A.B. Luthra

PHYSICS

J L Lewis O B E
Formerly Senior Science Master, Malvern College

G E Foxcroft O B E
Formerly Senior Science Master, Rugby School

LONGMAN

Acknowledgements

We are indebted to the following authors for permission to reproduce copyright material from the Longman Physics Topics Series:
A.R. Duff for extracts from *Pressures* (LPT 1969); D.W. Harding and Laurie Gibson (formerly Griffiths) for an extract from *Materials* (LPT 1969); R.D. Harrison for extracts from *Forces* (LPT 1968); Miss Maureen Hurst for an extract from *Crystals* (LPT 1969); A.J. Parker for an extract from *Heat* by Parker and Heafford (LPT 1969); G.W. Verow for an extract from *Magnetism* (LPT 1974).

We are grateful to the following for permission to reproduce photographs: Bryan and Cherry Alexander, page 82; Arabian American Oil Company, page 240 below; Barnaby's Picture Library, pages 4 below right, 5 above right (photo John Edwards), 123 above left (photo J. Palmeri), 131 below (photo M.G. Webb), 132 above left (photo USIS), 132 above right (photo New Zealand High Commission) and 205 (photo Camera Talks); British Geological Survey, page 10 above; British Rail, page 239; Camera Press, pages 2 (photo Zentralfoto), 5 below (photo Andrew Beard), 203 and 252 (photo Pierre Berger); J. Allan Cash Photolibrary, pages 4 above left, above right and below left and 6; Dunlop, page 57; Patrick Eagar Photography, page 123 below; Mark Edwards/Still Pictures, page 317; *Farmer's Weekly,* page 81; Geoscience Features, pages 12 above left and 102; Handford Photography, page 142; Harrap, Martin, *Thirteen Steps to the Atom*, pages 9 and 12 above right; P.E. Heafford, page 240 above; Heinemann Educational Books, page 41 below; Hulton Picture Company, pages 53 and 123 above right (photo Keystone); IBM, page 192 right; Ibstock Building Products, page 5 above left; Institute of Gelogical Sciences, London, page 8 below right; Kodansha, page 42; Andrew Lambert, pages 25 left and right, 41 above, 69, 159, 170, 202 and 235 left; John Lewis, pages 10 below, 12 centre, 13, 44, 46, 63 below, 73, 116, 131 above left and above right and 232; Lick Observatory, page 287 above; Longman Photographic Unit, pages 167, 179–88 and 303; National Power, pages 129, 132 below and 140; Panos Pictures, page 325 (photos Sean Sprague above left and below right, Ron Giling above right and below left); L. Phoenix, page 7 right; Quadrant, page 130; Ann Ronan Picture Library, page 90; Royal Observatory, Edinburgh, pages 280 and 287 below; Science Photo Library, pages 8 below left (photos Arnold Fisher), 12 below (photo Claude Noridsany and Marie Perennod), 63 (photo Harvard College Observatory), 192 left (photo Nelson Morris), 196, 274 below left (photo Hank Morgan), 274 right (photo John McFarlane), 281 (photo Dr. Fred Espenak), 285 (photo NASA), 288 (photo Royal Greenwich Observatory), 289 (photo Hale Observatory) and 323 (photo N. Feather); Tate and Lyle, page 7 left (photo H.E.C. Powers); Telegraph Colour Library, page 274 above left (photo Masterfile); Unilab, page 235 right.

The cover photograph of new aero-generators for energy production at Den Helder, Netherlands is from Tony Stone Worldwide.

Contents

Good laboratory practice

Danger

Biohazard

Flammable

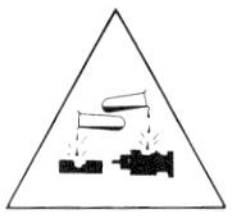

Corrosive

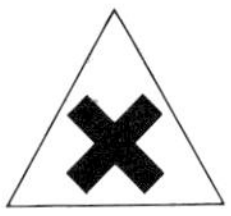

Irritant

Toxic

Wear eye protection

These are the safety symbols used in these books. You should get to know these so that you can recognise hazards that you might come across during your science lessons.

To avoid accidents you should:

- take special care when you see one of these symbols.
- always read through **all** the instructions given before you start doing your experiments.
- check with your teacher if you are not sure about any of the instructions.
- always check with your teacher before beginning any experiment that you have designed yourself.
- always wear eye protection when you see the eye protection symbol or when your teacher tells you.
- always wash your hands thoroughly with soap and water after handling pond water, soil, microbes, plants and animals.
- always stand when you are handling liquids so that you can move out of the way quickly if you spill anything.
- if you do spill anything on the bench wipe it up with a damp cloth, being careful not to get it on your hands.
- if you spill anything on your skin wash it off immediately and thoroughly with water. If you spill anything on your clothes tell your teacher.
- if you get anything in your eyes tell your teacher immediately.

Preface

While the authors approve that all pupils should study a balance of the sciences up to age 16, they recognise that not all teachers will necessarily want to follow an integrated science course; many will prefer to teach the three sciences separately. This book has been written for those who are studying physics up to Key Stage 3 of the National Curriculum in Science.

The authors were, at one time, closely involved in developing the Nuffield Physics Projects. Those courses were based on the belief that it was by doing experiments and carrying out investigations that a better understanding of physical principles would be acquired. This Nuffield spirit very much pervades the criteria for GCSE examinations, and also the National Curriculum. Our book therefore uses an experimental approach in order to involve pupils in physics.

It is not easy to write such a book if the text is not to ruin the spirit of enquiry by providing answers to experiments which should be done by the pupils. We hope that we have avoided this difficulty without leaving so much open-ended that pupils have no idea where they are going.

Yet another difficulty is to meet the needs of students with a wide range of ability. We were aware that, if we wrote for too wide an ability range, certain topics would require a very superficial treatment which could be dissatisfying for some. We have, therefore, tried to avoid too superficial a treatment in the hope of meeting the needs of average and above-average pupils.

A lot of material from the book Physics 11–13 has been incorporated. New chapters on electromagnetism, weather, light, the Earth in space, sound and electronics have been added, as required by the National Curriculum up to Level 7. Most of the chapters now include at least one piece of 'Background Reading' – short passages which we hope will interest readers and stimulate further study. Much can be learnt through answering questions and, for that reason, there are many scattered throughout the text.

An essential part of any physics course is the acquisition of laboratory skills and the ability to investigate. A particular feature is the inclusion of 'Explorations' – usually open-ended experiments to assist in the development of those skills and their assessment.

Finally, the authors wish to thank their many friends who have at some stage helped in the writing of the book. We are also grateful to members of CLEAPSS and the ASE for ensuring that the text meets with current safety recommendations. For our part, after many years of enjoyment teaching physics, we hope our book may help pupils to realise how worthwhile a study of physics is.

J L Lewis, G E Foxcroft, 1991

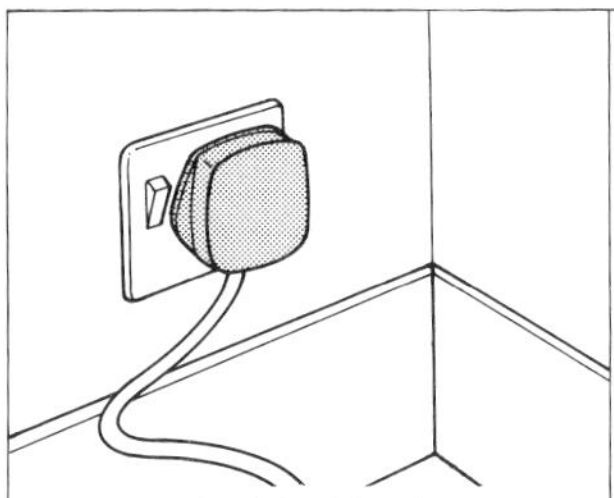

electrical leads

bicycle frame

glass window

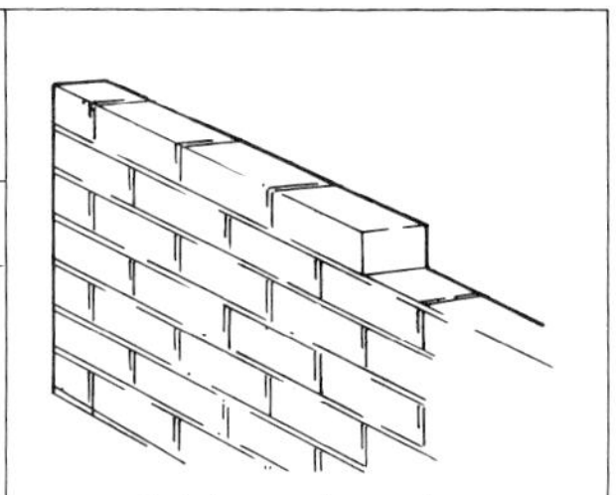

bricks and mortar

chairs and table made of wood

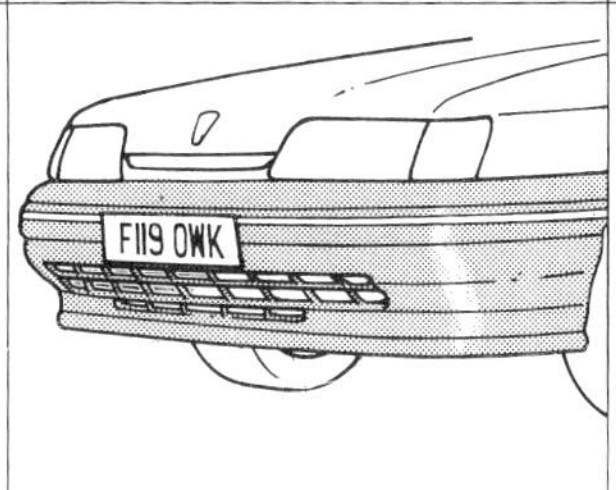

car bumper

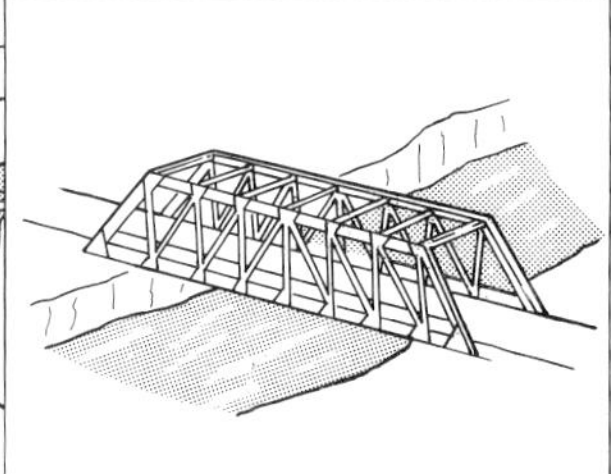

iron girder bridge

saucepans

a stone house

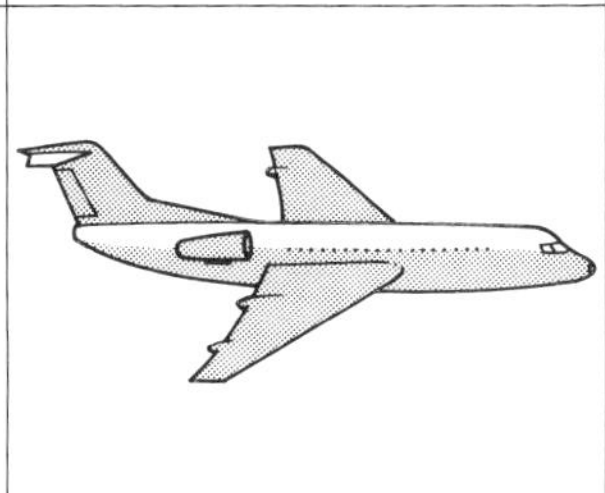

aeroplane

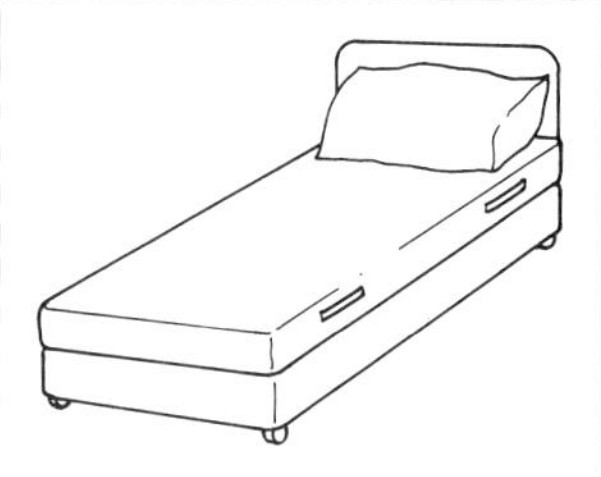

mattress and pillow

milk bottle and glasses

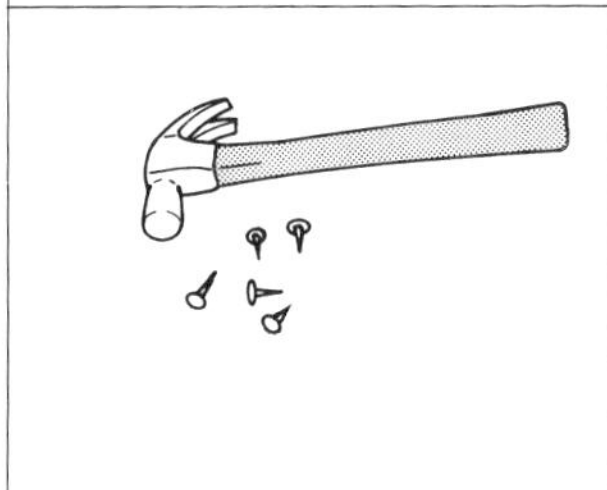

hammer and nails

curtains

coffee machine

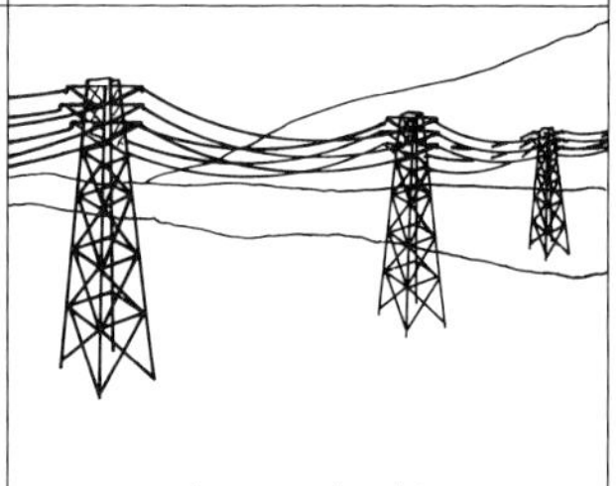

pylons and cables

porcelain

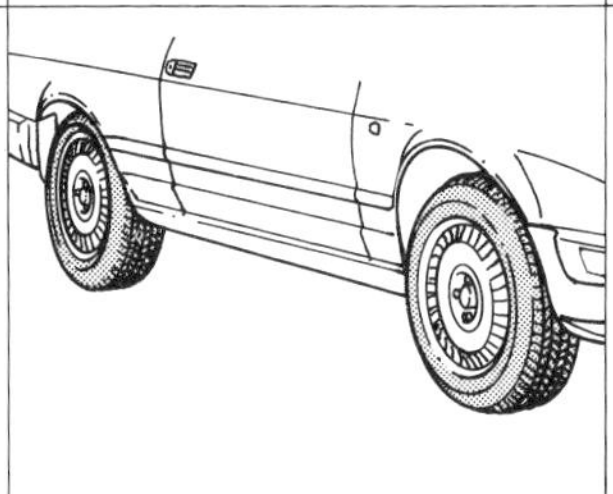

car tyres

paper

ship's cable

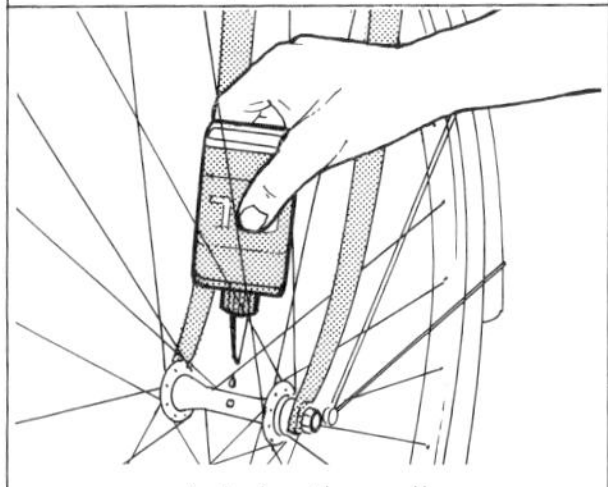

lubricating oil

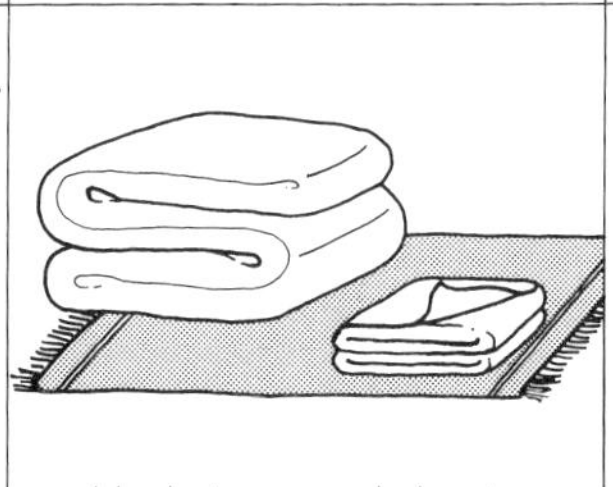

blanket, rug and sheets

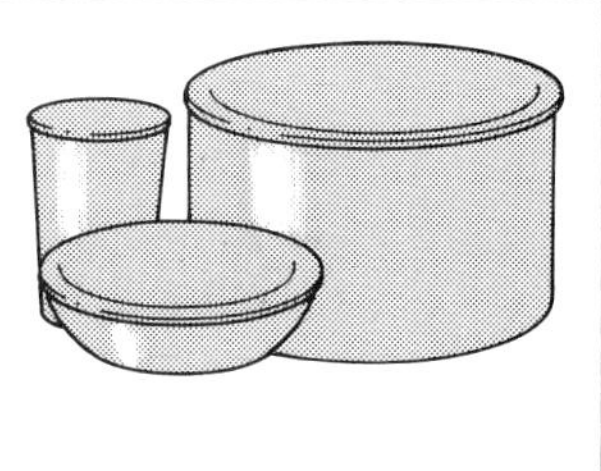

polythene bowls

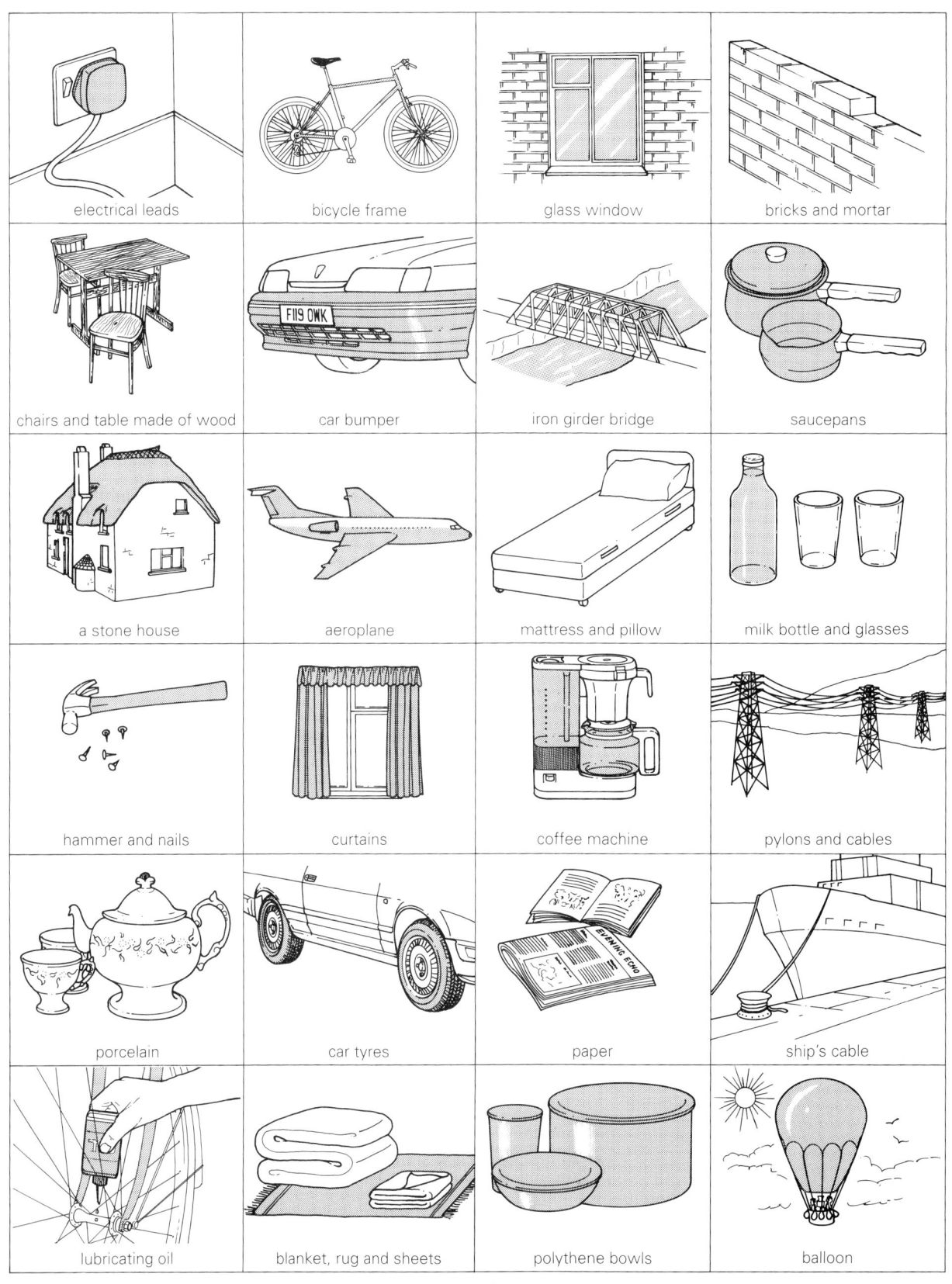

balloon

Chapter 1 Materials

The variety of materials

In our homes, in our schools, in the streets and in the countryside, we are surrounded by all kinds of things. There is also an immense variety in the materials from which these things are made. In your home there are tables and chairs made of wood, saucepans of aluminium, windows of glass, curtains of cotton, bowls of polythene, bicycle frames of steel, electrical wires of copper.

In your laboratory you will see a variety of different kinds of materials. Some of the substances are **solid**: iron, copper, aluminium, rubber, brick, lead, wood, polythene. Some of the solids consist of little bits, but the bits are none the less solid: sugar, salt, iron filings, sand. These substances have very varied properties. Some are hard, some soft; some bend easily, some break as soon as you bend them; some stretch; some can be dented by your finger, some spring back to their original shape when your finger is removed; they may differ in smell, in colour and in their feel or texture.

When we wish to make something we choose carefully the substance with the most suitable properties. Why do you think we make furniture from wood? Why do we not choose wood for a saucepan? Why use aluminium instead? Why is the handle made of a different substance from the rest of the saucepan? Why are car tyres made of rubber? Why is rubber unsuitable for a saw?

Some of the substances you have seen at home or in your laboratory are **liquid**: water, paraffin, mercury, treacle, vinegar. They too have varied properties: some for example flow very easily, others are very sticky, some smell, some are coloured.

Yet other substances are **gases**, like the air around us. Perhaps you have seen balloons filled with different kinds of gases: a balloon of hydrogen will rise upwards, a balloon of carbon dioxide will fall.

Solid, liquid and gas

How can you tell whether a substance is a solid or a liquid? One sensible answer would be to see if it will pour. Oil will pour, but a

Molten iron being poured

block of iron does not, and we therefore call oil a liquid, but iron a solid. Or will iron pour? The photograph shows molten iron being poured.

Why does it pour in this case? Because it has been heated, and this leads us to another important property of matter. Many solids when heated will become hot enough for them to melt and turn into a liquid. Heat the liquid and it will eventually boil and turn into gas. The most familiar example of this is ice; heat it and it becomes water, a liquid; heat it further and the water turns to steam, a gas. Normally iron is a solid, but heated sufficiently it turns to liquid and any iron in a star like the Sun would be a gas.

Sometimes it is a little difficult to decide whether a substance is solid or liquid. A piece of pitch may appear to be solid, but if a can of pitch is left on its side, it will eventually flow out. What about glass? If you examine glass in an old window in a church you may find that the glass is thicker at the bottom than at the top. What could be the explanation of this?

Questions for class discussion

1 Why are saucepans made of aluminium? Why are they not made of polythene or wood?

2 Why are electrical connections usually made with copper wire? Why do we not use silver? Why is the copper wire often surrounded with rubber or plastic?

3 Could the blade of a spade be made of lead or aluminium instead of steel?

4 Why are the handlebars of bicycles made of steel? Why are they usually chromium plated?

5 Why are tyres made of rubber?

6 What material is used for making aeroplanes?

7 Why are ships made of steel? Since a piece of steel sinks when it is put on water, how is it possible for a ship made of steel to float?

8 What kind of material is useful for clothing on a cold day?

9 Would gold be a good substance for a garden fork? Would there be any advantages in an iron fork with a thin layer of gold over the iron (gold plated)? What are the disadvantages?

10 Why are most houses in the Cotswolds made of stone?

11 Why is concrete used in building bridges, but not for motor car bodies?

A game to play

Put a block of each of the following on a tray: iron, brass, lead, glass, aluminium, softwood, hardwood, foamed polystyrene,

Perspex, paraffin wax. Feel each of them. Weigh them in your hand. Smell them. Find how hard they are by trying to dig your fingernail into them. When you think you know the materials, get someone to blindfold you and to move the blocks around on the tray. Then see if you can decide which block is which. You will be doing well if you get them all right.

There is a more difficult version of this game. Start with ten blocks as before and get to know them in the same way. But when you are blindfolded try to identify ten other objects, made of the same materials, but different in shape. Some might be tubes, some rods, some blocks, some wires and so on. This is much harder and really tests how well you got to know the original blocks.

The game described above is typical of what a scientist has to do. To decide what each article was made of you had to find clues – how hard it was, how strong it was and so on. A scientist is always looking for evidence (clues) and on the basis of that evidence forms an opinion.

Homework assignment: the right material for the job

The material chosen to make an object depends on what job the object is going to do. If it is to be used to boil water on a gas cooker, it would not be very sensible to make it out of wood. It would be much better to make it from a metal. Gold and silver are both metals, but no one uses gold or silver saucepans. Why is aluminium the metal most commonly used? On the other hand, aluminium is not a very good material for the handle of a kettle or saucepan. Why is that? Which materials would be suitable for the handle of a kettle or saucepan?

Make a list of ten different solid objects which you might find in a kitchen – or elsewhere in your home. Against each item, write down what material it is made of and suggest a reason why that material was chosen.

Background Reading

Building materials

Human beings have always needed somewhere to live and primitive people relied on caves to provide shelter. As soon as they wanted to choose where to live instead of living where the caves happened to be, they had to build their own homes. If a home was to last a lifetime, stone would be used because it was strong, it did not crush, it did not rot and it could be carefully split and shaped into convenient pieces. The cost of the material and the distance it had to be moved were always important considerations – they are still very important today when choosing materials for any job in a home or in industry. The builder therefore used the local stone found in the area. The examples

(a) Crofter's cottage (Scotland)

(b) Slate cottages (North Wales)

(c) Stone houses (Cotswolds)

(d) Flint houses (Norfolk)

here show (a) a crofter's cottage in Scotland, (b) cottages in North Wales made from slate, (c) typical Cotswold stone houses, (d) flint houses in Norfolk.

Most towns however make use of bricks to provide their buildings. Cost will have a lot to do with the decision to use bricks: there is probably little natural stone in the area and skilled craftsmen are needed to shape it. Bricks however have other advantages as they can be made to a standard size which makes them convenient for building. Most bricks are made from clay formed by the weathering of rocks into fine grains, which cling together when wet. The wet clay can be shaped in moulds and baked in the sun (as in many African countries) or heated more strongly so that it becomes hard and strong (the clay is then said to be 'fired'). Various substances can be added to the original wet clay to make the final appearance of the bricks different. The photograph shows a loaded kiln car.

Loaded kiln car leaving kiln after firing

Using mortar to build a wall

Cement is one of the most useful materials a builder uses. It is made from clay and limestone mixed together and finally ground to a bluish-grey powder. Mortar is made by mixing together cement, sand and water in the right proportions to make a thick paste which can be spread on to a brick. It becomes the builder's glue (see the photograph). As the mortar hardens, it forms a joint between the bricks.

A substitute for bricks and stone which is widely used these days is concrete. This is made from a mixture of cement (one part), sand (two parts) and broken stones or gravel, usually called aggregate (four parts), together with water to mix it. The wet concrete can be poured into a hole for foundations of a building or bridge, or between wooden boards (shuttering) when making concrete walls.

Building under construction

Sydney Opera House

The concrete can be made particularly strong if it is **reinforced** with steel wire or rods. The photograph on page 5 shows a building under construction. The steel rods are assembled, the wooden shuttering is placed around them and wet concrete is poured into the gap. The boards are removed when the concrete is hard and the 'reinforced concrete' remains.

Such concrete is particularly good for building bridges and high-rise buildings. The photograph above shows the Sydney Opera House which is made of concrete.

Homework assignment: a visit to a building site

Is there a chance for you to be taken to a building site near your school or home? Perhaps your father or mother could take you to see one.

Make a list of the materials in use on the site. What is each material used for in the building? Ask if you can start a collection of small samples of the building materials.

Not all the materials will necessarily be lying around the site on one afternoon. It would be much better if you could arrange to visit the site several times to see how things develop and what new materials are brought into it.

Chapter 2 Crystals

Looking at things

In this chapter we will start thinking about the structure of matter. To do this it would seem sensible to look at it closely and perhaps the first thing would be to use a hand lens (a magnifying glass).

Fig 2.1

Experiment 2.1 Using a hand lens

Hold the lens close to your eye with one hand and bring up the object towards the lens with the other hand until you can see it clearly (Fig 2.1). Look at your finger nail, your skin, a piece of material, a bit of nylon stocking. Look at the blocks you looked at earlier: blocks of iron, aluminium, wood and so on. Look at a picture from a newspaper: you may be rather surprised by what you see. Above all, look at some photographic hypo (solid), some common salt, some sugar. If possible, look at sugar of various kinds: granulated, castor, demerara and icing sugar. What do you notice about the grains of salt or the grains of sugar?

You will use microscopes in biology (see *Biology 11–14* page 36) and it is interesting to look at grains of salt and sugar using them.

Safety note

Never look at the Sun through a lens.

Granulated sugar

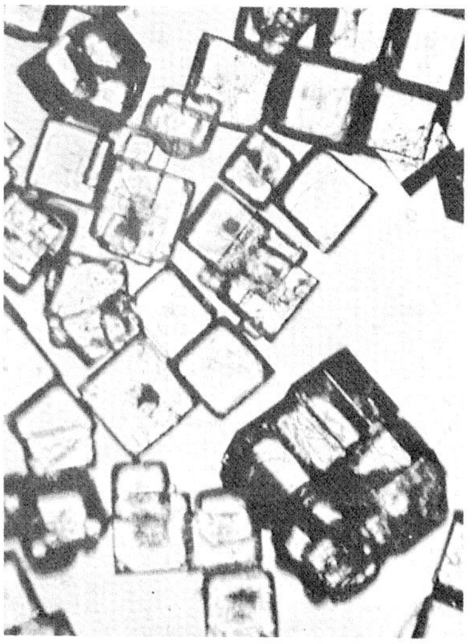

Salt crystals

Regularities

Alum crystal

The interesting thing about the salt, or the sugar, is the regularities between one grain and another. Not all the sugar grains are the same size, nor do they necessarily have the same shape, but they all appear to have the same angles between faces. These regular structures are called **crystals**.

The photograph shows a large piece of alum, in which there are a lot of alum crystals. Notice the regularity of the angles of these crystals even though the crystals are of different sizes and shapes.

Crystals

The photographs below show examples of crystals which occur in nature. They all show a lot of smooth faces with sharp edges. These flat surfaces can be very good reflectors when light shines on them.

If you hold a piece of granite so that light can be reflected off it and if you rotate it slowly, you should notice parts of the surface suddenly appearing bright or sparkling and thereby revealing the presence of crystals. Three main materials can be seen in a specimen of granite: shiny flaky mica, clear glass-like quartz and some felspar, usually pink or white in colour.

What is the reason for this regularity in crystals? The photograph on page 9 shows a model built with polystyrene spheres. It shows the same sort of regularity as the alum crystal at the top of this page.

Cubic crystals of rock salt

One possible explanation of the regularity of crystals is that they too are built up from many small 'building blocks'.

The 'building blocks' in one case are the polystyrene spheres. The 'building blocks' for the crystals might possibly be particles such as atoms or molecules. We will now see if there is any other evidence to support this idea of crystals being made of particles stacked up together.

Crystals growing

You have already looked at the regular shape of salt crystals. How do the crystals get like that? To find out, it is sensible to watch crystals growing by doing experiments.

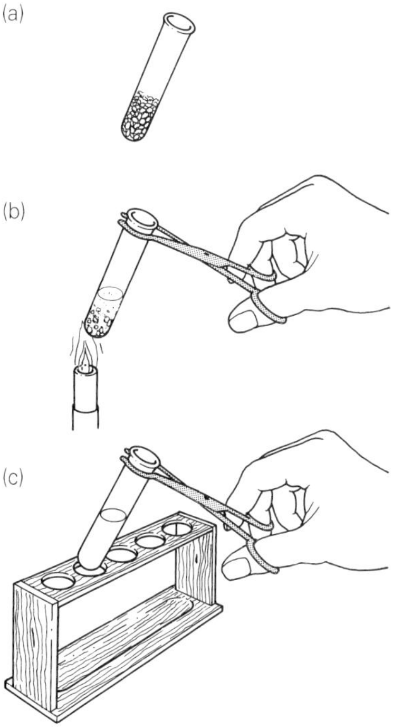

Fig 2.2

Experiment 2.2
Hypo crystals

Take a test-tube and fill about a third of it with hypo crystals (Fig 2.2a). Heat the tube gently and the crystals will dissolve (Fig 2.2b). Then let the test-tube slowly cool to room temperature by placing it in a test-tube rack (Fig 2.2c). Hold the test-tube in your hand and drop into it a small hypo crystal. Something immediately starts to happen. Did you notice any change in temperature?

The experiment can easily be repeated by once again warming the crystals so that they dissolve. You can cool the test-tube under a running tap, but if you try to hurry things up like this, you may find crystals suddenly forming even before you have put in a small hypo crystal.

This experiment shows crystals growing very quickly. It is possible to grow much larger single crystals and this is fun to do in your school laboratory or at home. It takes much longer to grow them, but it is well worth taking some trouble. Your teacher will give you details.

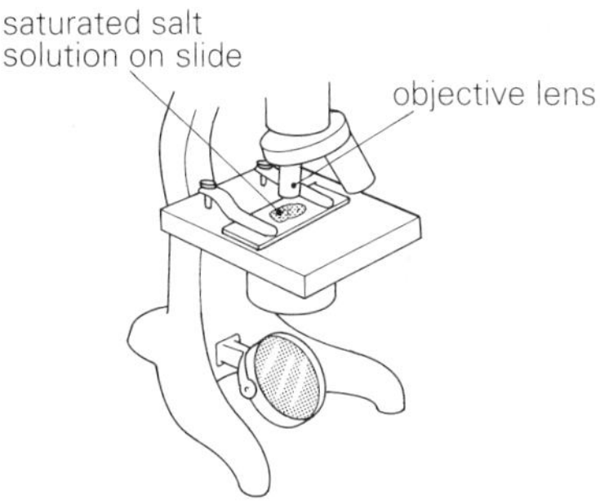

Fig 2.3

Experiment 2.3
Salt crystals

Put some common salt in a test-tube, then half fill it with water. Shake it until the salt has dissolved. Add more salt and shake again. Continue the process until no more can be dissolved. You will then have a saturated salt solution.

Warm a microscope slide over a low Bunsen flame and put a drop of the salt solution on the slide. Put the slide on the platform of a microscope (Fig 2.3) and then watch the crystals forming at the edge of the drop. (To get the drop in focus, it may help to put a pin on the slide and focus on that.)

Experiment 2.4
Growing salol crystals

A very good substance for showing crystals growing is salol. It is not expensive and can be bought at a chemist's shop. Put a little on a microscope slide and place it on a radiator or convector heater to melt it (a temperature of 43 °C is sufficient). Remove the slide and watch what happens to the liquid when a very small seed crystal is put in it. It is better if you watch with a hand lens or use a low-powered microscope.

You will see the crystals growing. They usually start at the edge of the liquid. When they grow they always have a particular shape. After·a while they 'bump' into each other and growth gets restricted in that direction.

The photograph shows a thin layer of copper sulphate crystals growing on a microscope slide. Some copper sulphate was dissolved and a few drops were put on the slide. Crystals started to grow and could be seen under the microscope. Notice that there are angles in the different crystals which are the same.

Copper sulphate crystals

Beads in a tray

Models of crystal growth

How do these experiments fit in with our idea that a crystal is made up of particles acting as building blocks?

If you take a lot of glass beads (all the same size) and pour them into a small tray so that they are one layer thick, you will notice that they settle into a definite pattern without your having to arrange them (see photograph).

Of course there are some irregularities, but you will notice that there is very often a hexagonal packing, one bead with six beads all round it.

Experiment 2.5
Crystal models

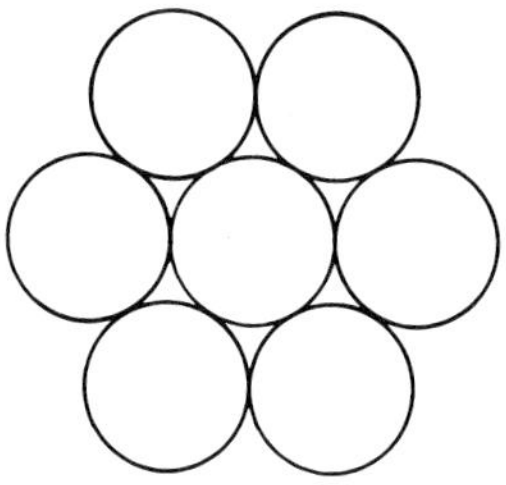
Fig 2.4

Arrange several polystyrene spheres in the hexagonal pattern as shown in Fig 2.4. (You can use marbles instead, provided they are all the same size.) Then add more spheres to the pattern so that it gets bigger. What do you notice about the angles as the shape grows? What do you notice about the way the other spheres pack together?

The spheres need not be arranged in this hexagonal way. They might be arranged in squares. (Fig 2.5a) shows nine spheres arranged in a square on a baseboard (the baseboard merely stops them from rolling about over the bench). What would happen if some more spheres were added on top of this?

Clearly we should get a pyramid (Fig 2.5b). The pyramid has a base of nine spheres, three in each side. The model could be extended so that there were four spheres in each side of the base, or five spheres as in (Fig 2.5c). And do you notice that on the side of that large pyramid the hexagonal packing appears again? Each sphere seems to be surrounded by six spheres. If you were to stick the balls together and go on adding more and more, you could build up the model on page 9. This has considerable similarities with the large crystal of alum on page 8. Perhaps this is good evidence that crystals are made of basic building blocks.

Fig 2.5

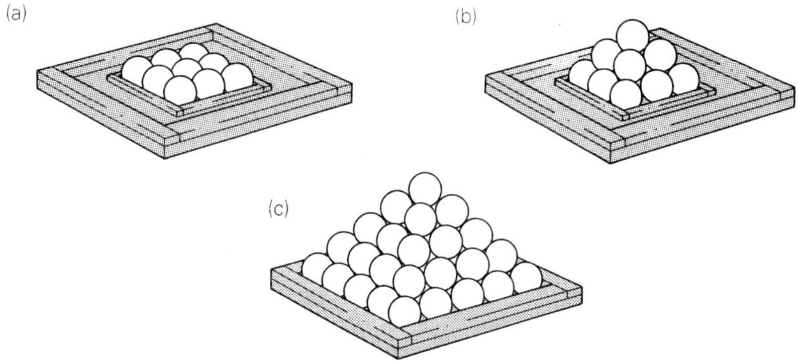

Of course, if all crystals were made of spherical particles packed in this way, then all crystals would have the same angles. And you can easily see that this is not so. Perhaps that is because the particles are not like spheres; and perhaps they pack together in some other regular way.

Cleaving crystals

Further evidence comes from the way some crystals can be cleaved. (Cleave is another word for split.) These crystals tend to split along certain flat surfaces. If the regular shape of crystals means that they are made up of layers of particles, we might

Cleaving calcite

Cleaving a polystyrene model

Galvanised iron

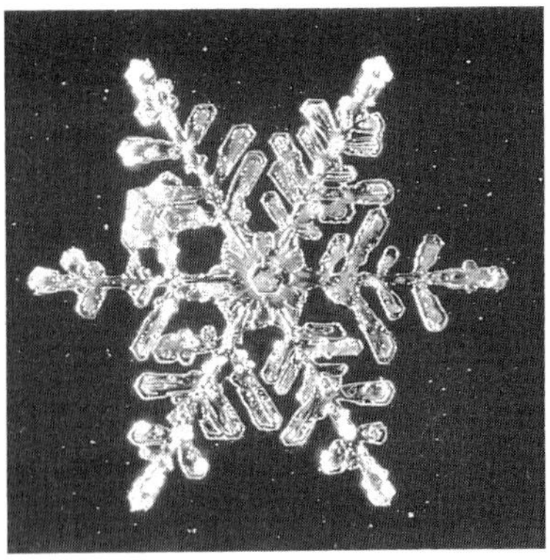

Snowflake

expect them to cleave along certain planes or flat surfaces and this is just what happens. The photograph on the left shows a calcite crystal being cleaved and the other shows what is happening in terms of our model.

Questions for class discussion

1 If you look at a particularly fine type of icing sugar with a hand lens or a microscope and you cannot see crystals, can you assume from this that the icing sugar is not crystalline?

2 You saw calcite being cleaved in your laboratory. It was very important to cleave it at the correct angle. What would happen if the angle were wrong? What would happen in your polystyrene sphere model if you tried to cleave it at the wrong angle?

3 If you take a cube of sugar and try to cleave it, you cannot do so: it shatters whatever angle is chosen. When you looked at sugar with a hand lens, each small piece looked like a crystal. Do these two things contradict each other?

4 You have seen hypo (sodium thiosulphate) crystals growing quickly, but growing an alum or copper sulphate crystal takes place slowly. Why do you think there is this time difference?

5 The photograph shows a piece of galvanised iron. What do you notice about it? How can you explain what you see?

6 The photograph shows a snowflake. Why does it show such a pattern?

Crystals in a cast metal bar

Some things to try at home

1 Try to find an old brass door knob. It may have been 'etched' by the sweat of people's hands. If so, what do you notice about it?

2 Try to find a piece of broken metal and examine the broken part for crystals (see the photograph). Look for evidence of crystals in galvanised buckets or inside fruit tins.

3 Try growing a crystal of alum or copper sulphate at home. Your teacher will give you details. Perhaps the most important thing is to keep the jar where the temperature does not change too much.

Conclusion

All the evidence you have seen suggests that it is reasonable to think of a solid as a regular arrangement of small particles. We will assume this until we get evidence to the contrary. Of course we do not know why the particles hold together: we shall have to wait until a later stage before we find out.

Background Reading

Gemstones

Diamond, ruby, sapphire and emerald are the most precious of gemstones and they are all crystals. They are used in crowns, tiaras, rings and bracelets. Their value depends first of all on their beauty and lustre, but if they were less hard (they are amongst the hardest substances known) they would soon be damaged, their beauty would be lost and they would be less precious. They often occur in regular attractive shapes, though their appearance is usually enhanced even more by cutting and polishing.

Diamonds were originally formed in volcanic pipes from patches of carbon, under conditions of intense heat and pressure, during eruptions billions of years ago. Many diamonds reached the surface as the rock weathered, and were carried away by rivers.

Although diamond is the world's hardest natural substance, a diamond crystal contains planes of weakness along which it can be cleaved, as you have seen calcite cleaved. A skilled craftsman in this way exposes natural faces of the diamond, and afterwards may cut extra surfaces. The reflection and refraction (bending) of light in a properly cut diamond is what gives its characteristic fire.

Diamonds have now been made synthetically by copying nature and applying very high temperatures and pressures to carbon in heavy presses. Although made from carbon, they cost just as much to make as to mine because of the high

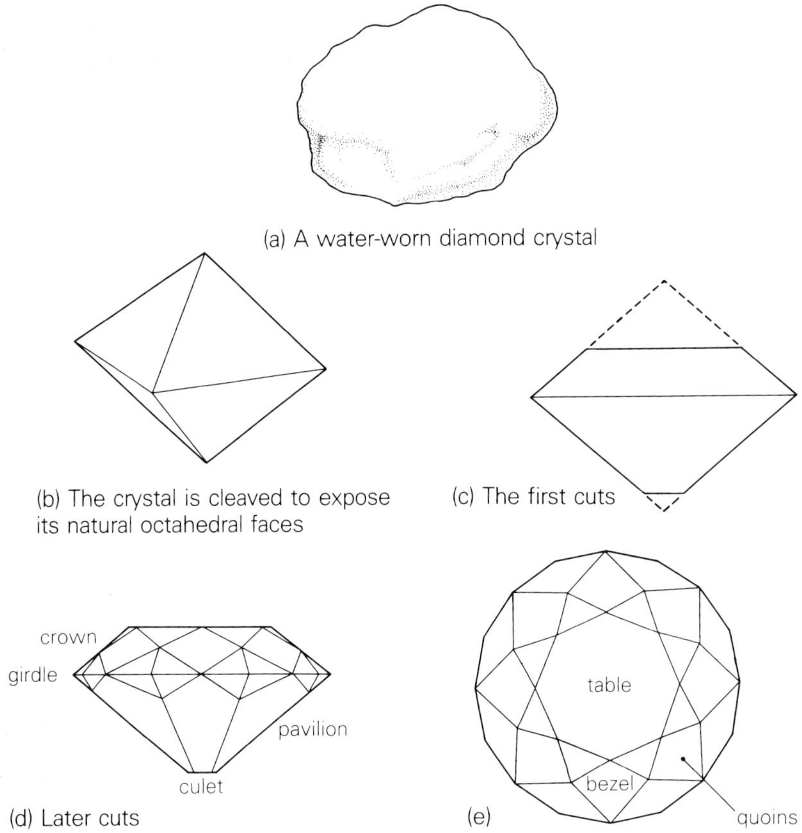

Fig 2.6 The cutting of a diamond

temperatures and pressures, so there is no hope yet of diamond tiaras appearing in bargain basements! Most of the world's output of diamond is used in industry, mainly for cutting, drilling, grinding and polishing. Some synthetic diamonds are set into the tips of saws and drills which will then cut and drill the hardest rock and concrete. The stylus of a record player is often tipped with a diamond.

Ruby and sapphire are both crystalline forms of aluminium oxide. The ruby contains small quantities of chromium oxide and iron oxide, which colour it red; sapphire owes its blue colour to a trace of titanium oxide.

(Adapted from Longman Physics Topic *Crystals* by MM Hurst.)

Chapter 3 Measurement

It is easy to make guesses about matter and its structure, but it is only possible to make real scientific progress when measurements are made. In no subject is measurement more important than in physics and it is now time to start thinking about it.

In Britain we have in the past measured road distances in miles, lengths of material in yards, sugar in pounds and milk in pints. However, the metric system is being used increasingly in our everyday life and it has for a long time been the system used in scientific work. We will use it throughout this book.

Measurement of length

The standard unit for measuring length is the metre. Your school laboratory will have some metre rules and these will help you to learn to judge lengths in metres.

The metre is too small a unit for measuring distances between towns and it is usual to use kilometres (1 kilometre = 1000 metres). For small distances centimetres and millimetres are both used: 100 centimetres = 1 metre and 1000 millimetres = 1 metre. There are standard abbreviations for these units and it is wise to become familiar with them.

Unit	*Abbreviation**
metre	m
kilometre	km
centimetre	cm
millimetre	mm

1 km = 1000 m	1 m $= \frac{1}{1000}$ km	= 0.001 km
100 cm = 1 m	1 cm $= \frac{1}{100}$ m	= 0.01 m
1000 mm = 1 m	1 mm $= \frac{1}{1000}$ m	= 0.001 m

You will come across kilo-, centi- and milli- a lot in the future. 'Kilo' means 1000 times, 'centi' means $\frac{1}{100}$th and 'milli' means $\frac{1}{1000}$th.

* It is the accepted scientific usage not to put a full stop after m, cm, mm even though they are abbreviations, unless of course they come at the end of the sentence. Another accepted usage is to omit commas when writing large numbers. For example 1,000,000 may be written as 1 000 000. Notice that there are no plurals in unit abbreviations. For example, 2 metres is written as 2 m and not 2 ms because s is the abbreviation for the second.

Quick exercises

1 Measure the length of this page in centimetres. Write down the length in metres and in millimetres.

2 Measure the thickness of this book in centimetres.

3 Measure your own height in centimetres and in metres.

4 Compare the length of a metre rule with one yard. Which is bigger? Is a 100 metre race longer or shorter than a 100 yard race? By how much would they differ?

Powers of ten

The distance from the Earth to the Sun is about 150 000 000 000 m. This involves a lot of noughts and it is not very convenient writing out such a large number. Scientists perfer to use a shorthand way of writing it. 150 000 000 000 is 1.5 multiplied by 10 a number of times. How many times? How many places would the decimal point have to be moved to change 1.5 into 150 000 000 000? The answer is 11 so we can write 150 000 000 000 as 1.5×10^{11}. This is much quicker to do and easier to understand.

Similarly,

$$50 = 5 \times 10 = 5 \times 10^1$$
$$500 = 5 \times 10 \times 10 = 5 \times 10^2$$
$$5000 = 5 \times 10 \times 10 \times 10 = 5 \times 10^3$$
$$5\,000\,000 = 5 \times 10 \times 10 \times 10 \times 10 \times 10 \times 10 = 5 \times 10^6.$$

You will study powers of ten in more detail in your mathematics course. We shall, however, use it as a convenient shorthand in this book.

Standard form

The distance from the Earth to the Sun can be written in the following ways.

150×10^9 m $\quad 15 \times 10^{10}$ m $\quad 1.5 \times 10^{11}$ m $\quad 0.15 \times 10^{12}$

All of these are equivalent mathematically, but scientists usually prefer to write such a quantity with one figure in front of the decimal point. The **standard form** would be 1.5×10^{11} m. It is useful to get into the habit of writing numbers like this.

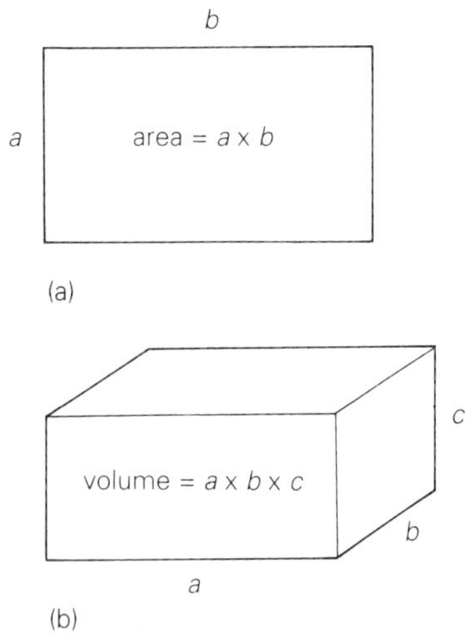

Fig 3.1

Measurement of area and volume

You have learned in mathematics how to find the area of a rectangle: you multiply the length by the breadth (Fig 3.1a). Similarly you have also learned how to find the volume of a rectangular block: you multiply the length by the breadth by the height (Fig 3.1b).

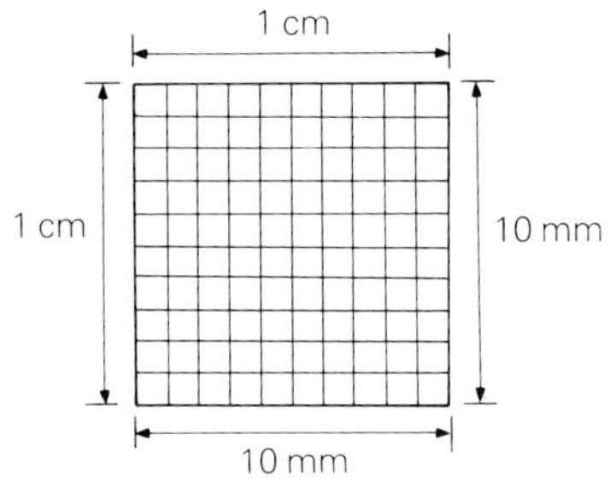

Fig 3.2

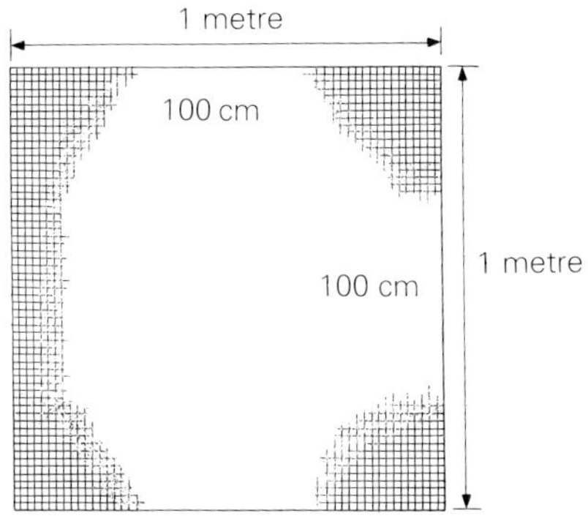

Fig 3.3

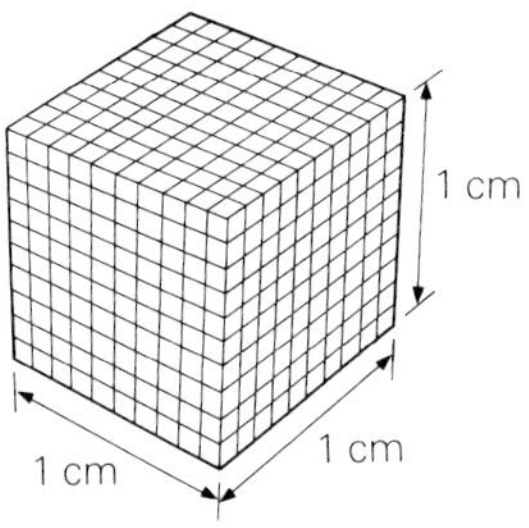

Fig 3.4

The units for measuring area and volume are also based on the metre: square metres for area, cubic metres for volume. There is a very convenient shorthand for writing square metres as m^2, and cubic metres as m^3 and these will be used in this book. Similarly cubic centimetres can be written as cm^3.

Suppose Fig 3.2 represents 1 square centimetre (not drawn to scale). As there are 10 millimetres in 1 centimetre, there will be 10 millimetres along each side. Each little square in the drawing is 1 square millimetre. How many little squares are there? It follows that

$$100 \text{ mm}^2 = 1 \text{ cm}^2.$$

Since there are 100 cm in 1 m, there will be 100 cm along each side of a 1 m square (Fig 3.3). Therefore there will be 100×100 small centimetre squares in 1 square metre. In other words

$$10\,000 \text{ cm}^2 = 1 \text{ m}^2.$$

Similarly, there will be $10 \times 10 \times 10$ cubic millimetres in 1 cubic centimetre (Fig 3.4). In a cubic metre, there will be $100 \times 100 \times 100$ centimetre cubes. In other words

$$1\,000\,000 \text{ cm}^3 = 1 \text{ m}^3.$$

Using our power-of-ten notation, we can write

$$1 \times 10^4 \text{ cm}^2 = 1 \text{ m}^2 \quad \text{and} \quad 1 \times 10^6 \text{ cm}^3 = 1 \text{ m}^3.$$

Questions for homework or class discussion

1 What is 10×10 in words?
Write down the answer in standard form.
Do the same for $10 \times 10 \times 10$ and for $10 \times 10 \times 10 \times 10 \times 10 \times 10$.

2 The mass of the Earth is about

5 980 000 000 000 000 000 000 000 kg.

Write this down using powers of ten. You may have written your answer as something $\times 10^{22}$, or as something $\times 10^{23}$, or as something $\times 10^{24}$. Now write it down in all three ways. Which of these three ways is the standard form?

3 What is the area of a rectangular field 220 m long by 50 m wide?

4 What is the area of a rectangular sheet of metal 3 m long and 20 cm wide? Give the answer in cm^2, and in m^2.

5 What is the volume of a block of metal 5 cm long, 4 cm wide and 3 cm high?

6 What is the volume of a metal cube with side 4 cm?

7 A wooden crate is 2 m long, 1 m wide and 1 m deep. How many 1 cm cubes could you fit into the crate? Express your answer in powers of ten.

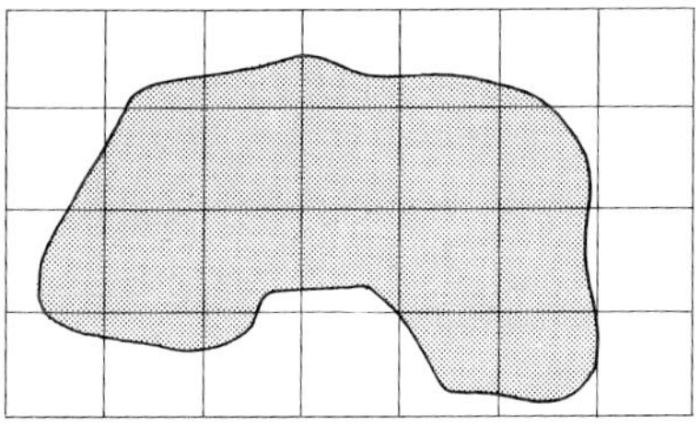

Fig 3.5

8 Fig 3.5 is a map of an island. The lines drawn on the map are each 1 km apart. Estimate the area of the island in square kilometres. (*Hint*: what is the area of one square; how many complete shaded squares are there; how many squares have more than half shaded; how many have less than half shaded?)

Measuring irregular areas

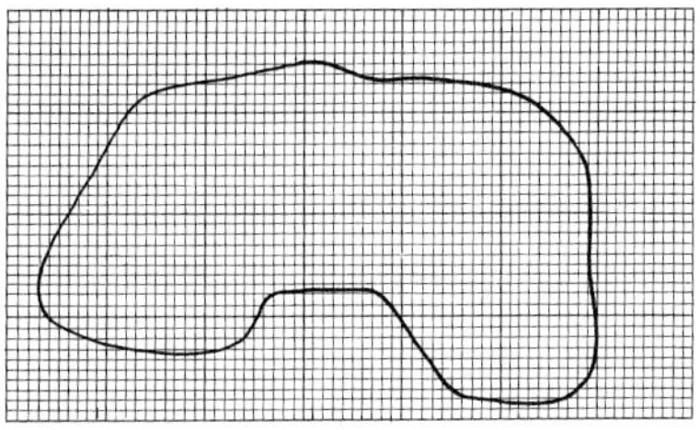

Fig 3.6

You can find the area of a rectangle by measuring the length and the breadth, and multiplying the two together. Question **8** above suggests how you might estimate the area of an irregular shape. But suppose you wanted to know the area with rather greater accuracy. What could you do?

You might divide the big squares into a hundred little squares (Fig 3.6), and then you could count first the number of complete big squares and then the number of little squares. Each big square will represent 1 km^2, each little square will be $\frac{1}{100}$ km^2. This will give you a more precise answer.

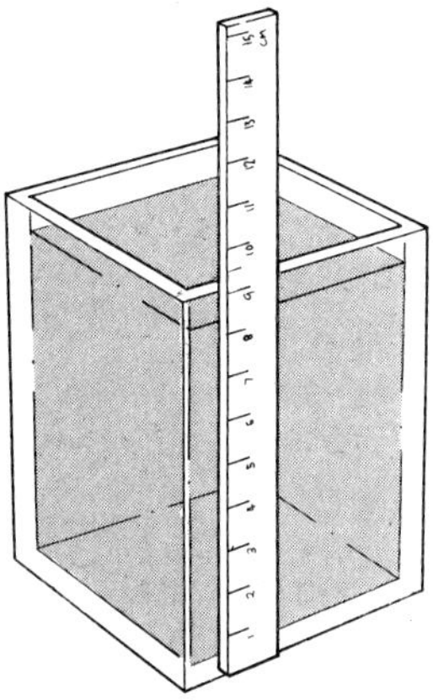

Fig 3.7

Measuring the volume of liquids

It was easy to find the volume of a solid rectangular block but what can we do about liquids? They do not have a definite length, breadth and height, but take the shape of the container in which they are put. But that gives us the clue how to measure their volume.

Pour the liquid, whose volume you want to know, into a Perspex container as shown in Fig 3.7. (Those recommended for school use have *internal* length and breadth of 5 cm.) Measure the internal length, breadth and height, and multiply all three together, and you have the volume.

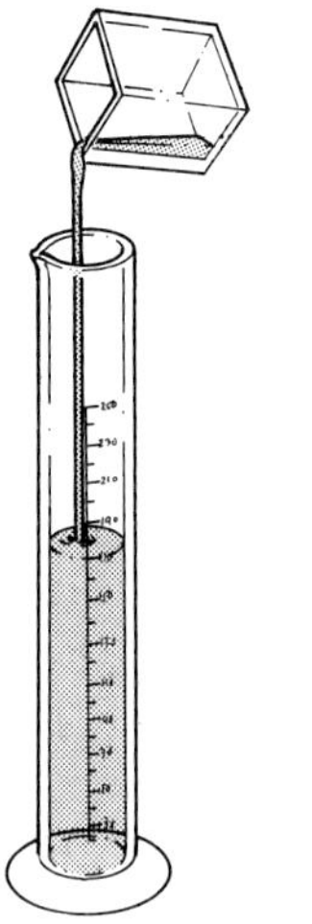

Fig 3.8

Experiment 3.1
Checking the scale on a measuring cylinder

A very convenient way of measuring the volume of liquids is to use a measuring cylinder. These have a scale marked on them (Fig 3.8). To check whether the scale is correct, put water into one of the 'square' perspex containers to a depth of 4 cm. The volume of water will be $5 \times 5 \times 4$ cm^3 or 100 cm^3. Pour this water into the measuring cylinder and see if it comes to the 100 cm^3 mark. How would you check the 200 cm^3 mark and the 250 cm^3 mark on a 250 cm^3 measuring cylinder?

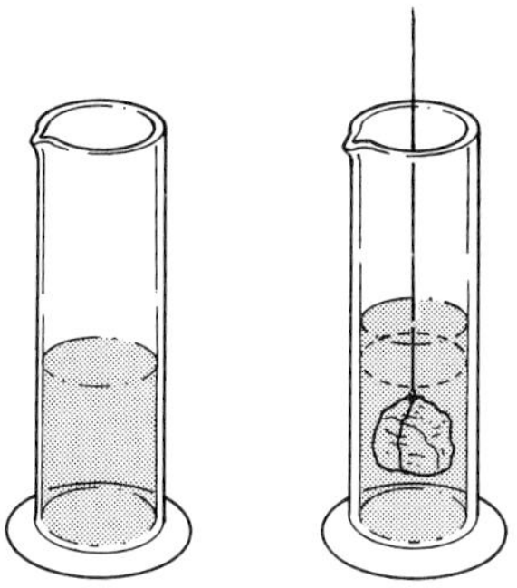

Fig 3.9

Experiment 3.2
Measuring irregular solids

It is easy to calculate the volume of a rectangular block. But many solids are irregular, for example a stone, a pair of scissors, a bicycle. Their volumes cannot be easily calculated from a few measurements.

One way to find the volume of a stone is shown in Fig 3.9. Pour water into a measuring cylinder so that it is about half full, and read the volume. Then lower the stone on a piece of cotton into the water. The level will rise. Read the new volume. The difference between the two readings gives the volume of the stone.

Accuracy

Suppose you measure the height of a book and you say it is 21 cm. What does this mean? To a physicist, it means that the length is neither 20 cm nor 22 cm, but that it lies between 20.5 cm and 21.5 cm. That sort of accuracy is sufficient if you are deciding whether the book will fit on bookshelves 25 cm apart. But it would not be sufficiently accurate for the length of a piece of wood to be used for a chair leg: to that accuracy the four legs might be very different. One leg may be 20.7 cm, another 20.8 cm, the third 21.1 cm and the fourth 21.4 cm, and you would not have a very satisfactory chair.

To describe something more accurately it is necessary to give further decimal places.

21.5 cm	means it is between	21.45 cm	and	21.55 cm
21.55 cm	means it is between	21.545 cm	and	21.555 cm
21.555 cm	means it is between	21.5545 cm	and	21.5555 cm

Similarly, 21.0 cm means the length lies between 20.95 cm and 21.05 cm.

It does not have any meaning to say that the height is *exactly* 21 cm, unless you mean 21 followed by a decimal point and an infinite number of noughts, and that, of course, is absurd.

Averages

Whenever possible a scientist does not rely on only one reading when making a measurement. It is preferable to take several readings and calculate the average. To do this the readings are added up and then divided by the total number of readings.

It can sometimes happen that an obvious mistake is made in one of the readings, in which case that reading should be omitted when working out the average, as will be shown in one of the following examples.

Questions for homework or class discussion

1 If you buy 8 m of rope, it is impossible to cut it at *exactly* 8 m. Strictly speaking, 8 m would mean anything between 7.5 m and 8.5 m, although the salesman is usually careful to make it over 8 m rather than under. Why does he do this? If you want it between 7.95 m and 8.05 m, how should you write the length?

2 A boy measures the length of a table and he says it is 1.231 42 m. If he measured it with a tape measure marked only in centimetres, is this a reasonable answer? What sort of accuracy could you expect with such a tape measure? If he had measured it with a tape measure marked in millimetres, write down what would have been a reasonable answer for him to give. Why is a fabric tape measure not a good thing to use for scientific measurements?

3 A girl walks round a field several times. On each occasion she times herself. The readings are 11 minutes, 12 minutes, 13 minutes, 12 minutes. What is the average time?

4 A boy uses a watch with a seconds hand to time a ball rolling down a slope. He takes three readings. They are 12 seconds, 12 seconds, 13 seconds. He adds them up and gets 37 seconds. He divides by 3 and says the average is 12.333 3 seconds. Is this a sensible answer?

5 Six girls are asked to measure the length of a piece of wood using a metre rule. The lengths they write down are the following.

63.2 cm	63.9 cm
63.8 cm	63.9 cm
36.4 cm	63.1 cm

One of the readings is a bit strange. Which one? What do you think that girl did wrong? Calculate an average value for the length of the piece of wood.

Estimations

Scientists often need to make estimates. An estimate may be only a quick rough guess, but it is not necessarily bad science because of that. Suppose you have a doorway 1 m wide and you want to know if a table will go through it. A rough estimate that the table is 80 cm wide is quite sufficient to tell you what you want to know, even though it may in fact be 75 cm or 84 cm.

Rough estimates also help us to avoid careless mistakes in arithmetic: they provide a valuable check on our work. Suppose a man wants to know the size of the floor of a large room. He measures the length as 12.4 m and the breadth as 9.3 m. He multiplies the two together and writes down the area as 1153.2 m^2. If he does a quick estimate taking the length as 10 m and the width as 10 m, he will realise the answer should be about 100 m^2 and he quickly sees his mistake in the position of the decimal point.

Question for homework or class discussion

1 Estimate the height of the room in which you are sitting. It may help if you start by thinking of the height of a person. Can one person stand up in the room? Could one person stand on the shoulders of another person and still be upright? Could more people stand on each other's shoulders? Give your answer in metres.

Estimate the length and breadth of the room in metres. Calculate the floor area. Give your answer in square metres.

Use your estimates of the length, breadth and height of the room to calculate its volume in cubic metres.

Making estimates

Making estimates may seem difficult at first, but you will find that it becomes much easier with a little practice. Making estimates can be rather fun once you realise that no one is looking for an exact or *correct* answer.

Suppose you want to estimate the contents of a box of drawing pins. At first you may think you have no idea. But is it one? No, many more than that. Ten? No, more than that. Is it a thousand? No, that is far too many. Five hundred perhaps? Less than that. Already, you have estimated that it is greater than 10 and less than 500. Perhaps 100 might be a sensible guess. But, you might say, it depends on whether the packet is large or small. All right. If it is small I shall change the estimate to 50 and if it is large to 200. These are not at all likely to be exact answers to the question, but they are not bad guesses.

You might want to estimate the number of words in this book. It is obviously more than 100 and more than 1000. Could it be a million or a thousand million? You have no idea. All right. How many pages are there? About 300. Good. How many words on a page? The trouble is there are a lot of pictures on each page. All right. What is the average number of lines on a page? You might count the number of lines on four or five pages in the middle of the book. Perhaps the average is about 40. And how many words to a line? An estimate might be about 10. That means about 400 words to a page and, with 300 pages, there must be about

120 000 words. To find the exact answer would take someone a very long time to count and would it really make any difference if you knew it was 113 475? Making sensible estimates is good science.

Questions for homework or class discussion

1 Estimate the length of a car in metres.
Estimate the number of cars, bumper to bumper, which could be fitted into 100 metres.
Estimate the width of a car in metres. How many cars could be put side by side in 50 metres?
Use this to decide the maximum number of cars that could be fitted into a field 100 metres by 50 metres.
In a car park it would not be possible to put cars as close as that. Estimate what you think would be the maximum number of cars you could conveniently park in such a field.

2 Estimate the diameter of an apple in centimetres.
Apples are usually sent to market in boxes. Estimate the length, breadth and height of a conveniently sized box. How many apples could be put along one side of the box? How many apples could you fit in the bottom layer? Estimate the total number of apples you could get into the box.

3 Estimate the size of a grain of salt. How many grains would there be in a pinch of salt? Estimate the number of grains there might be in a packet of table salt bought at a grocer's shop.

Background Reading

Taking measurements

Science is concerned with finding out about the world, why things happen, how things work. The most universal physical instruments for finding out are built into our bodies. We get most information through our eyes, but our ears are nearly as important. We also get information through touch: the delicate touch of our finger tips can tell us about texture and there are the muscular senses of pushing and pulling, the feel of hot and cold, and our sense of balance. There is also smell and taste, which give further information. These tools for gaining information are called our senses.

Early scientists, the ancient Greeks for example, relied almost entirely on their senses. They were good at observing and at suggesting explanations of what they saw, but they had no idea of 'doing experiments' as we do them today. They relied on their senses, and senses can be deceived. For example, Aristotle believed that heavy objects fall faster than light ones. Is that in fact correct?

The way the senses can be deceived is illustrated in these drawings.

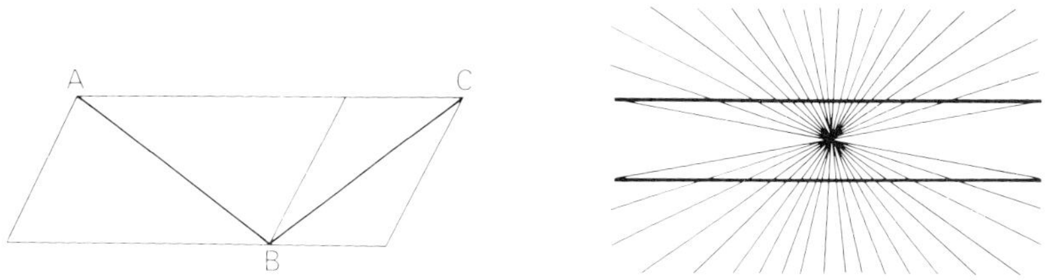

Which is longer, AB or BC? Are the two thick lines straight?

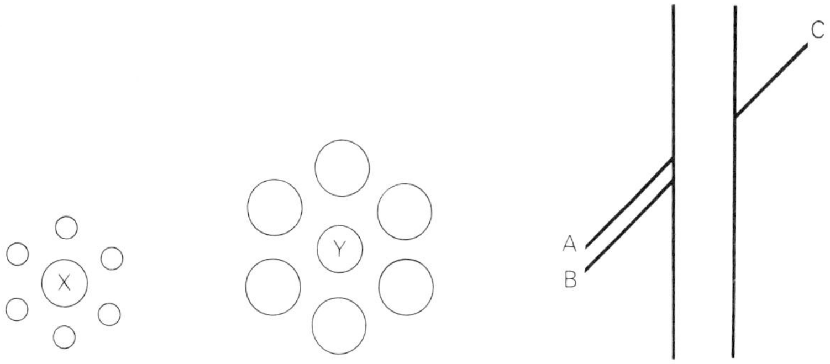

Which of the two centre circles X or Y is larger? Does A or B extend to C?

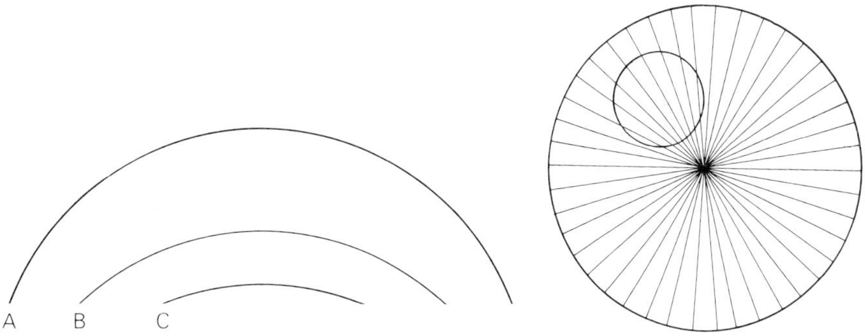

Which arc comes from the largest circle? Which circle is not perfectly round?

It is not only optical illusions where our senses let us down. The experiment with three bowls of water, one cold, one tepid and one hot, described on page 45, is another example. It is because our senses are not reliable enough that scientists make careful measurements. It is only by making measurements that you can convince yourself that AB and BC are the same length in the first example above.

Measurements have helped scientists and engineers to come to an understanding of motion, how aeroplanes fly, how satellites behave, how jet engines work. Measurements make it easier to describe observations. Measurements make observations easier for other scientists to use. Measurements help scientists to sort out and classify their observations. Measurements help scientists to look for patterns in their observations.

It is easy for the traffic police to see that there is a traffic jam at a busy road junction, but if careful measurements of traffic flow are made, the information can be fed into a computer and ways of avoiding the difficulty can be found.

Of course, like the senses, instruments can be deceived. They all have their limitations. Scientists will always repeat their measurements because they know they can make mistakes, but it is also necessary to test instruments and to cross-check to make sure they are reading correctly. Careful measurement is an important part of the work of any physicist.

Measuring mass

A young child soon discovers that, unless a ball or a brick is held up, it will fall to the ground. We explain this by means of the pull of gravity. It was Sir Isaac Newton who developed the idea of gravity, and you have probably heard the story (which may or may not be true) that he watched an apple fall and realised that it did so because of the gravitational pull of the Earth.

This pull or force due to gravity, acting on an object, is called the **weight** of the object. The pull of gravity is a lot less on the surface of the Moon than it is on the Earth. Even on the Earth the pull of gravity gets a little less the higher up you are, so that the weight of a brick is a little bit less on the top of a mountain than it is at sea level.

On the other hand, there is just the same amount of matter in the brick wherever it is. The quantity which measures the amount of matter in an object is called its **mass**, and it is measured in kilograms (kg). One kilogram of butter has exactly the same mass of 1 kg whether it is on the Moon, on the top of Mount Everest or on the breakfast table, because there is the same amount of butter in each situation. But it will have different weights in those different places because the pull of gravity is different.

The kilogram is sometimes too large a unit and the gram is used instead. 1000 grams is the same as 1 kilogram. It is sometimes convenient to use a milligram (abbreviated to mg), which is one thousandth of a gram.

Unit	*Abbreviation*
kilogram	kg
gram	g
milligram	mg

$$1\ \text{kg} = 1000\ \text{g}$$
$$1\ \text{g} = \tfrac{1}{1000}\ \text{kg} = 0.001\ \text{kg}$$
$$1\ \text{mg} = \tfrac{1}{1000}\ \text{g} = 0.001\ \text{g}$$

It is very unfortunate that people use the word *weight* when they mean *mass*. For example, you may find a packet of butter or groceries marked 'net weight 250 g' when it ought to say 'net mass 250 g'. And people use the word *weighing* when they mean *finding the mass* of something. In this book, we shall do our best to use these words correctly and only use the word *weight* to mean the pull of gravity on the object.

Measuring mass using a balance

The easiest way to find the mass of an object is to use an instrument called a balance. There are many different kinds of balance, and the photographs show two different kinds. The first is a lever-arm balance and the second is an electronic balance.

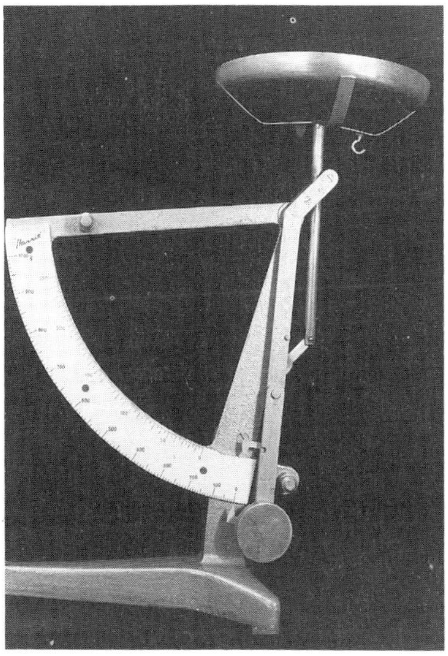

Lever-arm balance

Electronic balance

Experiment 3.3
Using a balance to measure mass

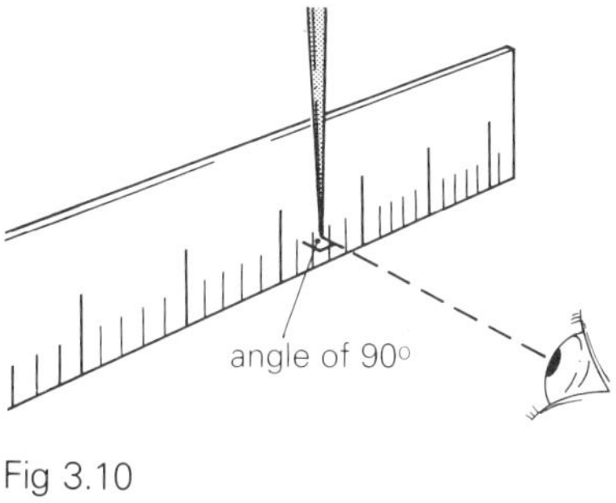

Fig 3.10

Your teacher will tell you how to use the kind of balance in your laboratory, and there will be different objects whose mass you can measure.

Make sure you know how to adjust the balance to read zero when there is nothing in the pan. Some balances have a pointer moving near a scale, and you will find that the reading can change if you move your head from one side to the other. The reading should be taken when your eye is directly in front of the pointer and scale, as shown in Fig 3.10. This helps to keep any errors as small as possible.

You may wish to check the accuracy of the balance you are using. How could you do this? (*Hint*: there will be some 100 g and some 1 kg masses in your laboratory which might help you.)

Background Reading

The meaning of words

In Lewis Carroll's *Alice Through The Looking Glass* Humpty Dumpty says that words often mean whatever we want them to mean. Certainly it must be very difficult for foreigners to learn to speak English when so many words have more than one meaning. The word 'fine', for example, can be used in different ways.

'The weather is fine'.
'I have got a parking fine'.
'This thread is very fine'.
'That is a fine building'.
'I am feeling fine'.
'She is studying Fine Arts'.

In science it is very important that people should be able to communicate together without misunderstandings. Scientific words should have only one meaning so that there is no misunderstanding.

In Israel the language spoken is Hebrew, a language based on that used in the Old Testament Bible. Unfortunately, when that language was developed, the knowledge of science was not great and the same Hebrew word was used for **force** and for **energy**. As these two are quite different things scientifically, it is very difficult for young people who have grown up using the same word for each.

In England, the confusing word, as already explained, is the word *weighing*. It is this word that makes it difficult for some people to understand the word *mass*. It is unfortunate that we do not use the word *mass* as a verb, for then we could drop the word *weighing* and talk about 'massing an object' instead. That would be better as far as science is concerned, but it is never easy to change the meaning of words used in everyday language.

Chapter 4 Sorting things out

Let us think again about why certain substances are chosen to do certain jobs in everyday life. Consider what is used in the making of an easy chair. The framework is probably made of wood held together by steel screws and glue. This framework is often covered with a web of woven fabric strips, fixed to the frame with steel tacks and padded with foam plastic. The seat cushion often rests on several long steel springs or strong rubbery strips to make it more comfortable to sit on. The whole is covered with some decorative material which will not wear out quickly.

It is easy to think of the reasons why different substances are used. A strong material like wood is chosen for the frame so that the chair does not collapse when someone sits on it. Steel screws are used because steel is harder than wood, so they can be screwed into the wood. What are the special properties which the padding needs? To design and make a chair successfully, you need to select the best substances for the various parts. And that means knowing about what substances are available and how they behave.

In order that a designer can choose the best substance, it is necessary to sort them out: which are hard and which are soft; which are strong and difficult to stretch and which are weak and stretch easily; which are flexible and bend easily and which are stiff and difficult to bend. Some substances can be seen through (they are said to be transparent). Some float in water, some sink, yet others will dissolve in water. Some melt very easily as the temperature rises, and some are very difficult to melt.

Experiment 4.1
Exploring materials

In this experiment you are asked to compare materials to find out, if you can, which are strong and which are weak, which are stiff and which are flexible, which are hard and which are soft.

The first stage of such an experiment is to think about how you are going to do the investigations. You will need to plan on paper what you are going to do, and what measurements to make, and the tests must be fair. It is not much use deciding aluminium is flexible and wood is stiff if you compare a piece of aluminium foil with a log of wood!

At the end of the experiment, put those substances you try in an order of hardness or flexibility or strength.

Questions for class discussion

1 Use the words strong, weak, stiff, flexible, hard, soft to compare
- **a** a biscuit with a piece of cooked spaghetti;
- **b** cotton thread with thin copper wire;
- **c** sellotape with a strip of paper;
- **d** steel with aluminium.

2 Make a list of some of the materials used in building a car and write a sentence or two to say why each is chosen for the job it does.

3 a Why is concrete used for building bridges but not for motor car bodies?

b Could the blade of a spade be made of lead or aluminium instead of steel?

Density

Suppose you were to build a metal bridge. What metal would you use? Most would say steel – and that would be correct because steel is hard, strong and stiff. Those are also the reasons why steel is used to build cars. But why is it not used to build aeroplanes? Why is aluminium used instead?

It is a great temptation to say that aluminium is lighter than steel, but you must be careful. A large block of aluminium is a lot heavier than a small steel screw!

You have probably heard the old riddle: 'Which is heavier, a kilogram of lead or a kilogram of feathers?'. What a temptation it is to say lead, but of course the lead and the feathers both have the same mass of 1 kilogram and therefore weigh the same, although the volume of the feathers is much greater than the volume of the lead. If we were to compare the masses of the *same* volume of lead and of feathers, the mass of the lead would be much greater.

This gives us another useful way of comparing substances. We can find the mass of a certain volume of the substance, and a convenient volume to use is 1 cubic centimetre. Actually, it is more usual to use 1 cubic metre as the unit of volume, but the arithmetic will be easier if volumes are measured in cubic centimetres.

Suppose a block of aluminium measures 3 cm × 4 cm × 5 cm with a mass of 150 g. How do you find the mass of 1 cm^3? The volume of the block is 60 cm^3, that is to say, it is the same as 60 cubes each of 1 cm^3. So the mass of a cube of size 1 cm^3 is found by dividing 150 by 60 to give 2.5 g.

The mass of a 1 cm^3 or 1 m^3 volume is called the **density** of the substance and, in this case, the unit is grams per cubic centimetre or g/cm^3. Notice how we found the density by dividing the mass of the substance (150 g above) by the volume (60 cm^3).

$$\text{DENSITY} = \frac{\text{MASS}}{\text{VOLUME}}$$

Experiment 4.2
Measuring the density of a solid

Measure the masses of various blocks of different substances using a balance. Draw a table similar to the one below and write down the masses in the second column.

Material	Mass g	Length cm	Breadth cm	Height cm	Volume cm^3	Density g/cm^3
aluminium						
iron						
softwood						
hardwood						
lead						
glass						
marble						
Perspex						
paraffin wax						
...						
...						

Blocks of material

Measure each block using a ruler and put the measurements in the table (you need only measure to the nearest centimetre). Then calculate the volume and put the value in the next column. Finally, work out the density and put the value in the last column.

Now make a list of the materials you have measured, sorting them in order of increasing density.

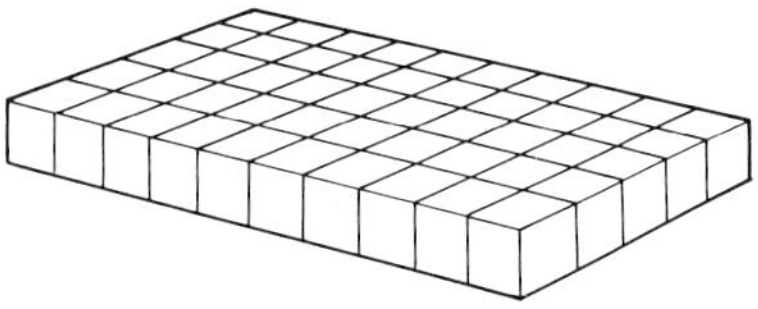

Fig 4.1

Try to measure the density of a glass stopper or of a stone. Look back to Experiment 3.2 if you have forgotten how to find the volume of an irregular solid.

What problem might you have if you were trying to find the density of a copper sulphate crystal? How could you get over this difficulty?

Try to find the mass of a small piece of expanded polystyrene. It is a bit difficult to get an accurate answer. Can you think of a way of getting over this difficulty? Perhaps Fig 4.1 gives a clue.

Questions on density

1 A block of iron is 5 cm × 5 cm × 4 cm. It has a mass of 750 g. What is the volume of the block? What is its density?

2 If the density of lead is 11 g/cm^3, what is the mass of **a** 1 cm^3 **b** 10 cm^3 **c** 100 cm^3 of lead?

3 The density of balsa wood is 0.2 g/cm^3. What is the mass of **a** 1 cm^3 **b** 5 cm^3 **c** 10 cm^3 **d** 50 cm^3 of balsa wood?

4 The density of a substance is 3 g/cm^3. What is the volume of a piece of the substance which has a mass of **a** 27 g **b** 270 g?

5 Water has a density of 1 g/cm^3. Ice has a density of 0.9 g/cm^3. Which has the bigger volume: 10 g of water or 10 g of ice?

6 Write down your own mass in kilograms. What is your mass in grams? The density of the human body is about the same as that of water. Use this fact to calculate the volume of your body.

7 The density of lead is 11 g/cm^3. The density of iron is 7.5 g/cm^3. Which has the greater mass: 16 cm^3 of iron or 10 cm^3 of lead?

8 (Difficult) The density of gold is 19 g/cm^3. A block of gold is measured and the volume is found to be 10 cm^3. It is found that the mass of the gold is only 170 g. Can you suggest a possible explanation of this?

Measuring the density of a liquid

You are not allowed to pour liquid on to the pan of a balance in order to find its mass. It would soon ruin the balance if any liquid got into the works! To find the mass of liquid you have to use a separate container as explained in the experiment below.

Experiment 4.3
Measuring the density of a liquid

First, using a measuring cylinder, measure out 100 cm^3 of liquid. Then take a beaker and find its mass when it is empty and dry.

Pour the liquid into the beaker and find the combined mass of the beaker and the liquid. Calculate the mass of the liquid by subtracting these readings. Knowing the mass of the liquid and its volume, calculate the density.

Use this method to find the density of several different liquids.

It is suggested that the experiment could be simplified a little if a polystyrene cup were used instead of a beaker. What do you think of that suggestion?

Floating and sinking in liquids

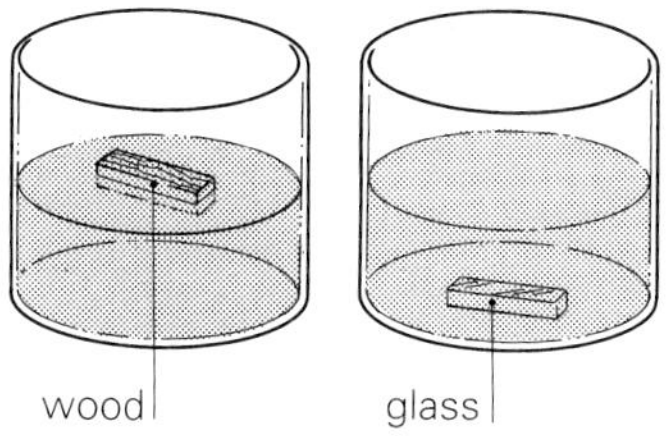

Fig 4.2

Some solids will float when placed in a liquid, others will sink.

If a dish is half-filled with water as shown in Fig 4.2, what happens when a block of softwood is placed in the water? What happens when a similar shaped glass block is placed in it?

The density of water is 1 g/cm^3. The density of softwood is about 0.5 g/cm^3, but the density of glass is about 2.5 g/cm^3. It looks as though the solid floats when its density is less than that of the liquid, but sinks when it is greater. We can test this theory with other liquids.

Mercury is a liquid metal which has a density of 13.6 g/cm^3. A glass block would therefore float in it. A lead block with density 11.4 g/cm^3 would also float in it, but a block of gold (density 19.3 g/cm^3) would not.

When a wooden block floats on water, does the pull of gravity on the block disappear? If it does not, why doesn't the block fall to the bottom of the beaker? It would be silly to think the Earth stops pulling on the block just because it is in water. Being in water certainly did not stop the Earth pulling on the glass block. When the wooden block is placed in the water, the water pushes upwards on the wood (Fig 4.3). The block sinks until the upward push of the water equals the downward pull of gravity, and the block then floats. With the glass block, the upward push is never as big as the downward pull and so it sinks.

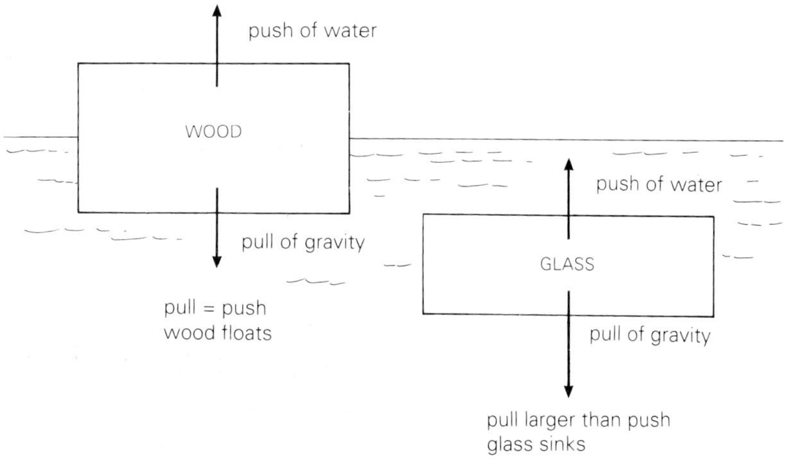

Fig 4.3

Questions for class discussion

1 An iceberg will float in the sea with some of it above the surface (though a lot more is below the surface). What can this tell us about the density of ice compared with the density of water?

2 When a boy floats in water, there is not much of him above the surface. What does this tell us about the density of the boy?

3 If the same boy floats in the Dead Sea, he will lie with rather more of his body out of the water. What does this tell us about the water in the Dead Sea?

4 Three liquids were poured into a glass jar, shaken up and left to stand. When they had settled the appearance was as shown in Fig 4.4. Which liquid, A, B, or C, has the greatest density? Which liquid has the least density? If A is water and C is mercury, why cannot B be a solution of copper sulphate?

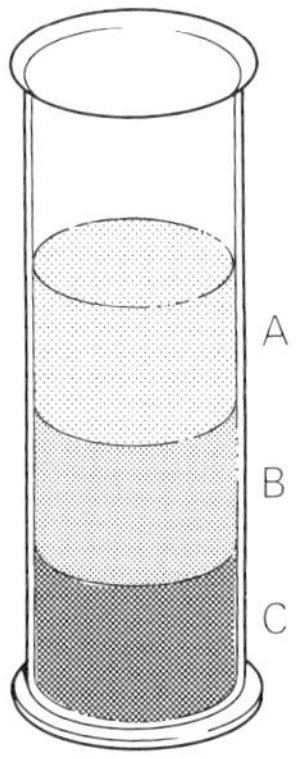

Fig 4.4

Air

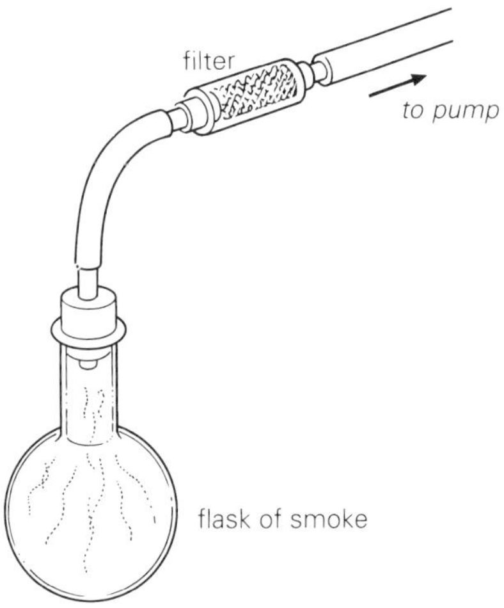

Fig 4.5

We have found the density of solid substances and of liquids, but what about gases? How could we measure the density of air?

We speak of a bottle as being half full of water. Does that mean that it is also half empty? No, because there is air in the other half. In the exhibition of materials in your laboratory at the beginning of this course there were two bottles, both of which looked empty, but one was labelled 'air' and the other 'vacuum'. How do we know that the second bottle had nothing inside it? You cannot open it and look inside. Some demonstration experiments will help to answer this.

Demonstrations with a vacuum pump

1 To show the effect of a pump, some smoke should be placed in a round flask which is connected to the pump through a filter (Fig 4.5). When the pump is switched on, the smoke disappears because the air in the flask and the smoke particles are removed. Can you think why a filter is advisable?

2 To show the effect of atmospheric pressure, the pump may be connected to a large polythene bottle or a tin can (Fig 4.6). When the air is pumped out, the bottle or can collapses because of the air pressing on the outside but not the inside.

3 To investigate two glass bottles, one labelled 'air', the other 'vacuum', the two bottles should be similar, and one is attached to the vacuum pump so that the air is drawn out. The clip on the tubing is closed before it is disconnected from the pump. If each bottle is opened with the end of the tubing

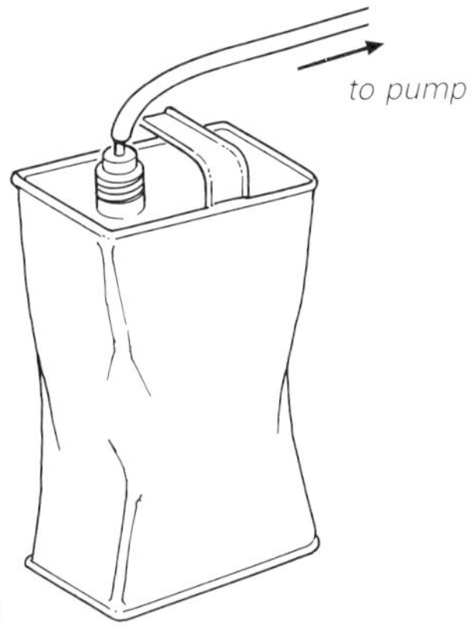

Fig 4.6

under water (see Fig 4.7), water enters one of the bottles but not the other. Can you explain the difference?

Although water enters the bottle labelled 'vacuum', the water does not fill the whole bottle because some air remains. A good vacuum pump removes most of the air but not all of it.

Fig 4.7

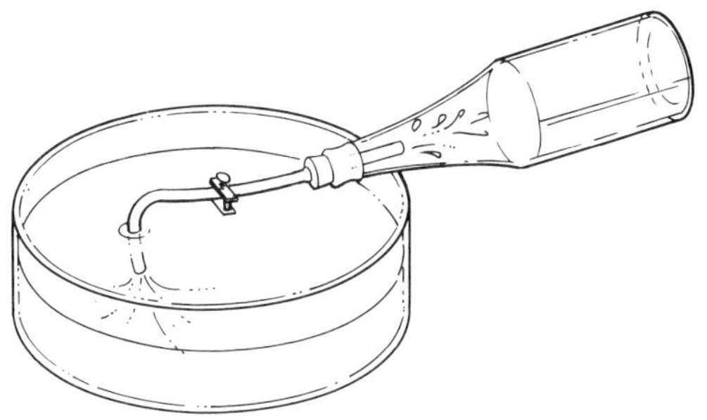

Experiment 4.4
Measuring the density of air

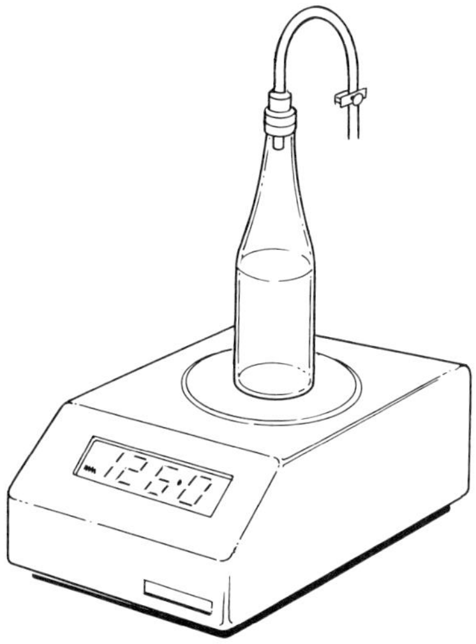

Fig 4.8

When finding the mass of the liquid in Experiment 4.3, you measured the mass of an empty container and then the mass of the same container with liquid in it. This time, you find the mass of the bottle plus the air in it, and then the mass of the bottle with the air removed by a vacuum pump. The difference is not very large, but a sensitive electronic balance (Fig 4.8) will allow you to do this experiment easily and with accuracy. Think hard about the best way to measure the volume of air which the pump extracted. Could you use water and a measuring cylinder to do this? Why is it better to measure the mass of the bottle full of air first rather than the evacuated bottle first?

If a sensitive balance is not available, it is possible to obtain a good estimate of the density of the air using the method of Question **5** overleaf.

Under normal conditions, the density of air can be taken to be 0.0012 g/cm^3.

Questions for class discussion

1 Estimate the length, breadth and height of the room in which you are now sitting. What is the volume of the room, in cubic metres and in cubic centimetres?

2 What is the mass of the air in the room?

3 What volume of water would have the same mass?

4 If a balloon is filled with hydrogen and let go, it rises upwards. Why is this?

5 In an experiment to measure the density of air (see Fig 4.9), it was found that the large cubical plastic container (measuring 30 cm × 30 cm × 30 cm) originally had a mass of 460 g. When air had been pumped in under pressure, the container had a mass of 466 g. When the 'extra air' was let out and collected as shown in Fig 4.9, it was found that there was enough to fill a cubical Perspex box, 10 cm × 10 cm × 10 cm, five times.
a What is the mass of the 'extra air'?
b What is the total volume of the 'extra air' released?
c Calculate the mass of 1 m^3 of air at normal pressure.

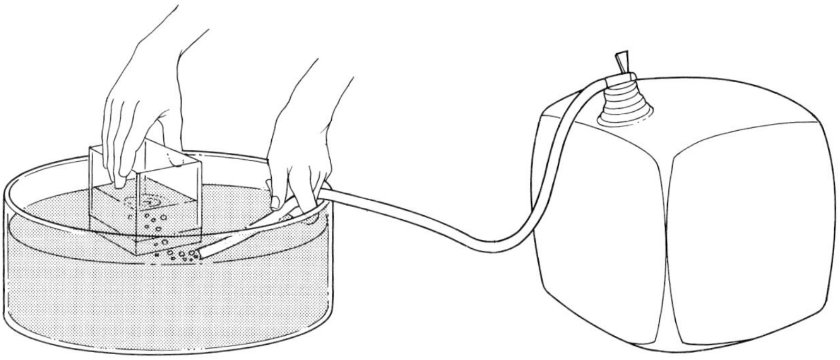

Fig 4.9

Values for the densities of different substances

Substance	Density in kg/m^3
aluminium	2 700
copper	8 940
gold	19 320
iron	7 860
lead	11 350
nickel	8 900
platinum	21 450
silver	10 500
zinc	7 130
brass	8 300
expanded polystyrene	15
glass	2 800
marble	3 200
paraffin wax	950
Perspex	1 200
softwood	500
hardwood	700
methylated spirits	790
mercury	13 550
turpentine	860
water	1 000
air in a room	1.2
hydrogen in a balloon	about 0.09

In this chapter we have measured our masses in grams and our volumes in cubic centimetres because it was convenient to do so. But scientists prefer to use kilograms for their standard unit of mass and cubic metres instead of cubic centimetres. Sometimes the densities are measured in grams per cubic centimetre, sometimes in kilograms per cubic metre.

It is not difficult to change from g/cm^3 to kg/m^3 and vice versa. Think about water.

1 cm^3 of water has a mass of 1 g.
There are 10^6 cm^3 in 1 m^3.
Thus, 1 m^3 of water has a mass of 10^6 g.
There are 10^3 g in 1 kg.
Thus, 1 m^3 of water has a mass of $\frac{10^6}{10^3} = 10^3 = 1000$ kg.
The density of water is 1000 kg/m^3.

1 g/cm^3 is exactly the same density as 1000 kg/m^3.

To change g/cm^3 into kg/m^3, you need to multiply by 1000.
To change kg/m^3 into g/cm^3, you need to divide by 1000.

The table on the left gives densities of common materials in kg/m^3.

Chapter 5 Further measurements

In Chapter 4 we learned how to measure densities. In this chapter we shall think about some more ways of making measurements. Sometimes an experiment will not be very successful at first; then it is our task as scientists to think of ways of improving it. If possible, you should try all the experiments yourself; you can always compare your results afterwards with those of other people – scientists like doing that as well.

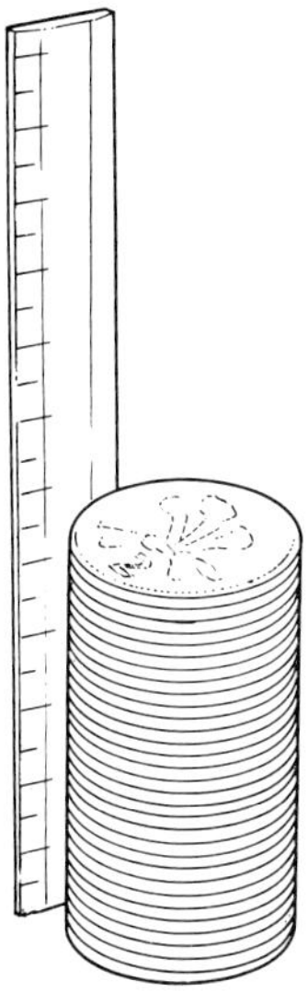

Fig 5.1

Experiment 5.1
Finding the thickness of a coin

First guess the thickness of a coin. Then try to measure it using a plastic ruler marked only in centimetres.

Neither of the above are very precise. It would be better to use a ruler with a millimetre scale on it. How precise can you be now? How can you improve the experiment further? Fig 5.1 gives a clue.

What is the height of ten coins? What does that tell you about the thickness of one coin?

See if you can measure the height of 100 coins. What does that measurement tell you about ten coins? About one coin?

Would there be any point in going further?

Experiment 5.2
Finding the thickness of a piece of paper

Measure the thickness of a piece of paper as accurately as you can.

The thickness of paper in books and magazines varies considerably. Investigate how they differ. Find the thickness of a page in this textbook and compare it with the thickness of a page in a glossy magazine.

Measuring mass again

Most balances are not very useful when it is necessary to measure small masses such as the mass of a hair or a grain of sand. The next experiments will show you how you can make your own sensitive balance for measuring small masses.

Experiment 5.3
A simple balance to find the mass of a parcel

The object of this experiment is to find the mass of a small parcel using a simple balance. Set up the balance as shown in Fig 5.2.

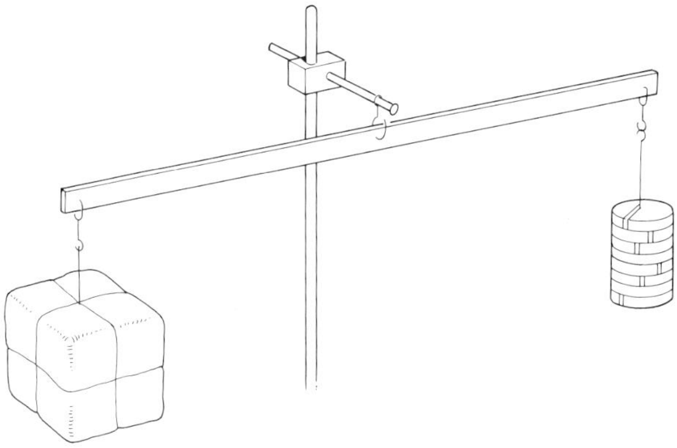

Fig 5.2

Put the parcel on one side of the balance and the hanger on the other side. Add 100 g masses to the hanger and try to work out the mass of the parcel.

It will probably be very difficult to get it to balance exactly. Does this matter? Can you still estimate the mass of the parcel?

Experiment 5.4
Using the simple balance to find the mass of a letter

Use the same balance to find the mass of a letter (Fig 5.3). It is much lighter than the parcel so you will need to use a hanger with 10 g masses instead of the 100 g masses.

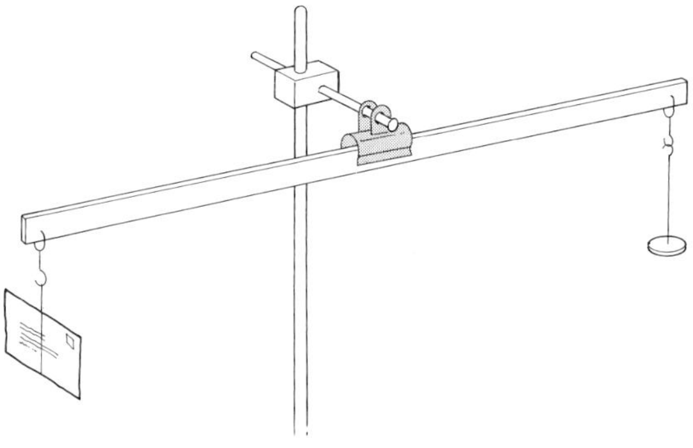

Fig 5.3

It may be very difficult to get it to balance. If so, use a bulldog clip instead of the central hook. Can you estimate the mass?

Suppose you need to know if the mass is greater or less than 25 g, but your masses on the hanger only give 10 g, 20 g, 30 g and so on. Can you think of a way of doing this? Perhaps Fig 5.4 will give you a hint of how it might be done.

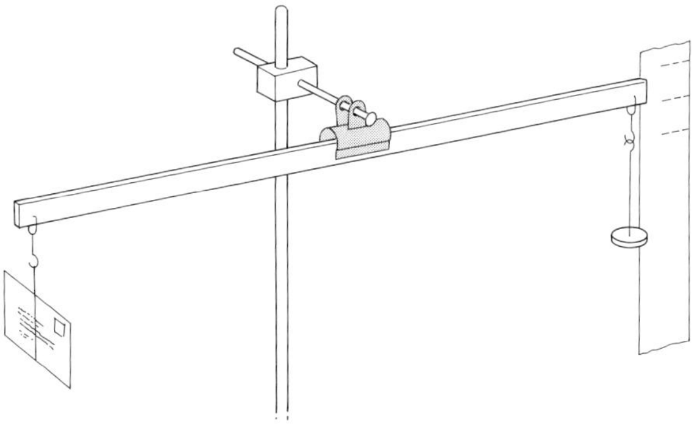

Fig 5.4

The above apparatus is quite good for finding the mass of an envelope, but it would not find the mass of a dead fly or a hair. It looks as though we need something different for that.

Experiment 5.5
Making a microbalance

You will need the equipment shown in Fig 5.5.

Put the screw in the end of the drinking straw and fit the other pieces together as shown in Fig 5.6.

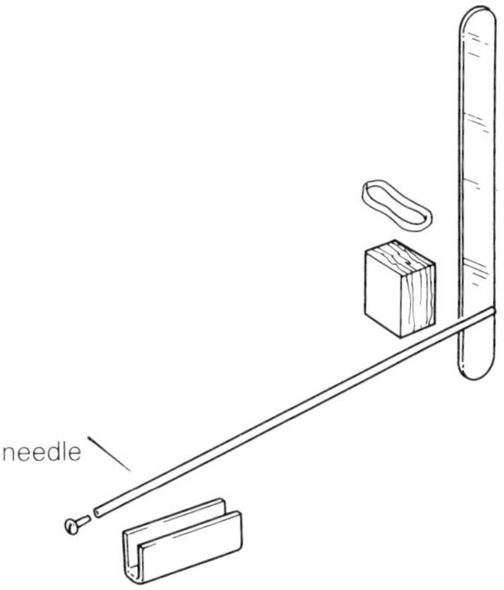

Fig 5.5

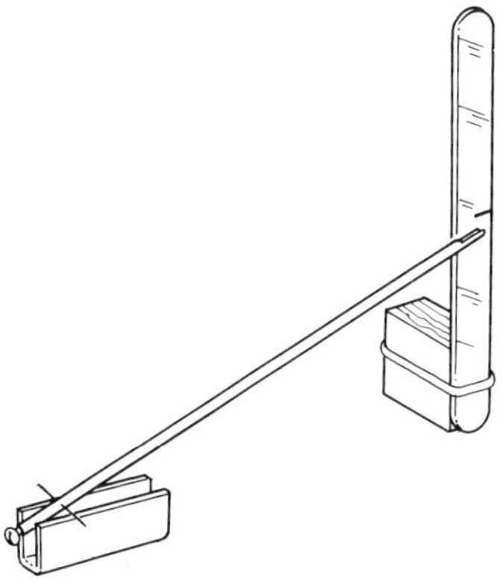

Fig 5.6

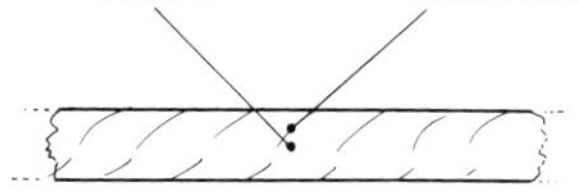

Fig 5.7

Before sticking the needle through the straw the approximate balance position can be found by balancing the straw on the needle. The needle should not go through the centre of the straw, but above it (see Fig 5.7). The screw can be used to adjust the balance. Squash the other end of the straw to make a little platform on which to put things to be measured.

Do not worry if you damage your straw: you can always have another one.

When the apparatus is assembled and balancing, put a grain of sand or a hair on it, and see if it is working. You may find that draughts are a nuisance; if they are, make a shield with a screen or a large book.

Experiment 5.6
Using the microbalance

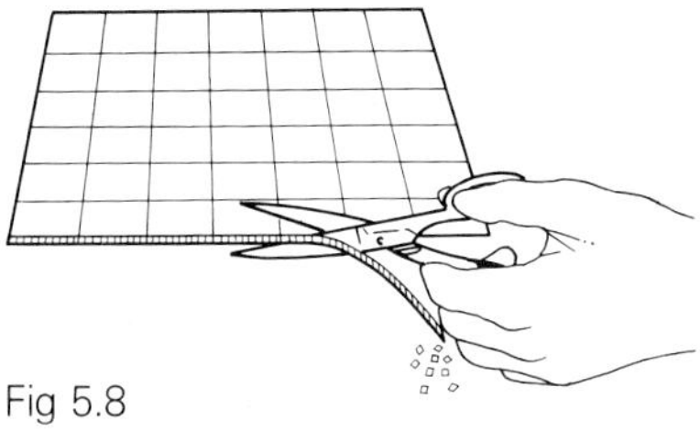

Fig 5.8

To do any measuring with your microbalance, you will need some masses. You have some graph paper and some scissors with your apparatus. You can cut up some small squares from the graph paper (see Fig 5.8) and these will make very useful masses.

Mark the position of the end of the straw on the upright when there is no mass on it. Put on one small square and mark the position, put on a second square and mark that position, and so on. Go on putting on more masses and mark all along the scale. If you put the hair on the balance instead of the squares, you will be able to read off the mass of the hair, measured in 'small squares'.

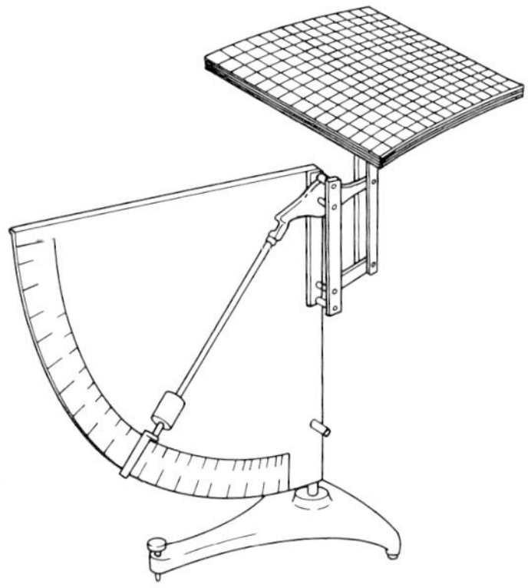

Fig 5.9

How can you find the mass of the hair in grams? You need to know the mass of each 'small square'. The easiest way to find this is to put 100 sheets of the graph paper on an ordinary balance and find their mass (Fig 5.9). It is then easy to find the mass of 1 sheet. Count up the number of large squares on the sheet and you can calculate the mass of one of them. But there are 100 'small squares' in one large one. So you can find the mass of one small square. It is usually convenient to measure it in milligrams (1 mg $= \frac{1}{1000}$ g).

Can you now find the mass of the hair in milligrams?

Sensitivity of a microbalance

We call a balance very sensitive if a small mass added to it moves the pointer a large distance. The sensitivity of a microbalance depends to some extent on where the needle is put through the drinking straw. The sensitivity is low if the needle is put near the edge of the drinking straw; it gets higher the closer the needle gets to the centre of the straw. If you have time, try changing the sensitivity of your microbalance (Fig 5.10).

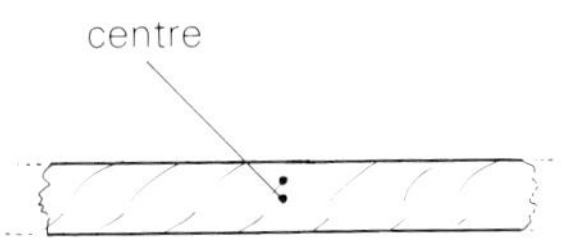

Fig 5.10

Experiment 5.7
Measuring the thickness of aluminium leaf

This is a difficult experiment and will require some thought. The aluminium leaf is very thin and it is not possible to pile it up and measure the thickness of the pile as was done for sheets of paper.

Start with a square piece of leaf, 5 cm × 5 cm. We know its length and breadth, but want to know its thickness.

We cannot measure its thickness directly. Is there anything else we can measure? Its mass. This is very small, but it can be found with the microbalance.

If we know its mass, can we find its volume? Yes, because we have already found the density of aluminium in an earlier experiment. As we already know the length and breadth of the original piece of leaf, we can now use the volume to find the thickness.

This experiment can be represented by the chart shown below.

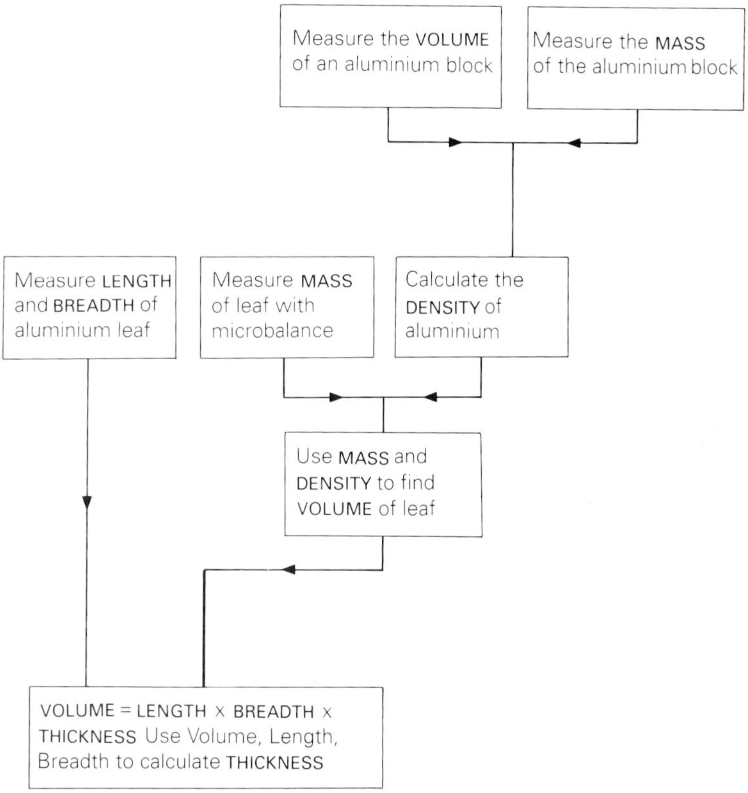

Measuring time intervals

Time is another quantity that scientists need to measure. The usual unit is the second (abbreviated to s) though there will, of course, be occasions when it is necessary to measure in minutes, hours, days or years.

The time between two events can be measured with a clock, preferably one with a second hand. Stopwatches are particularly useful for timing.

Questions for class discussion on measuring time intervals

1 When you are timing the interval between events which occur regularly, such as your pulse, why is it more accurate to time ten intervals and divide the answer by ten than it is to time one interval?

2 When counting ten intervals, why must you be careful not to count the first event at which you start timing as one, the second as two and so on up to ten? Why should you count the first event at which you start timing as zero, the next as one and so on up to ten? (In practice, it helps sometimes to have a countdown to zero.)

Timing things

You should use a stopwatch to time different things: the time between hand-claps, the time for a book to fall, the time it takes to measure the mass of an aluminium block, the time to walk round the room or to run round the building.

Try estimating time intervals. You can become quite good at counting seconds with a little practice. One way is to say 'apples and pears one', 'apples and pears two' and so on at a normal speaking speed. (Some people have their own words, for example 'Mississippi one', 'Mississippi two', and so on.) It does not matter what you use, but you should practise with it and you will soon find the right speed at which to say the words, so that you can count quite accurately to 30 seconds.

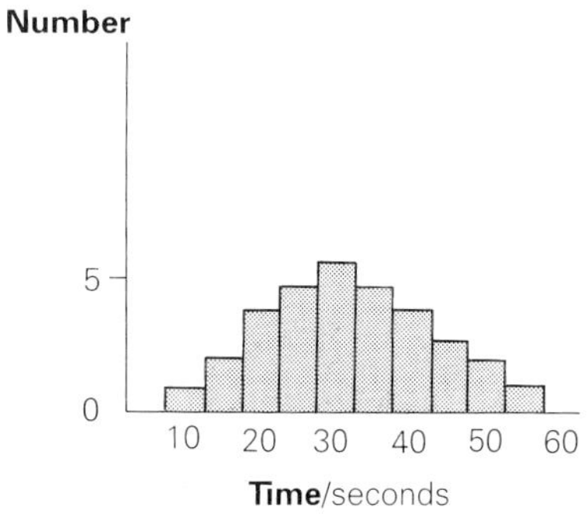

Fig 5.11

Experiment 5.8
Class experiment estimating 30 seconds

An interesting game can be played in which everyone in the class estimates 30 seconds. The teacher has a watch and starts everyone together. As soon as you think 30 seconds have gone by you raise your hand quietly. (It is better to do it with your eyes shut and to raise your hand so that others do not hear.)

The first time it is likely that someone will put their hand up after 20 seconds or even earlier, and some may not raise theirs for 50 seconds. There might be a spread like the first chart (Fig 5.11).

After you have played the game several times, everyone will be much better at estimating. Perhaps the spread will be more like the second chart (Fig 5.12).

Can you see the difference?

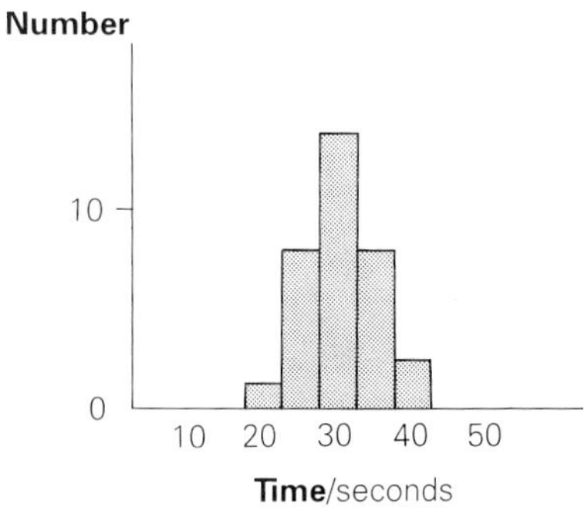

Fig 5.12

Background Reading

Measuring speed

Clocks and watches can be used to measure the time between the start and end of a journey, in other words what we usually call a **time interval**. If you know the distance travelled by a vehicle in such a time interval, you can easily calculate its speed. Speed is calculated by dividing the distance an object travels by the time it takes to do so. It is the distance the object travels in one unit of time and it has a unit such as miles per hour or metres per second. Of course if you find how far a car travels on a long journey, you will only get an average speed as the actual speed will not stay the same the whole time.

Stroboscope

$$\text{average speed} = \frac{\text{distance travelled}}{\text{time taken}}$$

In the laboratory, speed can be measured by a photographic method using a flashing lamp, called a **stroboscope** (see the photograph).

To obtain the photograph below the toy car was illuminated every tenth of a second by the stroboscope and a time-exposure photograph was taken as it moved past a half-metre rule. A series of pictures was then obtained on the same negative.

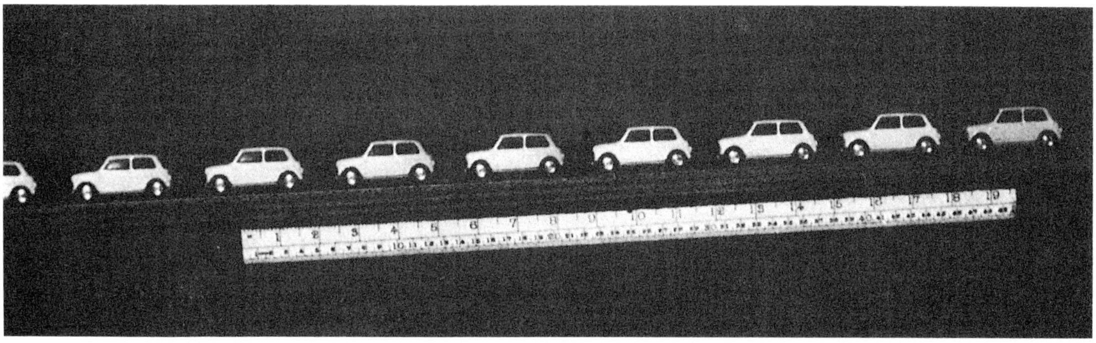

It is possible to see from the photograph that the toy car was moving at a steady speed. How do we know this? As the flashes occurred every tenth of a second, it took the car six-tenths of a second to travel about half a metre and from that we can calculate its speed.

The photograph of a tennis player serving (page 42) was taken using an even faster stroboscope. Can you tell when the racket is moving slowly, when it is speeding up and when it is slowing down? Do you notice the difference in the speed of the ball when it is being thrown upwards and after it has been hit?

Two other stroboscopic photographs are shown. The first shows a ball projected upwards: the second shows three balls projected horizontally at the same moment with different speeds. Do you notice anything about how they are falling?

Stroboscopic photograph of a tennis player serving

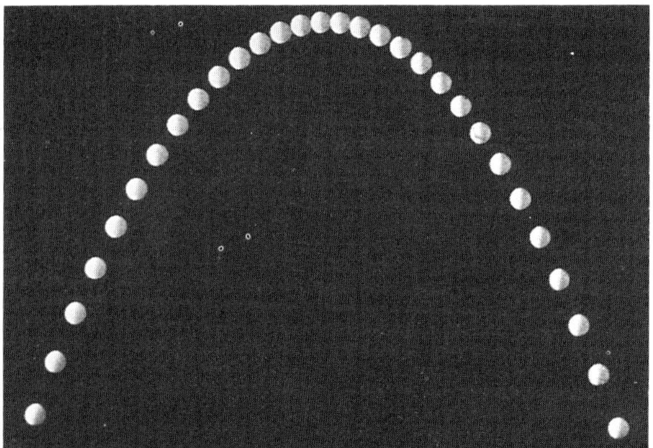

Path of a ball projected upwards

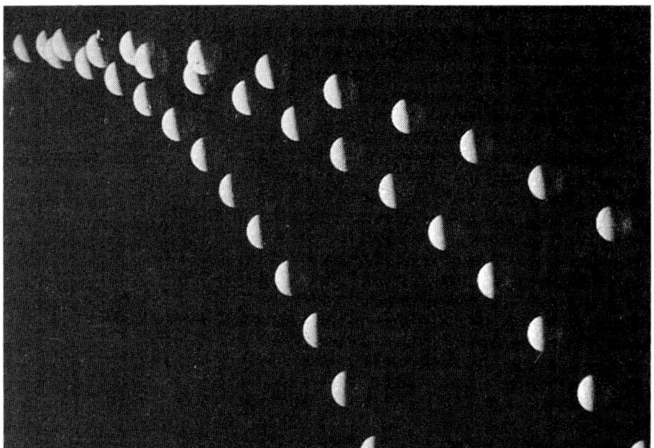

Paths of a ball projected horizontally, at three different firing speeds

Statistics

A statistics frame (see Fig 5.13) is a useful way of showing a spread of results. The following experiment will illustrate its use.

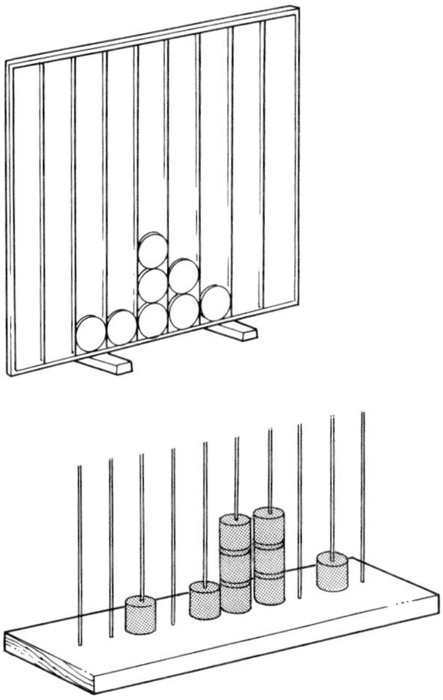

Fig 5.13

Experiment 5.9
Class experiment with statistics frame

Everyone in the class should find their mass in kilograms. (Some bathroom scales are necessary for this.) Every student should be given a disc to place in the statistics frame. Suppose the first slot in the frame represents masses up to 10 kg, the next slot masses between 10 kg and 20 kg, the next between 20 kg and 30 kg, and so on up to 90 kg. (See Fig 5.14.)

Each student then puts their disc in the slot appropriate to their mass. You will see that we get a spread. But you will notice we soon get one or two slots over-full. What can we do about it?

Let us try a different spread. Suppose the first slot is 35–40 kg, the next 40–45 kg, the next 45–50 kg, and so on. (See Fig 5.15.) This spreads out the commonest masses.

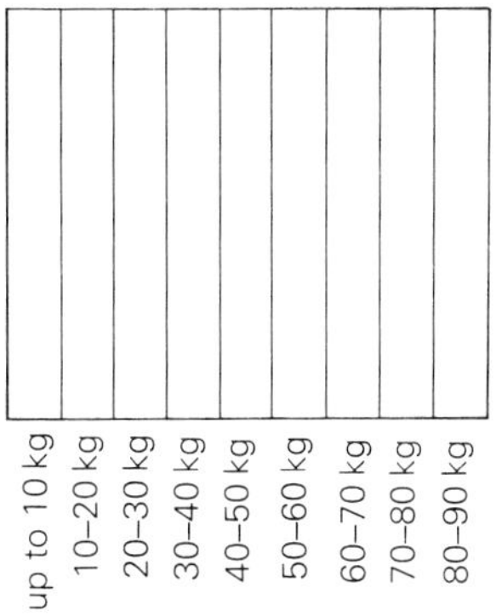

Fig 5.14

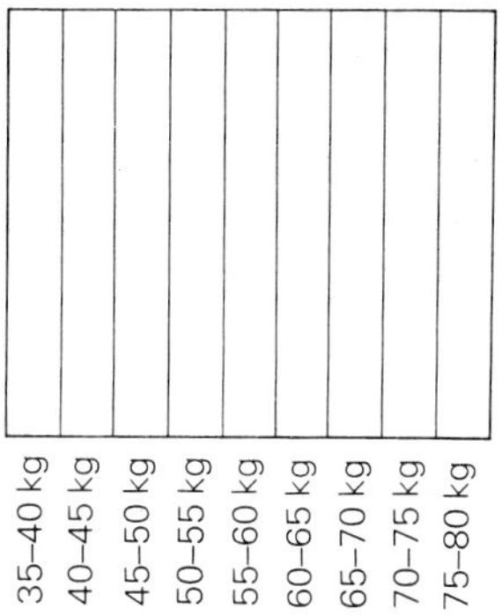

Fig 5.15

Other statistical experiments

1 Let everyone in the class measure their height (or the length of their foot, or the size of their waist). Put the results on the statistics frame.

2 Let everyone in the class measure the mass of the same block of aluminium. Not everyone will get the same answer: there will be a spread of readings. Show these on the statistics frame.

3 If your school has a lot of 500 g masses (or 1 kg or 200 g masses), measure the mass of each of them on a balance. It is unlikely that all the readings will be the same. Show the spread on the statistics frame.

Averages

Statistical experiments **2** and **3** on page 43, show that we must be very careful about using phrases like 'absolutely accurate' when making measurements: there is usually a statistical spread. The best you can do is to get an average value.

The usual method for finding an average is to add all the measurements up and divide by the number of measurements. Thus the average of 501 g, 507 g, 502 g, 502 g, 503 g is obtained as follows:

$$\text{average} = \frac{501 + 507 + 502 + 502 + 503}{5}\,\text{g} = \frac{2515}{5}\,\text{g} = 503\,\text{g}$$

The average value is therefore 503 g.

There is in fact an easier way to work out the average of those five numbers. Can you think what it is?

Estimating games

Being able to estimate distances and times is very helpful in scientific work. You can get practice in this when out for a walk. Before you go, use a metre rule to measure the length of your pace. What distance in metres will be the same as ten paces? What distance will be the same as 100 paces?

1 Choose something ahead of you on your walk, like a lamp-post or a tree, and estimate how many paces you think it will be until you reach it. Then pace it out and see how close your estimate was.

2 Then start to make estimates in metres. Try it with objects which are quite near and ones which are much further away.

3 Another game is to estimate the time it will take to walk to the tree or the lamp-post, if you walk at a steady speed.

4 Along the side of a railway track there are usually quarter-mile posts (see the photograph) and $\frac{1}{4}$ mile is very close to 400 m. If you look out of a train and estimate the time taken between two of these posts, you can find the speed of the train. This table will help.

Quarter-mile post

Time between $\frac{1}{4}$ mile posts (in seconds)	Time to travel 1 mile (in seconds)	Speed (in miles per hour)	Speed (in kilometres per hour)
10	40	90	144
15	60	60	96
20	80	45	72
30	120	30	48
60	240	15	24

5 A more difficult game is to work out the speed of a car on a road. You must first pace out 100 metres along the road, and then estimate the time a car takes to travel that distance. From this you can work out its speed.

Measuring temperatures

Before we leave measurements, we ought to mention one more: temperature measurement. You will be studying heating later: this is merely a first look at one aspect of it.

In scientific work, temperature is often measured on a scale called the Celsius scale. If you have a thermometer marked in degrees Celsius (°C), measure the temperature of melting ice and the temperature of boiling water.

- On the Celsius scale the temperature of melting ice is 0 °C.
- Pure water boils at 100 °C under normal conditions.
- If the temperature is below 0 °C, water will freeze and turn to ice.
- 5 °C is the temperature of a very cold day, 35 °C would be sweltering.
- −88 °C is the lowest temperature recorded in Antarctica.
- 59 °C is the highest temperature ever recorded in the shade (in Algeria).

Find out what the temperature is of the room in which you are reading this. What is the temperature of the water from a cold tap? What temperature do you like the water for a bath?

You are probably quite good at telling whether a bath is hot or not by putting your toe in it. But actually your body is not very good at estimating temperatures. Try the following experiment.

Experiment 5.10
Measuring temperature

Safety note

Take care that the water is not too hot.

Take three bowls of water, one nearly full of very cold water, one containing tepid water and the third full of hot water (Fig 5.16). Put one hand in the cold water, one hand in the hot water, and keep them there for a minute. Then put both hands in the tepid water. How does the tepid water feel to each hand? One hand tells you it is warm, the other tells you it is cold!

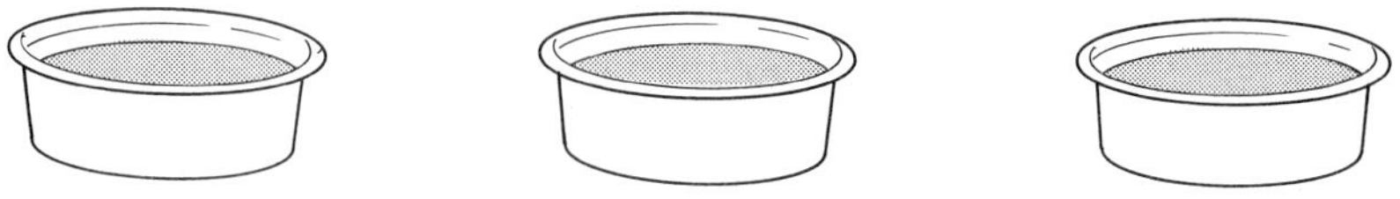

Fig 5.16

To measure temperature we use a thermometer. Put a thermometer in the cold water and another in the hot water and see what they read. Then put them both in the tepid water and you see that they read the same. This is one reason why thermometers are preferable to hands when measuring temperature.

Thermometer

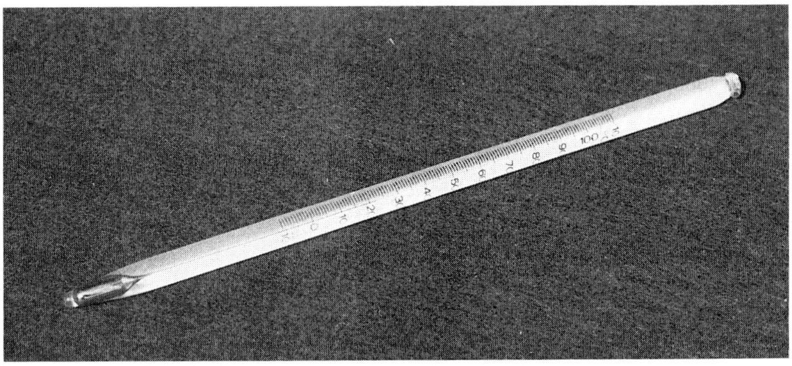

Background Reading

Your temperature

If you have ever been in hospital, you will know how carefully doctors and nurses measure your temperature in order to know what progress you are making. They measure the temperature of your blood by putting the thermometer under your tongue or under your arm. They may take the temperature several times a day and plot it on a chart like the one shown in Fig 5.17.

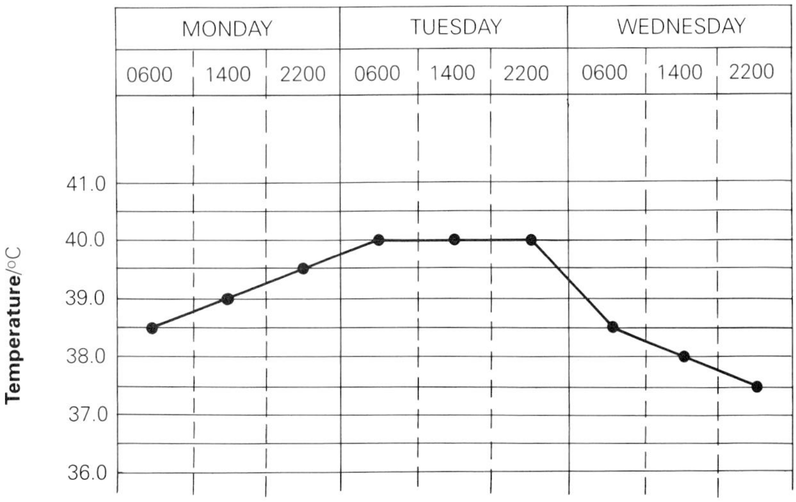

Fig 5.17

The 'normal temperature' is 37.0 °C, though most of us have temperatures below that when we are perfectly well. With a temperature of 38.0 °C or 39.0 °C you will not be feeling very well.

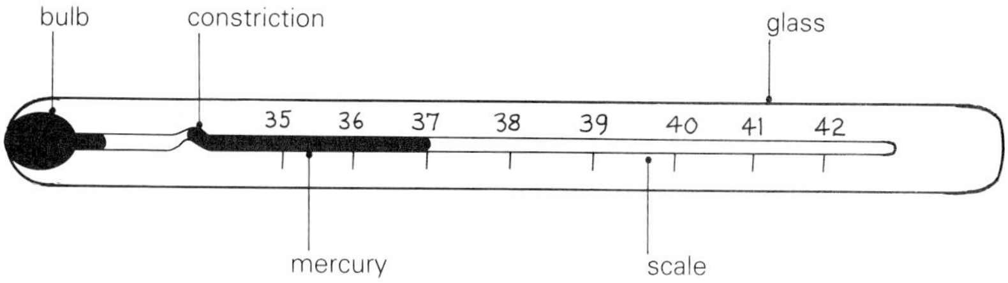

Fig 5.18

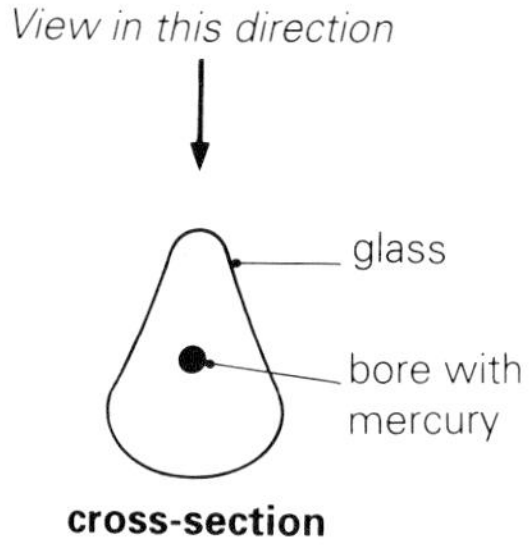

Fig 5.19

Because it is only necessary to read the temperature over a small range (from 35 °C to 42 °C), a special thermometer called a clinical thermometer is used (Fig 5.18). At one end there is a bulb containing mercury which expands when heated. Inside the stem is a very thin 'bore' tube up which the mercury moves as the mercury expands, and there is a scale on the side to say what the temperature is. The bore is very fine (Fig 5.19), and you have to look into the thermometer in the right direction in order to see the mercury inside. The shape of the glass causes it to magnify the tube and makes it easier to see the mercury if the thermometer is rotated into the correct position.

There is a special feature about a clinical thermometer. When doctors or nurses take the temperature of a patient, they do not want the mercury to contract and go back to the bulb until they have had time to read it. There is a narrow constriction in the bore and when the bulb cools, the mercury to the right of the constriction in the diagram does not move back. It can be shaken back after the temperature has been read.

Electronic thermometers are slowly replacing mercury thermometers in medicine. They give the reading in °C on a small screen. After being placed in the mouth or under the armpit, a buzzer sounds when the reading of the temperature has become steady.

plank sagging
bow and arrow
tug of war
hammering a nail
flag pole
cycling
apple falling
rowing
suspension bridge
sitting
pulley and load
catapult
stopping a plane
weight lifting
parachute
chest expander
mowing machine
pushing a truck
clothes line
fishing
sail
wind blowing a tree
crowbar
horse and cart

Chapter 6 **Forces**

We are always pushing and pulling. We have a special name for a push or a pull: we call it a **force**. Forces are everywhere around us and a few examples are shown in the drawings opposite. Look at each picture and think about what forces are acting.

How do we recognise a force? If you put a football in the middle of a field, it will remain there without moving until a force is applied to it. The force may be caused by the wind blowing it, by someone kicking it or by someone picking it up. In each case a force causes motion.

A force can also decrease motion or stop it. A moving football can be stopped by the force exerted when it hits a player. A parachute attached to a jet plane when it is landing causes a force to be exerted on it and this slows it down. We can recognise a force as something which changes motion.

If a football hits a wall, it will bounce off it. If it hits the wall at an angle, it will bounce off at an angle; the force of the wall has in this case *deflected* the motion. The direction of the force is clearly important. When you kick a stationary ball, it moves in the direction of the force. When it is picked up, it moves in the direction of the upward force applied.

To summarise, a force is a push or a pull which changes the motion of an object by speeding it up or slowing it down or changing its direction of motion.

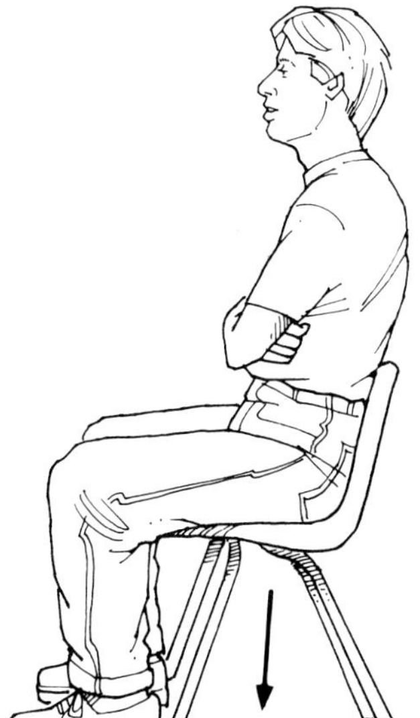

Fig 6.1

Forces in equilibrium

We can recognise a force by its ability to start motion, or to change motion. But if you sit on a chair there is more than one force acting on you and yet you do not move (Fig 6.1). In this case the forces just balance and produce no motion. The forces are said to be **in equilibrium**. Sitting on the chair, there is a downward force on you due to gravity (your weight) and this is exactly balanced by the upward force exerted on you by the chair. (There are also forces to stop you sliding off.) This is similar to what happened when a small block of wood floated in water (page 31).

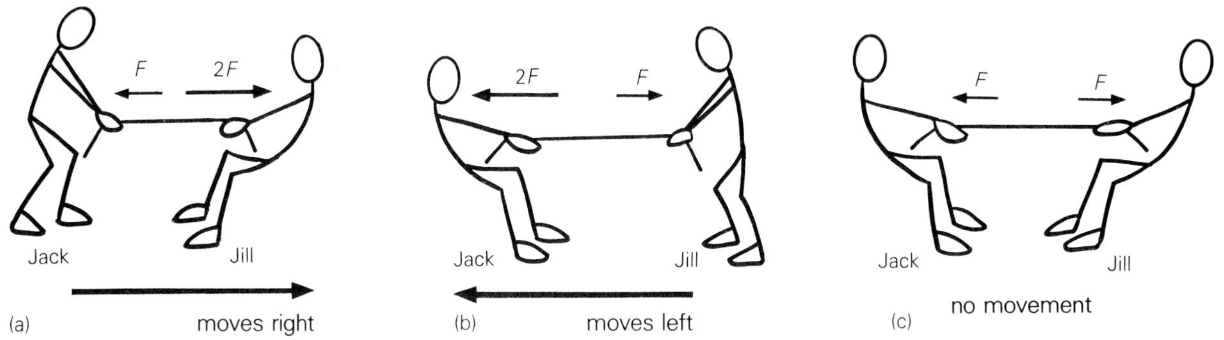

Fig 6.2

Fig 6.2 shows two people pulling on a rope. In the first drawing, Jill pulls with twice the force of Jack: she will pull Jack to the right. In the second, Jack exerts the bigger force so the motion is to the left. But in the third drawing the forces are equal, they are in equilibrium and there is no motion.

Kinds of forces

The most obvious forces of which you are aware are pushes or pulls exerted through the actions of your muscles, but there are other ways in which forces can be produced.

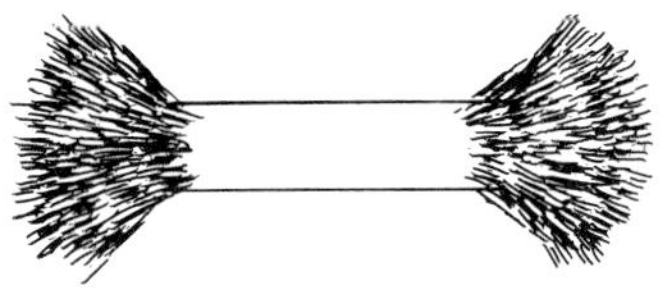

Fig 6.3

Magnetic forces

Magnetic forces can be felt by holding the ends of two magnets together. Sometimes the forces are attractive forces (pulls), sometimes they are repulsive forces (pushes). To find out about these it is best to do your own investigations as described in the next experiment.

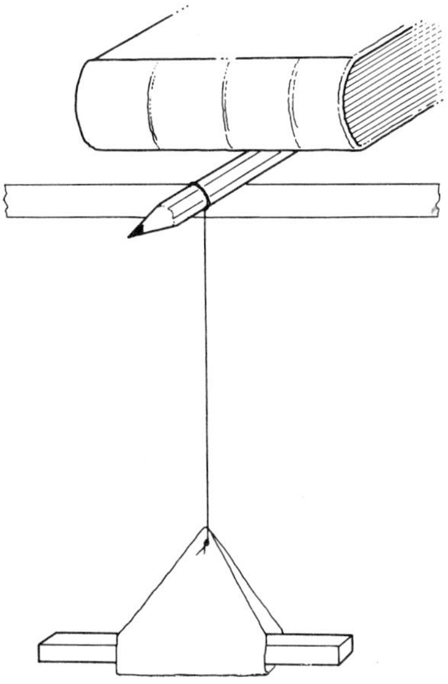

Fig 6.4

Experiment 6.1
Investigation of magnets

a Take a small magnet and find out what things are attracted to it. Pieces of paper? A paper clip? A piece of cloth? A plastic comb? A screwdriver? A piece of wood? The metal end of a hammer? What sort of things are attracted to it?

b Sprinkle some iron filings on to a piece of paper. Try picking up the iron filings with the magnet. Are the filings attracted to both ends of the magnet? They should be like the drawing shown in Fig 6.3. The two centres of attraction are called the **poles** of the magnet.

c Hang up the magnet by a cotton thread so that it hangs freely (Fig 6.4). When it has come to rest, you may notice that the magnet is lying approximately in a north–south direction. A freely suspended magnet always comes to rest like this.

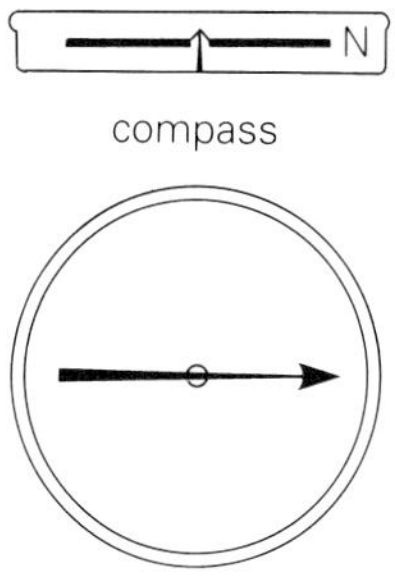

Fig 6.5

The pole at the end which points towards the north is called the **North-seeking pole** or, for short, a North pole; the other pole, the **South-seeking pole** or South pole for short. This is an important property of a magnet and enables it to be used as a compass. In a simple compass, a small magnet is balanced on a sharp point (Fig 6.5).

d Take two cylindrical magnets and first arrange them on the bench so that they attract each other (Fig 6.6). What happens if you turn one of the magnets round? What happens if you then turn the other round as well?

e Hold the magnets, one in each hand. Bring two of the ends together. What happens? What happens if you bring the other two ends together? Turn one of them round. What happens when the ends are brought together now?

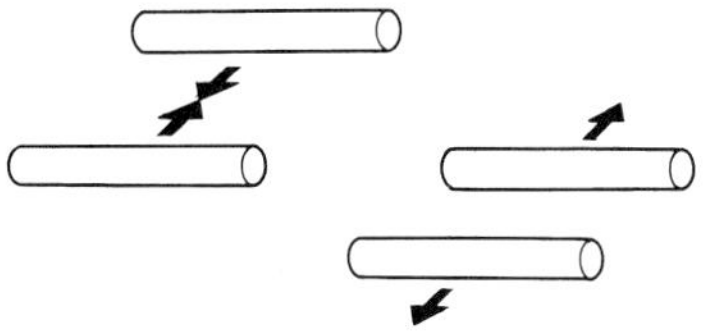

Fig 6.6

The rule for magnetic forces is that like poles repel, unlike poles attract. Can you see what this means? Did your experiments agree with it?

You will have noticed that magnets influence one another even when they are not touching, and that iron filings jump on to a magnet when the magnet is brought near.

Magnetic effects can be detected in the region around a magnet (see Fig 6.7): we say that the magnet is surrounded by a **magnetic field**.

A small compass will show the existence of a field. If you place a magnet on a table and move a small compass round it, the direction in which the compass points changes.

Horseshoe magnets also have poles. Can you think of a way to show that one end is a North-seeking pole and the other a South-seeking pole? (*Hint*: You already know that one end of your small cylindrical magnet is a North-seeking pole and 'like poles repel'.)

A good demonstration to show magnetic forces is given by mounting two horseshoe magnets on trucks as shown in Fig 6.8. Push the trucks together and see what happens. What would happen if one of the magnets was put the other way up? What would happen if they were both turned the other way up?

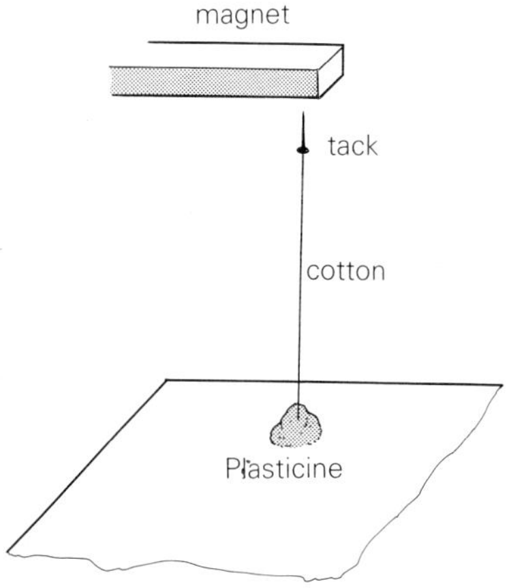

Fig 6.7

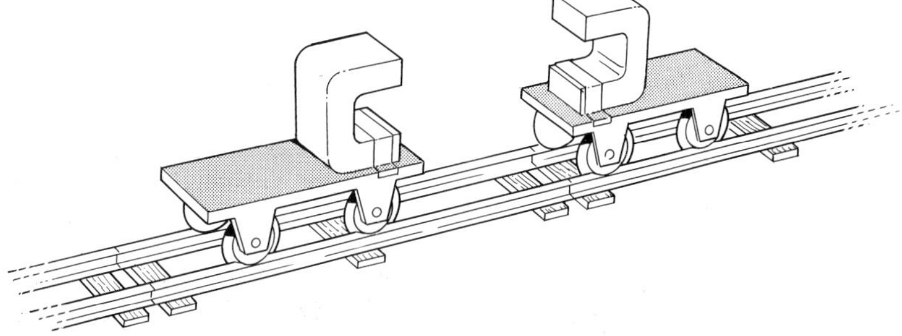

Fig 6.8

Experiment 6.2
Mapping the magnetic field of a bar magnet

a Place the magnet on a sheet of paper. Outline it with a pencil (so that you can put it back where it was if you nudge it) and make a dot near one of the magnet's poles. Then place the compass so that one end of it is close to the dot (position A in Fig 6.9) and make another dot opposite the other end of the compass. Move the compass to position B and repeat what you did in position A. Continue doing this until you reach the magnet again. Join the series of dots with a line, and put arrowheads on the line to show which way the compass was pointing.

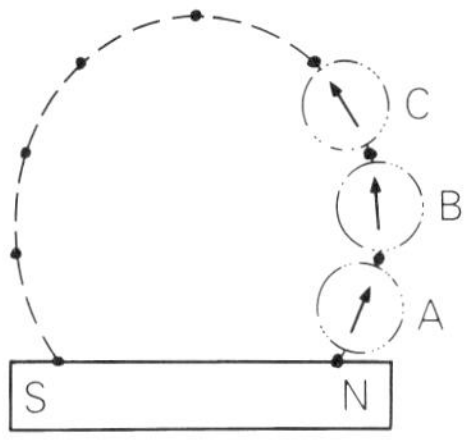

Fig 6.9

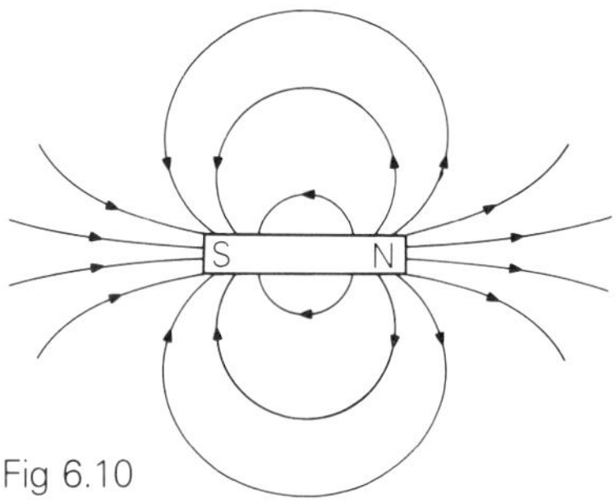

Fig 6.10

b Obtain a number of these lines to map the field round the magnet (Fig 6.10) by starting from a different point near the magnet.

The lines of field which you have obtained show the direction in which a compass points at various places in the field. Can you explain why the line always points in the end to the South pole of the magnet?

c Magnetic field maps can also be obtained by using iron filings. This time, two wooden strips are used to support a piece of card placed over the magnet (Fig 6.11).

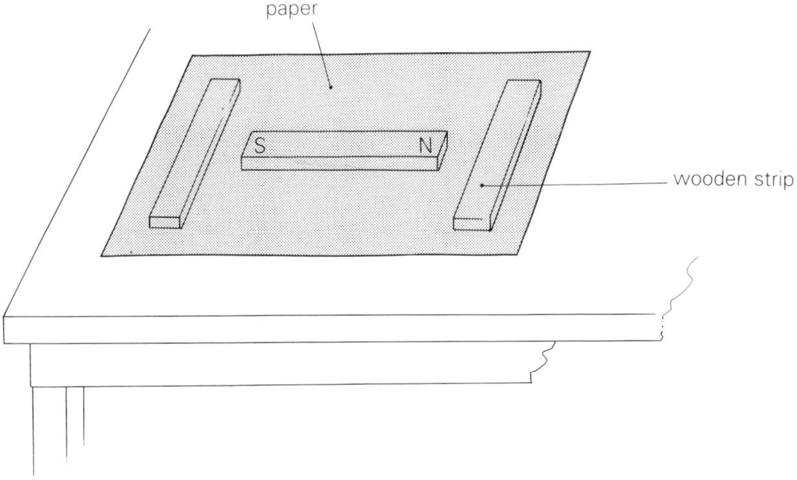

Fig 6.11

Scatter iron filings evenly over the card and then tap the card gently with your finger or a pencil. The pattern of the magnetic field appears as the iron filings set themselves along the field lines.

You might like to get other patterns by using two magnets. Fig 6.12 shows four different ways in which they could be arranged.

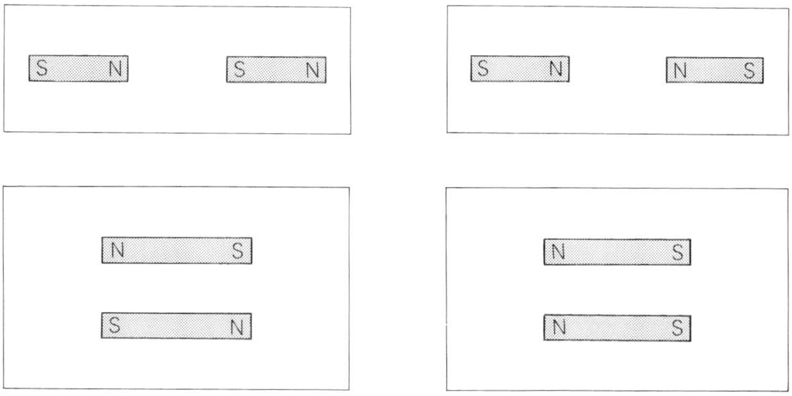

Fig 6.12

Background Reading

Early history of magnets

Very early in history a particular mineral was known to have the property of attracting iron. This material, or stone, is a dark ore of iron now known as magnetite because it was found in Magnesia in Asia Minor. Fanciful stories were woven around this property of attraction in the *Arabian Nights* and later during the Middle Ages.

Gilbert demonstrating magnetism to Queen Elizabeth I

Theories to explain the attractive force suggested that the gods were responsible or that the stone had a soul.

It was Dr Gilbert, court physician to Queen Elizabeth I, who discounted superstitious tales of the magical powers of magnetite. He did this by experiment. In 1600 he produced one of the earliest scientific works when he published *De Magnete* ('about magnets, magnetic bodies and the great magnet the Earth; a new discussion of their nature') written, of course, in Latin. Gilbert's method of testing theories and statements by experiment is now taken for granted in science and is recommended to anyone studying physics.

Another property of magnetite, known certainly by the Chinese over 2000 years ago, gives it the alternative name of lodestone, from the Anglo-Saxon 'lode' meaning way, course or journey. When hung up by a fine thread, a piece of magnetite settles down pointing in a constant direction. This led to its use as a simple form of compass, and to the name 'lodestone'. In studying this property, Gilbert came to the conclusion that the Earth behaved like a great magnet.

(Adapted from Longman Physics Topic *Magnetism* by G W Verow.)

Electric forces

Later in your course you will learn about electric forces. You will find there are two kinds of electric charge which for convenience we call '*positive*' and '*negative*'. There is a repulsive force between two positive charges, and there is also a repulsive force between two negative charges, but an attractive force between a positive and negative charge. We must leave detailed discussion of this until later, though it would be interesting to do one experiment to show it.

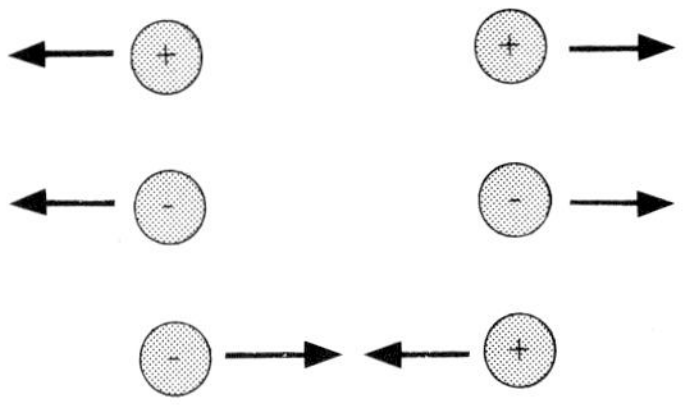

Experiment 6.3
Electric forces shown with balloons

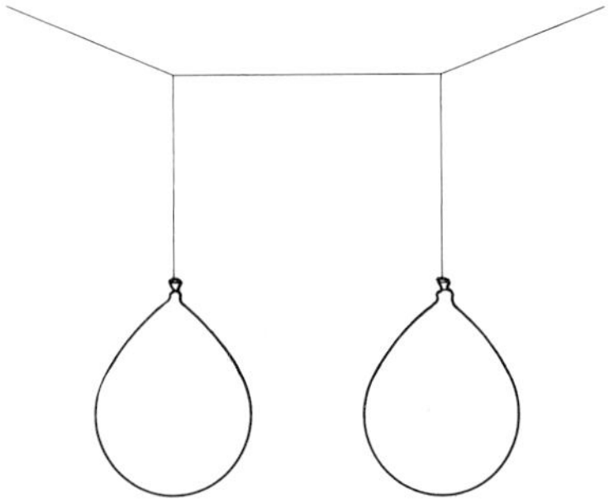

Fig 6.13

Blow up two rubber balloons. Attach a long thread to each and hang them up from a horizontal thread as shown in Fig 6.13.

Rub one of the balloons with your sleeve. Then rub the other balloon similarly. What happens when the balloons come near to each other? What happens when you put your sleeve near either of the balloons?

This is only a brief look at electrical forces. We will return to them later.

Gravitational forces

We are all aware of the gravitational force pulling everything towards the centre of the Earth. It is this gravitational force which causes an apple to fall from a tree or a brick to fall to the floor when we let go of it. It is this gravitational force between a body and the Earth which we call the **weight** of the body.

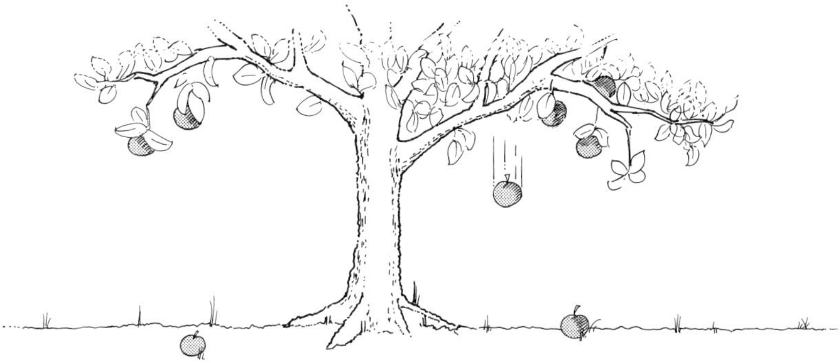

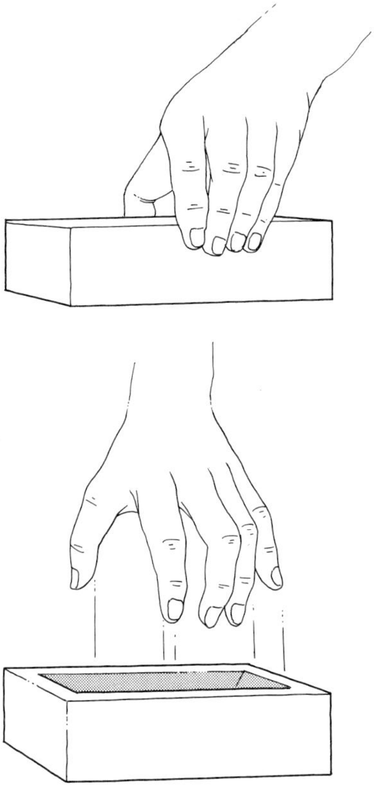

At a very young age, a small child realises that to hold up a ball – to keep it in equilibrium – an upward force must be exerted on it to balance the weight.

It was Sir Isaac Newton who first realised that there is a force of attraction between any two masses. It is usually a very small force, but if one of the masses is the Earth then the force is very much greater and we are all aware of it.

Since the mass of the Moon is much smaller than that of the Earth, the gravitational force on objects is less on the Moon even though they are nearer to its centre. This means that the weight of a given mass will be less on the Moon than it is on the Earth: the weight of anything on the Moon is about one-sixth of its weight on the Earth.

Impact forces

A stationary body can be set in motion if a moving body collides with it. The moving body therefore exerts a force on it, and such a force is called an **impact force**. A moving hammer can knock a nail into a wall, a stream of water can wash away its banks, the jet of water from a hose pipe can move soil or a pile of sand, a strong wind can blow trees over. Thus solids, liquids and gases can all exert impact forces.

When a pea from a pea-shooter hits a target, it exerts a small impact force on it. This lasts for a very short time, but if there is a continuous stream of peas it will seem like a continuous force (see Fig 6.14). This is how a stream of water or a gust of wind exerts a force. Each molecule exerts an impact force. As there are many such molecules, all the impact forces will add up to give a large force.

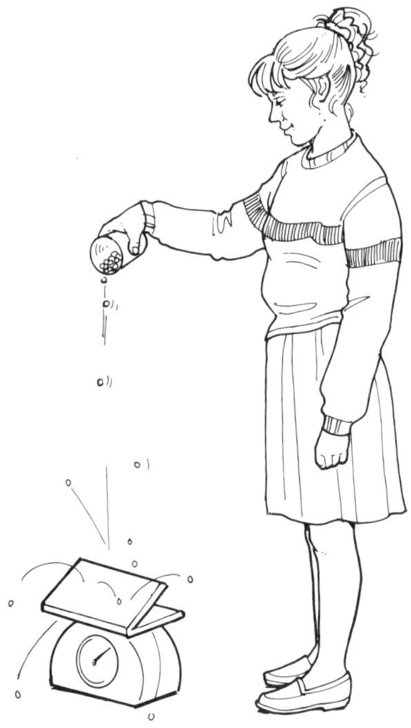

Fig 6.14

Strain forces

When you pull a piece of rubber (for example, an elastic band) by applying a force with both hands, there is a small movement and the rubber becomes longer until equilibrium is reached (Fig 6.15a). You become aware of a force or **tension** in the rubber. If you pull a bit harder, it stretches a bit more and the tension becomes greater. The more you pull, the more it stretches and the greater the tension in it, provided of course you do not pull too much and break it. If you let go, there is no longer equilibrium, the tension force sets the parts of the rubber in motion again so that it returns to its original length.

Such strain forces are felt not only by stretching. If you take a piece of rubber and twist it, you will feel similar forces (Fig 6.15b). When you let go, the rubber goes back to its original shape.

If you take a strip of metal, like a hacksaw blade, or a long piece of wood, like a metre rule, and bend it, you will feel the strain force (Fig 6.15c). The more you bend it, the greater the force.

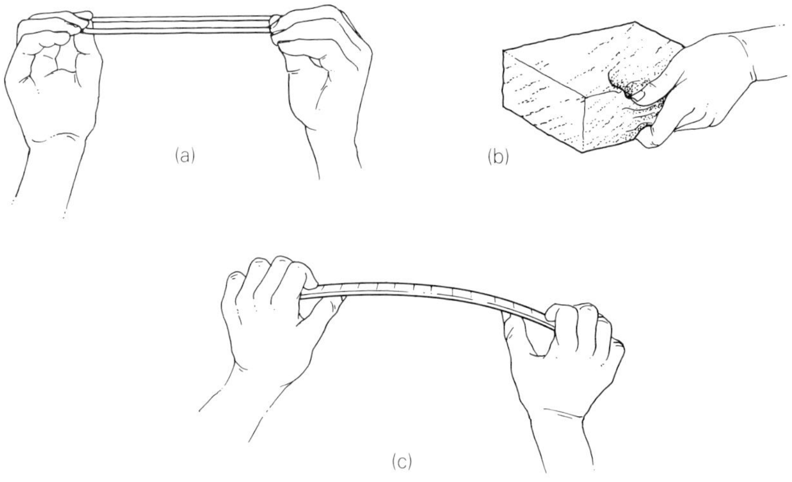

Fig 6.15

Forces are also experienced when squashing takes place. This can be shown by mounting two springs on the railway trucks as illustrated in Fig 6.16. What happens when they are pushed together?

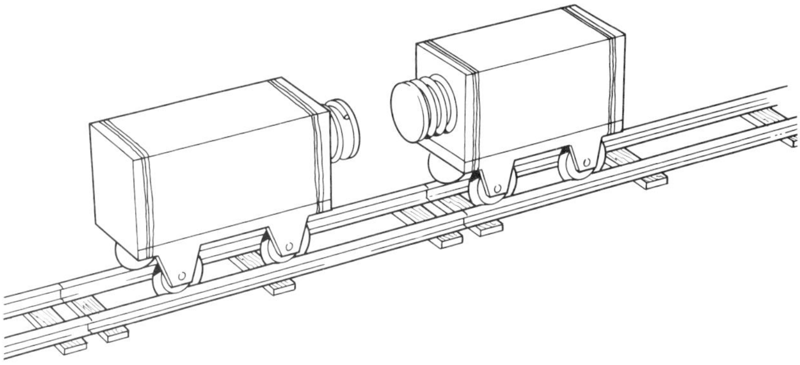

Fig 6.16

Bodies which return to their original shape when the forces are removed are called **elastic** bodies.

It is these strain forces which explain why a ball bounces. When a moving ball hits the floor, there is an impact force which deforms the ball, flattening it slightly. The impact force ceases and the strain force restores the ball to its original shape causing the ball to move away from the floor.

Golf ball squashed on impact

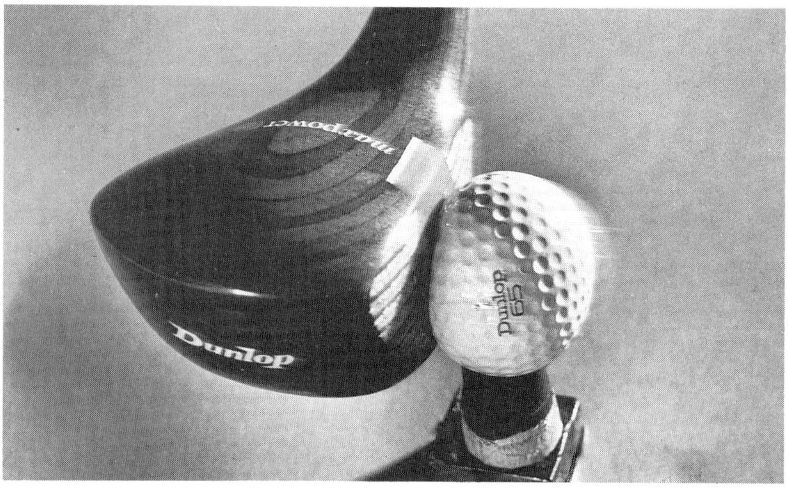

Frictional forces

Probably the most common force is the force of friction. Friction has some strange properties. It cannot start a body moving, it can only oppose motion and the size of the force can vary automatically. Try the following experiment.

Experiment 6.4
Investigating frictional forces

Tie a thin rubber band to a block of wood on a table (Fig 6.17). Pulling the band will cause it to stretch and to pull on the wood.

The more the rubber stretches, the bigger the pull on the wood.

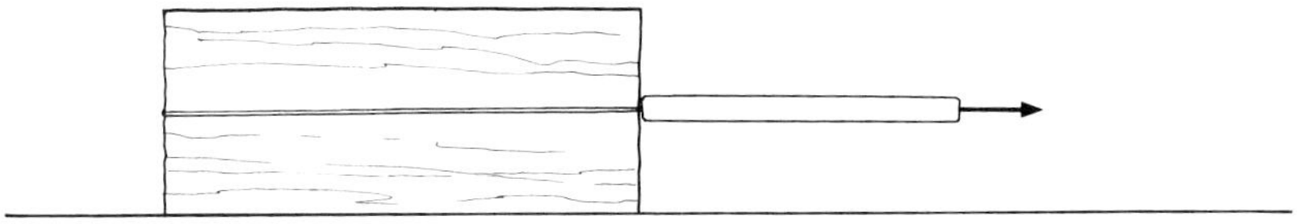

Fig 6.17

Try pulling the rubber band gently. Then slowly increase the pull (Fig 6.18). At first, the block does not move. So what is balancing the pull of the rubber band on the block? We say there is a force of friction between the block and the table. This force acts so as to prevent motion and, if the block does not move, it equals the pull.

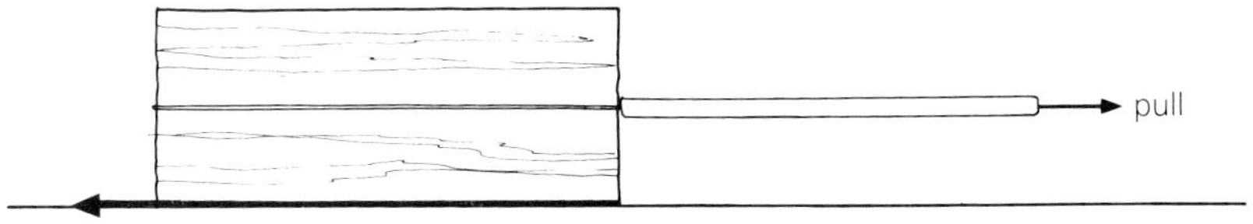

Fig 6.18

As the pull on the rubber band grows, so does the force of friction, until, at some stage, the block suddenly moves. This is because the size of the frictional force is limited – it can only grow up to a certain size and when the pull is greater than that, the block moves.

Friction therefore has some strange properties. It cannot start a body moving, it can only oppose motion and the size of the frictional force can vary up to a certain maximum.

Frictional forces on moving bodies

If a body is in motion and there is a frictional force acting on it, it can be brought to rest by that force. If there is no force on it, the moving body will go on for ever. It may be surprising to learn that a body could go on for ever without any force acting on it, but think of a block of wood moving over the ground. Frictional forces soon bring it to rest, but put it on smooth ice where the frictional force is much smaller, and it will travel much further before the friction stops it. Now imagine the frictional force getting smaller and smaller; the block will travel further and further. Imagine then what happens when there is no frictional force: the block would go on for ever.

A moving body will have its motion changed by forces acting on it, but if the forces on it are in equilibrium then it will continue to travel at a steady speed as if no forces were acting.

For example, if the forward thrust due to the engines of an aeroplane is equal to the backward force due to air resistance, the

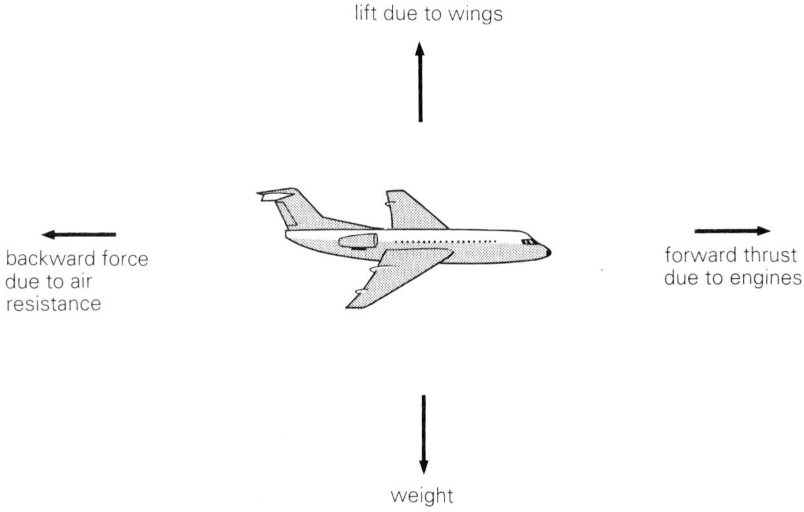

Fig 6.19

forces are in equilibrium and the aeroplane moves at a constant speed (Fig 6.19). If they were not in equilibrium, the aeroplane would speed up or slow down depending on which force was greater.

We usually think of friction as a nuisance: we oil machinery to reduce it. But most of the time it is essential to us. Think how difficult it is walking on a slippery surface. In walking, there has to be a forward force on the body. When we push our feet backwards, there is a forward force due to friction and it is this force which moves us forward. If there were no friction, there could not be any forward force and we could not walk.

Questions for class discussion

1 Suppose you put your finger through the ring in each of the drawings in Fig 6.20. What force will you feel? In what direction will it act? Will there be any change in the force if you move your finger very slowly upwards? Or downwards?

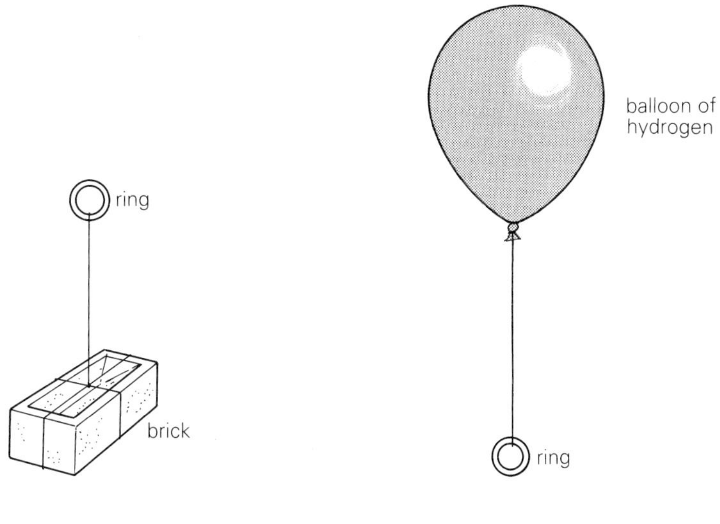

Fig 6.20

2 If you put your finger through the ring in each of the drawings in Fig 6.21, what force will you feel? In what direction will it act this time? Does it make any difference to the force if you move your finger slowly upwards? Or downwards? What is the effect of the pulley?

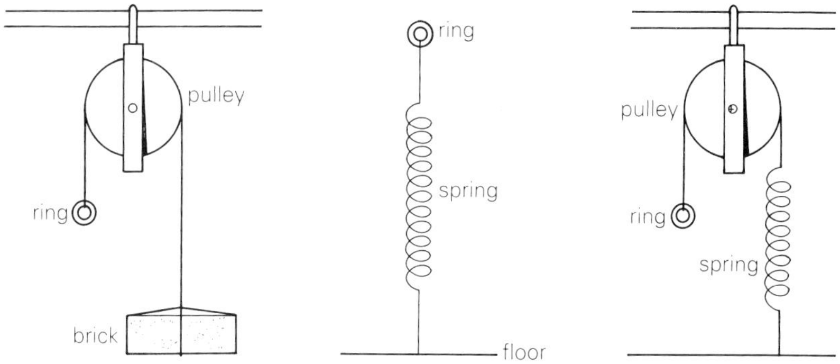

Fig 6.21

3 Give examples of each of the following:
a a force exerted on something which starts it moving;
b a force exerted on something which stops it moving;
c a force which changes the direction of something moving;
d a frictional force exerted on a moving body;
e a frictional force exerted by a liquid;
f a frictional force exerted by a gas;
g a strain force when something is stretched;
h a strain force when something is compressed;
i a gravitational force on a solid body at rest;
j a gravitational force on a moving body.

4 When a skater is moving on ice, there is very little friction. How is it possible for the skater to turn a corner?

5 When an aeroplane is flying horizontally at a steady speed, the forces on it are in equilibrium. The aeroplane has a downward force on it, its weight. What does this tell us about the upward force on the aeroplane?

6 Look at the drawings on page 48 and describe what forces are acting in each picture.

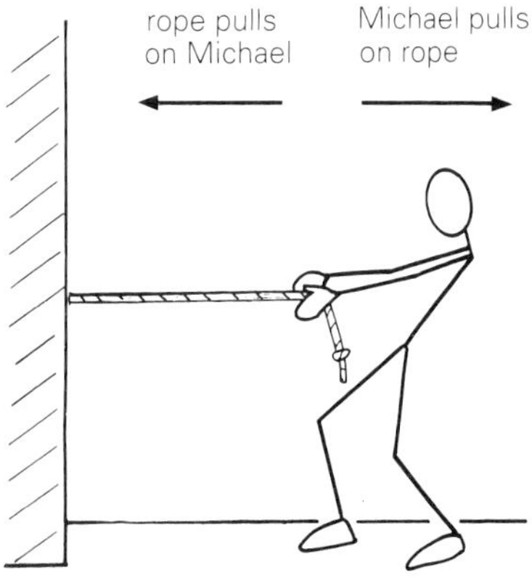

Fig 6.22

Forces occur in pairs

If Michael pulls on a rope, the rope also pulls on Michael (see Fig 6.22). If a block of wood rests on a table, the block pushes down on the table and the table pushes up on the block. When you carry a bag of shopping, the bag pulls down on your hand and you pull upwards on the bag.

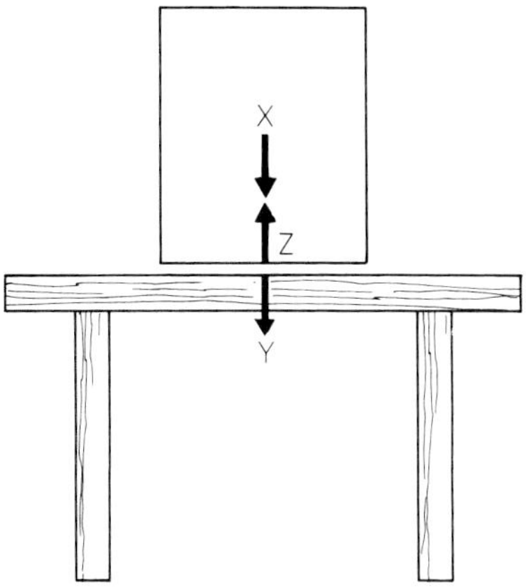

Fig 6.23

Forces always come in pairs like this. If an object, A, exerts a force of any kind on object B, then object B exerts an equal force in the opposite direction on A. But if you have to decide whether an object will be in equilibrium or not (in other words, whether the forces balance or not) you have to think *only* about the forces acting on that object. For example, think of an object on a table (Fig 6.23). The diagram shows the three forces acting:

X is the force of gravity on the object (its weight),
Y is the force of the object on the table,
Z is the force of the table on the object.

Y and Z are *always* equal and opposite in direction. The forces *on the object* are X and Z. The object will be in equilibrium if X and Z are equal and opposite in direction.

Questions for class discussion

1 The Earth pulls on an apple and this force causes the apple to fall to the ground. Forces occur in pairs. The same force must therefore act on the Earth due to the apple. Why does not the Earth appear to move towards the apple instead of the other way round?

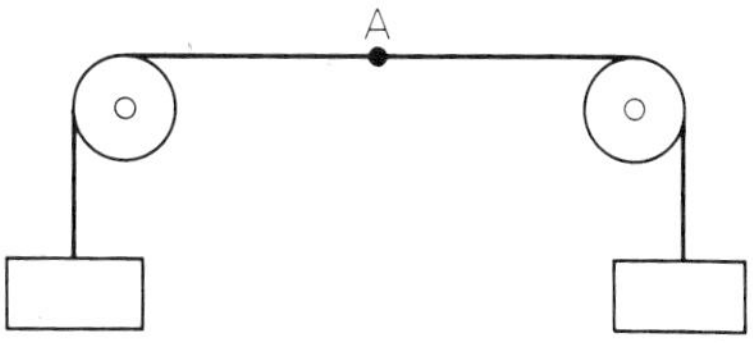

Fig 6.24

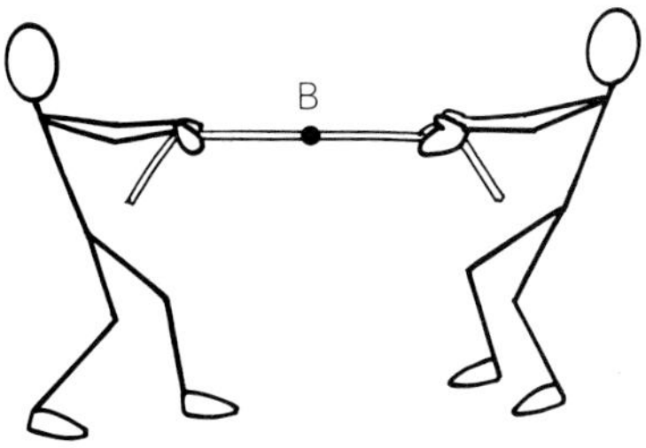

Fig 6.25

2 Explain the following.

a Two equal masses are joined together by a string which is hung over two smooth pulleys (Fig 6.24).
The string suddenly breaks at A and both masses fall in the same direction to the ground.

b Two girls are pulling each other with a rope (Fig 6.25). Suddenly the rope breaks at B and they both fall over backwards in opposite directions.

Background Reading

Universal gravitation

One of the most mysterious features of gravity is the fact that it acts across space. There is nothing connecting the body attracting and the body attracted – nothing between the Earth and the falling apple. This feature is termed action at a distance and the early scientists were very puzzled about it. Eventually Sir Isaac Newton avoided the problem by pointing out that, even if we do not know how it happens, there is no doubt that it does and it is more profitable to concentrate on finding out as much as we can about how gravity behaves – that is, to describe the laws of gravitation – than to wonder how it works. Newton himself found out most of what we know on the subject and published it in 1687 in one of the greatest books on physics, *Principia Mathematica.* It is only in the present century that we have learned a little more about it, and even now we do not know how it acts. Gravitation is regarded as one of the fundamental forces of nature which cannot be explained in terms of anything simpler.

Perhaps the most surprising of Newton's conclusions was that *all* bodies attract each other gravitationally. It is not only that the Earth attracts small objects on its surface. It also attracts the Moon and deflects its path into a more or less circular orbit around the Earth. The Sun attracts the Earth and deflects its path into a more or less circular orbit around the Sun. Also the Moon attracts the Earth and moves the oceans over its surface, causing the tides.

Gravitational attraction between Jupiter and its 'moons' (see the photograph) causes them to go round Jupiter just as the Moon goes round the Earth.

Jupiter and its moons

Distant stars attract each other so that they cluster in huge galaxies, like the Milky Way. The photograph on the next page shows the Andromeda nebula, a collection of about 100 000 000 000 stars held together by gravitational forces.

Even insignificant objects like two apples attract each other, but the force between them is far too small to be detected. Only the theory tells us it is there.

There is a very famous story that when Newton was trying to explain why the Moon goes round the Earth, he sat in an orchard and saw an apple falling from a tree. Suddenly he realised that the Moon *falls* round the Earth under the force of gravity, which stretches all that way out in space. No-one is quite sure whether this story is true or not.

Nobody quite knows how people get new ideas. Certainly one has to think very hard about a problem for a long time before one can solve it. Then very often the solution comes quite suddenly, when one is thinking of something else. A chance happening, like

Andromeda nebula

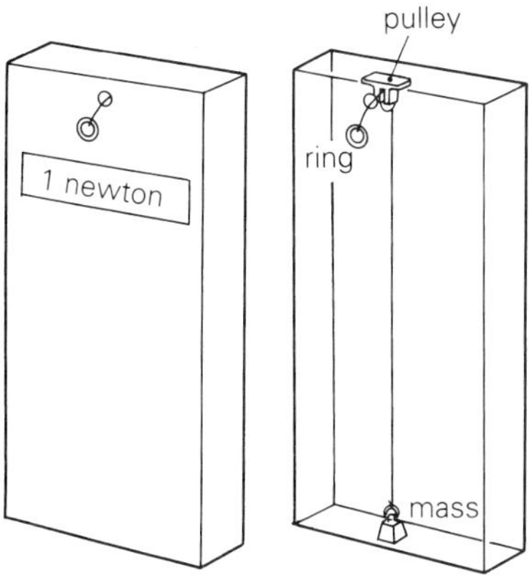

Fig 6.26

an apple falling, may provide the vital clue. So even if the story of Newton and the apple is not true, it could be true. It tells us something about how people think.

(Adapted from Longman Physics Topic *Forces* by R D Harrison.)

Newton spring balance

Measuring forces

We have already seen that scientists like to measure those things with which they have to deal. Forces are no exception, and they are measured in **newtons**. You should experience for yourself what a force of 1 newton is like by pulling on the forces demonstration box (Fig 6.26). (The abbreviation for 1 newton is 1 N.) 1 N is just about the weight of a 100 g mass.

(The drawing above shows how the box is made. The string passes over a pulley and on the lower end is a suitable mass of weight 1 newton. So pulling on the ring on the front of the box gives the feel of 1 newton.)

Forces can often be conveniently measured using a newton spring balance (see the photograph). This has a scale on the side marked in newtons and the position of the pointer tells you the force applied.

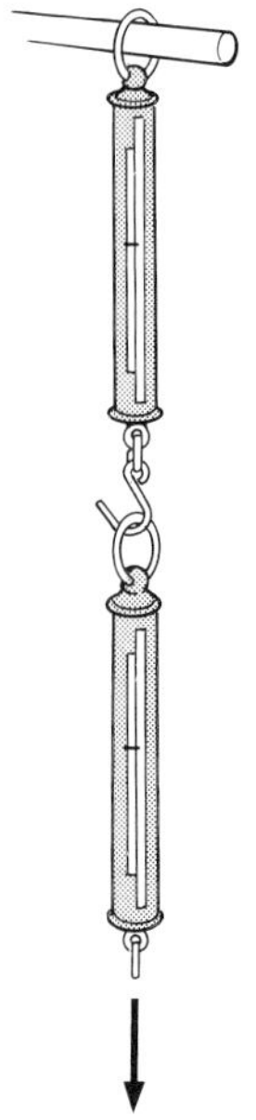

Fig 6.27

Experiment 6.5
Using a spring balance

Use a spring balance, marked in newtons, to measure a lot of different forces.

Find out what size of force is necessary for the following:

a to pull a block of wood along a table;

b to pull a table or chair across the room;

c to pull a drawer open;

d to pull a door open.

Use the newton spring balance to measure the gravitational force on a shoe, on a book, and on other objects in your laboratory.

What happens if you join two spring balances together? Fix one of them so that it hangs vertically (Fig 6.27). Attach the other spring balance to it and pull with a force of 5 newtons. What is the reading on each spring balance?

By hanging a 1 kg mass on the spring balance, measure the gravitational force on it. Do the same with a $\frac{1}{2}$ kg mass if one is available.

Measuring the strength of the gravitational field

We say there is a gravitational field anywhere a gravitational force is felt. We are in a gravitational field everywhere on the Earth's surface.

The previous experiment will have shown that when you hang up a mass of 1 kg the downward force on it due to gravity is approximately 10 newtons.

Similarly, the downward force on 2 kg is 20 newtons; on 3 kg it is 30 newtons and so on. We can say that the strength of the gravitational field is 10 newtons on every kilogram, or 10 newtons per kilogram (10 N/kg).

On the Moon, the gravitational field of the Moon is about one-sixth of this. In other words, the force on 1 kg would be only 1.6 newtons. This downward force is what we call the weight of the body, so the weight of an object on the Moon would be less than that of the same object on the Earth. As weight is a force, it should always be measured in newtons. Unfortunately, suppliers of groceries do not always use the correct scientific terms: a packet of sugar may be labelled 'weight 250 g', whereas it should read 'mass 250 g'.

You now know that a mass of 1 kilogram will have a weight of approximately 10 newtons on the Earth, and its weight on the Moon will be approximately 1.6 newtons. The mass is always the same; the weight will be different in different gravitational fields.

Experiment 6.6 Investigating friction

The frictional force between a wooden block and a wooden plank can be demonstrated with the apparatus shown in Fig 6.28. If the crank is turned, the wooden plank can be pulled steadily over the rollers. The spring balance will measure the frictional force on the block.

Measure the frictional force when the plank is pulled at a steady speed. Then put a second block on top of the first and measure the frictional force again. It will be greater. Add more blocks and each time measure the force.

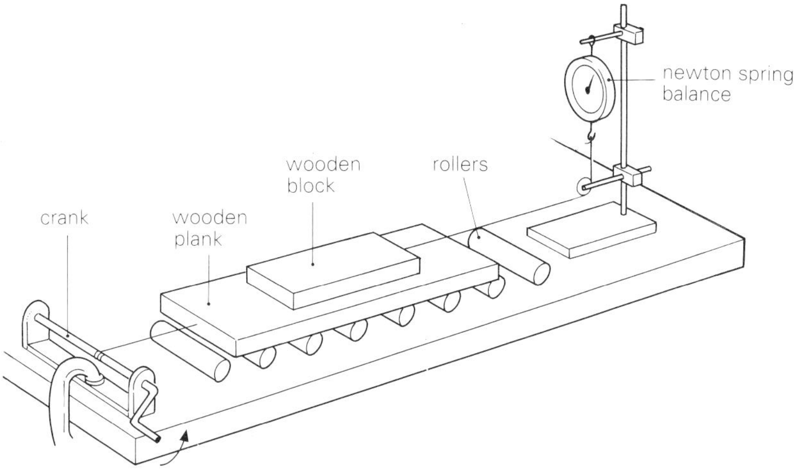

Fig 6.28

You can also investigate the frictional force when you press down on the block with your hand. What happens if you turn the block on its side so that the area of contact is smaller?

Background Reading

A theory of friction

Quite recently it has become possible to explain frictional effects as a result of cohesive forces between atoms in the two surfaces in contact. We must remember that even the smoothest surface would look very rough and, indeed, mountainous if we were only about the size of an atom. So when we press two surfaces together it will be something like Fig 6.29.

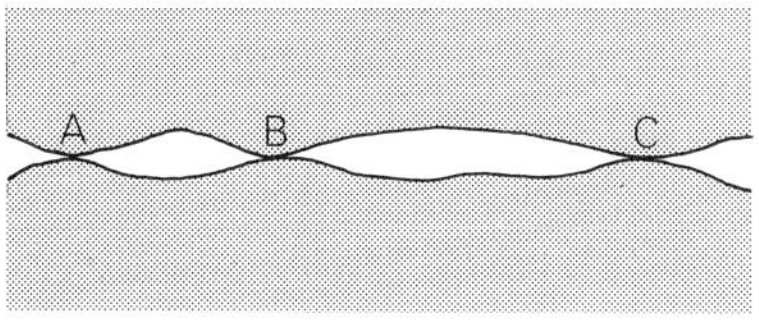

Fig 6.29

Only high spots like A, B and C will actually touch. So of course the pressure at these points will be quite enormous, and the tiny 'hills' which are in contact will tend to squash flat. The harder we push the surfaces together, the more these points squash. They will go on squashing until the area in contact has become big enough to support the force. This squashing under pressure is really quite drastic (almost as bad as the squashing of the atoms at the centre of the Earth). The two bits in contact become almost liquid and flow into each other and stick together – this process is known as cold welding and is sometimes used by engineers to stick things together.

Now, if we want to make the surfaces slide past each other we have got to break all these tiny little welds, and as fast as we do so, new points will come into contact and weld together.

If we want to make friction as small as possible, the theory tells us that we must either prevent the high spots from sticking together or make sure that it is very easy to tear them apart again after they have stuck.

A film of oil can prevent the two surfaces ever quite sticking together. That is why lubrication is so important in all the moving parts of a machine. It not only reduces friction, but also wear, which may be even more serious. Nowadays a very thin layer of air at high pressure is sometimes used for lubrication instead of oil. It serves the same purpose of keeping the surfaces apart, and allows sliding to take place even more easily than oil. The same principle is used on a big scale in a hovercraft.

Another way of reducing friction is to make one of the surfaces of quite a soft material, so that even if they do stick together, they can easily be pulled apart again. This is the principle used by engineers in white metal bearings which are used in motor car engines and other pieces of machinery.

Yet another way is to look for a material whose atoms are so strongly joined to one another that they do not easily join up to the atoms of a different material. Polytetrafluorethylene, or PTFE for short, is such a material. It slides over almost all other substances with very little friction and is used, among other things, for the best and most expensive skis.

The theory of friction is a valuable guide when it comes to looking for new materials which will give low friction.

(Adapted from Longman Physics Topic *Forces* by R D Harrison.)

Chapter 7 Investigation of springs

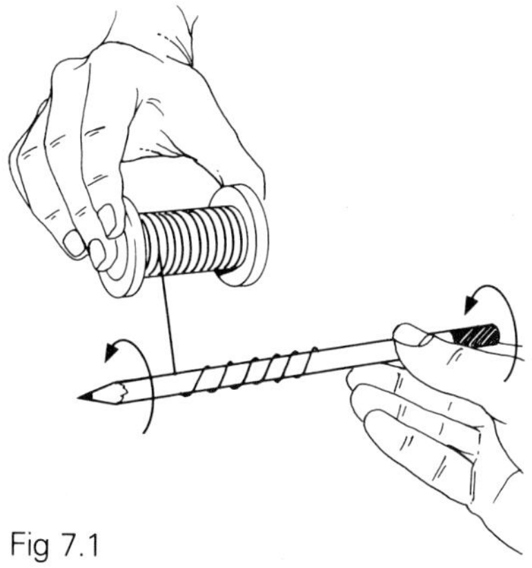

Fig 7.1

In this section you will be finding out how springs behave. At the end you should know quite a lot about them and be able to predict how they will behave in any future experiments you do. In order to help sort out what you find, keep a careful record of your investigations in your notebook. Good scientists always do that.

Experiment 7.1
Home-made springs of copper wire

To begin this investigation of springs, start by making your own spring with copper wire. You will need about 80–90 cm of wire which should be wound into a spiral on a pencil. The spring should be made *by turning the pencil round*, not by twisting the wire, as shown in Fig 7.1. Make a twisted loop at both ends so that you can hang up your home-made spring using a nail.

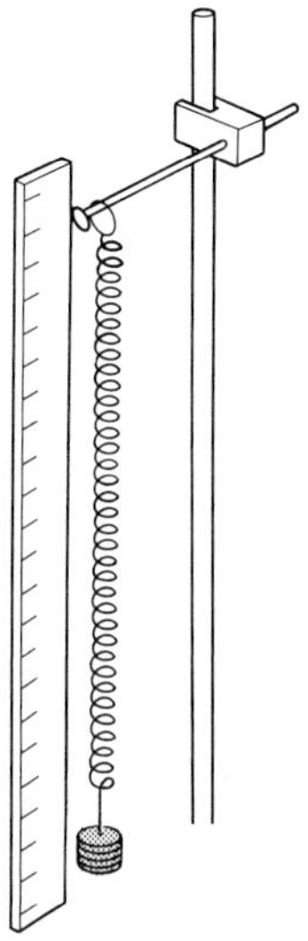

Fig 7.2

a Use a metre rule (or a half-metre rule, whichever is more convenient) to measure the length of your spring before anything is attached to it.
b Attach a small hanger to the lower end of the spring (Fig 7.2). What happens? Is there anything you can now measure? What is the best way to hold the rule? What is the most convenient way to measure the length?
c Add a 10 g mass to the hanger. What is the total load now? Is it 10 g or more? (*Hint*: What is the mass of the hanger?) What is now the length of the spring?
d Take off the 10 g mass. What does the length go back to? What does it go back to if you take off the hanger as well?
e Put the hanger back on again. Measure the length once more. Add 10 g and re-measure. Add more masses, keeping a record of the lengths.
f After you have loaded up with several masses, find out what happens to the length when you take the hanger off.
g Go on adding more masses. Is there any limit to the number you can add? Describe what happens.
h If time allows, make another spring, winding it on the same pencil, but using twice the length of copper wire and making the spring twice as long. Load up this spring in the same way. What difference does this new length make?

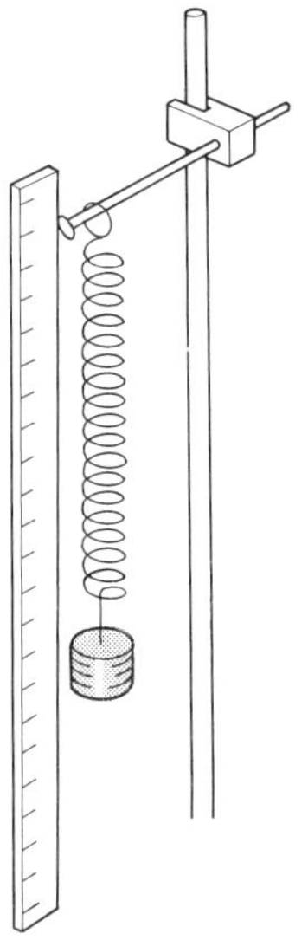

Fig 7.3

Experiment 7.2
Investigation of steel springs

Carry out a similar investigation using the steel springs provided instead of your home-made copper spring.

This time you should use larger hangers, which have a mass of 100 g (Fig 7.3). Add masses which are each 100 g. You may find it makes it easier if you fix the rule in a clamp, but that is for you to decide.

Do not worry if the springs get damaged, your teacher will always supply you with another.

When you have finished the investigation, write down a description of how the behaviour of the steel spring differed from that of the copper spring.

Plotting results

It is often much easier to see what is happening at a glance if you plot the results on a graph instead of just making a list of measurements. You can plot the length of the spring against the load, as shown in the first graph of Fig 7.4.

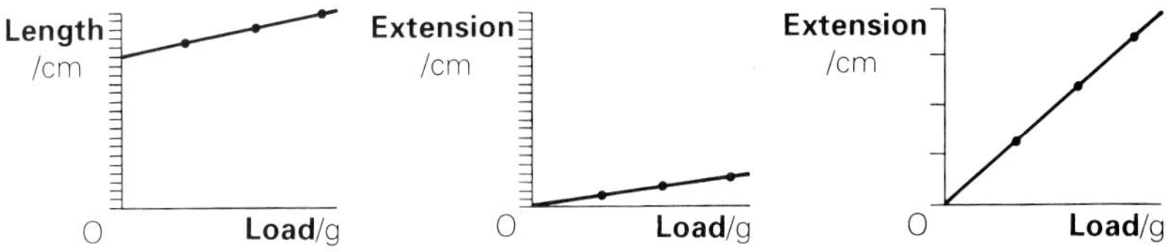

Fig 7.4

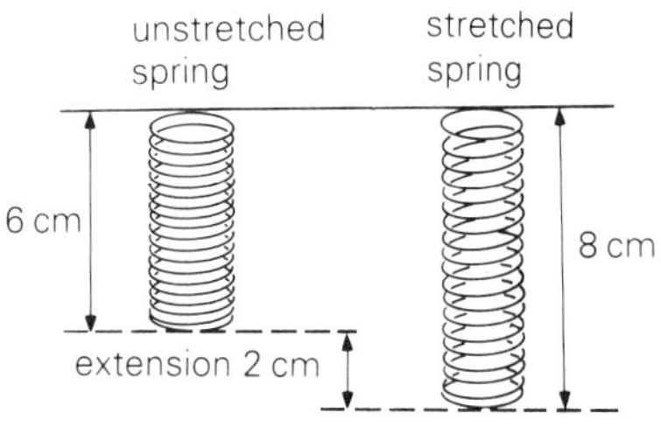

Fig 7.5

Sometimes it is more convenient to plot the **extension** against the load instead of the actual length. By extension we mean the amount the length has increased or extended. If a spring has an unstretched length of 6 cm, and a load attached to it stretches it so that the new length is 8 cm, then the extension is (8−6) cm or 2 cm (Fig 7.5). If you plot extension against load, you will get the second graph of Fig 7.4.

An advantage of plotting extension against load is that you can now change the vertical scale and get a graph like the third one in Fig 7.4.

A simple rule for stretching springs

Use the readings you obtained for your steel spring in Experiment 7.2. to plot a graph of extension against the load applied. Can you see any simple rule about the way the spring stretches? (*Hint*: What happens to the extension when you double the load? What happens if you treble the load?)

The fact that the extension goes up steadily with load (that it is proportional to the load) was discovered by Robert Hooke about 400 years ago and it is therefore often called Hooke's law.

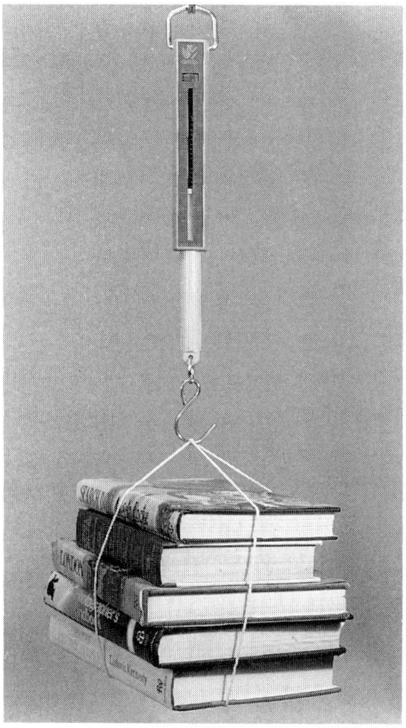

Newton spring balance

Further experiment

When you applied a small force to a steel spring (by hanging a small load on it) and you removed the force, you found that the spring went back to its original length. But will this happen however large the load? Investigate what happens when you apply much larger loads. What happens to your plot of extension against load when these bigger loads are applied? You will certainly damage the spring by having to test it like this, but you will have to do this if you want to know all about how a spring behaves.

The newton spring balance

In the last chapter you learned to use a spring balance to measure forces. Your experiments in this chapter will have shown you how they work. The balance contains a spring which stretches when a force is applied. The rule for the stretching of springs tells us that the extension increases steadily as the force is steadily increased. You can therefore put a scale on the side which tells you directly what the force is. Most spring balances have a stop which prevents the spring inside being overpulled. We usually refer to the stage at which the spring does not return to its original length as the **elastic limit** of the spring. The stop prevents the spring being stretched to its elastic limit.

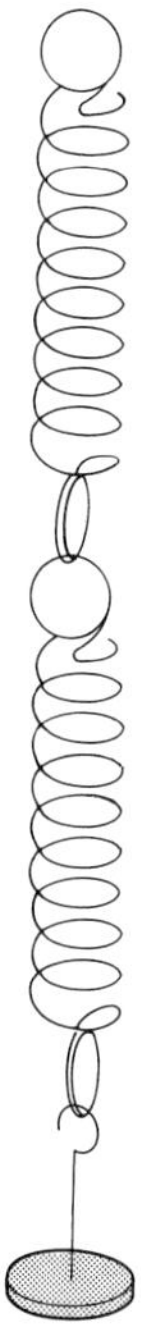

Fig 7.6

Experiment 7.3
Investigation of springs in series

Take two similar springs and measure how much each stretches when a 100 g load is attached to them. As the springs are similar, the extension should be the same in each case.

Then join the two springs together as shown in Fig 7.6. (We call this joining them **in series**.) How much does each spring stretch when a 100 g load is attached to the bottom end? Each spring is now stretched by the same force. How does this compare with the extension when a 100 g load was added to each spring separately?

On the same sheet of graph paper, plot the graph of extension against load, first for one spring on its own and then for the two springs in series.

Make a guess what the graph might look like if there were three springs in series. Then do an experiment to see if your guess was correct.

Experiment 7.4
Investigation of springs in parallel

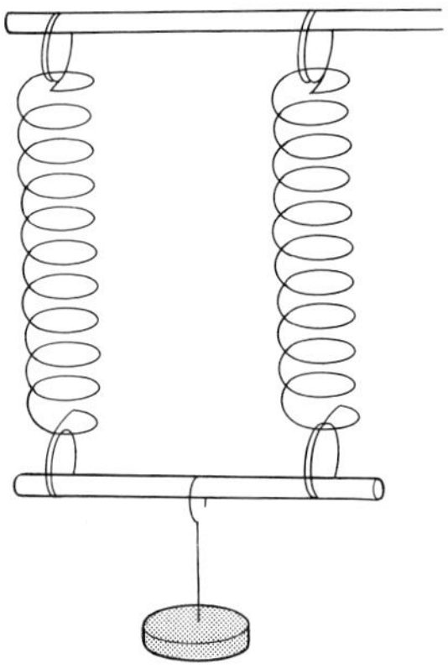

Fig 7.7

Hang two similar springs side by side as shown in Fig 7.7. (We call this joining them **in parallel**.) Investigate how the extension varies with the load in this case. The springs now share the load.

Plot the graph of extension against load on the same graph paper as a similar plot for one of the springs on its own. How does this differ from the result obtained in Experiment 7.3? Why does it differ from that result?

What do you think would happen if the load were not attached at the midpoint, but closer to one of the springs? Try it and see if it agrees with your answer.

Of course there will also be upward forces on the springs at the support. These forces balance the downward forces on them.

Stretching copper wire

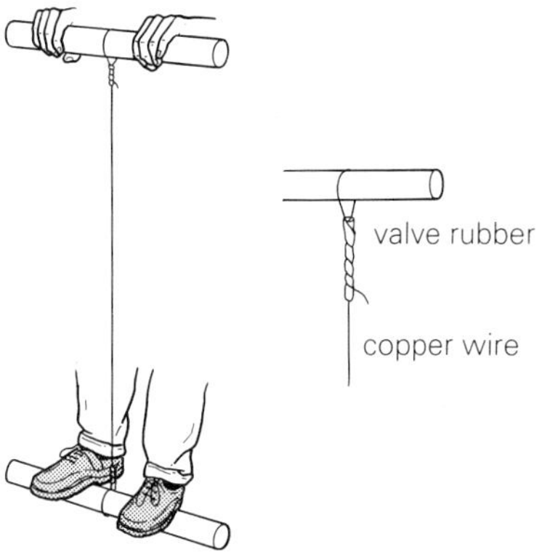

Fig 7.8

In this chapter we have so far only considered springs, but most things stretch a bit when force is applied. You can feel this for yourself if you take some very thin copper wire; just over a metre will do. If you try to pull it as it is, it will probably pull through your fingers. A good way to hold it is to wrap each end round a pencil and twist the ends round a piece of valve rubber as shown in Fig 7.8. Hold one pencil between your feet and the other in your hands. Pull on the wire and see if you can feel it stretching. You can probably feel the 'cheesy' effect as it stretches.

If you break your wire, look at the broken ends with a hand lens. What do you see?

Questions for homework or class discussion

1 A spring is 4 cm long. When a load of 200 g is attached to it, the new length is 6 cm and it returns to 4 cm when the load is removed. What will the length be when a load of 100 g is attached instead of the 200 g load? What will it be for a load of 50 g?

2 A 6 cm spring extends 4 cm with a load of 400 g. The spring returns to 6 cm when that load is removed. What is the extension for a load of **a** 100 g, **b** 200 g, **c** 600 g, **d** 60 kg?

3 A spring has an extension of 4 cm for a load of 800 g. What will be the extension when a second similar spring is connected in series with it and a load of 800 g is attached at the bottom?

4 What would be the extension if the two springs in the previous question were connected in parallel and the 800 g load were attached to the middle of a light rod connecting their lower ends?

5 Spring X is similar to spring Y except that X is twice as long as Y. If a load of 1 kg stretches Y by 10 cm, by how much will the same load stretch spring X?

6 Spring A is 10 cm long when no load is attached, but it stretches by 8 cm when 400 g is attached. It returns to its original length, and an unknown load is then attached to it (instead of the 400 g). The new length of the spring is 13 cm. What was the unknown load?

7 A spring stretches 12 cm when a load of 600 g is attached to it on the Earth. The gravitational field on the Moon is one-sixth of its value on the Earth. How much would the spring stretch if 600 g were attached to the same spring on the Moon?

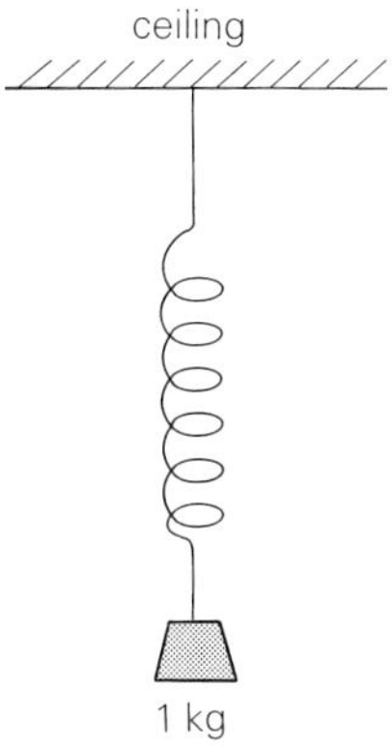

Fig 7.9

8 A mass of 1 kg is attached to the end of a spring so that the spring stretches 2 cm (Fig 7.9). The weight of the mass pulling downwards on the spring is a force of 10 newtons and it is this force which causes the spring to stretch.
- **a** How big is the force exerted by the spring on the ceiling?
- **b** What is the direction of the force which the spring exerts on the ceiling?
- **c** If the experiment were done on the Moon, would you expect the spring to stretch more, less or the same amount? Give the reason for your answer.

9 A girl fixes the top of a spring to a hook. To the other end she attaches various masses. Every time she changes the masses, she measures the length of the spring. The following is a record of her measurements, in the order in which she made them.

Load(g)	Length(cm)
0	10.0
200	12.4
400	14.8
600	17.2
400	14.8
0	10.0
600	17.2
800	20.4
1000	26.0
400	19.0

- **a** What would have been the length if the third reading had been 300 g?
- **b** The first time there was a mass of 400 g, the length was 14.8 cm. It was the same the second time. Why was it different the third time?
- **c** If she had taken the mass off after the last reading so that the load was once again 0, would the length have been 10 cm? Give a reason for your answer. What do you think the reading might be?

Background Reading

The dynamometer car

Any new product has to be tested to see if it comes up to the designer's expectations, and railway locomotives are no exception. The engineers need to know how well a locomotive performs when it is pulling a train of coaches and particularly how much force it can exert and how powerful it is.

To do this, the engineers use a special coach – called a dynamometer car – which is coupled between the locomotive and the train of passenger coaches. In the dynamometer car, there is a large, strong spring, one end of which is firmly fixed to the framework of the car, whilst the other end forms the coupling to the locomotive (Fig 7.10). In practice the spring is a stronger spring than a coiled spring.

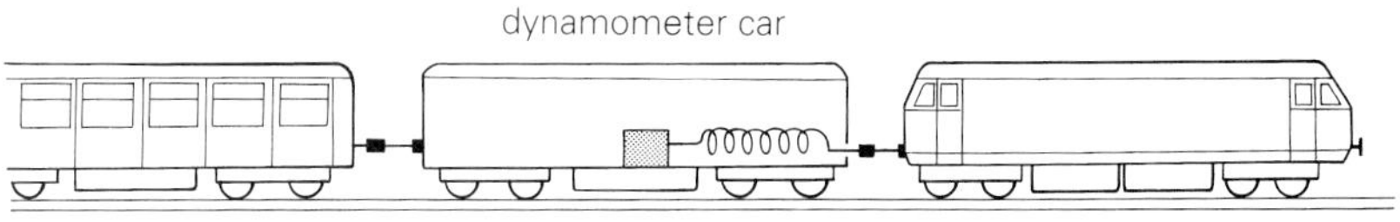

Fig 7.10

Thus, as the locomotive pulls its train, the force of the locomotive causes the spring to stretch, and the force on the train can be found from how much the spring stretches. As the speed of the train or the gradient changes, the extension of the spring changes, and so it is necessary to make a continuous record of the length of the spring. To discover how powerful the locomotive is, it is also necessary to know its speed (as you will learn later) and that is measured (and recorded) from the rate at which the wheels are turning.

So, the dynamometer car is really a spring balance on wheels, with a small laboratory built round it!

Chapter 8 The turning effect of forces

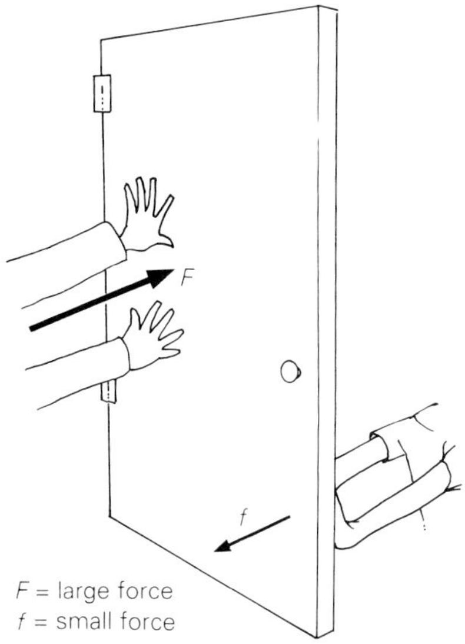

Fig 8.1

It is very difficult, if not impossible, to loosen the nut on the hub of your bicycle wheel with your fingers. It is necessary to use a spanner. If it is very difficult to loosen, where do you hold the spanner? In the middle or at the end?

We have already discussed forces. We know that it matters what their size is and in which direction they are applied, and we will see in this chapter that it also matters at what point they are applied.

The turning effect of a force depends on where the force is applied. For example, if a man pushes on an open door, near the hinges, with a large force, the door can easily be kept open by a small child pushing in the opposite direction near the edge of the door (see Fig 8.1). The small force has the same turning effect as the large force because of the places at which the forces are applied. Just a small increase in the child's force would cause the door to open wider.

It is because of this effect that a claw hammer can be used to extract a stubborn nail, and that nutcrackers can be used to crack a hard nut. Fig 8.2 shows the forces on the tools used.

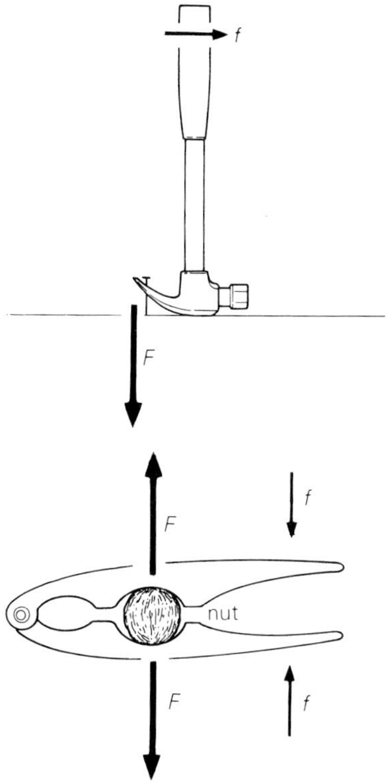

Fig 8.2

Imagine a painter, standing on a plank of wood supported on two trestles while painting the ceiling. All is well when he stands on the plank between the trestles, but if he moves along the plank to the part which overlaps the end, what may happen? In this chapter we shall look further at the turning effect of forces.

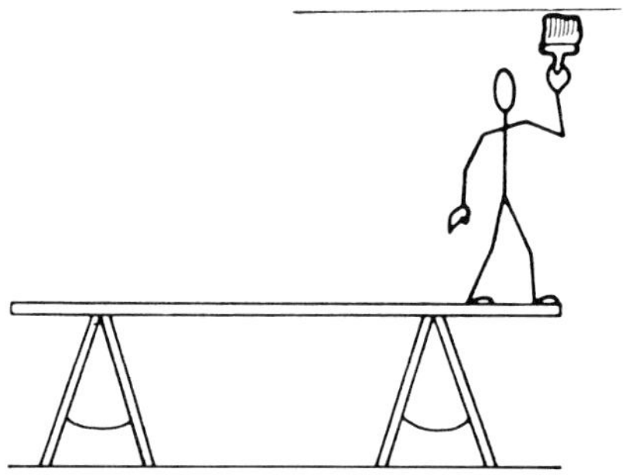

Experiment 8.1
Investigation leading to the law of the lever

Set up the apparatus shown in Fig 8.3. The wooden beam should be supported with the central groove resting on the top of the triangular support. (Such a support is sometimes called a **fulcrum**.) If the beam does not balance, put a lump of Plasticine underneath at some position so that it does balance.

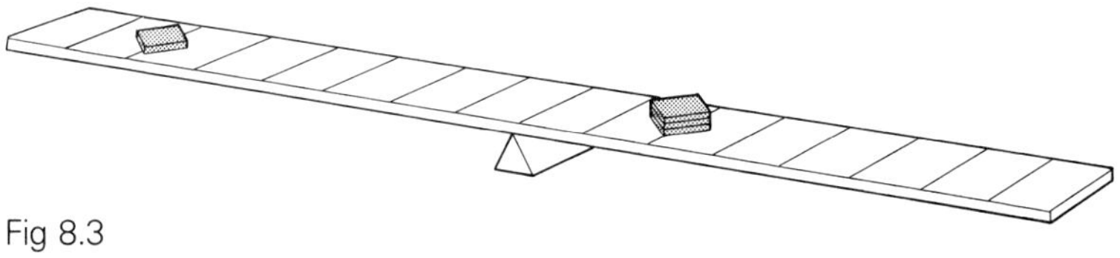

Fig 8.3

Put some metal squares or coins (all similar) on one side and some on the other so that the beam balances. You should put the loads so that they are one, two, three divisions out and not say, two divisions and a bit, because those fractions will make it more difficult to find out any law. It is probably best to put the squares diagonally on the beam.

Start by making the beam balance with two piles of squares, one on each side. You will not be able to get it to stay exactly balanced in mid-air, but the beam will tip over to one side or the other. It will be much like 'weighing sweets'. Find the point at which a very little more will tip the scale one way or the other.

Find out what you can about balancing the beam. Get four squares on one side balanced by two on the other; then get four squares balanced by one or by three. Keep a record of how many squares on the left at different distances are balanced by how many on the right, and at what distances. See if you can find some rule which tells you in advance whether two loads will balance. You may find that a table helps.

Number of squares	Distance

balanced by

Number of squares	Distance

Questions for class discussion

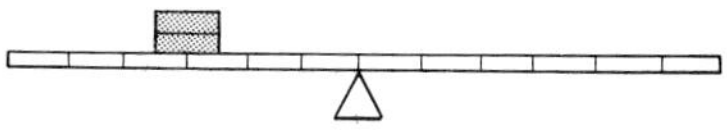

Fig 8.4

1 Two squares are placed on the left-hand side of a beam three divisions out (Fig 8.4).

a Where could two squares be placed on the right to balance the beam?

b Where could three squares be placed to balance it?

c Where could one square be placed?

2 Three squares two divisions out can be balanced by two squares three divisions out (see Fig 8.5). But that is not the only position where the two squares can be put on the right.

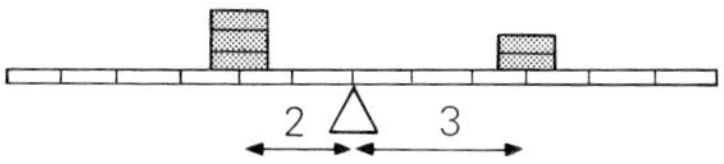

Fig 8.5

case (a)

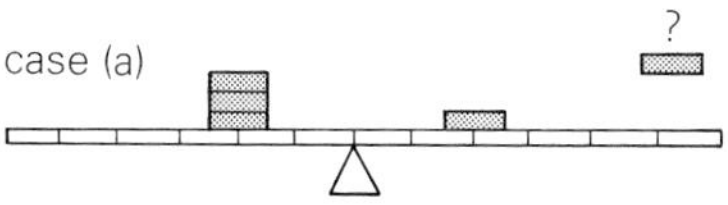

case (b)

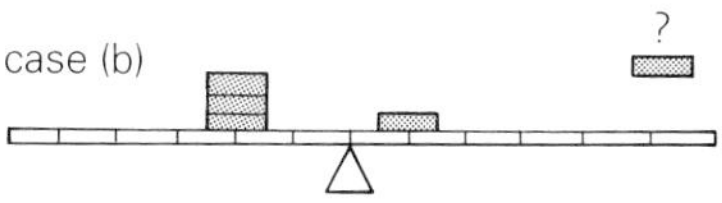

a Suppose one of the squares on the right is put two divisions out, where must the other be put?

b Suppose one of the squares on the right is put one division out, where must the other be put?

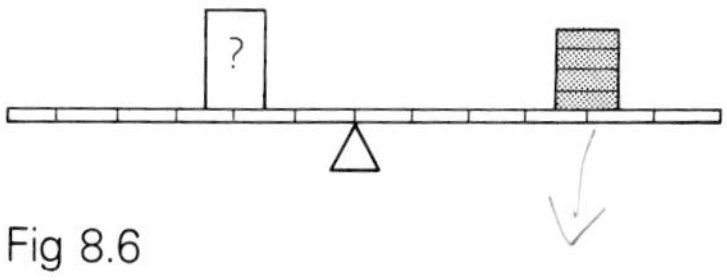

Fig 8.6

3 A pile of squares two divisions out on the left is balanced by four squares four divisions out on the right (see Fig 8.6). How many squares are there in the pile on the left?

The turning effect of a force

These experiments have shown that the tipping effect of a force depends on the size of the force and the distance of the force from the point about which the turning takes place. (This tipping or turning effect, the product of force multiplied by distance is sometimes called the **moment** of the force, but there is no need to memorise this name.)

When the beam balances, the total turning effect of all the forces trying to turn the beam one way equals the total turning effect of all the forces trying to turn it the other way. This is the *law of the lever* or the *law of moments*.

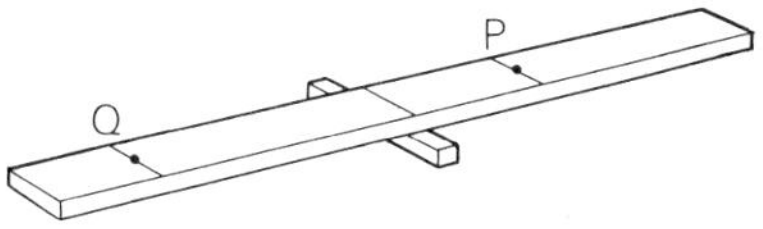

Fig 8.7

Experiment 8.2
Using a see-saw to measure your own mass

For this experiment you need a long, strong plank of wood. Balance it on a piece of wood. Mark a point P on one side of the plank which is 20 cm from the centre of the support (see Fig 8.7). On the other side mark a point Q which is 80 cm from the support. Stand on the plank at P. Get someone to put 1 kg masses on the plank at the point Q until a balance is reached. Then you can calculate your mass.

If the plank is not long enough, or if you have not got enough 1 kg masses, you may have to choose different distances. We shall leave that to you as an exercise.

More about forces

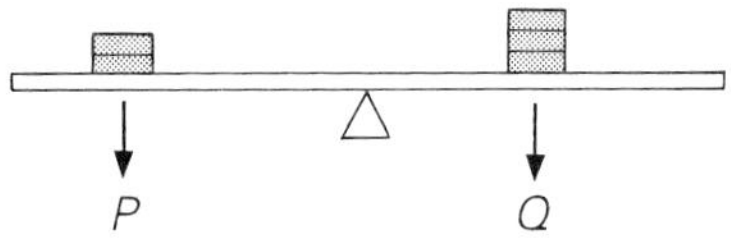

Fig 8.8

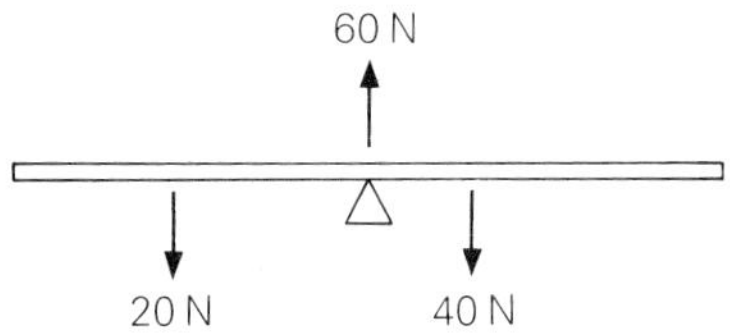

Fig 8.9

When two piles of squares balance each other as in Fig 8.8, it is of course the gravitational forces acting on the masses which matter. The turning effect of the force P about the pivot is exactly equal to the turning effect of Q about the pivot.

The forces P and Q are acting downwards, but the beam does not move. There must therefore be an upward force to balance them. Where is this force? It can only be at the support and it will be equal to $(P + Q)$, but upwards, as in Fig 8.9. This assumes that the beam is light, that its mass is small. The actual upward force will be $P + Q +$ the weight of the beam.

The forces acting on a lever need not be gravitational forces. In Fig 8.10, the force trying to turn the lever anticlockwise is provided by a spring attached to the floor. In Fig 8.11 the force trying to turn the lever clockwise is provided by the load whilst the force trying to turn it anticlockwise is provided by the spring pulling upwards.

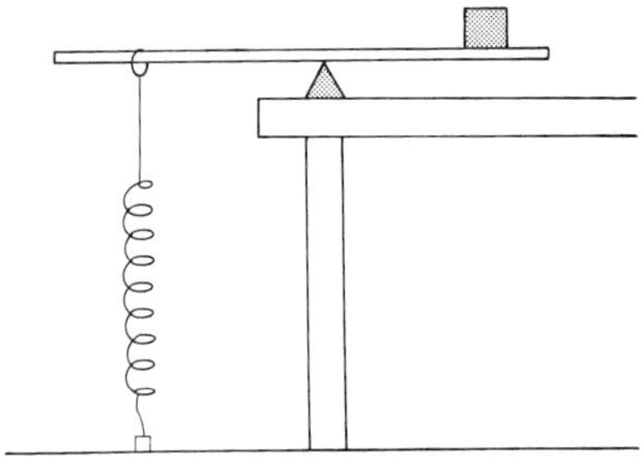

Fig 8.10

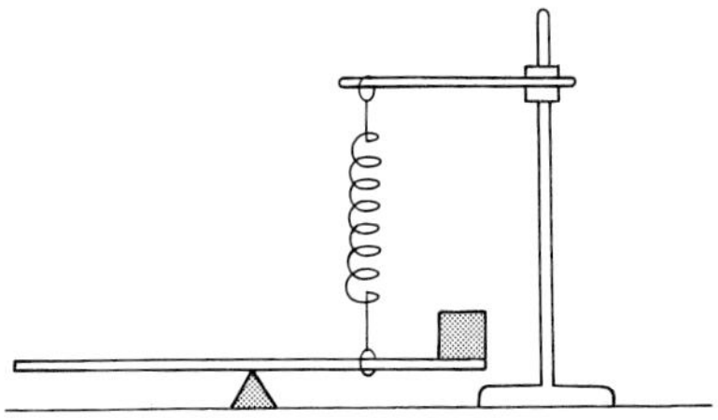

Fig 8.11

If a plank is balanced on a brick, there will be a downward force, the weight of the plank acting on it. As the plank does not move, it must be balanced by an upward force exerted by the brick. We can conveniently think of the weight of the plank acting downward through the balance point even though we know it is really distributed over the whole plank. If the plank is uniform (that is, it has the same thickness and width all the way along and the density is everywhere the same), then the balance point is in the middle. We call this balance point the **centre of gravity** of the plank.

Questions for homework or class discussion

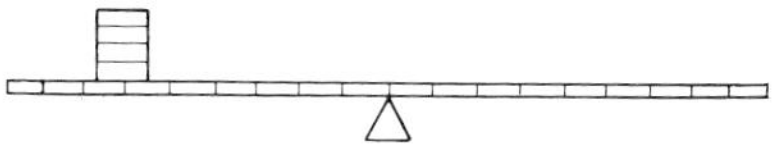

Fig 8.12

1 Four squares are placed six divisions out from the centre of a beam as shown in Fig 8.12. How many squares must be placed four divisions to the right to balance it?

2 The see-saw shown in Fig 8.13 is balanced at its centre, and equal divisions are marked on it outwards from the centre. Various masses are put on it. Say in each of the following cases whether they balance or not. If they do not balance, say which end will go down.

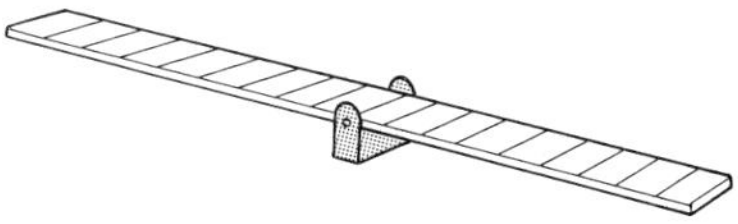

Fig 8.13

a 10 kg four divisions out on the left; 20 kg eight divisions out on the right.
b 10 kg four divisions out on the left; 20 kg two divisions out on the right.
c 2 kg eight divisions out on the left; 4 kg four divisions out on the right.
d 10 kg four divisions out on the left; 10 kg two divisions out and 20 kg one division out on the right.
e 2 kg two divisions out and 4 kg five divisions out on the left; 8 kg two divisions and 2 kg four divisions out on the right.

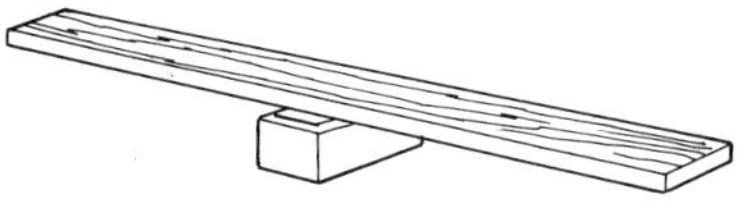

Fig 8.14

3 A plank of wood is balanced on a brick (Fig 8.14). A boy stands at a point 30 cm from the centre. Ten 1 kg masses are placed on the plank 1.5 m from the centre on the opposite side and the beam balances. What is the mass of the boy?

4 A light beam is supported at its centre and two forces of 5 newtons and 10 newtons act on it (Fig 8.15).

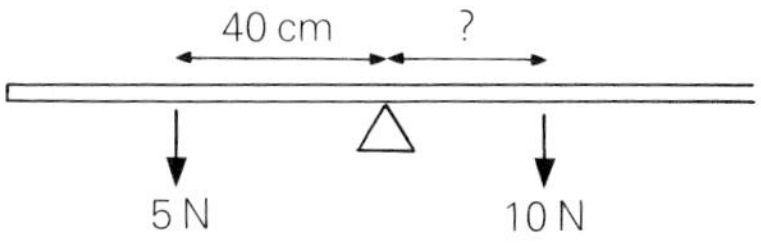

Fig 8.15

a If the beam balances, at what distance from the centre must the 10 newton force act?
b What upward force is exerted by the support on the beam?

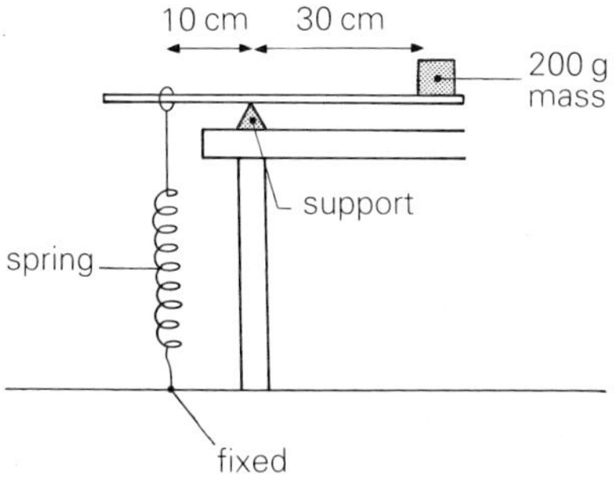

Fig 8.16

5 A light beam is balanced as shown in Fig 8.16 by the force due to the spring acting on one side of the balance point and by a mass of 200 g on the other side. The gravitational force is 10 newtons on every kilogram.
a What is the downward gravitational force acting on the 200 g mass?
b What downward force must be exerted by the spring?

Explorations 1

A A lorry is loaded with wooden crates by hauling them up a ramp.

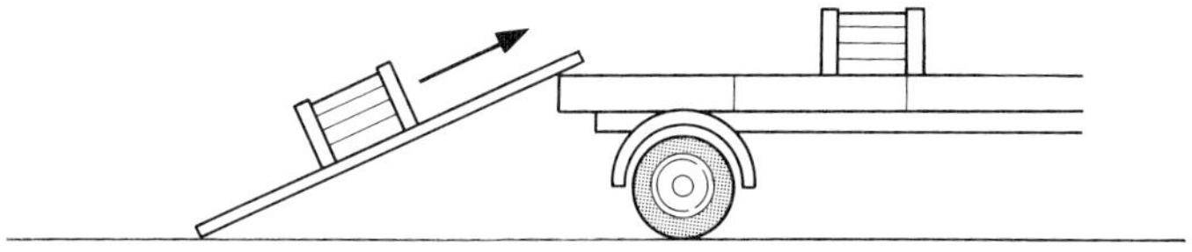

If the loaders stop for a rest, the crates may sometimes slip down again.

Do some experiments to find out the smallest angle at which this happens.

Does the angle depend on how heavy the crate is?

Does it depend on how big the crate is?

Does it depend on what the ramp is made of?

B Pirates used to make their prisoners walk the plank, and sometimes the plank was such that the unfortunate prisoner would slip off the plank as he reached the end.

Using a lath of wood or a metre rule as the plank, and a brass weight to represent the prisoner, explore how long the plank should be for 'prisoners' of different weights to slide off the end of the plank.

Can you discover what effect the width of the plank has? What effect does the thickness have?

C A child swings to and fro taking a certain time for each swing.

Use the apparatus provided to find out whether the time it takes for each swing depends on the mass of the child.

Does the time of swing depend on the length of swing?

Does it depend on how hard the child is pushed at the start?

What pulls the child back after it has been pushed outwards?

Can you guess what the results might be if the experiment were done on the Moon?

D The speed which is reached by a toboggan as it slides down a snow-covered slope depends partly on how steep the slope is.

Explore how speed depends on slope by rolling a ball down a ramp.

Measure how long it takes to cover a measured distance and calculate its average speed. See how this changes as the height *h* is increased.

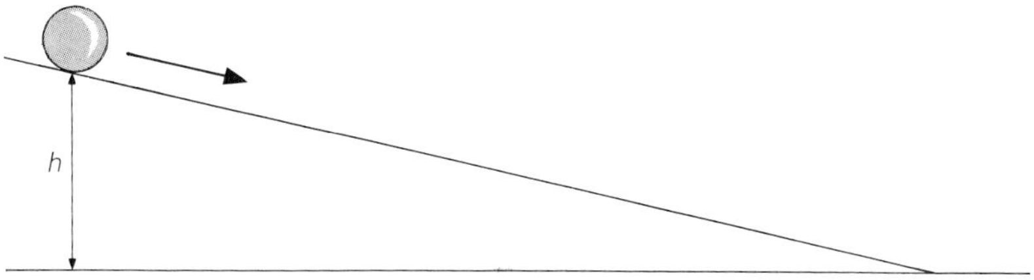

E The force on the pedals of a bicycle results in a force between the tyre of the rear wheel and the road.

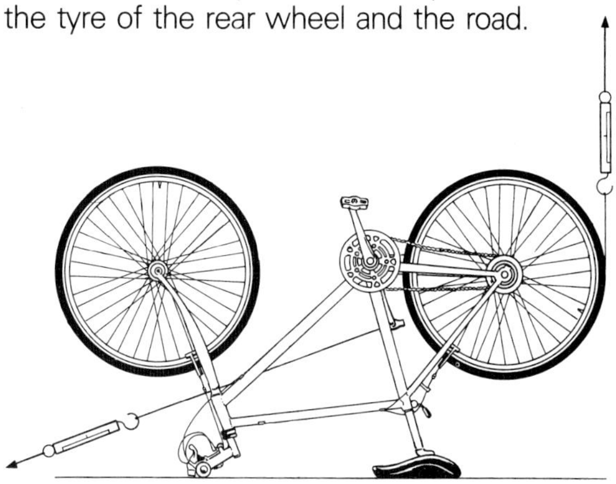

Using spring balances to measure the forces, explore how these two forces are related.

Explore what differences are made by the use of gears.

F You are given a number of tins. They are all sealed so that they cannot be opened, but each has something different inside.

Explore how each tin behaves when you lift it, shake it, roll it or do any other experiment. Decide what is inside each tin.

Write down your answer giving the reason for your choice.

G **a** You are given a piece of paper or cardboard, about 30 cm by 21 cm (A4 size). If you make a bridge with it across a gap, it will take very little weight to make it collapse. Can you make a stronger bridge, perhaps by folding the paper or cardboard? Investigate how to make the strongest 'girder' to bridge the gap.

b Use a sheet of paper (30 cm by 21 cm) to make a paper tower which will hold a 1 kg mass 21 cm above the ground without collapsing. Investigate what arrangement supports the greatest weight.

H For this investigation you need a number of wooden blocks similar in shape. To the side of each fix different surfaces, such as rubber, felt, lino, sandpaper, plastic, carpet, *etc.*

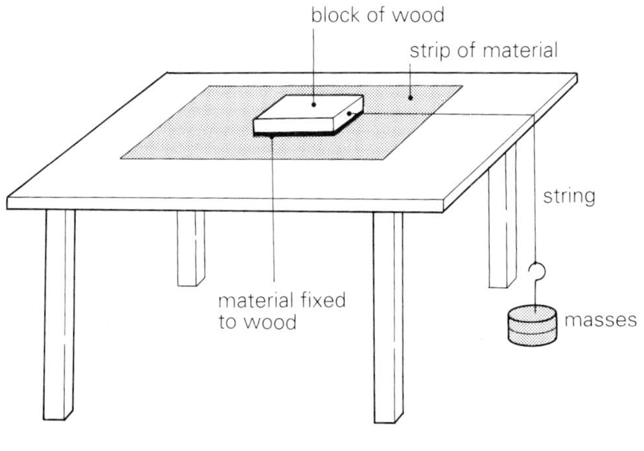

Arrange the blocks so that they can move on different surfaces made of strips of various materials (hardboard, metal, plastic, lino, sandpaper, carpet, *etc.*). Add masses as shown in the diagram to find what load is necessary so that the block just starts to move. Keep a careful record of what you find.

Does it make any difference if a 1 kg load is put on top of the block? Is the friction increased?

Does it make any difference if the 'road surface' is made wet?

Chapter 9 Pressure

In earlier chapters, you have learned about forces, and you know that it matters in what direction they are applied and where they are applied. In this chapter, you will learn that the effect of a force can also depend on the area over which the force is applied.

Here are some things for you to think about and discuss.

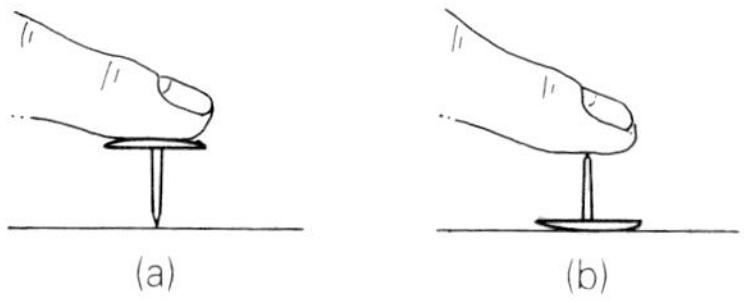

Fig 9.1

1 Suppose you push a drawing pin against a very hard surface. How does pushing the pin as in Fig 9.1(a) differ from pushing as in Fig 9.1(b) using the same push in each case?

2 It is more painful to stand barefooted on small pebbles or gravel than it is to stand barefooted on a smooth floor. Why?

3 Farm tractors have wheels with large, wide tyres. Why?

4 For moving on snow, skis are used rather than skates. Why?

Your discussions should have brought out the fact that the effect of a force is more noticeable when the area of contact is smaller. To push on the point of a drawing pin is painful; to support your own weight on a small area of contact (as on pebbles or gravel) is painful; tractors with small narrow tyres will sink more easily into soft ground; skates will simply sink into the snow whereas skis will not. Can you think of other examples which show that the effect of a force is less when the area of contact is large?

Experiment 9.1
The effect of area

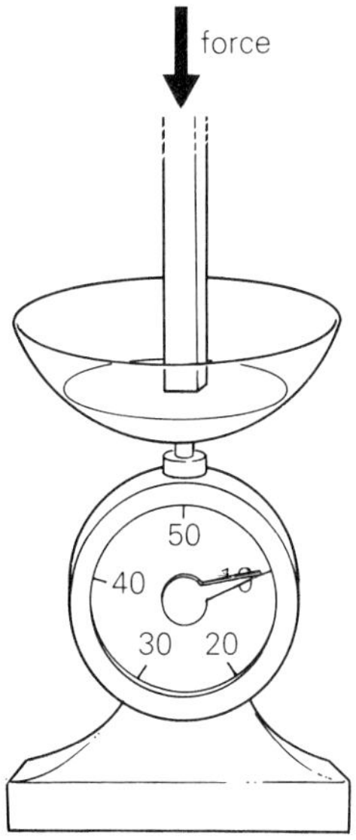

Fig 9.2

Take some modelling clay. It needs to be reasonably soft for this experiment, so work it in your hands beforehand and get it pliable. Put a layer of it, at least 2–3 cm thick, into the pan of the scales.

For the experiment you will need a rod of wood of cross-section 1 cm × 1 cm and another 2 cm × 2 cm. Hold the first rod over the scales and push it down on to the modelling clay until the reading of the scales is, say 10 newtons (or 1 kg if they are scales for measuring mass). Hold the rod at the reading for a short while (see Fig 9.2).

Repeat the experiment using the second rod applying the same 10 newton force. You will find the dent is less deep in the second case.

Repeat the experiment with the 2 cm × 2 cm rod, but this time increase the force to 40 newtons. This time the dent will be about the same as when 10 newtons were applied on the 1 cm × 1 cm rod. In one case there were 40 newtons acting on 4 square centimetres; in the other case 10 newtons were acting on 1 square centimetre. In other words the same dents were obtained when the force on each square centimetre was the same.

We use the word **pressure** for **force per unit area**. In both the above experiments, when the dents were the same, the pressure was 10 N/cm^2.

$$\text{PRESSURE} = \frac{\text{FORCE}}{\text{AREA}}$$

Questions for class discussion

Snow shoes

1 In snowy regions, why do people wear cumbersome snow shoes?

2 Suggest one reason why footballers have studs on their boots.

3 If a boy falls through some ice on a pond, why is it better for a man to crawl across the ice when he goes to the rescue? The best thing for the man to do is to put a plank of wood or a ladder on the ice and to crawl along that. Why is that?

4 Skis are very awkward things to carry around. Why is it necessary to have them so large?

5 Why do grocers use a fine wire for cutting cheese?

6 You can carry a suitcase comfortably by the handle. But if you have no handle and you tie the case with string, why is it very painful carrying the case with your hand through the string?

7 Make an estimate of your mass. What is the downward gravitational force on you in newtons? Estimate the area of the flat top of a drawing pin. What is the pressure when you sit on the flat side of the drawing pin? Estimate the area of the pointed tip of a drawing pin. What is the pressure if you sit on that, assuming that all your weight is supported by the point?

Experiment 9.2
Water pressure

Connect a small and large syringe together as shown in Fig 9.3, first filling the tubing and the large syringe with water. There should be no air in the system.

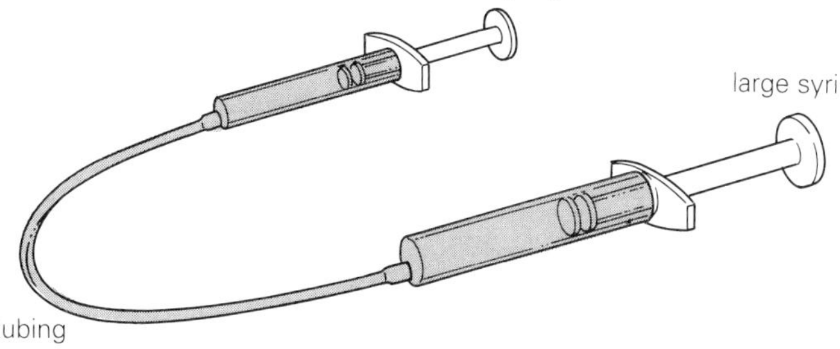

Fig 9.3

Hold the small syringe while pushing the water from the large syringe into it. Feel the forces. Then push the water from the small syringe to the larger one. Again feel the forces. You will find the forces different because of the different areas involved.

Pressure of liquids

The above experiment has shown that liquids also exert a pressure. Liquids spread over a surface; any material which spreads over an area will exert a pressure.

You can feel the pressure due to a liquid by turning on a tap and putting your thumb over the outlet. You might try this with a tap at the top of a house and then with a tap at ground level or in a basement. Do you notice any difference? (You may not notice any difference if the taps are connected directly to the water mains. It should be very noticeable with hot water taps.)

Water pressure is obviously very important when building a dam or the wall of a reservoir. The water will exert a considerable pressure on the wall. Why do you think engineers build dams with the wall thicker at the bottom than at the top (see Fig 9.4)?

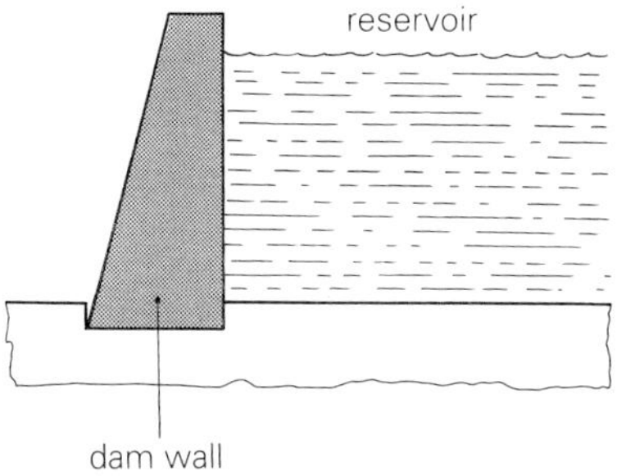

Fig 9.4

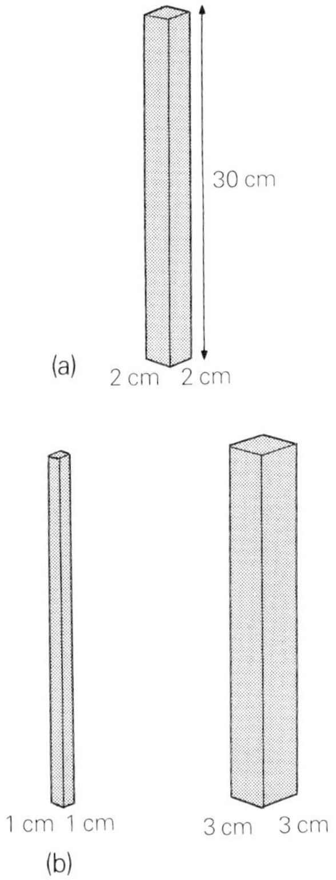

Fig 9.5

Questions for class discussion

Imagine a column of water with height 30 cm and cross-section 2 cm × 2 cm (Fig 9.5a).

1 What is the volume of the water?

2 The density of water is 1 g/cm^3. What is the mass of the water?

3 The gravitational field has strength 10 N/kg. What is the downward force due to gravity on this mass of water?

4 What is the area of the base on which this force acts?

5 What is the pressure on the base, that is, the force on 1 cm^2?

6 In just the same way, work out what the pressures would have been if the base had been 1 cm × 1 cm or 3 cm × 3 cm (Fig 9.5b).

7 What do you notice about the results?

8 What would happen to the pressure if the height of the column were 60 cm and not 30 cm?

9 What would happen to the pressure if the density were 2 g/cm^3 and not 1 g/cm^3?

These examples have shown that the pressure due to the liquid does not depend on the area of the base: the pressure was the same in each of the cases in Questions **5** and **6**. It also showed that the pressure depends on the height: if you double the height, you double the pressure; if you treble the height, you treble the pressure and so on. Similarly it depends on the density: if you double the density, you double the pressure.

So far, we have considered the pressure on the base. But does a liquid exert a pressure in any other direction?

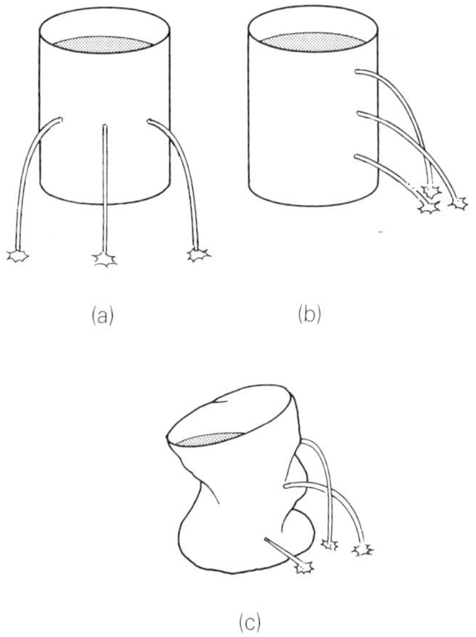

Fig 9.6

Experiment 9.3
Investigating the direction and strength of water pressure

a Take a tin can and make holes in it with a round nail. It helps when making these to put a block of wood inside as an anvil. Put the holes at different places round the can but at the same level (Fig 9.6a). Take care to make the holes the same size.

b Fill the can with water and watch how the water spouts out. If you wish, you can put the can under a tap so that water flows in at the same rate as it flows out.

c Take another can and make equal holes, one near the bottom, one near the top and one in the middle (Fig 9.6b). Watch how the water flows out of these holes.

d Finally, take a third can and batter it into an irregular shape. Make holes in three or four different places and watch what happens this time (Fig 9.6c).

This experiment should show that the pressure is certainly greater at the bottom than it is near the top because water comes out faster where the pressure is greater. You will also notice that the water comes out at right-angles to the surface. Of course it does not continue in that direction because gravity acts on the jet and it falls towards the Earth, but it always starts at right-angles to the surface.

Optional experiment for the bath

Fill a balloon or polythene bag with water. Make some holes in it with a pin (one near the top, one in the middle, one near the bottom) and watch what happens. The jets should all start at right-angles to the surface: the jet from the one near the top will start upwards.

Experiment 9.4
Water finds its own level

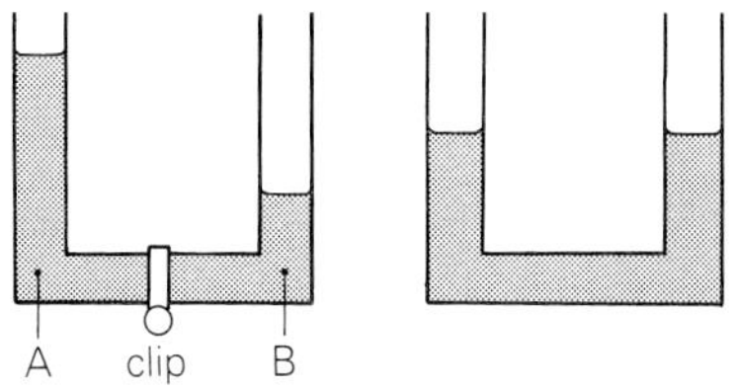

Fig 9.7

It is often said that 'water always finds its own level'. What does this mean? And what is the cause of it?

Two glass tubes should be held vertically and joined at the bottom by rubber or polythene tubing. Put a clip in the middle so that water cannot flow from one side to the other (see Fig 9.7). Put some coloured water in the left-hand tube so that it is nearly full (a few drops of ink will colour the water) and put less coloured water in the right-hand tube.

We know that the pressure depends on the depth of the water. It is therefore greater at A than it is at B. But A and B are at the same level. As a result, when the clip is opened, water will flow from A to B. It will continue to flow until the pressures are the same, in other words until the water levels are the same.

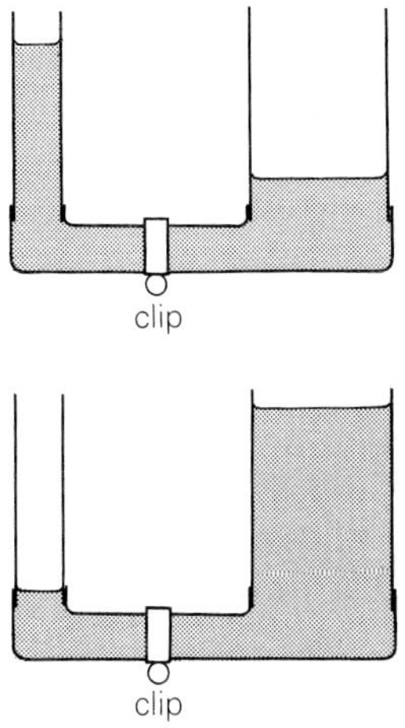

Fig 9.8

Suppose the two tubes are unequal in size as shown in Fig 9.8. What would happen in these cases when the clip was opened?

From the above we see it is not the mass of water on the two sides that must be the same. It is merely the pressure that is the same and that means that the heights must be the same. The shape of the tubes does not make any difference, as is shown in the teapot or the odd-shaped glass container in Fig 9.9.

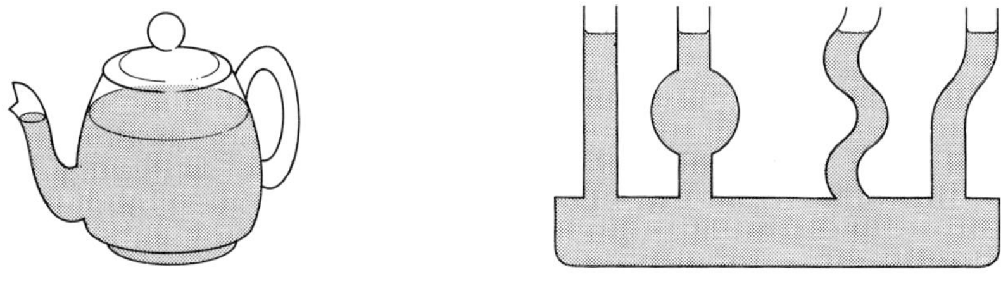

Fig 9.9

Background Reading

Control of pressure differences

When a high-pressure liquid or gas is connected to one at low pressure, there is a flow from the high pressure to the low pressure. Many of our everyday appliances depend on pressure difference and we need to be able to regulate this flow. We need to use taps and valves. The tap enables us to stop or reduce the flow when we want to; the valve enables us to control the direction of flow.

The domestic gas tap consists of a cylinder with a hole through it, which fits into a circular opening in the gas supply tube. When the tap is closed, the hole is not in line with the gas supply. When the tap is open, the hole connects the gas main to the fire or cooker, and gas flows (see Fig 9.10).

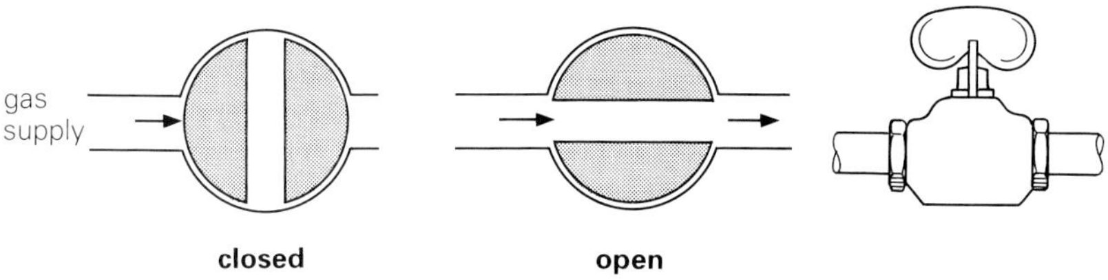

Fig 9.10

The gas tap in Fig 9.10 would not be very satisfactory as a domestic water tap. Can you suggest why not?

When the domestic water tap shown in Fig 9.11 is opened, the washer is lifted up and water flows. When the tap is closed, the washer is pressed against the seating and the flow stops. The washer is made of either leather or rubber; it can get worn and has to be replaced occasionally.

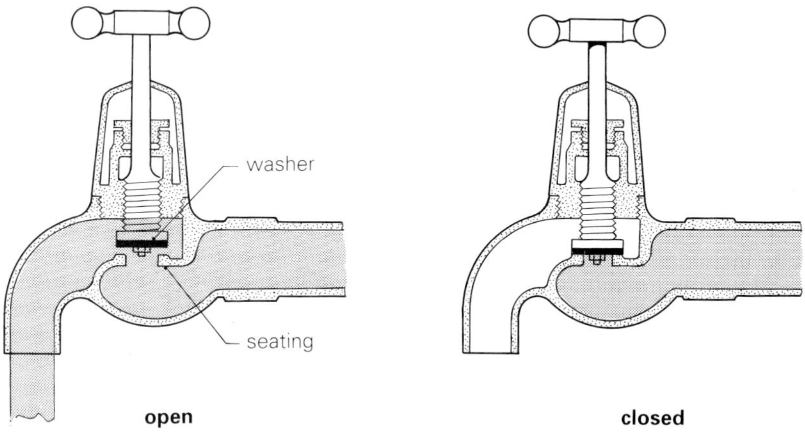

Fig 9.11

The ballcock is a special kind of tap used in water tanks in the attic and in the lavatory cistern. When water is used from the tank, the level of water goes down. The ball floating on the water descends as the water level falls (see Fig 9.12a). The tap then

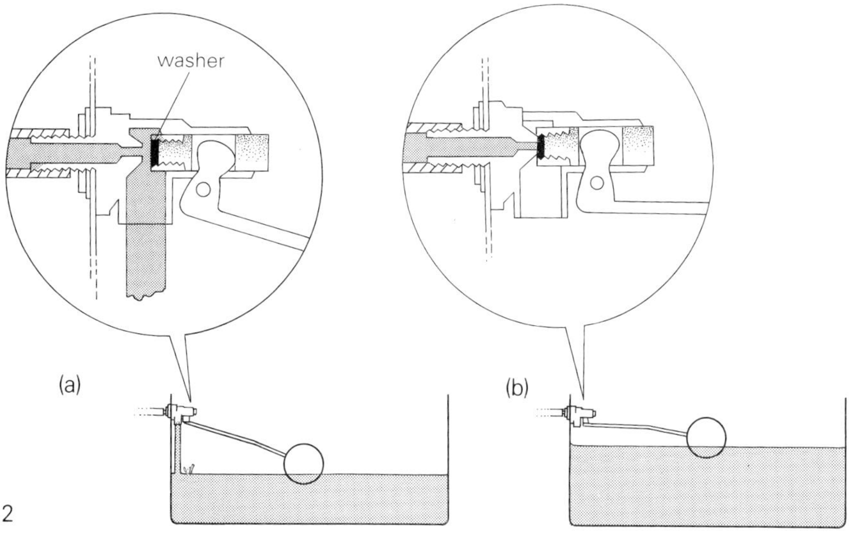

Fig 9.12

Fig 9.13

opens to let water flow into the tank again. As the water from the inlet pipe fills the tank, the floating ball rises and closes the inlet (see Fig 9.12b). This ingenious device fills the tanks automatically, and always to the same level.

A non-return valve is a device which will allow a flow in only one direction. There are valves for gases, valves for liquids and even valves for electricity, all of which allow flow one way only.

Perhaps the easiest way to understand the idea of a non-return valve is to make a very simple one. Take a sheet of paper, moisten your lips, open your mouth wide and press the paper against your lips, all round, as shown in Fig 9.13. Blow air out from your mouth; you will find it comes out quite easily. Now try drawing air into your mouth: it cannot be done because the paper is forced against your lips, blocking the entry. You have made a simple non-return valve for your mouth. It allows air to go out, but not to come in.

A simple non-return valve for liquids, like that shown in Fig 9.14, can be put within a pipe or at its opening.

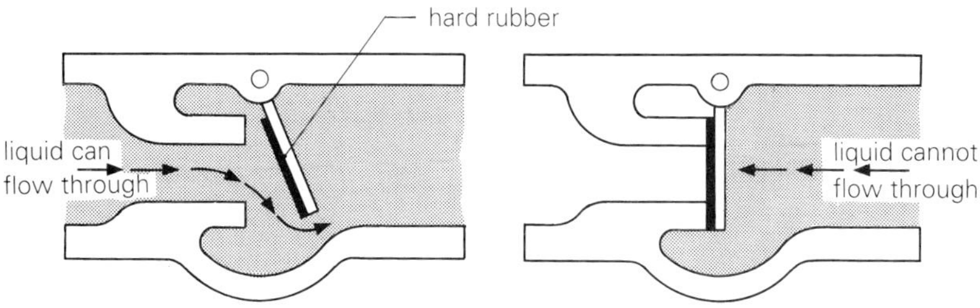

Fig 9.14

If the liquid is at a higher pressure on the left, it pushes open the hinge and liquid flows from left to right. If it is at a higher pressure on the right, it presses the rubber washer against the metal and liquid cannot flow. A similar arrangement can be used for gases as well as for liquids.

(Adapted from Longman Physics Topic *Pressures* by A R Duff.)

Pressure of gases

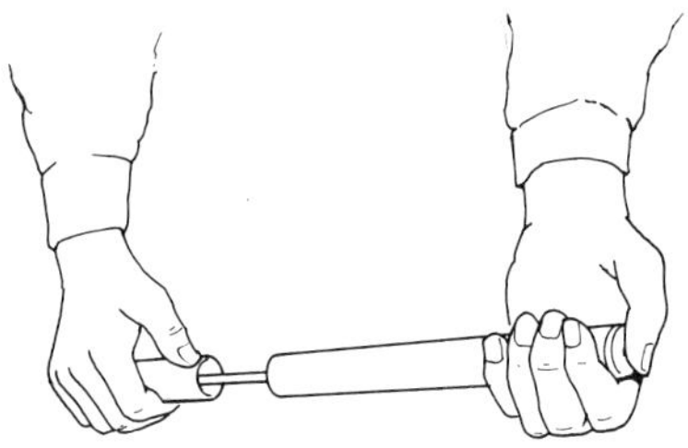

Fig 9.15

Gases also exert a pressure. If you put your finger over the outlet of a bicycle pump and push the handle in (Fig 9.15), you will feel the force produced by pressure. If you blow up a balloon with a small hole in it, the air will come out of the hole whichever direction the hole is pointing. What about area? The next experiment will tell us something about that.

Experiment 9.5
Feeling forces with two syringes of different sizes connected together

Repeat Experiment 9.2, but this time have only air in the syringes and tubing, and not water.

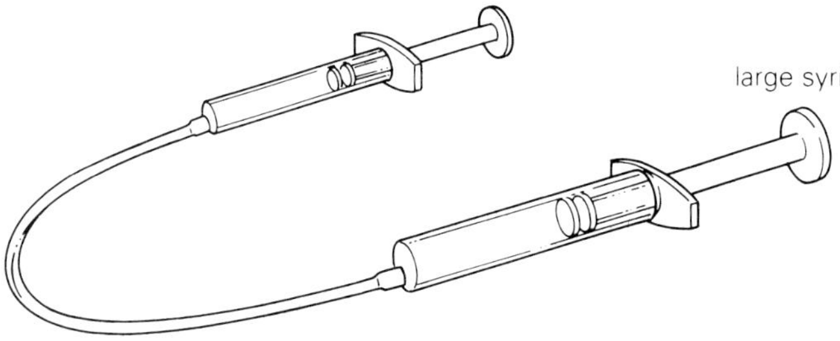

Fig 9.16

Hold the small syringe while pushing on the larger one so that air goes from one syringe to the other. Feel the forces on the two syringes. Then push on the small syringe while holding the large syringe. Feel the forces again. The force is greater on the plunger with the larger area than it is on the plunger with the smaller area.

Experiment 9.6
The pressure box

The apparatus shown in Fig 9.17 has a plastic bag inside it. There are two moveable platforms resting on the top; one is four

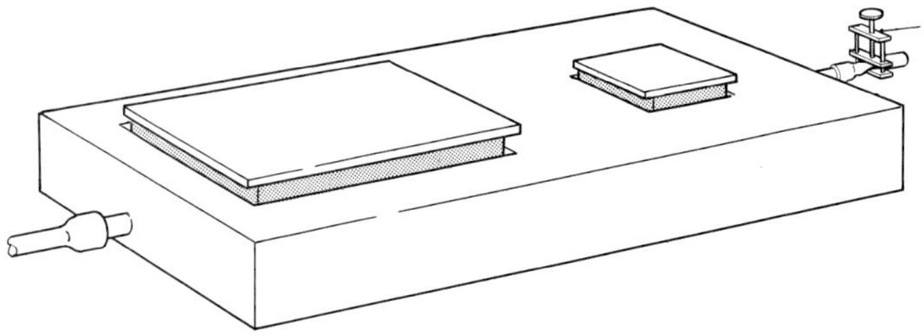

Fig 9.17

times the area of the other. Close the outlet tube and blow air into the inlet tube until the bag is full of air.

Place a 1 kg mass on each platform and increase the air pressure inside by blowing into the box. Which of the platforms rises first? Why does this happen?

Repeat the experiment, keeping one 1 kg mass on the small platform but with two, three and four 1 kg masses on the large platform. Which rises first on each occasion? Do these results agree with your knowledge of pressure?

Atmospheric pressure

Earlier in the course, you measured the density of air. 1 cm^3 of air has a mass of 0.0012 g which is not very large, but our atmosphere contains a great deal of air extending upwards for many kilometres.

We are in fact at the bottom of a great sea of air and this exerts a large pressure on us, about 10 N/cm^2. We are so used to this pressure that we do not ordinarily notice it, but it is there none the less. The following demonstration experiments should convince you of its existence.

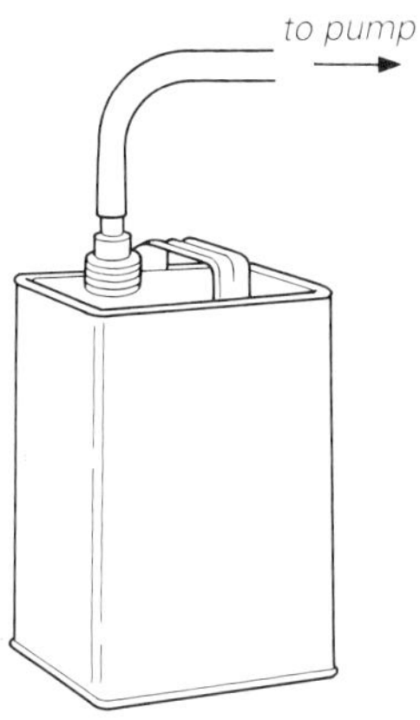

Fig 9.18

Experiment 9.7
Collapsing can

A can with a well fitting bung is connected by pressure tubing to a vacuum pump (Fig 9.18).

At first the pressure of the air inside the can is the same as the atmospheric pressure pressing on the outside of the can. The pump is then used to draw the air out of the can and the can collapses because of the atmospheric pressure on the outside.

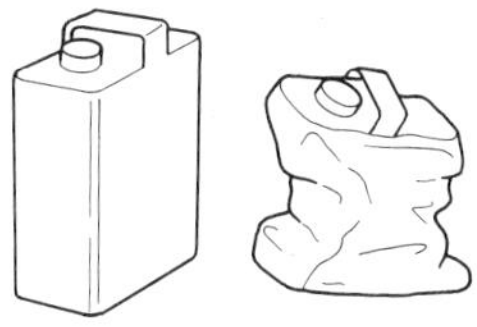
Fig 9.19

Experiment 9.8
Another collapsing can

For this experiment you need a can with a good bung or airtight cap. Put a little water in the bottom of the can and boil it vigorously with the top open. The water vapour will drive much of the air out of the can. Then close the top. Let the can cool. The water vapour inside will condense back to liquid.

The pressure inside will be much less than the atmospheric pressure outside and the can will collapse (Fig 9.19). (If you try this experiment at home, it is wise to use a rectangular can. Some cylindrical cans will stand up to big pressure differences and will not collapse. Do not use one which has contained petrol, paraffin, white spirit, or any other flammable liquid.)

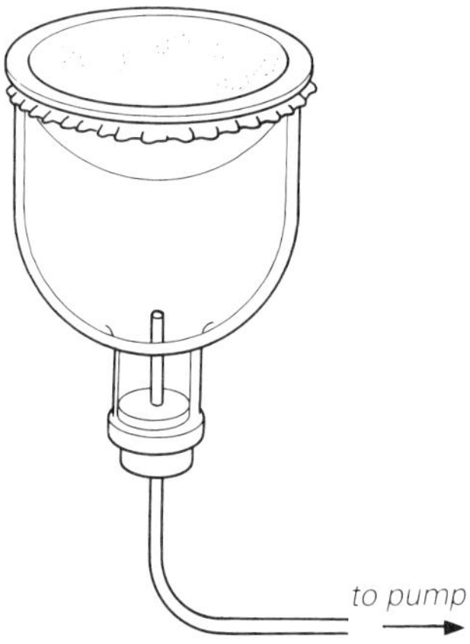

Fig 9.20

Experiment 9.9
Demonstration of atmospheric pressure

A rubber sheet is tied over the open end of a bell-jar (so called because it is shaped like a bell). The other end of the jar has a well fitting bung and is connected by tubing to a vacuum pump (Fig 9.20). At first the atmospheric pressure on one side of the sheet is balanced by the pressure due to the air inside the bell-jar. The pump then takes away the air inside and the effect of the atmospheric pressure is seen.

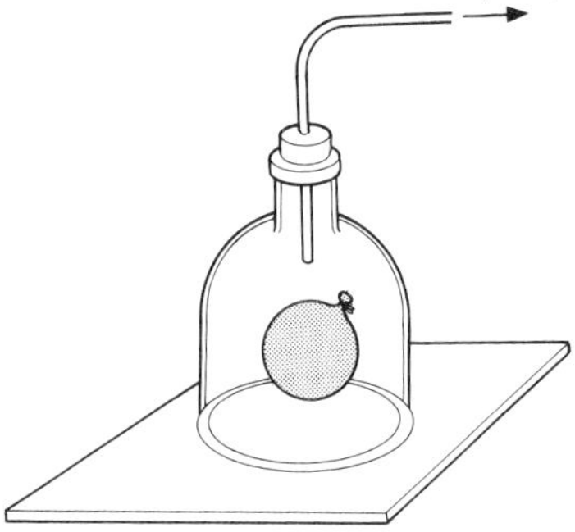

Fig 9.21

Experiment 9.10
Balloon in a bell-jar

Put a partially filled balloon inside a bell-jar which is placed with its open side on a thick glass plate (Fig 9.21). Some vacuum grease is probably necessary to seal it and so make it airtight. Connect the top of the jar to a vacuum pump. At first the atmospheric pressure on the outside of the balloon balances the air pressure inside, but when the vacuum pump removes the air and there is no longer the same pressure on the outside, the balloon expands.

Background Reading

The Magdeburg experiment

Magdeburg experiment

One of the first vacuum pumps was made by Otto von Guericke. He used it to do a famous experiment in the city of Magdeburg in Germany in the year 1657.

He had two hollow hemispheres which fitted together to form a sphere. A greased leather washer between the hemispheres ensured an airtight joint.

He used his vacuum pump to extract as much air as possible from inside so that the atmosphere outside pushed the two hemispheres together. Then he harnessed two teams of horses, one team to each hemisphere, and drove them to pull the hemispheres apart (see picture on page 90).

It needed eight horses in each team to do that. The hemispheres were about 35 cm in diameter and the atmospheric pressure was about 10 N/cm^2. You can therefore show that the force needed to pull the hemispheres apart was about 10 000 N.

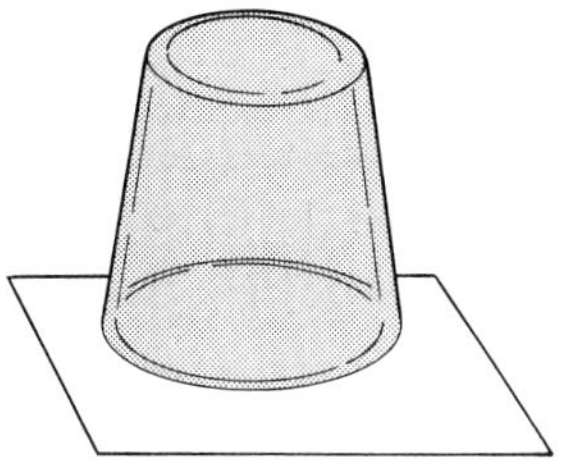

Fig 9.22

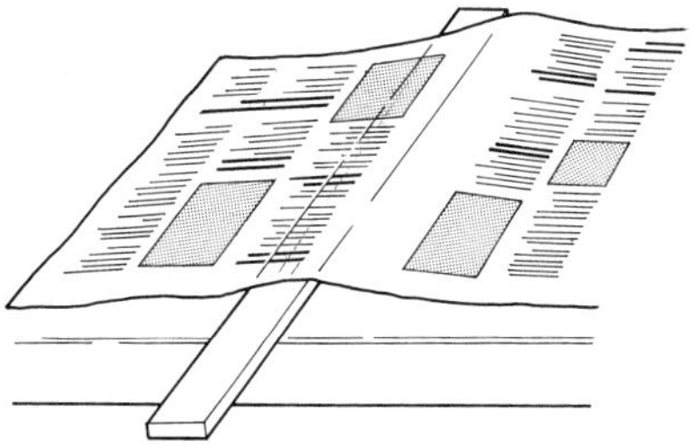

Fig 9.23

Class discussion about further experiments

1 A tumbler is filled with water and a piece of paper is put on the top. It is important that there are no air bubbles in the water. With your hand holding the paper on the tumbler, turn it over and remove your hand (Fig 9.22). Why does the water stay in the glass?

2 A sheet of newspaper is placed over a thin piece of wood on a table (Fig 9.23). If a small steady force is applied to the end of the wood, what happens? If you give a sudden sharp blow to the wood, what happens? Explain the difference.

3 A drinking straw is filled with liquid. Place a finger over both ends and hold the straw vertically. When you take your finger off the bottom, the liquid stays in the tube. Why is this?

4 A sucker pressed on to a flat surface can support quite a heavy mass (Fig 9.24). Why is this?

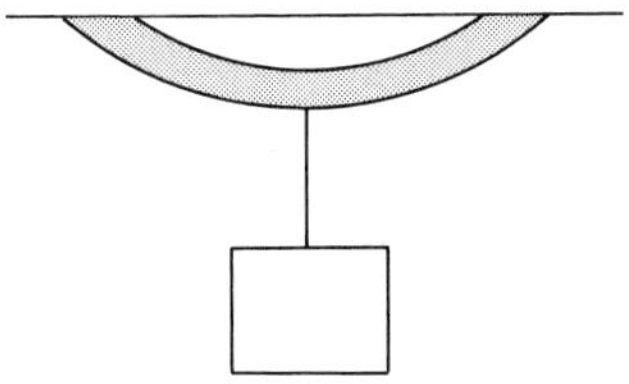

Fig 9.24

Fig 9.25

5 A piece of cotton wool is wedged in the bottom of an empty tumbler. The tumbler is turned upside-down and then pushed below the surface of the water in a deep bucket (Fig 9.25). The wool does not get wet. Why is this?

6 Take a milk bottle and fill it with water. Put a drinking straw in it. Fix Plasticine round the drinking straw at the top of the bottle and make sure you have an airtight joint (Fig 9.26). Can you suck the water out of the bottle? Explain what happens.

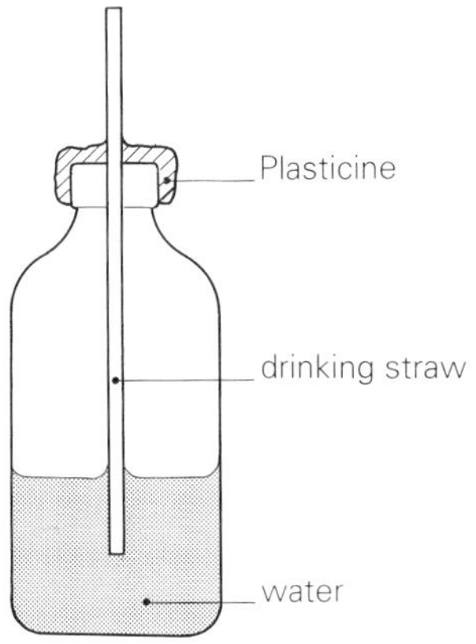

Fig 9.26

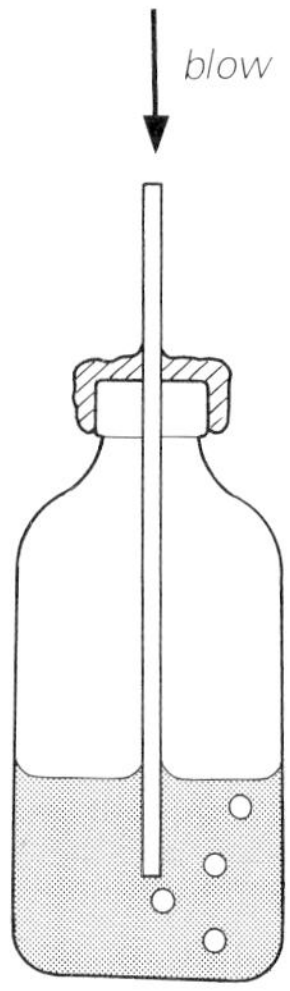

Fig 9.27

7 Using the same bottle, drinking straw and Plasticine, see how many air bubbles you can blow through the drinking straw (Fig 9.27). When you have got as much air as possible into the bottle, take your mouth away from the drinking straw and see what happens. (It is wise to do this near a sink!) Explain what occurs.

Background Reading

Pressure due to gases

Like liquids, gases also exert a pressure on their containers. If you try to blow up a balloon with a small hole in it, you will find the air comes out of the hole. The air comes out of the hole no matter in which direction you hold the balloon.

Air pressure plays a most important part in the tyres of your bicycle and the tyres of a motor car. It manages to support the bicycle or car, and to prevent the road rubbing on the rim of the wheel. The car or bicycle is supported because of the extra pressure of the air in the tyre. If the pressure is low, then there has to be a bigger area of tyre in contact with the ground in order to support the weight of the car or bicycle (Fig 9.28).

If the extra pressure in the tyre were only half of what it should be, the total area of contact would need to be doubled to support the weight.

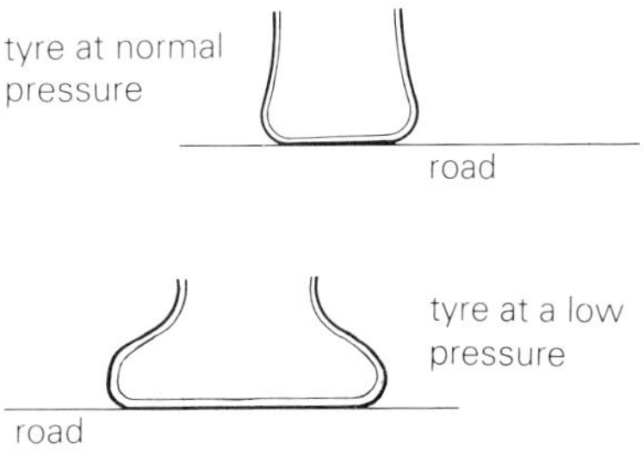

Fig 9.28

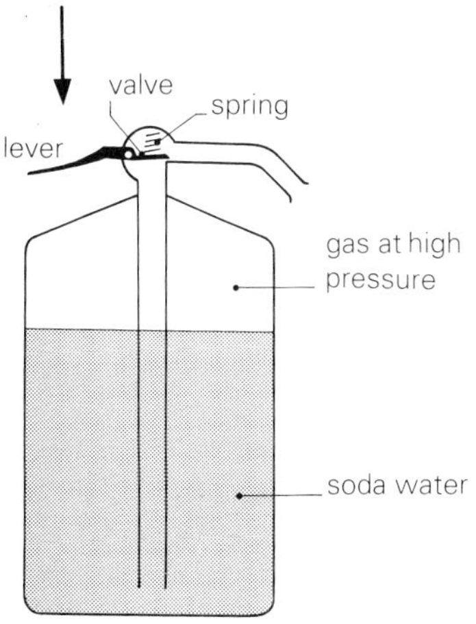

Fig 9.29

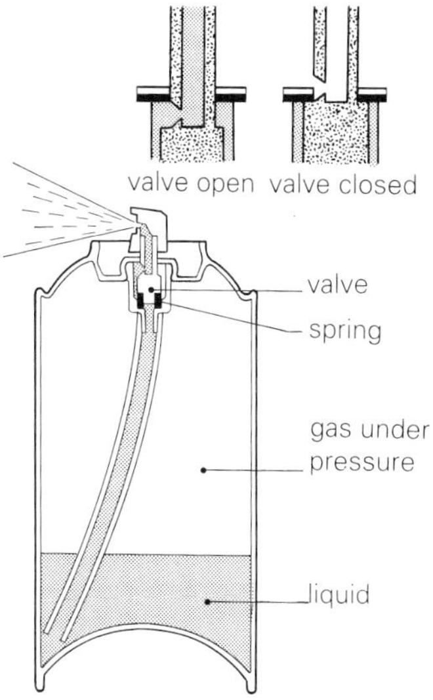

Fig 9.30

A soda siphon (Fig 9.29) is a good example of a gas exerting a pressure. The space above the liquid is filled with gas at high pressure. When the valve is raised by the depression of the lever, the gas pushes down on the liquid and forces it up the tube and through the outlet. Another example is the aerosol spray (Fig 9.30) which also contains gas under pressure. The gas forces liquid out when the valve is opened.

(Adapted from Longman Physics Topic *Pressures* by A R Duff.)

Measuring pressure

Manometers

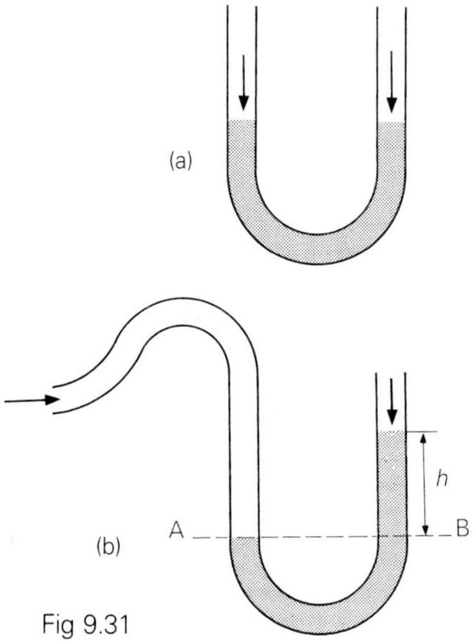

Fig 9.31

A useful device for measuring pressure differences is a manometer. This is a U-shaped tube of glass or clear plastic, partially filled with liquid (see Fig 9.31).

If the tube is open at both ends, as shown in diagram (a), the level of the liquid will be the same on both sides. The pressure of the atmosphere is of course pressing down on the liquid surfaces, but as it is the same on both sides the levels are the same. If you raise the pressure on one side by blowing into a tube attached to that side, as shown in diagram (b), the levels will be different. When the liquid is at rest, the pressure at A will be the same as the pressure at B. The pressure at B will be equal to the pressure of the atmosphere plus the pressure due to the column of liquid, *h*. We say that the excess of pressure on the left-hand side is '*h* cm of liquid'.

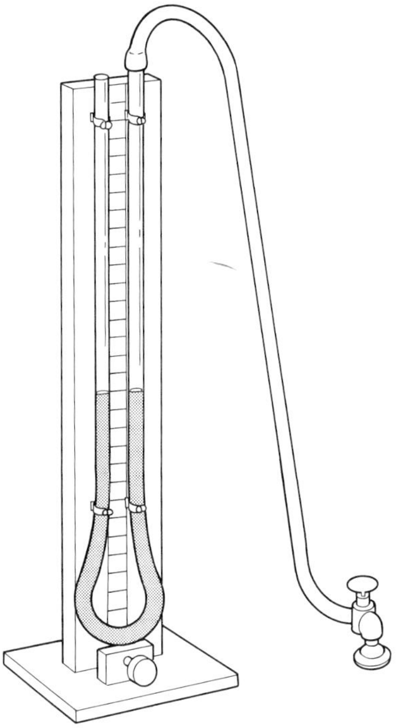

Fig 9.32

Experiment 9.11
Measuring the pressure of the gas supply

Connect the manometer to the gas supply (Fig 9.32). Turn on the gas supply and leave it on. Measure the distance between the water levels. How much bigger than atmospheric pressure is the gas pressure, measured in centimetres of water?

Experiment 9.12
Using a manometer with unequal tubes

In this experiment use a manometer in which one of the tubes is much wider than the other (Fig 9.33).

Join the narrower tube to the gas supply and measure the difference in levels. Then join instead the wider tube to the supply and measure the difference in levels. What do you find? Is this what you would expect?

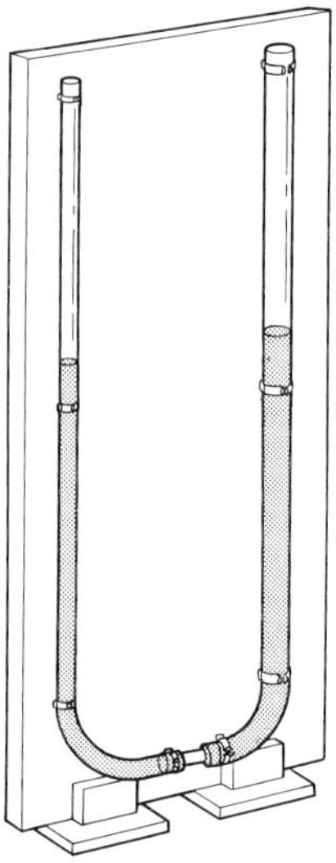

Fig 9.33

Experiment 9.13
Comparing a water manometer with a mercury manometer

A large water manometer and a mercury manometer are connected through a three-way connector to a bicycle pump with a non-return valve. Use the bicycle pump to increase the pressure inside the tube. How do the levels compare in the two manometers? The density of mercury is almost fourteen times the density of water. Does this explain the difference?

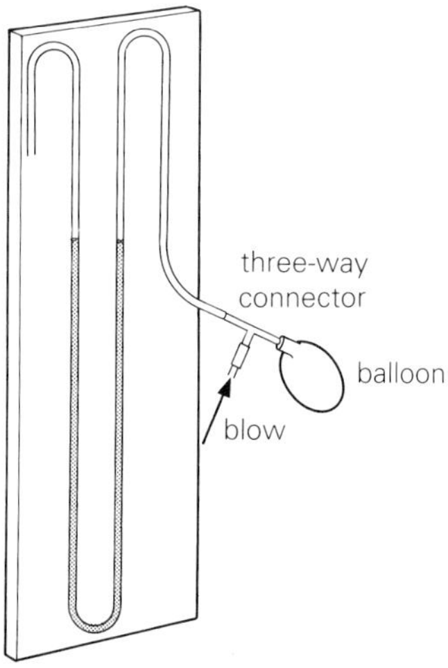

Fig 9.34

Experiment 9.14
Measuring the pressure necessary to blow up a balloon

A balloon is tied securely to a three-way connector, which is connected to a large water manometer as shown in Fig 9.34. Blow up the balloon through a tube connected to the three-way connector. Get your partner to watch how the pressure changes. Does it stay the same as the balloon is blown up?

The Bourdon gauge

Fig 9.35

Another device for measuring pressure is the Bourdon gauge. You are doubtless familiar with the toy, shown in Fig 9.35, used at parties. The harder you blow (in other words, the greater the pressure you exert) the more it uncurls.

The Bourdon gauge works on the same principle: the greater the pressure applied the more the curved tube inside straightens out and this moves a pointer across a scale (see Fig 9.36).

Bourdon gauges can be calibrated to measure the pressure directly in N/cm^2.

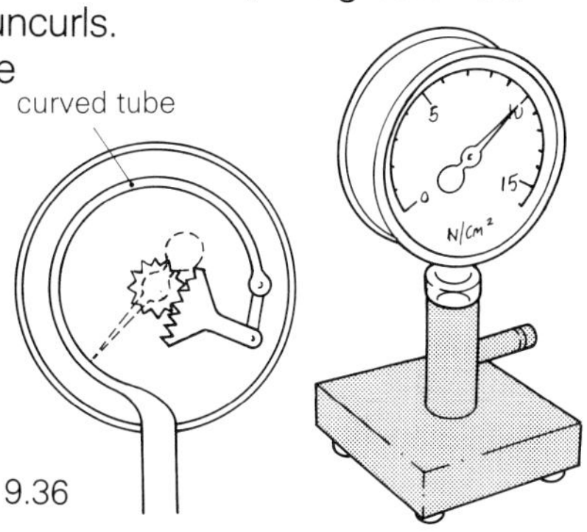

Fig 9.36

Experiment 9.15
Using a Bourdon gauge to measure pressure

Connect a tube to the Bourdon gauge and blow into it to measure your lung pressure. Also try sucking to see how low a pressure you can reach. You could also try this experiment using the large water manometer instead of the Bourdon gauge.

Experiment 9.16
Using a Bourdon gauge to measure the pressure of a balloon as it is blown up

This is a repeat of experiment 9.14, but this time using a Bourdon gauge instead of a manometer (Fig 9.37).

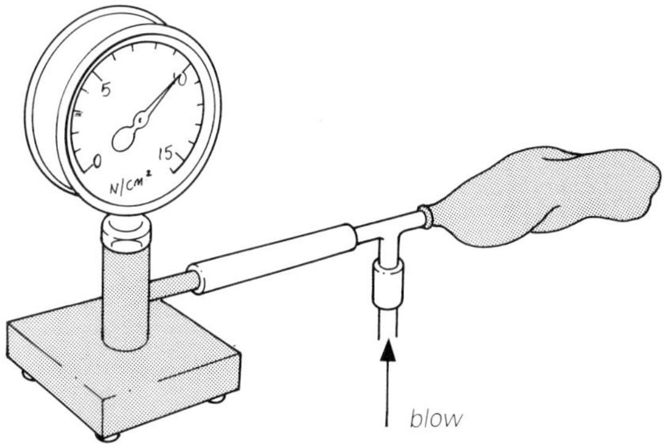

Fig 9.37

Demonstration experiment 9.17
Using a Bourdon gauge to measure low pressures

Connect a Bourdon gauge directly to a vacuum pump. Switch on the pump and see how low the pressure falls. Great care must be taken in this experiment when letting air back in again: if it rushes in too fast, it can damage the gauge.

Measuring atmospheric pressure

Atmospheric pressure, usually about 10 N/cm^2, varies from day to day and from place to place. It plays a very important part in predicting the weather and for that reason it is necessary to be able to measure it.

Safety note

Care must be taken as mercury is an extremely toxic substance.

Demonstration experiment 9.18
Height of a mercury column

A glass tube, 1 metre long, is held vertically with its lower end below the surface of some mercury in a trough. (The trough should be in a tray in case any of the mercury gets spilled.) The top end of the tube is connected to a vacuum pump. As the pump might be seriously damaged if mercury gets into it, it is necessary to include a round-bottomed flask, as shown in Fig 9.38, as a trap for any mercury.

Fig 9.38

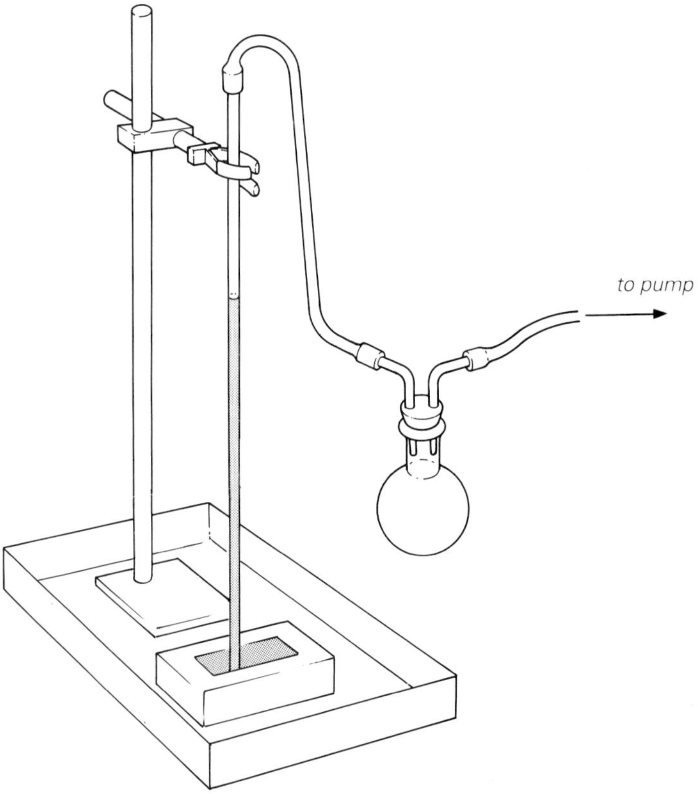

The mercury in the tube will start at the same level as the mercury in the trough, as the pressure both outside and inside the tube is atmospheric.

If the pump is switched on for a short while, the air pressure in the tube will be lowered and some mercury will rise up the tube.

In Fig 9.39, the pressure at A is the atmospheric pressure, P, due to atmosphere pushing down on the mercury in the trough.

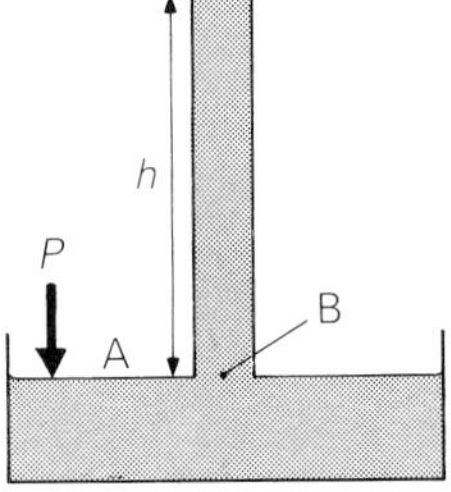

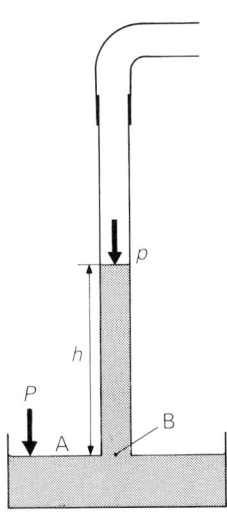

Fig 9.39

If the mercury is not moving, the pressure at B must be the same as the pressure at A. The pressure at B, however, will be equal to the pressure, *p*, due to the air left at the top of the tube plus the pressure, *h*, due to the column of mercury.

If the pump is switched on again, more air will be taken away, the pressure, *p*, will be less and the mercury in the tube will rise further (see Fig 9.40).

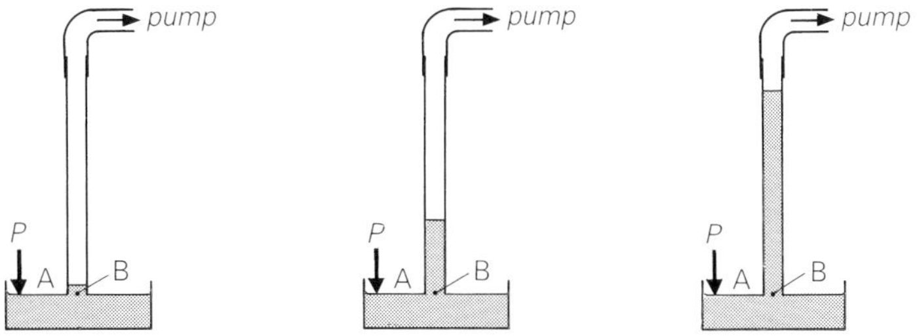

Fig 9.40

Will it be possible to draw the mercury right up to the top of the tube? That might be serious as we have said it is possible to damage a pump if mercury gets into it. Let us see what happens when you pump continuously.

You will find that the mercury rises to a height of almost 76 cm and then will rise no more.

The pressure at A is the atmospheric pressure and this is the same as the pressure at B. Assuming the pump is a good one, the pressure at the top of the tube is now almost zero. (It would be zero if there were no air or vapour at the top of the tube, so that there was a perfect vacuum there.) So the pressure at B is merely the pressure due to a column of mercury 76 cm high.

This is the reason why you will sometimes find pressures measured in centimetres of mercury. If you read that the pressure is 76 cm of mercury, it means that the pressure is the same as that exerted by a column of mercury 76 cm high.

Questions for homework or class discussion

1 A boy has a mass of 50 kg. The gravitational field exerts a force of 10 newtons on each kilogram. If the total area of his shoes in contact with the ground is 100 cm^2, what is the pressure exerted on the ground?

2 A car has four tyres at an excess pressure of 15 N/cm^2. The mass of the car is 1000 kg. What is the weight of the car in newtons? What area of each tyre is in direct contact with the ground?

3 A car has a mass of 1200 kg. It is supported by four tyres which have an excess pressure of 20 N/cm^2. After several weeks' use, the pressure falls to 16 N/cm^2. How is it possible for this lower pressure to support the car?

4 Is there a maximum length for a drinking straw used for drinking milk out of a bottle?

Barometers

It is possible to measure the atmospheric pressure more simply by setting up a barometer, as in the following experiment.

Safety note

Care must be taken as mercury is an extremely toxic substance.

Demonstration experiment 9.19
Setting up a simple barometer tube

The barometer tube is usually made of thick glass for strength, and it is sealed at one end. First, it must be completely filled with mercury: this is done in a tray in case of spillage (Fig 9.41a). Wearing gloves, a finger is then placed over the open end of the full tube. This end is then lowered and held below the surface in a trough of mercury (Fig 9.41b). (Even wearing gloves it is wise to wash hands after the experiment. Mercury is a dangerous substance.) Can any air get into the barometer tube when the finger is taken away (Fig 9.41c)? Clearly there is no way in which it can.

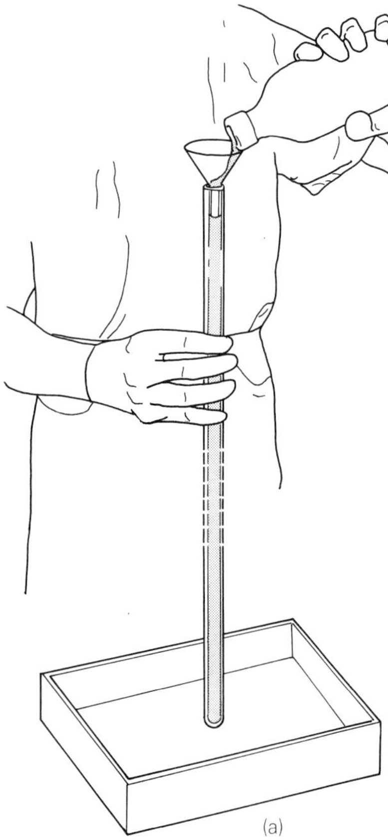

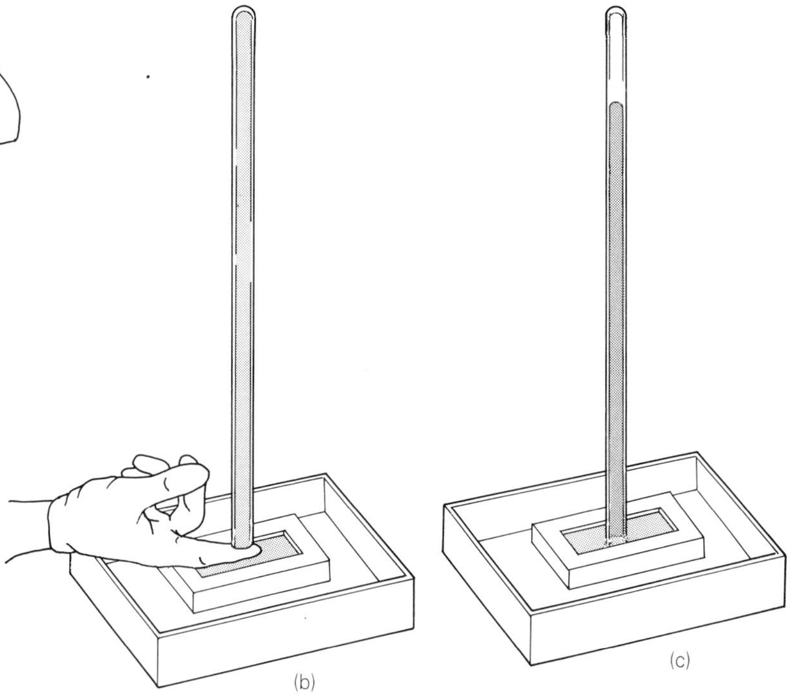

Fig 9.41

When the finger is removed, the mercury level in the tube will fall until the height, h, is such that the pressure at B, due to the column of mercury, is the same as the atmospheric pressure at A (see Fig 9.42).

As no air got into the tube, what will there be in the top of the tube at C?

You can now measure the atmospheric pressure in centimetres of mercury by measuring the height, h.

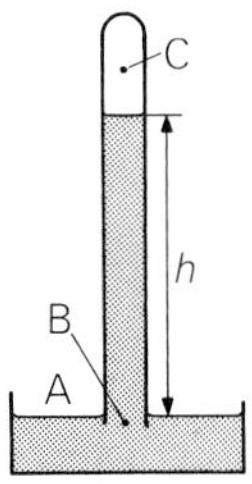

Fig 9.42

Would it make any difference to the height, h, if a barometer tube with a larger diameter were used? To find the answer to this, two barometer tubes of different diameter may be set up side by side (Fig 9.43). What you already know about the pressure due to a column of liquid should tell you the answer.

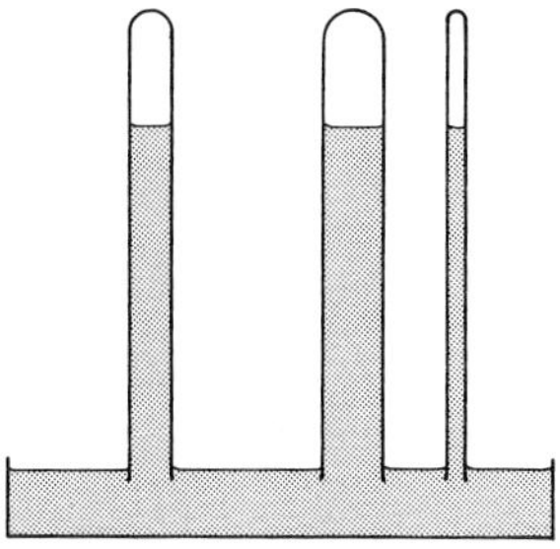

Fig 9.43

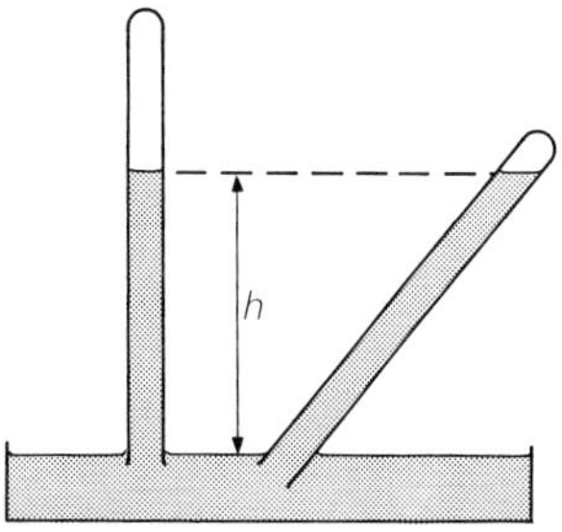

Fig 9.44

Finally, what happens to the height, h, if the barometer tube is inclined at an angle (Fig 9.44)? Find the answer by experiment, although once again you should be able to predict it.

Questions for homework or class discussion

1 A gas tap is connected to a manometer containing mercury. When the tap is turned on, the pressure difference is 1 cm of mercury.

a What would be the difference in levels if a water manometer were used? The density of mercury is 13.6 g/cm^3.

b If the difference in levels is 3.4 cm when a manometer with another liquid in it is used, what would be the density of this liquid?

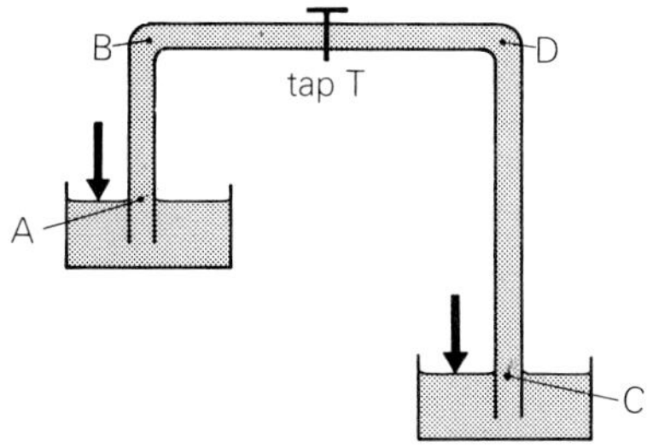

Fig 9.45

2 This question is about the way in which a syphon works (Fig 9.45). Initially the tap T is closed.

a What will be the pressure at A?

b B is a short distance above A. Will the pressure at B be less than, equal to or greater than the pressure at A?

c What will be the pressure at C?

d D is a large distance above C. Will the pressure at D be less than, equal to or greater than the pressure at C?

e Which will be greater, the pressure at B or the pressure at D?

f Will water flow from B to D, or from D to B, when the tap T is opened?

g At what stage will the water cease to flow from the upper bowl to the lower one?

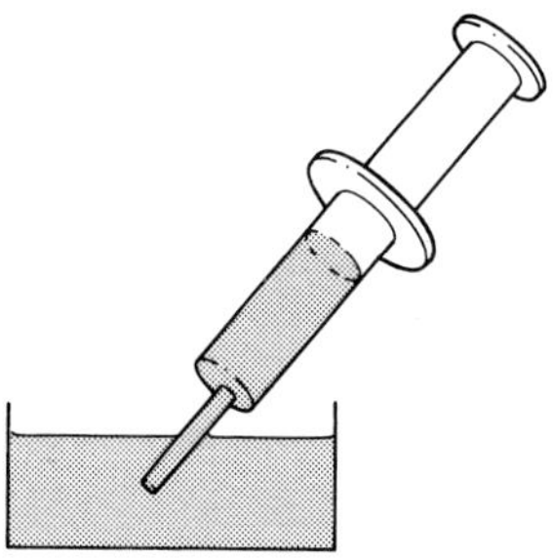

Fig 9.46

3 To fill a syringe, the plunger is pushed in and the end of the syringe is put in the liquid. As the plunger is pulled out, liquid enters the syringe (Fig 9.46). Explain why this happens.

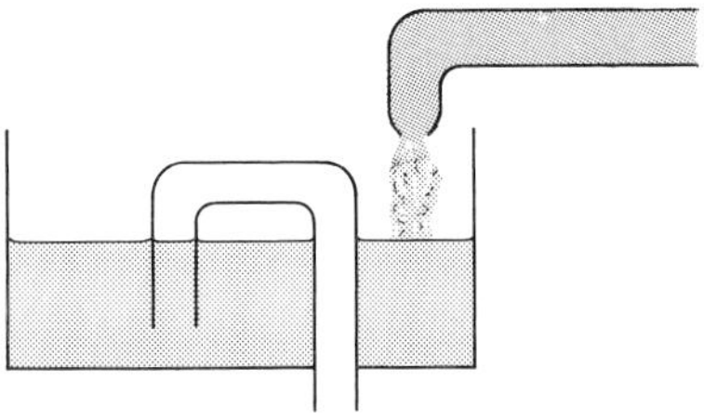

Fig 9.47

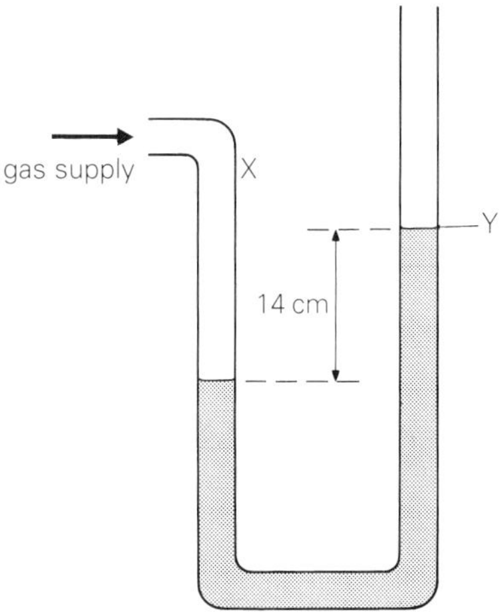

Fig 9.48

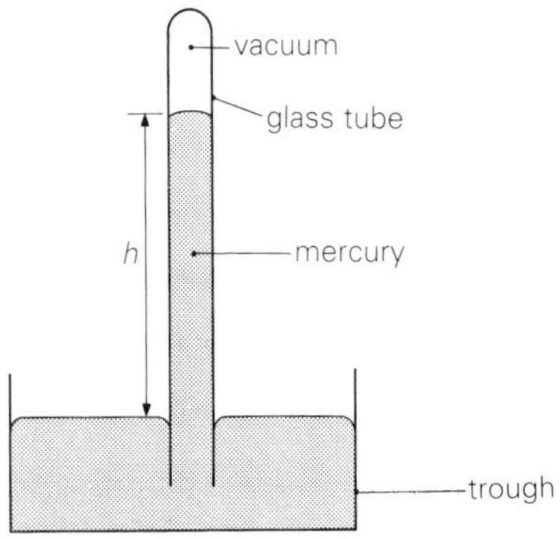

Fig 9.49

4 The inside of an aeroplane is usually 'pressurised'. What does this mean? Why is it necessary to do this pressurising?

5 If the atmospheric pressure is 10 N/cm^2, what will be the total force due to the atmosphere on a flat roof which measures 12 m × 9 m?

6 Fig 9.47 is a drawing of an automatic flushing cistern. Water flows into the tank from the tap which is left running permanently. When will it start to flush? When it has started, at what stage will it stop flushing?

7 On a certain day the pressure due to the atmosphere is equivalent to the pressure exerted by a column of mercury 75 cm high.

a The density of mercury is 13.6 times the density of water. What would be the height of a column of water which exerted the same pressure as the atmosphere?

b The density of water is about 1000 times greater than the density of air. If we assume that the density of air is always the same, what would be the height of a column of air which would give the same pressure as the atmosphere?

c This gives a possible figure for the height of the atmosphere. In fact, the height of the atmosphere is greater than this. Has anything gone wrong in the calculation? Explain why the actual height is greater.

8 When a water manometer is connected to a domestic gas supply the difference in levels is 14 cm (Fig 9.48).

What would be the difference in the levels of water in tubes X and Y if

a tube Y were twice as wide as tube X,

b tube Y were half as wide as tube X,

c mercury were used instead of water? (Suppose mercury is 14 times as dense as water.)

Could astronauts on the Moon use this method to measure the pressure of their oxygen supply? Give reasons for your answer.

9 Fig 9.49 shows a barometer.

a What, roughly, is the distance marked *h*?

b What does this tell us about the atmosphere?

c Give a reason for using mercury rather than water.

d If we pushed the tube 2 cm down into the trough, what would happen to the distance *h*?

e What would happen if there were a very small hole in the glass at the top of the tube?

10 What does a barometer measure?

How, if at all, would the reading of a mercury barometer be altered if

a the tube were to have double the diameter,

b the barometer were taken to the top of a mountain? Give a reason.

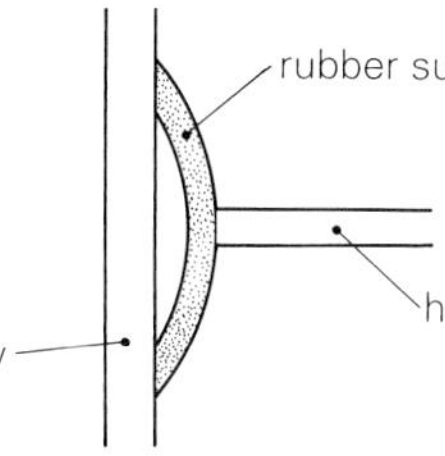

Fig 9.50

11 A rubber sucker is moistened and pressed against a window thereby pushing out most of the air inside the sucker (Fig 9.50).

a Explain why it is difficult to remove the sucker from the window.

b Why is it preferable to moisten the rubber sucker?

12 Fig 9.51 is a diagram of a simple lift pump for raising water out of a well. A non-return valve is a regulator which allows fluids to flow through it one way but not the other. A and B are non-return valves.

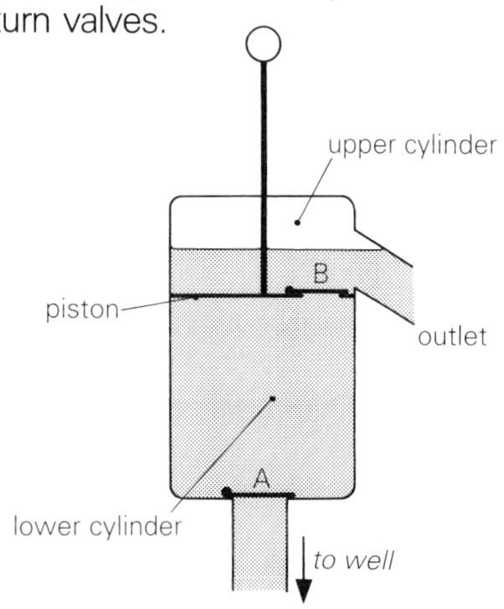

Fig 9.51

a Describe the action of the valves and water when the piston is pushed down from its present position.

b Describe what happens when the piston is raised again.

c What makes the water come up the tube out of the well?

d Why can it not take water out of a well more than 10 m below the pump?

Background Reading

The aneroid barometer

The most common type of domestic barometer, the aneroid barometer, does not contain mercury but has a corrugated metal box with a partial vacuum inside (Fig 9.52). This box is kept from collapsing by a strong spring. When the atmospheric pressure on the box changes, the spring moves slightly. This movement is magnified by a series of levers and moves a pointer round a scale. The movement of the pointer is a guide to the weather.

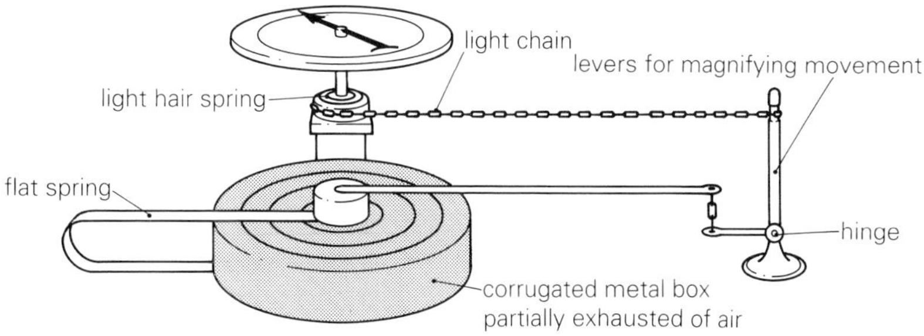

Fig 9.52

The barograph is similar to an aneroid barometer but it is used for making a continuous record of the pressure. The needle is replaced by a pen, and a drum with paper on it slowly rotates.

Barograph

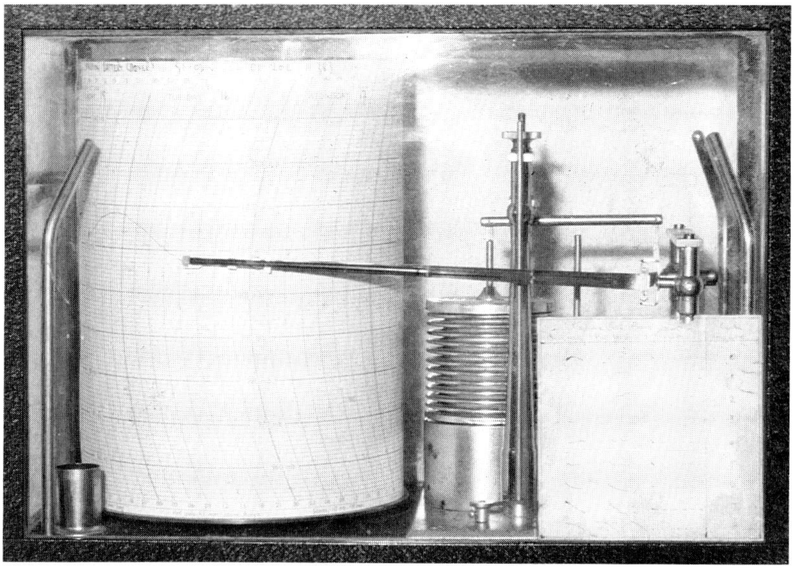

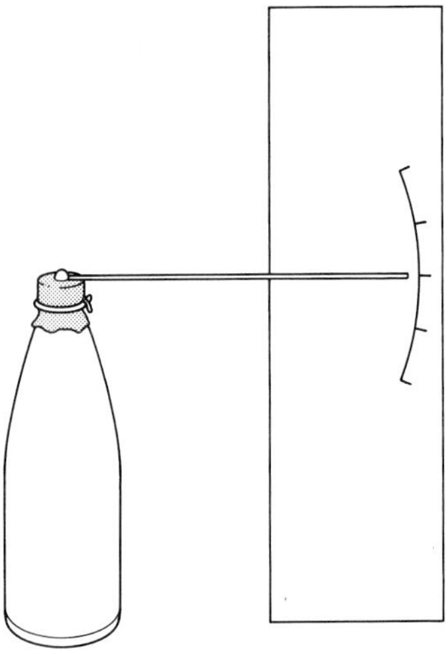

As something to do at home, you can make your own barometer as follows. Cut the end from a round toy balloon. Then stretch it smoothly over the mouth of a milk bottle and tie it in place (Fig 9.53). Fasten one end of a drinking straw to the *centre* of the rubber cap with a drop of wax. For a scale, mark some cardboard and prop it up beside the straw. (This will give a true result only if readings are always taken at the same temperature.)

When the pressure goes up, the cap is pushed in and the straw rises. When the pressure falls, the reverse happens.

Fig 9.53

(Adapted from Longman Physics Topic *Pressures* by A R Duff.)

Chapter 10 Particle model of matter

In Chapter 2 we looked at crystals, we noticed the regularities of shape and we saw what happened when crystals grew. We found we could begin to understand this if we imagined that matter was made up of particles.

We found support for the model when we saw crystals of calcite being cleaved. If the regular shape of crystals meant that they were made up of layers of particles, we might expect them to cleave along certain planes and that is just what happened.

Models play a very important part in science, but models do not have to be correct in all respects to be useful. We found a model of polystyrene spheres useful, but it certainly does not mean that all matter is made of little polystyrene spheres! You go on believing a model is good until you find evidence that contradicts it. Then you have either to abandon the model or to modify it to fit the new evidence. You cannot ever prove that a model is correct. All you can do is to collect more and more evidence which makes it likely.

Solids, liquids and gases

In this chapter we will extend our model further, and in particular we will see how it can be applied to liquids and gases. To do this we start by thinking about what happens to substances when they are heated.

Demonstration experiments 10.1
Turning solids into liquids and liquids into gases

a Put some ice in a beaker. Stand the beaker on a tripod over a Bunsen flame and watch what happens to the ice.

b Put a little water in a tin lid and heat it until the water boils. Then go on heating. What happens to the water?

c Hold a Bunsen burner at an angle of 45° so that the flame is directly over a fireproof mat. Use tongs to hold the end of a short piece of solder in the flame. The flame should be held at an angle so that the molten solder falls on to the mat and not into the burner. What happens to the solder on the mat?

d Use tongs to hold a short length of lead strip in the flame, with the Bunsen burner still at an angle, and watch it melt. What happens if a piece of iron wire and a piece of copper wire are held in the flame? (It is probably best to hold the pieces of wire in a pair of pliers. Why do you think this is a good idea?)

e Put 1 g of crushed roll sulphur in a test-tube and heat it gently, moving it in and out of a low flame. As soon as it begins to change, take it away from the flame to watch what happens. Heat it a little more if necessary to turn it all into a golden-yellow liquid. Put it in a test-tube rack to cool. Do you see crystals forming?

f Put some naphthalene in a test-tube to a depth of about 2 cm. Then hold the test-tube in a beaker of boiling water. What happens to the naphthalene? It is better to melt naphthalene in hot water rather than hold it over a Bunsen flame, since it gives an unpleasant smoke if it catches fire. Watch what happens to the molten naphthalene when it cools. Do you see crystals forming?

You know that ice melts when it is warmed and that if you boil water it turns to steam. You will have seen other examples of solids melting and turning to liquids in the experiments above, although you were not able to melt iron wire in the Bunsen flame. Iron does not melt until it reaches a much higher temperature. The melting points of various substances are given in the list below.

copper	1083 °C	lead	327 °C	silver	961 °C
ice	0 °C	mercury	−39 °C	sulphur	113 °C
iron	1535 °C	naphthalene	80 °C		

Heating liquids causes them to turn to gases as happens when water turns to steam. The boiling points of some substances are given below.

copper	2595 °C	mercury	357 °C	silver	2212 °C
iron	3027 °C	methylated spirit	79 °C	sulphur	445 °C
lead	1744 °C			water	100 °C

If substances which are gases at room temperature are cooled, it will be found that they turn to liquids and further cooling will turn them into solids. For example, oxygen at normal pressure liquefies at −183 °C, nitrogen at −196°C, hydrogen at −253°C. They turn from liquids to solids at still lower temperatures: oxygen −219 °C, nitrogen −210 °C, hydrogen −259 °C.

Molecules and atoms

So far we have referred to a solid as being made of particles which are too small to see. We ought now to consider what these particles are.

A lot of important evidence about the particles of which matter is made comes from chemistry. It tells us that matter is generally made up of **molecules**. A crowd is composed of people, a library is composed of books, a forest is composed of trees. Matter is composed of molecules. A molecule is the smallest part of the substance which is still that substance. It is possible to cut up trees or tear up books, but to do so changes the tree or the book. It is the same with molecules. Molecules can be broken up into **atoms**; for example, a molecule of water can be broken up into two hydrogen atoms and one oxygen atom, but if you do this it becomes hydrogen and oxygen and it is no longer water.

Chemistry tells us that there are many different kinds of atoms which occur in nature. These are the **elements**, of which hydrogen and oxygen are two. Other elements of which you will have heard are nitrogen, carbon, lead, copper, zinc, iron, sulphur, mercury and so on.

Quite often atoms do not like going around alone and they prefer to be in pairs. For example, hydrogen atoms usually pair off together and so do oxygen atoms. A molecule of hydrogen therefore consists of two hydrogen atoms, and a molecule of oxygen consists of two oxygen atoms.

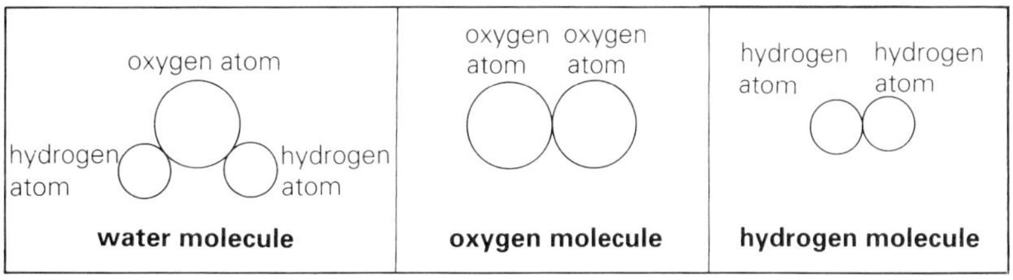

Fig 10.1

The drawings in Fig 10.1 represent the atoms of oxygen and hydrogen as spheres. This is merely another example of a model which we find very convenient. It does not mean that they are necessarily spheres. We have to go a lot further in our study before we can say more precisely what an atom is like, and even then we will merely be replacing one model by another.

Questions for class discussion on the particle model of a solid

1 The study of crystals led us to a model of a solid made of particles. If you take a piece of metal and pull it, it is very difficult to stretch it. What does this tell you about the forces between the particles?

2 It is very difficult to squash a piece of metal. What does this tell you about the forces between the particles?

3 (A harder question.) If you take two pieces of metal and hold them close to each other, they are not attracted to each other. Furthermore, if you touch the two pieces together, they do not stick to each other. What does this tell you about the forces between the particles?

4 If you pull a piece of steel wire so that it stretches a little and then let go, it returns to the original length. What does this tell you about the forces between the particles?

5 (Much harder.) If you pull on a piece of copper wire so that it stretches, it does not always go back to the original length. How can you explain this?

The particle model

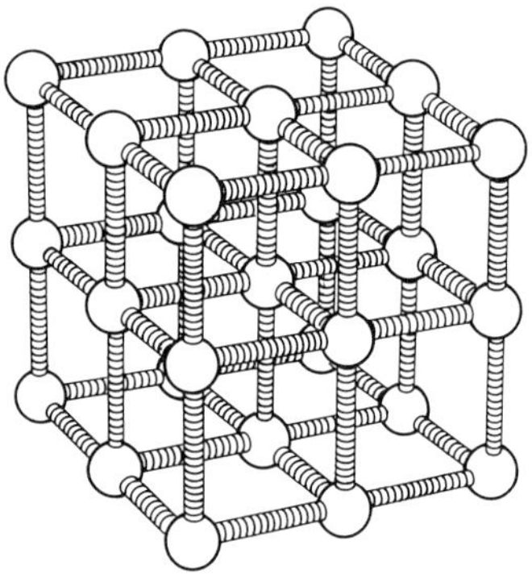

Fig 10.2

We can now suggest a model of a solid made of particles with relatively strong forces between them. Of course the force between a pair of particles is really very small, but there are so many particles in even the tiniest piece of material that it may take a considerable force to stretch or squash the material.

What can be said about the forces between our particles? When the particles are pulled apart by a small amount, a force of attraction tries to pull them back. When the particles are pushed together, the force between them is a repulsive force. When the material is neither stretched nor squashed, the net force on the particle is nil. A convenient model for our solid is shown in Fig 10.2 – polystyrene spheres joined by springs.

Fig 10.3 shows particle Y with particles X and Z next to it. If Y moves a little towards X, X will push Y away and Z will pull Y towards it. Y will then move towards Z, and when it gets nearer to Z than to X the forces will push it back towards X. Particle Y will wobble to and fro, and we call that a *vibration*. All the particles have a fixed average position but they will be vibrating. The model made of spheres and springs shows this happening if a sphere is moved a bit.

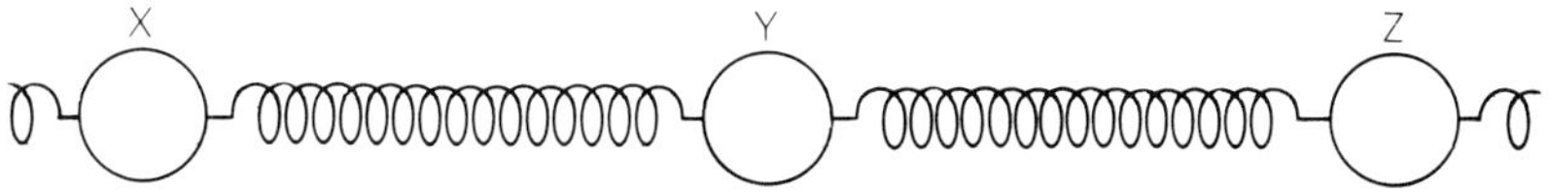

Fig 10.3

When a solid is heated the particles vibrate more and more but the general arrangement remains the same. If more heating took place, the particles might vibrate so much that some of the links between them would break, and so the solid would turn into a liquid. In a liquid there are still forces between the particles, but they are not as strong as before and the liquid is therefore able to flow.

In the previous chapter, we learned about the pressure exerted by a gas. Perhaps now we can explain this in terms of our model. If the liquid is heated, the molecules may be able to break away from each other entirely and, if the liquid is heated enough, all the molecules will be freed and will move around as a gas. If the gas is in the container, the moving molecules will bounce against the wall and will exert a force on it (in the same way that a ball bounced against a wall exerts a force on the wall). It is this impact force which is the cause of the pressure we have already discussed.

The particle model of gases and the use of the model to describe their properties is usually referred to as the **kinetic theory** of gases: kinetic comes from a Greek word meaning motion. Some further models may help to illustrate it.

Experiment 10.2
Two-dimensional kinetic model

(a)
(b)
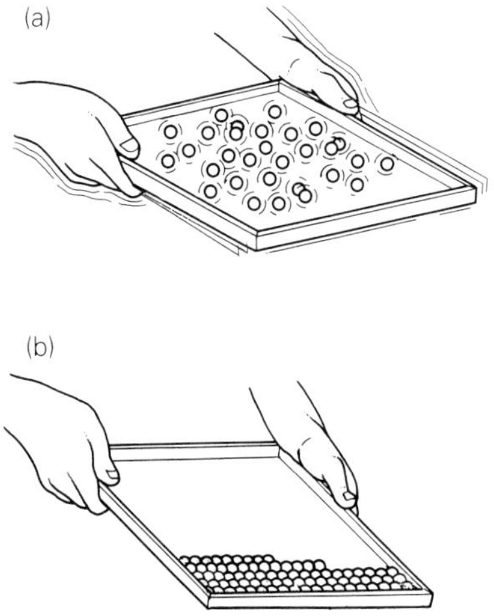
Fig 10.4

Put about twenty coloured marbles in the tray provided, and place the tray flat on the table. Agitate the tray, keeping it on the table, with an irregular shaking movement (Fig 10.4a). Watch the marbles moving about.

Try to use marbles which are of similar colour and have one which is distinctive and different, say red. Watch the motion of the red marble as the tray is agitated.

Look at the collisions that occur. You should be able to see two kinds: collisions when a marble hits a wall and collisions between the marbles themselves. The collisions with a wall cause a force to be exerted on it. This supports the idea that the pressure exerted by a gas is due to the collisions with the walls.

Tilt the tray and agitate it in the tilted position with most of the marbles at one side (Fig 10.4b). Why do you think this gives a possible model for a liquid?

Questions for class discussion

1 What path does a single marble take when the tray is being agitated?

2 Is the length of the path taken by one marble between collisions with other marbles always the same, or does it vary?

3 If, in Experiment 10.2, you agitate the tray rather more violently (in other words, 'heating the molecules' of the gas so that they move faster), what do you notice about the collisions with the walls? (From this model, we might expect the pressure exerted by a gas to be greater if the molecules were travelling faster, because there would be more impacts per second and each impact force would be greater. And this is exactly what we do find.)

4 If some more marbles are put in the tray, the model now represents a gas with a greater density. How does this affect the collisions with the wall if the tray is agitated so that the marbles have the same speed as before? Does this agree with the fact that the pressure of a gas increases with density if the temperature is kept constant?

Experiment 10.3
Three-dimensional model

In a gas, the molecules can move in all directions, so perhaps a better model is provided by the apparatus in Fig 10.5. The small ball-bearings in the plastic tube represent the molecules of a gas which move in three dimensions. An electric motor drives the vibrating rod under the rubber base and this sets the ball-bearings in motion. A cap on the top of the plastic tube prevents any of the balls escaping.

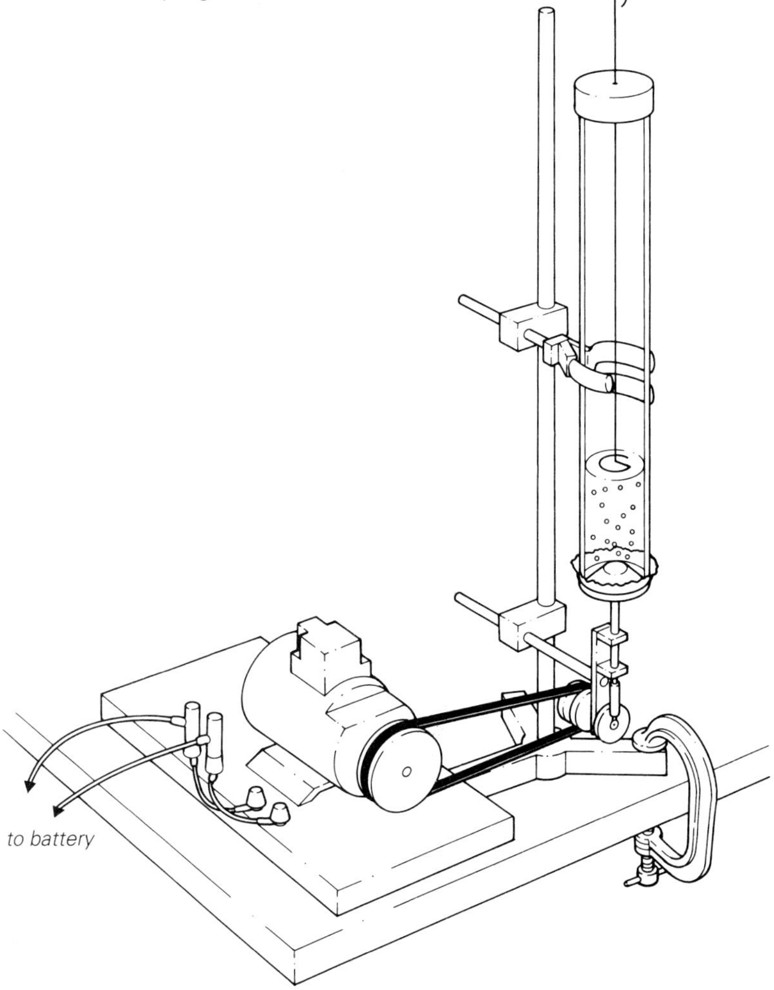

Fig 10.5

If the speed of the motor is gradually increased, the ball-bearings will fly around faster and faster, colliding both with each other and with the walls of the tube.

If a loose-fitting cardboard disc is lowered inside the tube, it will be bombarded by the ball-bearings. It will settle in a position when the downward force on the disc due to gravity (its weight) is balanced by the average upward force due to the collision of the balls on the disc. What happens if an extra cardboard disc is added to the top of the other disc?

Support for the particle model: change in volume

If the kinetic theory of matter is a good one, we ought to find some more evidence to support it. In our model of a solid the atoms or molecules are very close together, but in a gas they are a long way apart. We should therefore expect that the density of solids and liquids would, on the whole, be much greater than the densities of gases. That is exactly what we found earlier.

This merely compares the densities of different substances. What would you expect to happen to the volume of a substance when it changes from liquid to gas? The following experiment shows what happens.

Safety note

Petrol is flammable.

Demonstration experiment 10.4
Change of volume when a liquid changes to a gas

This experiment shows that a small drop of petrol has a larger volume when it turns into a gas.

A measuring cylinder (100 cm^3) full of hot water is placed inside the large beaker which has hot water in it (90 °C). Flexible tubing, full of water leads through the bung to the reservoir (Fig 10.6). The bung also has a short glass tube going through it. The bottom of the tube is closed with a rubber cap.

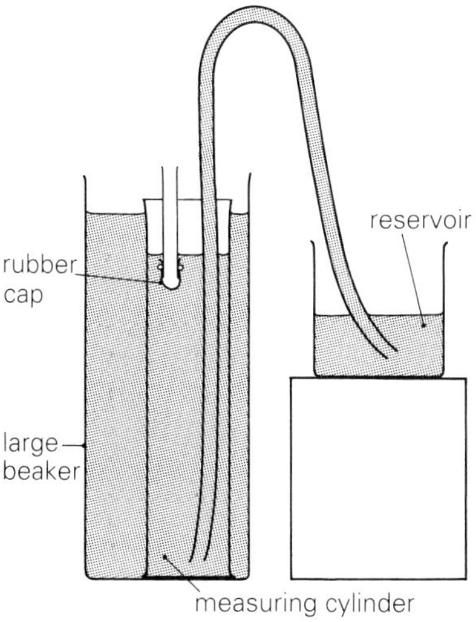

Fig 10.6

A small amount of petrol (0.1 cm^3) is injected through the rubber cap into the hot water in the cylinder. As the temperature is above the boiling point of the petrol, it turns to gas and this gas nearly fills the measuring cylinder (80 cm^3).

Water behaves in a similar way when it changes to steam. In fact 0.1 cm^3 of water will form about 160 cm^3 of steam, a change of volume of 1600 times.

In experiment 10.4, it is interesting to see what happens as the water cools. The gas turns back to liquid petrol, and the atmospheric pressure pushes water from the reservoir back into the measuring cylinder.

Safety note

Dry ice can burn.

Demonstration experiment 10.5
Change of volume when 'dry ice' turns to a gas

This experiment is possible only if your teacher can get some solid carbon dioxide, 'dry ice'.

Two people should hold open the neck of a rubber balloon. Put into the balloon a spoonful or two of solid carbon dioxide. Flatten the balloon quickly and tie a firm knot in the neck. What happens to the solid carbon dioxide as it warms up?

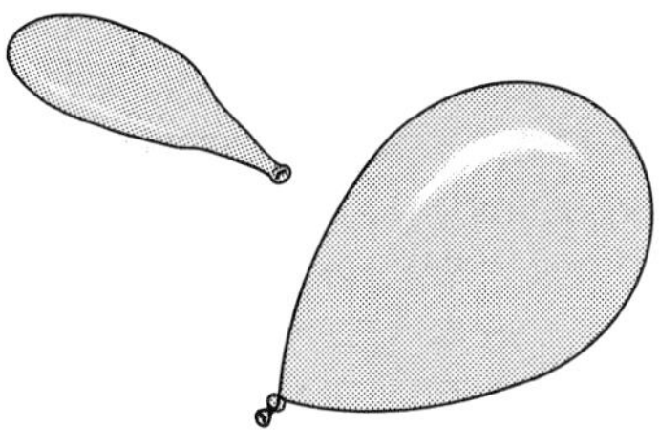

Fig 10.7

The change in volume is about 1 to 600. This experiment is also interesting since it shows a solid turning straight into a gas without going through a liquid stage.

Support for the particle model: Brownian motion

There is a very important experiment which supports the idea that a gas consists of molecules moving irregularly.

The trouble is that molecules are much too small for us to see, even with a very powerful microscope, and in any case they would all be moving very fast. However in this experiment we can see a direct result of these fast moving molecules.

To understand the experiment, imagine a lot of people all sitting round a football suspended by a string from the ceiling. Suppose they start throwing marbles at the ball. When it is hit by a marble it will move, though not very much because it is much more massive than the marble. Sometimes it will be hit on one side, then on another side and so on. The football will move first in one direction, then in another. Even when many marbles hit it at once, it is still likely that it will move because more are likely to hit it on one side than the other. This irregular motion of the football would be a direct result of the collisions.

In the experiment, you will look at smoke particles in a little cell containing air. You cannot see the molecules of the air, but you can see the light scattered by the smoke particles as bright points of light. The movements of these show the irregular motion of the smoke particles as each one jiggles around when the air molecules bump into it. But before you look at the actual experiment, two other experiments will help you to understand it.

Experiment 10.6
Two-dimensional model for Brownian motion

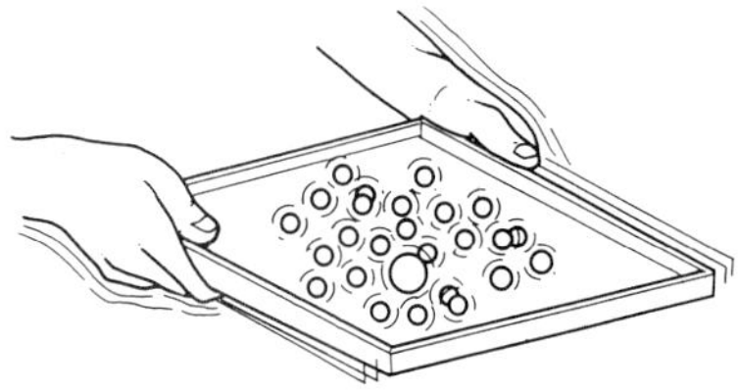

Fig 10.8

For this experiment, use the same apparatus as in Experiment 10.2. This time, put a much larger marble in the centre of the tray. Once again agitate the tray when it is flat on a table and watch the heavier marble (Fig 10.8). You will notice it has a much slower irregular motion.

Then put in the tray a piece of expanded polystyrene. It can be any shape as long as its base is flat. Watch the polystyrene as the marbles knock it around. It moves irregularly.

Experiment 10.7
Three-dimensional model for Brownian motion

This time, use the same apparatus as in Experiment 10.3. Put an expanded polystyrene sphere (about 1 cm in diameter) among the small ball-bearings. Set the vibrator in motion and watch the polystyrene sphere being knocked around in an irregular motion by the ball-bearings as they collide with it (see Fig 10.9).

We should, however, be careful. These experiments do not tell us anything about gases since they are only models. The ball-bearings or the marbles are *not* molecules of gas; they

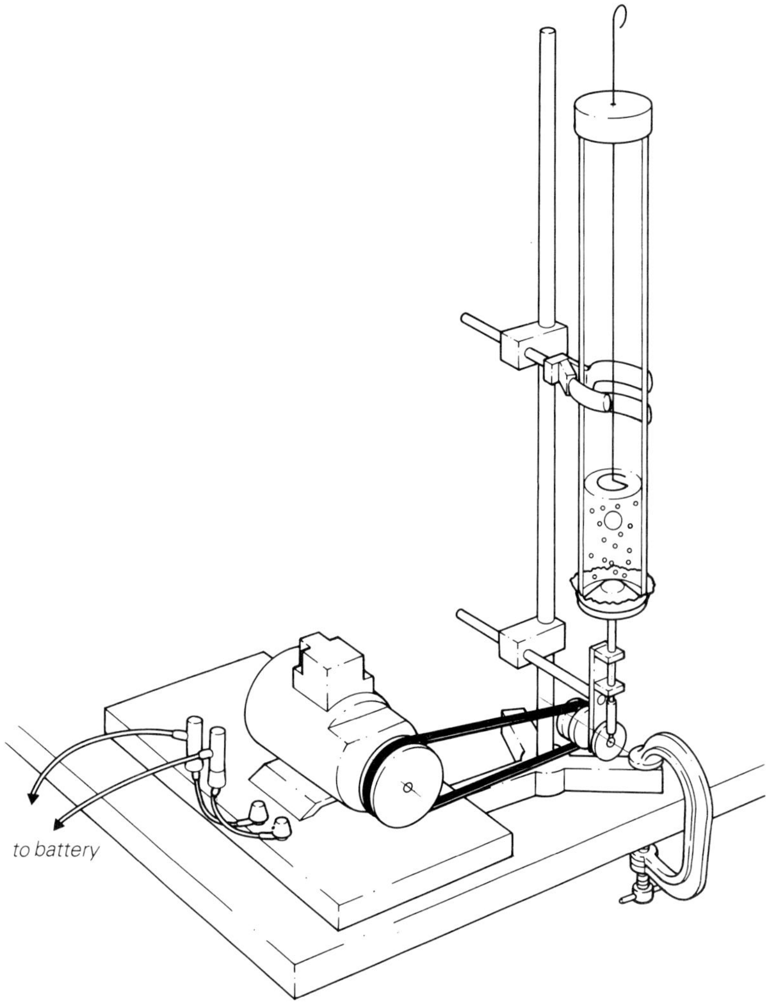

Fig 10.9

merely represent the molecules in our model. To get real evidence about the molecules, we must do the actual Brownian motion experiment.

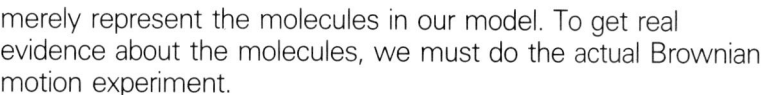

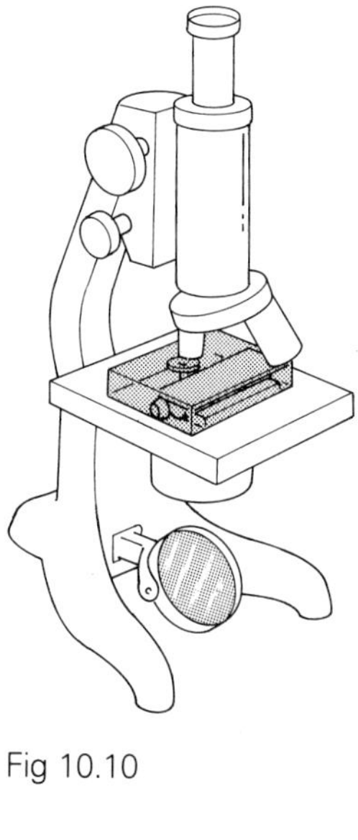

Fig 10.10

Experiment 10.8
Brownian motion in a smoke cell

The apparatus consists of a small glass cell which you are going to look into under a low-power microscope (Fig 10.10). The smoke particles must be illuminated. So the apparatus includes a small lamp and a glass rod which acts as a lens to concentrate the light on the middle of the cell.

First, remove the cover from the cell and fill the cell with smoke. The simplest method is to hold a paper drinking straw almost vertically over the cell. Then light the *top* end of the straw so that the smoke pours down inside the straw and into the cell. When it has plenty of smoke in it, put back the cover to prevent the smoke escaping and place the apparatus on the platform of the low-power microscope.

Connect a 12-volt supply to the lamp so that the light shines on the cell. Focus the microscope on the cell until you can see the light scattered by the smoke particles as bright points of light in it. Watch for a while and you will see the irregular motion. This irregular motion is called **random** motion. Random means that no-one can ever predict what is going to happen next to any particle. No-one can tell which direction it will move in or how fast it will go or when it will change its motion.

The random motion of the smoke particles is due to the air molecules hitting the smoke particles. (The smoke particles are *not* smoke molecules; they are tiny specks of soot.) And the random motion of the particles must occur because the *air molecules* are moving randomly and hitting the particles randomly. This is exciting for it is the first direct evidence for the existence of molecules and for their random motion in a gas.

Questions for class discussion

1 When you look at the smoke particles, some of the specks will suddenly become rather larger patches of light. What is the reason for this?

2 Sometimes one of the specks of light will disappear altogether. What has happened to the smoke particle?

3 What would be the difference if rather larger and more massive smoke particles were used?

4 If the temperature were raised, we think the air molecules would move faster. How would this affect the motion of the smoke particles? What might happen if the air were cooled?

5 How do you know that the motion of the smoke particles is not due to the light shining on them?

Brownian motion is so called because it was first observed by a Scottish botanist called Robert Brown. He observed it when watching pollen grains in water and he thought they were alive. The motion was due to the pollen grains being knocked around in an irregular way by the water molecules. It is possible to see the Brownian motion of carbon particles in water if a very small speck of graphite (Aquadag or Indian ink) is added to a few cubic centimetres of distilled water. Light should be shone through the water, and the particles will have to be viewed with a higher power microscope than was necessary in Experiment 10.8. You might like to try this if you have time.

Support for the particle model: diffusion

Support for the particle model also comes from diffusion. You will have noticed that if fish and chips are cooked in the kitchen and the door is left open, it is not long before the smell has reached all parts of the house. Does this support the idea that gases consist of particles in continual and rapid motion? If your teacher puts on some after-shave or perfume, how long is it before you can smell it everywhere else in the room? Further examples of diffusion in gases and liquids are given in the following experiments.

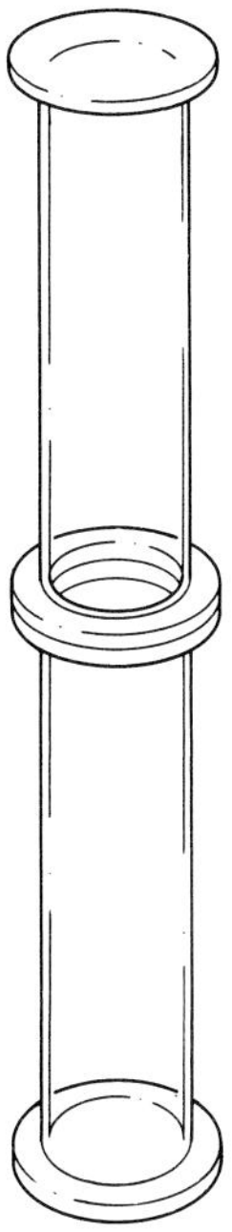

Fig 10.11

Demonstration experiment 10.9 Diffusion of nitrogen dioxide into air

Your teacher will prepare nitrogen dioxide gas by putting a mixture of equal volumes of concentrated nitric acid and water on some copper turnings at the bottom of a gas jar. The gas produced has a distinctive brown colour. When the action has stopped and the gas has cooled to room temperature, another gas jar containing air is inverted over the top of it (Fig 10.11). Watch what happens. It will help if a sheet of white paper is held behind the jars.

Does this show that molecules of nitrogen dioxide have diffused into air? Is there any evidence that air has also diffused into the nitrogen dioxide?

This experiment was done with the brown nitrogen dioxide underneath. (Nitrogen dioxide is denser than air and that seemed the sensible way in which to do the experiment.) What might have happened if the gas jar of air were underneath and the gas jar of nitrogen dioxide on top? Try it and see. This will show that diffusion occurs as before.

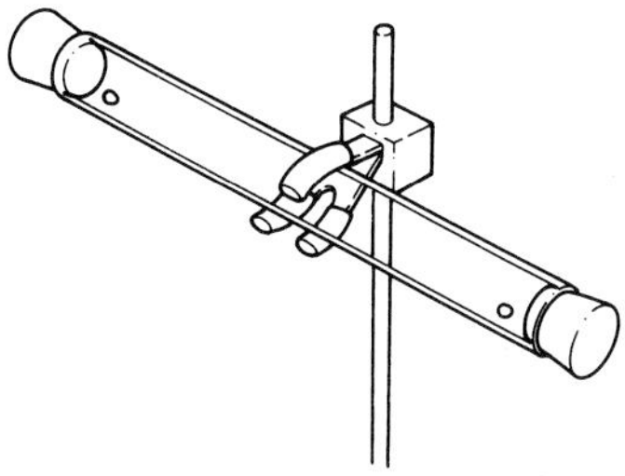

Fig 10.12

Demonstration experiment 10.10 Speed of molecules diffusing in air

When molecules of hydrochloric acid meet with molecules of ammonium hydroxide, they react and form a white powder or smoke. In this experiment a few drops of hydrochloric acid are put at one end of a glass tube while a few drops of ammonium hydroxide are put at the other end (Fig 10.12). A rubber stopper is inserted at each end of the tube.

Notice how long it takes before any smoke appears. Notice also where the smoke first appears. What does this tell you about the speeds at which hydrochloric acid molecules and ammonium hydroxide molecules diffuse through air?

There may appear at first sight to be some contradiction here. The Brownian motion experiment suggested that the air molecules were moving very fast and in the above experiment it took a few minutes for the hydrochloric acid molecules to diffuse down the tube. But there is a simple explanation for this. When the hydrochloric acid molecule diffuses through the air, it does not travel in a direct straight line. Its motion is an irregular one in which it is repeatedly colliding with air molecules. Its path will be

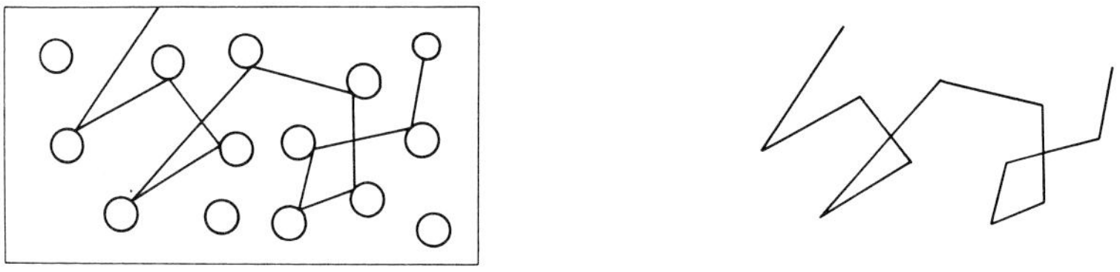

Fig 10.13

Safety note

Copper sulphate is harmful and must not be taken internally.

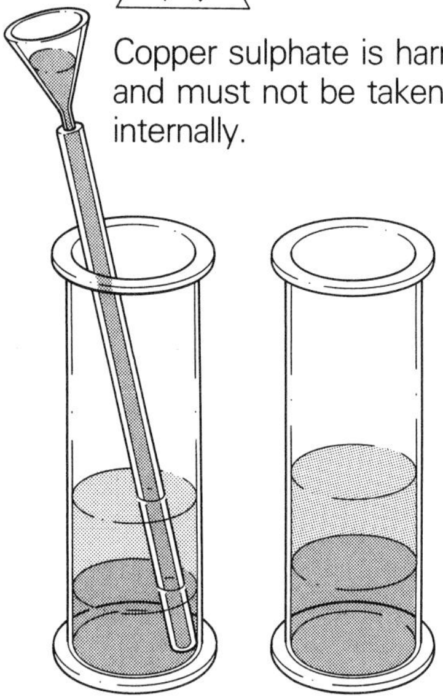

Fig 10.14

much like the path of the single marble you watched in the two-dimensional kinetic model (see page 110). In other words it is a random motion (Fig 10.13).

A man can run at speed from one end of a street to another if the street is empty. But if the street is full of people, all moving in different directions, and if he runs at the same speed he will be buffeted all over the place in an irregular way. It will take him much longer to travel the same distance. Molecules are like that. They travel at the same speed in both cases, but get buffeted about by air molecules and take longer to move down the tube when air molecules are also present.

Experiment 10.11
Diffusion of copper sulphate solution into water

Put water at the bottom of a gas jar. Add a concentrated solution of copper sulphate very slowly down a funnel and tube (Fig 10.14). The copper sulphate is more dense than water and will go to the bottom. Great care should be taken so that the liquids do not mix: there ought to be a distinct line of separation between the liquids. Diffusion in liquids will be seen after a few hours.

Another version of this experiment uses three different liquids (Fig 10.15). Put a strong sugar solution in the bottom of the glass jar. Put the concentrated copper sulphate solution on top of that. Finally, add water. After an hour or two, you will notice that copper sulphate molecules will have diffused both up into the water and down into the sugar solution.

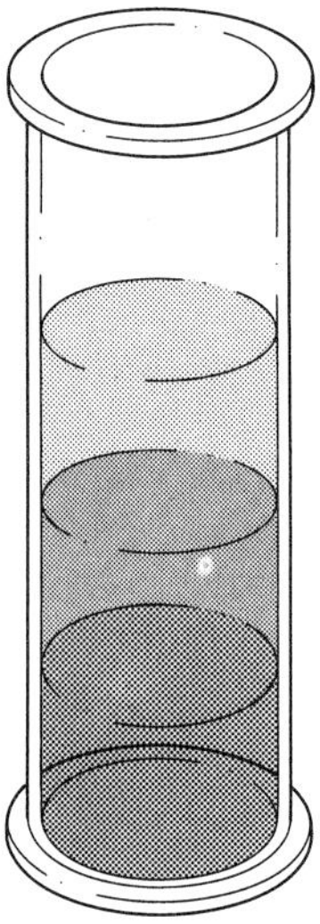

Fig 10.15

A summary

All the evidence in this chapter has supported a model of matter made up of very small particles or molecules.

In a **solid** the molecules are held in a regular array. They can vibrate to and fro, but do not move around. There are relatively strong short-range forces which make it hard to pull the molecules apart, and there are also relatively strong short-range repulsions which make it difficult to compress a solid.

In a **liquid**, the molecules are still fairly close together (there is not much change in volume when a solid melts). A liquid does not have a definite shape, but the forces keep the molecules sufficiently together to ensure that the liquid has a definite volume. The motion of the molecules is mainly vibration but some movement within the liquid occurs.

In a **gas**, the molecules are much further apart. The forces of attraction between them are now very small, perhaps zero, and so they move around freely filling the space available. They have a random motion and travel with high speeds. They exert a pressure on the walls of the container by colliding with them.

The size of molecules and atoms

How small are atoms or molecules? So far we have no idea except that they must be much smaller than the smoke particles we used in the Brownian motion experiment. You will shortly do an experiment in which you will estimate the size of a molecule, but first we must look at the skin effect of liquids (usually referred to as 'surface tension').

Safety note

Mercury vapour is toxic.

Experiment 10.12
Experiments on surface tension

a Look at the shape of drops of water forming on the end of a slowly dripping tap. Try to watch the shape of the drops as they fall.

b Look at the shape of small drops of mercury on a glass surface. Compare them with the shape of drops of water on the glass surface. Put a little wax from a lighted candle on to the glass and then put a drop of water on the wax. What is the shape of the drop this time?

c Make a frame of wire about 5 cm in diameter. Dip the frame into a dish containing a soap solution (50 % water, 50 % liquid detergent can be used) so that a soap film is formed on the frame.

Tie a short length of cotton to form a continuous loop. Put it in the soap film as shown in Fig 10.16. Then puncture the film inside the loop using a piece of chalk. What is the shape of the loop now?

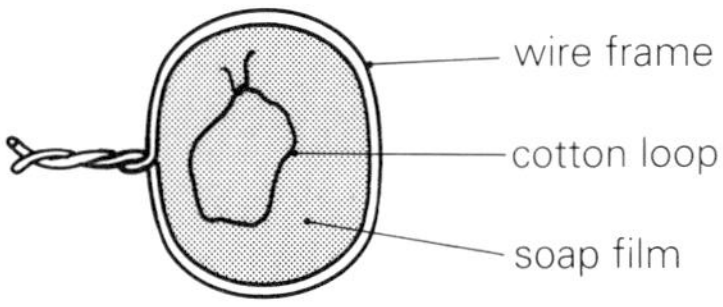

Fig 10.16

d Bend a piece of copper wire 12 cm long into the shape shown in Fig 10.17. Tie a piece of cotton thread between the ends so that it is slack and then tie a short length of cotton from the middle of the first piece.

Hold the frame at A and dip it into a dish containing a solution of liquid detergent in water so that a film is formed.

The cotton takes up a circular shape which shows the film is pulling on it evenly and in a direction such as to shrink the area of the film.

Now pull gently on the thread B. The film stretches – almost like a sheet of rubber – and when you release B, the film shrinks again.

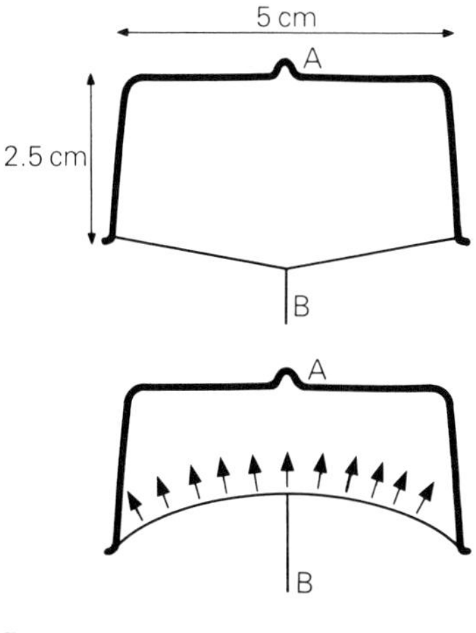

Fig 10.17

These experiments all suggest that the surface of a liquid behaves as though it had an elastic skin trying to keep the area as small as possible. There is not really a skin, but liquids behave as if there were.

This surface tension effect can be explained as follows. The molecules of the liquid attract each other (unless they are very close indeed, when they repel). For a molecule A inside the liquid (see Fig 10.18), the attractive forces pull in all directions. But a molecule B at the surface will have no liquid molecules above it to attract it. It will therefore tend to be pulled into the liquid and this means that the surface tends to shrink whenever possible. It therefore behaves as though it had a thin elastic skin.

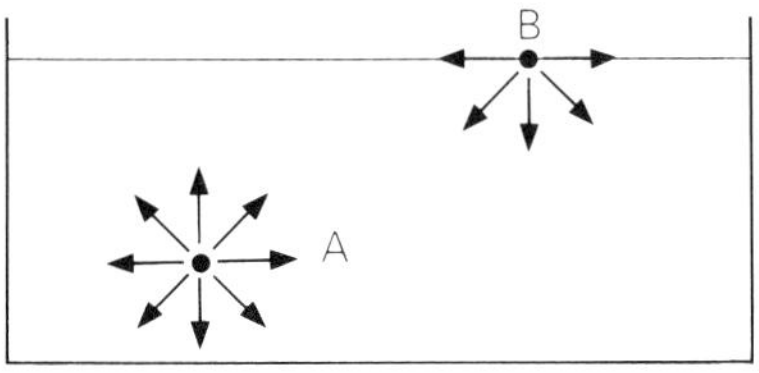

Fig 10.18

Measuring an oil molecule

How small are atoms? If we lined up atoms side by side, how many would there be in 1 cm? We have already seen that it can be useful to make estimates, but in this case it is a bit difficult to guess. Might it be 1000? Or a million? Or a million million? We really have no idea, so it would be good if some measurement could be made which would give an idea of the size. We shall choose a molecule of olive oil, which is in the shape of a long chain of atoms, about twelve atoms long. Unfortunately it is much too small for us to measure directly, but you will try to measure it by a roundabout method and it will be a great achievement if you can get some answer.

As a thought experiment, imagine you pour lead shot from a small beaker on to a tray so that it lies in a pile. You then spread it out into a flat layer. What is the thinnest layer you can get? Obviously one lead shot thick.

Suppose the volume of the lead shot is 100 cm^3. If the area covered by the lead shot is 100 cm^2, what is the thickness of the layer of lead shot? Suppose the lead is spread out into a layer one lead shot thick, and the area covered is now 1000 cm^2. What is the size of the lead shot?

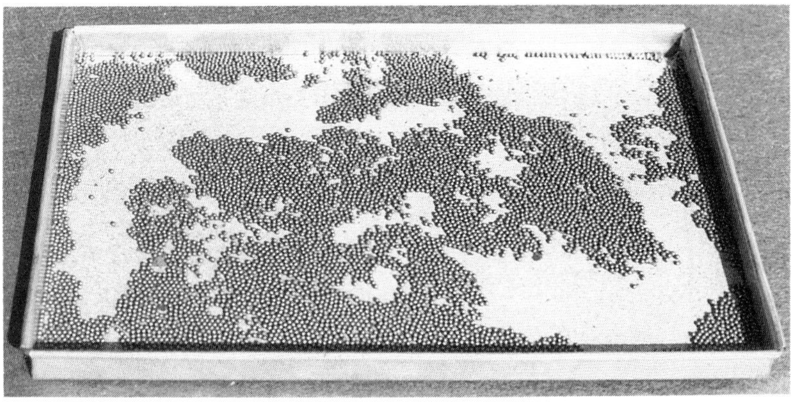
Lead shot in tray

If you did the above experiment spreading out lead shot until it was one layer thick, the result might be rather like the photograph on page 116. There are no forces of attraction between the lead shot and there would probably be empty patches.

But if instead of lead shot you spread oil on top of a water surface, the oil would float on the water and spread out. The forces between the molecules would however keep the oil film together. What would be the smallest possible thickness for the oil film?

How would the volume of the oil film compare with the original volume of the oil? This should give you a clue about how you are going to measure the length of an oil molecule.

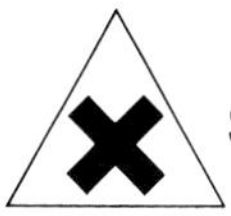

Safety note

Lycopodium can be an irritant.

Experiment 10.13 Preliminary experiments

In all these experiments it is extremely important that everything is very clean. Thoroughly wash the crystallizing dish with detergent and then rinse it several times in clean water to remove all the detergent.

Put water in the dish and lightly sprinkle the water surface with lycopodium powder. Put a drop of alcohol on the powdered surface and watch what happens (Fig 10.19). (If very little happens, or nothing at all, it means that the dish was not properly cleaned.)

Fig 10.19

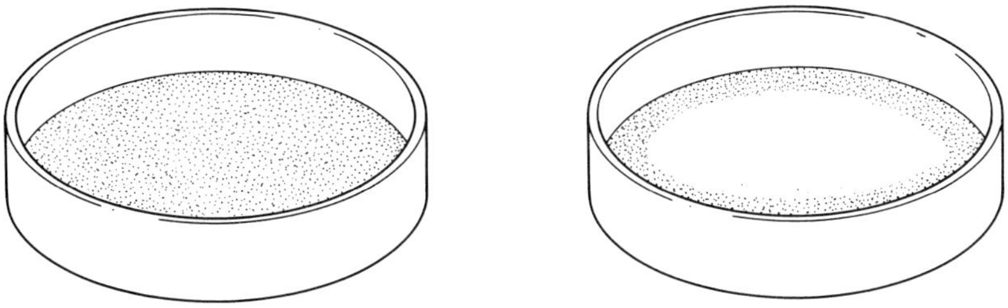

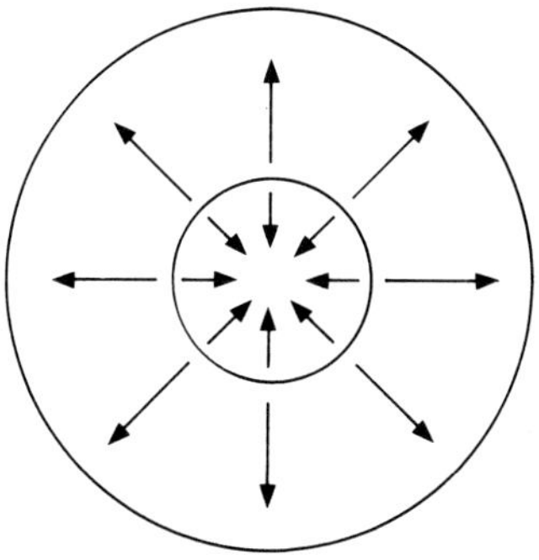

Fig 10.20

A clear circular patch appears in the powder as the alcohol spreads over the surface. The alcohol spreads because the surface tension of the water is greater than the surface tension of the alcohol. So the water surface shrinks and stretches the alcohol until it forms a thin film on the water surface (Fig 10.20). If the dish is really clean, the patch is circular, but it soon disappears because the alcohol evaporates and dissolves in the water.

Dip the end of a matchstick in some olive oil and wipe it clean. Lightly sprinkle some more lycopodium powder on a fresh, clean water surface. Touch the water with the end of the matchstick. Again a circular patch appears. (The matchstick was not wiped as clean as you may have thought!) The circular patch appears because the surface tension of the water is greater than the surface tension of the oil, the water surface shrinks and pulls the oil into a circular film, as happened before with the alcohol.

However, the oil does not dissolve in the water, or evaporate quickly, and so the circular patch remains. The oil cannot go on spreading and spreading: it can only spread until the film cannot get any thinner, when it will be one molecule thick. (The oil film does not break up because the surface tension of water is not big enough actually to pull the molecules of oil away from their neighbours.)

Finally, clean the dish again and put in some more water. Dust the surface again with lycopodium powder. Then dip a clean finger in the surface. Perhaps you will be surprised at the amount of oil on an apparently clean finger: that is why it was necessary to get everything so clean.

Now we are ready for the actual experiment. The previous experiment will have shown you how oil spreads on a surface. There have been examples of oil spreading on the sea when a leak has occurred in an oil tanker and the oil patch has spread over a very large area.

In this experiment you will take a very small drop of oil, find its volume and then let it spread on a water surface. You will then find the area of the oil film and calculate the thickness of the oil film.

The chemists tell us that the olive oil molecule is such that one end is strongly attracted to water and the other is not. The oil molecules are therefore upright on the water, much like the bristles in a brush. Our experiment will therefore give the length of the oil molecule assuming the oil film is only one molecule thick.

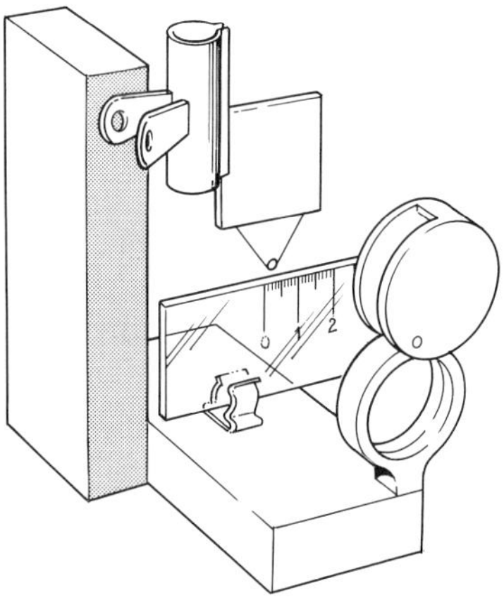

Fig 10.21

Experiment 10.14
Oil film experiment

For this experiment, a tray should be filled to the brim with clean water and dusted lightly with lycopodium powder.

It is necessary to prepare a drop of oil which is $\frac{1}{2}$ mm across. Put a transparent scale marked in $\frac{1}{2}$ mm (see Fig 10.21) in the holder provided and in front of it fix a hand lens. Dip the loop of steel wire in olive oil and take up a drop. Fix the card in the holder so that the drop and $\frac{1}{2}$ mm scale are clearly seen through the lens. Use a second loop dipped in the oil to 'tickle' the first drop until it appears to be $\frac{1}{2}$ mm across. It is often helpful to use the second loop to run small drops together until you have one of the right size.

Fig 10.22

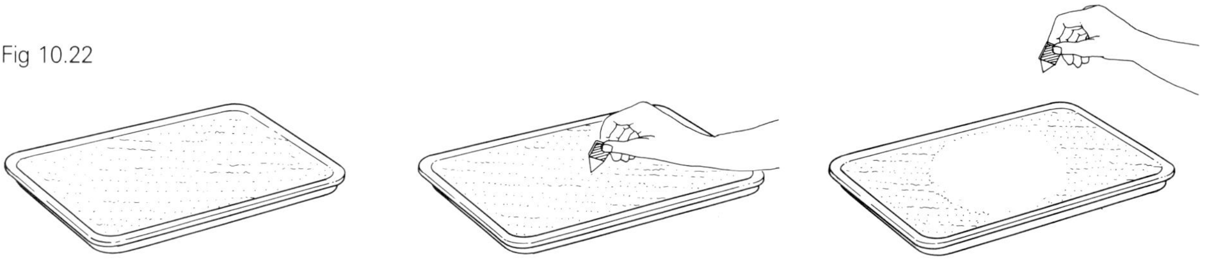

Carefully carry the $\frac{1}{2}$ mm drop to the tray and lower it into the centre of the water surface (Fig 10.22). It will immediately spread in a circle over the surface. Measure the maximum diameter of the film using a metre rule.

Calculating the size of the oil molecule

The drop was $\frac{1}{2}$ mm across. It will make the calculation easier if we assume the drop is a cube. We know the volume of a cube is the length multiplied by the breadth multiplied by the height. What is the volume of the oil drop? Give your answer in cubic millimetres. When the drop is put on the water surface it spreads out to form a film one molecule thick. What is the volume of the film in cubic millimetres?

Again to keep the arithmetic simple, let us suppose the oil film was a square patch. Then the volume of the film is also equal to the length of the film × breadth of the film × the thickness. You have measured how many millimetres it was across, so that you know the length and the breadth. You can calculate the area and then you should be able to work out the thickness of the film. When you have done this, you will have made your first measurement on the atomic scale.

The chemists tell us that the olive oil molecule consists of twelve carbon atoms along its length, so that if you divide your answer by twelve, you should get the size of a carbon atom. Of course, there were approximations in your calculation and it was not easy to be very accurate in your measurements, but it is a fine achievement to have got an answer of the right sort of magnitude.

The generally accepted value for this is a little more than

$$\frac{1}{10\,000\,000}\text{ mm} \quad \text{or} \quad \frac{1}{10^7}\text{ mm.}$$

This means that in a millimetre, there might be approximately 10 million atoms lying side by side.

Chapter 11 Energy

What is energy?

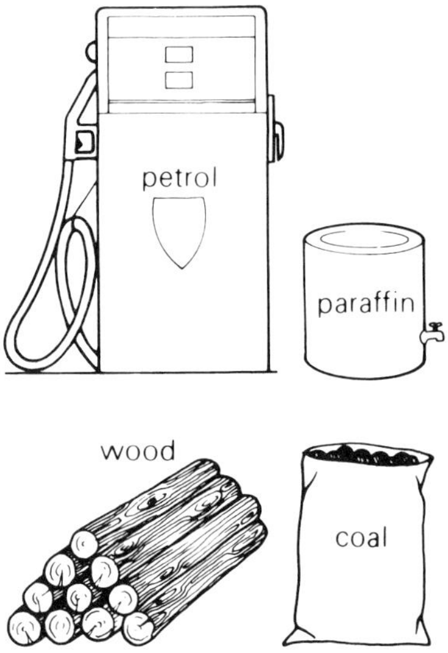

Some sources of energy

You have heard phrases like 'feeling energetic' or 'having no energy', and you will have seen advertisements for food which 'give you energy'. What is meant by this word *energy*?

If you say you 'have no energy', you mean that you do not feel like doing anything. If you have lots of energy, then you can bustle about doing lots of jobs. You need energy to lift things, you need energy to run round a field or to ride a bicycle, you need energy to saw wood or to hammer nails.

Everything you do needs some energy. If you do a hard morning's work lifting things, and going up and down stairs many times, or if you play a hard game of hockey and run about a lot, then you feel tired and hungry at the end of it. And you need to eat before you can do much more.

Jobs of work can also be done by engines of various kinds. Lorries can carry loads up a hill, cranes can lift masses, electric motors can drive drills and washing machines. Do these engines need a supply of energy? Of course they do. A car engine must have petrol. A diesel engine requires diesel oil. In the same way that human beings need to be fed, engines need to be fed with fuel. Energy is stored not only in food, but also in petrol, diesel oil, wood, coal, coke and gas.

To use the energy stored in the fuel or food, it has to be combined with oxygen – which is what happens when the fuel is burned in the engine or the food is digested in your stomach. The energy stored in fuels or food is sometimes called **chemical energy**.

Electric motors are driven by electric current flowing from a power station or battery. A power station has to have fuel in order to produce the electric current, and a battery uses up chemical energy from the chemicals inside it.

Questions for class discussion

1 Think about each of the following jobs. Which of them require fuel and which need none?

a Lifting a pile of books on to a shelf.
b Watching a pile of books on a shelf.

c Kicking a football.
d Hitting a post into the ground.
e A post holding up a clothes line.
f The sea keeping a boat afloat.
g A sailing boat moving across the sea.
h An ocean liner crossing the Atlantic.
i Sleeping in bed.
j Reading a book.
k Climbing a mountain.
l Winding a clock.
m A train travelling from station to station.
n Holding up a pile of books.

2 Make a list of five things different from those in Question 1 which require fuel and five things which do not.

Uphill energy

Look at a brick lying on the floor. Has it got any energy? Can it do a job of work? It does not seem very likely.

Suppose the brick is picked up and put on the edge of a table. The brick looks much the same, but if it falls from the table to the ground, it is capable of doing a job of work. It could certainly knock in a nail placed on the floor or it could make a dent in your foot.

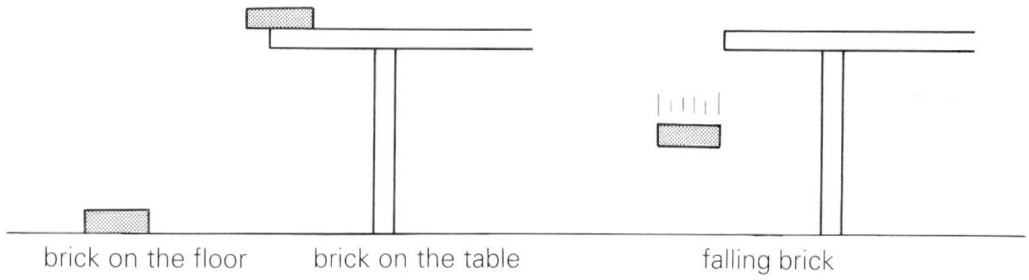

Fig 11.1

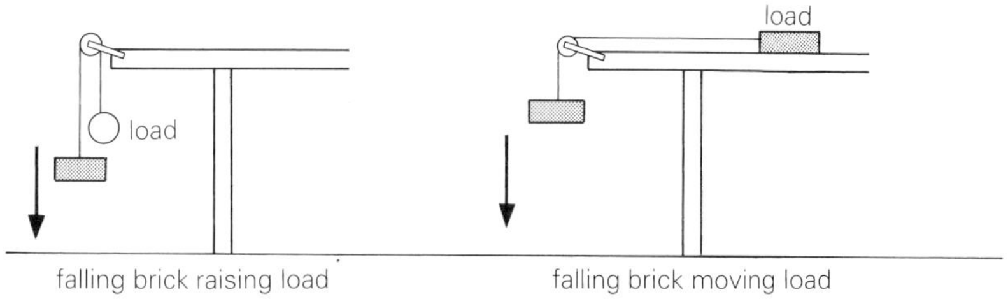

Fig 11.2

Suppose the raised brick is attached to a mass over a pulley as shown in Fig 11.2. The falling brick can lift the load off the floor if the mass of the brick is great enough. In the second diagram, the falling brick can pull the load along the table top. In both these ways the raised brick is capable of doing a job of work; in other words it has energy. But when the brick has reached the floor, it cannot lift the load any higher or pull it further across the table.

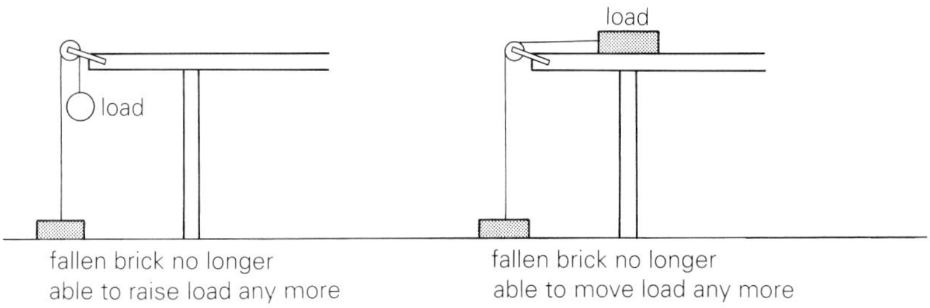

Fig 11.3

Since this energy depends on the height of the brick, we will call it **uphill energy**. The higher the brick is lifted the more uphill energy it has got. Of course to raise it, we have to use some of the chemical energy stored in our bodies. What we have done is to transfer that chemical energy to the uphill energy stored in the brick.

Later in the course (page 221) we will call this energy **gravitational potential energy** or more briefly **potential energy**. But at this stage it is probably clearer if we call it uphill energy.

Motion energy

Put a thin plank of wood on the floor at the foot of a table, as shown in Fig 11.4. (You may prefer to use balsa wood or hardboard.)

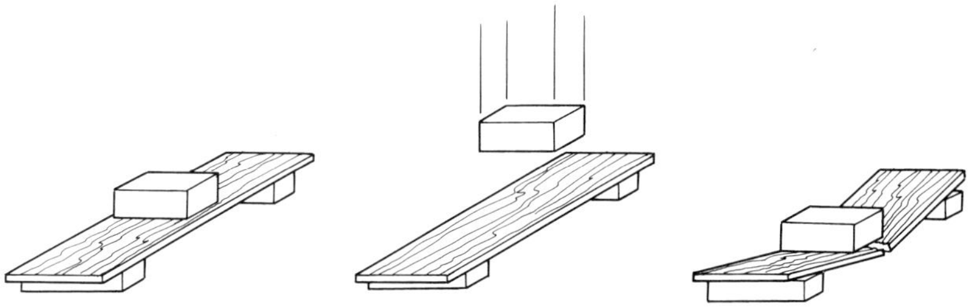

Fig 11.4

Put the brick on the wood. It is not able to break it.

Now raise up the brick to the top of the table so that it has got uphill energy. Then gently push the brick off the edge of the table. It falls on to the wood and this time it does break it.

The brick has uphill energy when level with the table, and as it falls, its uphill energy becomes less. At the moment it reaches the wood, its uphill energy is least, but it is now moving and it breaks the wood and comes to a stop. It must therefore have some kind of energy if it can do a job of work. This is energy due to the fact that it is moving, and we can call this **motion energy**. Notice that uphill energy has been changed into motion energy.

In a bowling alley, it is *motion energy* that causes the pins to be knocked over; a moving cricket ball can knock over the stumps; a moving car can do a lot of damage if it hits something.

Pins being knocked over in a bowling alley

Smashed car after a collision

Stumps knocked over by a cricket ball

Obviously the faster the body is moving, the more motion energy it has got. (Later we will call this **kinetic energy**, but at this stage it will be easier just to call it *motion energy*.)

Energy transfer

If you lift up a brick and put it on the edge of a table, it has gained some uphill energy. In order to lift it, you have to use some of the chemical energy stored in your body: if you spend a lot of time lifting bricks you certainly feel hungry at the end. Energy is transferred from one form to another. Chemical energy in the food is transferred to chemical energy in your body, which is transferred to the uphill energy in the brick. If the brick is then allowed to fall, the uphill energy will transfer to motion energy just before the brick hits the ground. This is shown in the diagram below.

Whenever anything happens and a job of work is done, energy is transferred from one form to another. Diagrams like this are useful for showing energy transfers and there will be more examples later in the book.

Getting hotter

If you take a block of wood with a flat surface and place it on a plank of wood, there is no rise in temperature of the wood however long you leave it. It makes no difference if you increase the forces between the block and the plank by loading the block.

Now, take the block in your hand and rub it a short distance backwards and forwards on the plank (Fig 11.5). If you do this vigorously for a minute or two, pressing down hard as you rub, you will find that both the block and the plank become warm. (A good way to feel this is to hold the block against your cheek before you start rubbing and then to touch your cheek again after rubbing.) The particles in the block are vibrating faster and so have more energy. Do you perhaps feel a bit tired after doing the rubbing? Have you used up some chemical energy?

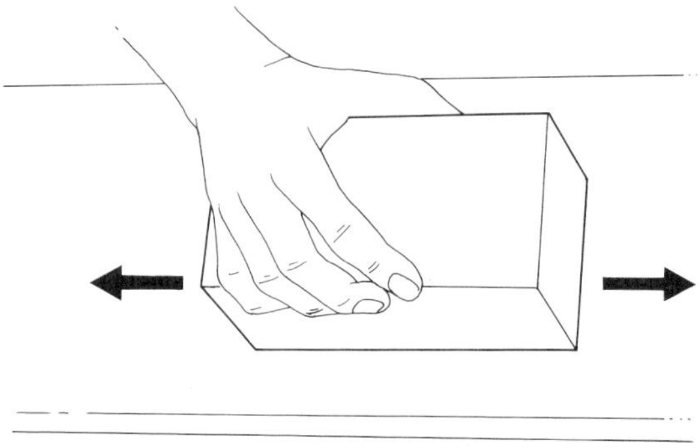

Fig 11.5

Rub your hands together vigorously. You are using some of the chemical energy in your body to give your hands some motion energy. What do you notice after doing this? Perhaps that you are a bit tired and certainly your hands will feel hotter. Some of the chemical energy has been transferred to energy of the particles in your hands.

We know that burning coal, wood or paraffin oil produces energy. The chemical energy in those fuels can cause heating. Heating results in increased motion of the particles of a substance and we call that **internal energy**.

Now think about the brick falling off the table again. When the brick falls off the table, we have already seen that the uphill energy turns to motion energy as it falls. Just before it hits the ground, the uphill energy has become motion energy. When the

Fig 11.6

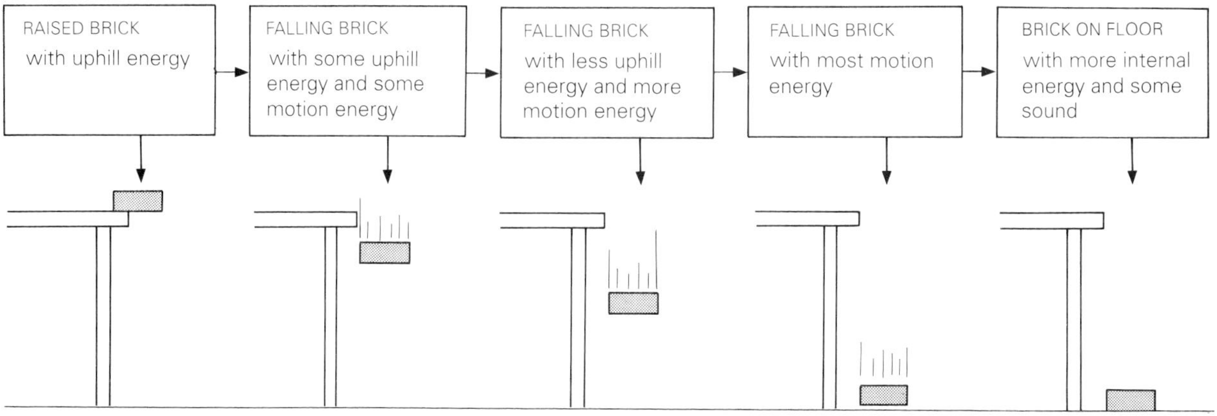

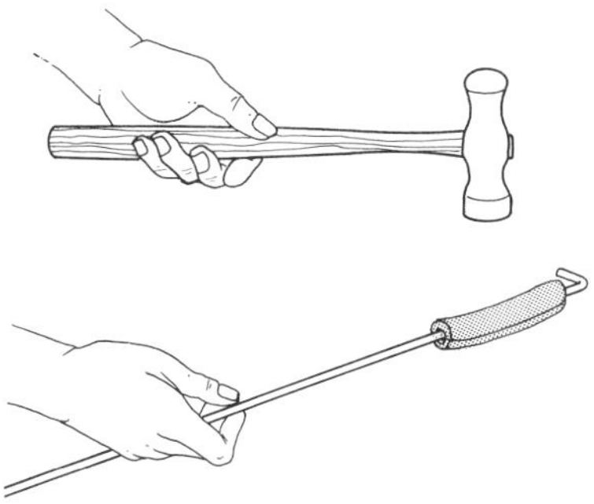

Fig 11.7

brick hits the ground, what happens to this motion energy? It turns to internal energy. Both the brick and the ground will be a little hotter after the collision. The flow diagram in Fig 11.6 shows how the energy is being transferred.

Another experiment to show that motion energy can be turned into internal energy is to hammer a small piece of lead sheet (Fig 11.7). It is convenient to fold the lead sheet round a piece of iron wire to hold it and then hammer it on a hard surface.

Heating caused by friction

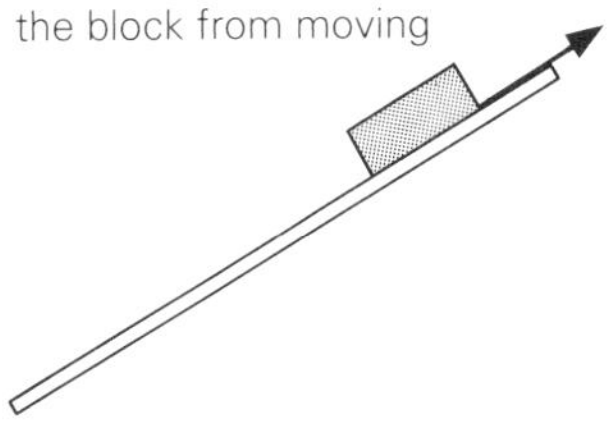

Fig 11.8

It is important to remember that friction is only a force, and forces on their own do not cause a transfer of energy. To show this, put a block of wood on a rough plank. If you tilt the plank (Fig 11.8), the block does not slide down (provided the slope is not too great). It is stopped from sliding down by friction, the frictional force between the block and the plank. But however long you leave the block in this position, it does not get hot. The frictional force does not cause any heating.

Earlier you rubbed a block of wood over a plank and the block became hotter. You were moving the block against the frictional force. There has to be movement against a frictional force for energy to be transferred to internal energy so that the block becomes hotter. If your bicycle is not moving and you apply the brakes, nothing gets hot. But when the brakes are applied to a moving wheel, the brake blocks can get very hot: the motion energy in the bicycle is transferred to internal energy in the brake blocks.

When a ball flies through the air, there is a small frictional force on it due to the air. As the ball is moving against this force, the air and the ball will be warmed a little. Some of the motion energy of the ball is transferred to heating the air and the ball, and the ball slows down slightly.

In most transformations some energy gets transferred to heating the surroundings. For this reason, the flow diagram for a batsman hitting a ball would be more correct if we drew it like this.

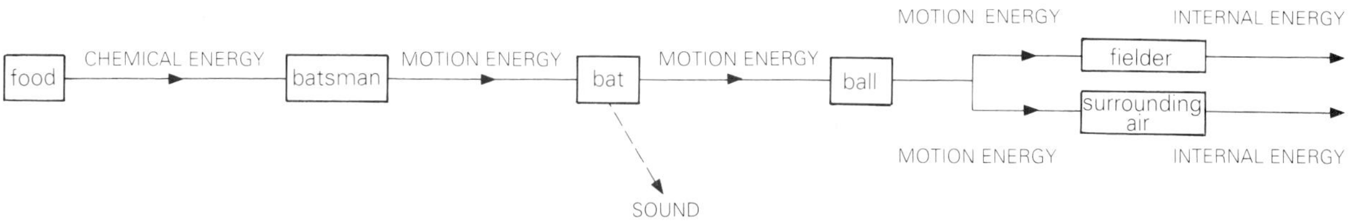

Questions for homework or class discussion

1 A ball is at rest at the top of a hill. It starts to roll down it, getting faster as it rolls further. At the bottom of the hill, it hits a wall and comes to rest. Draw an energy flow diagram for this.

2 It is said that when the scientist Joule was on his honeymoon in Switzerland he spent time trying to show that the temperature of the water at the bottom of a waterfall was higher than the temperature at the top. Do you think it might be true that there is a temperature rise? Give a reason.

3 When a lathe is used in a metal workshop, there is usually a flow of liquid on to the cutting tool. What is the reason for this?

4 Chemical energy from the petrol is necessary to start moving a car from rest.
  a Some of the chemical energy is transferred to motion energy of the car, but not all of it. Suggest what else the chemical energy gets turned into.
  b When the car is travelling at a steady speed, it still needs a supply of petrol to keep it going even though the motion energy stays the same. Can you suggest why this is necessary?

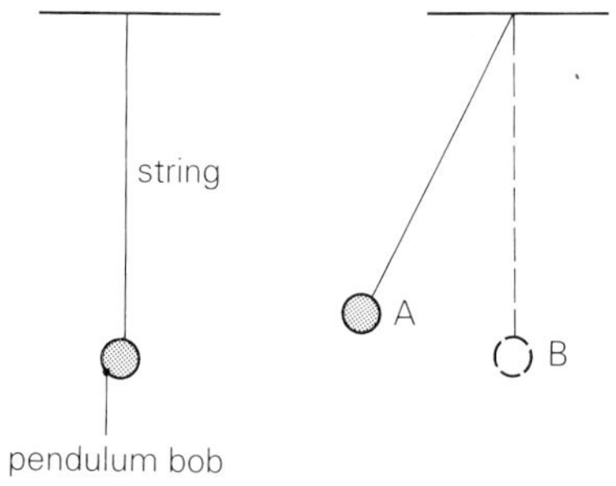

Fig 11.9

5 Fig 11.9 shows a simple pendulum hanging at rest. The bob is pulled aside until it is in position A. It has gained some uphill energy.
  a Where did this uphill energy come from?
  b The bob is released and the pendulum swings back. When it gets to position B, what has happened to the uphill energy?
  c Where will the pendulum swing to after passing B? What happens to the energy?
  d After some time, the pendulum stops swinging, coming to rest with the bob in position B. Where has all the energy gone?

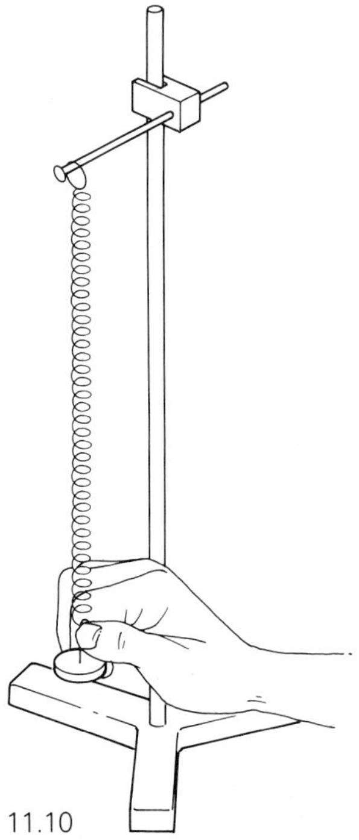
Fig 11.10

Strain energy (spring energy)

Fix the top end of a spring and pull down on the other end so that the spring is stretched (Fig 11.10). If you attach a mass to the bottom of the spring and let go, the spring will do a job of work and will raise the mass. In other words the stretched spring must have energy stored in it. We will call this **strain energy** or **spring energy**.

Further examples of strain energy can be seen in Fig 11.11. The stretched catapult, the arched bow, the bent branch of the tree all have *strain energy* stored in them; they are all capable of doing a job of work.

Here is another simple experiment showing an energy transfer. Blow up a rubber balloon and hold the neck tightly to keep the air in the balloon. Has the air inside any energy? Release the balloon and see what happens. The balloon flies about as the air inside escapes. Both the balloon and the air have gained motion energy, but where did it come from?

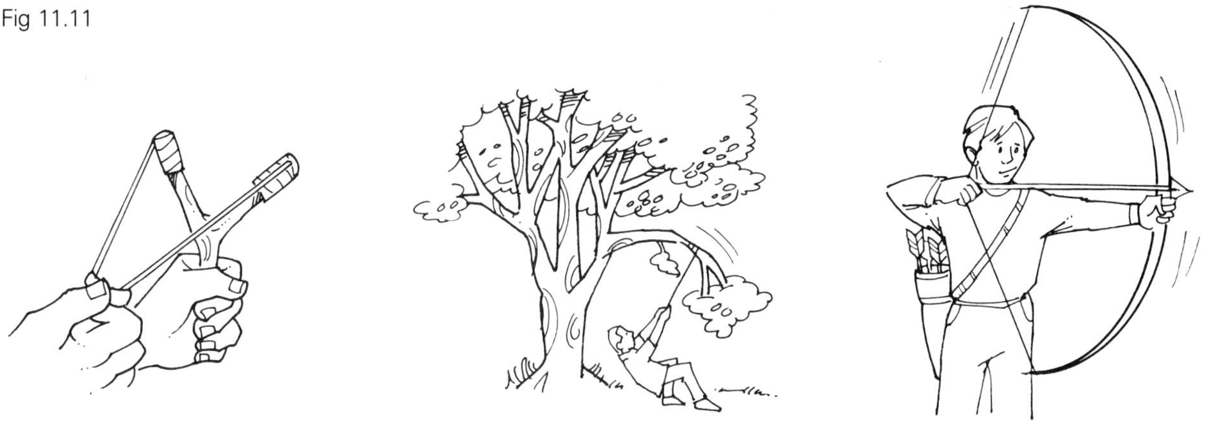

Fig 11.11

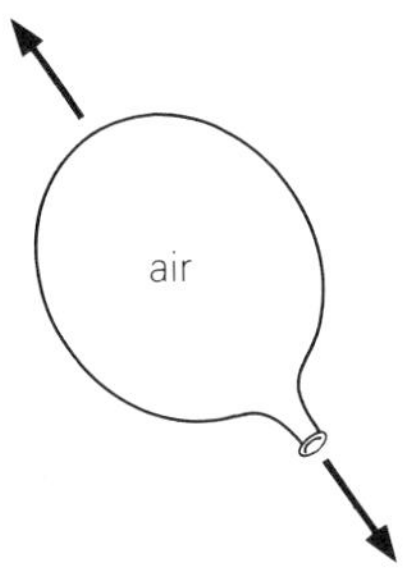

Fig 11.12

To blow the balloon up, we forced air into the balloon, the rubber stretched and the air was squashed at a higher pressure. When the air was allowed to escape, some energy came from the stretched rubber, but most came from the air as it expanded. A water rocket or a toy driven by a CO_2 capsule shows this energy transfer even more clearly, because the container does not stretch like the skin of the balloon does.

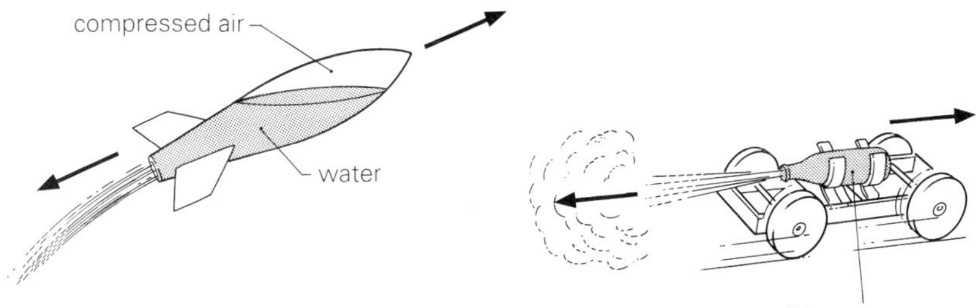

Fig 11.13

A liquid or a gas which has been squashed to a higher pressure has energy stored in it. It is very similar to strain energy and we shall call it the strain energy of the liquid or the gas. Sometimes it is called 'pressure energy'.

Other forms of energy

We know that a loud bang can cause windows to rattle and this tells us that sound is another form of energy. When a brick falls on the floor, most of the motion energy of the brick turns into internal energy, but some will become **sound energy**.

Another form of energy of which you will have heard is **atomic energy**, or **nuclear energy** as it should really be called. There is energy stored in the nucleus, the central part of the atom. The size of the atom as a whole is about 10^{-10} m in diameter with the

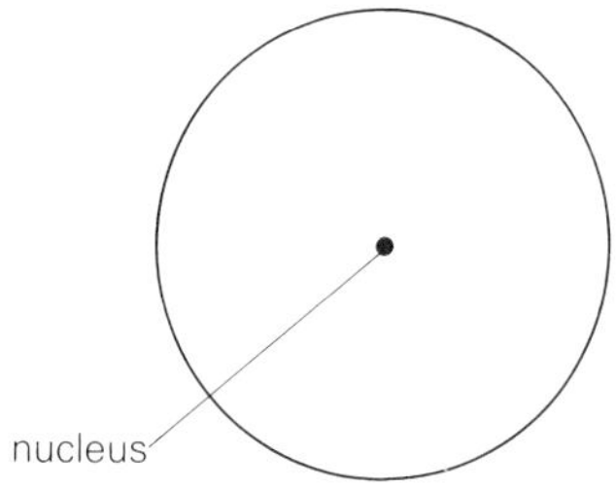

nucleus in the middle (about 10^{-15} m in diameter). Particles are shot out of the nucleus at very high speed, and this is evidence for the energy stored in the nucleus. Energy can also be released in a nuclear reactor when an uranium nucleus is bombarded by neutrons (neutral particles over 200 times lighter than the uranium nucleus). The uranium nucleus captures a neutron and then breaks into two parts, as shown in Fig 11.14 (notice that when the nucleus breaks up there is the release of extra neutrons).

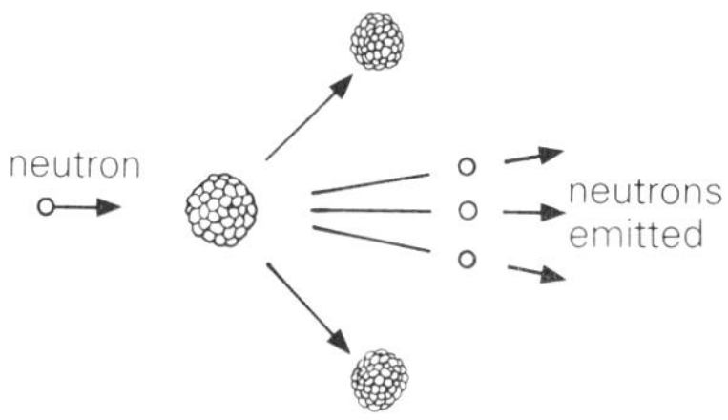

Fig 11.14

This breaking up is called *fission*. The amount of energy released in a single fission is very small indeed, but if the supply of uranium can be arranged so that the neutrons released in the fission can go on to bombard other uranium nuclei, it may be possible to keep the process going so that a much greater amount of energy is released, as shown in Fig 11.15.

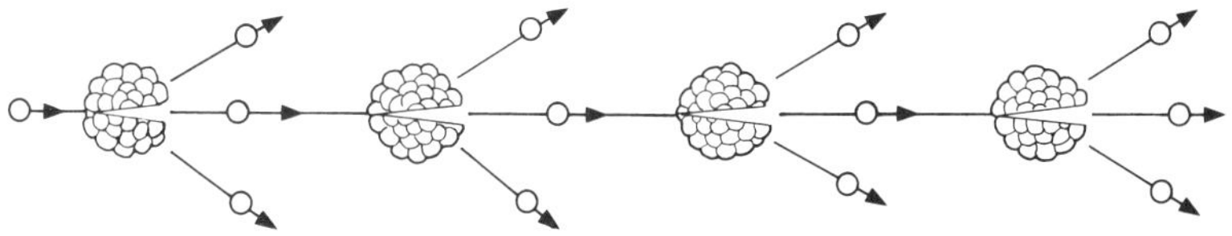

Fig 11.15

The energy released in this way in a nuclear reactor in a power station is used to generate electricity as will be described later. The photograph opposite shows a nuclear power station.

It is the release of nuclear energy within the Sun which enables it to send out so much energy to the Earth, though the reactions occurring in the Sun are different from those that have been used on Earth in nuclear power stations.

But how does the energy from the Sun reach us? The energy clearly travels through the empty space between the Sun and the Earth. There is therefore another kind of energy which we will call **radiation energy**. You have certainly felt this radiation reaching you whenever you have been sunbathing.

It is this radiation energy which gives us warmth, heating the land and the seas; it is this energy which is absorbed by plants enabling them to grow; without it, we would have no food. The stored chemical energy we have already talked about has in fact all come originally from the Sun.

It is radiation energy from the Sun which gives energy to the water in the sea, causing evaporation. The water rises and forms

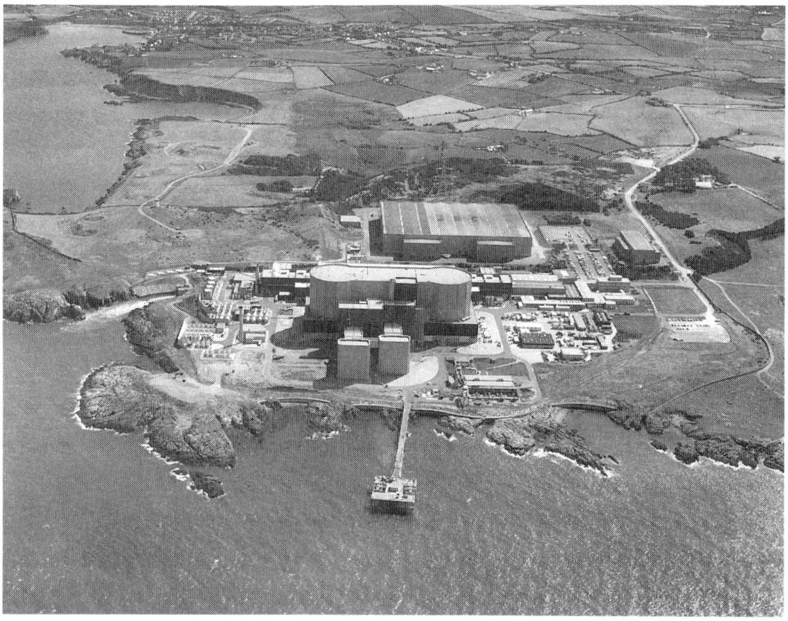

Nuclear power station at Wylfa, Anglesey

clouds, and then the water falls as rain, forming lakes and rivers. In other words the radiation energy from the Sun has given uphill energy to the water.

One form of radiation energy is **light**. How do we know that light is a form of energy? It will certainly do jobs of work: it will affect a photographic film so that we can take a picture with a camera; it will cause coloured material to fade.

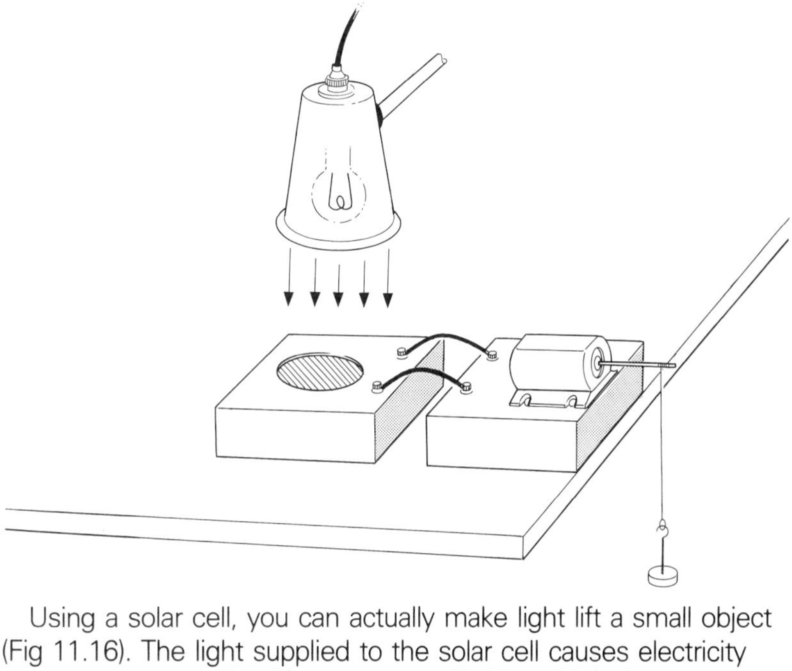

Fig 11.16

Using a solar cell, you can actually make light lift a small object (Fig 11.16). The light supplied to the solar cell causes electricity to turn a motor which lifts a very small object. Light is transferred into uphill energy of the object.

Radio waves are another form of radiation energy. Some of their energy is absorbed by the aerial of a radio and very small currents flow into the receiver. These cause the batteries of the receiver (or the mains) to supply a much larger amount of energy, some of which is the sound energy from the loudspeaker. X-rays are yet another form of radiation energy, as are gamma-rays from a radioactive source (see page 323).

Electrical energy

Electric trains can do a useful job moving people and goods from one place to another (see photograph). Electric motors can be used to lift masses or make all sorts of devices work. Electric fires will warm a room, electric kettles will boil water. In each case energy is needed to do the jobs and this is obtained from

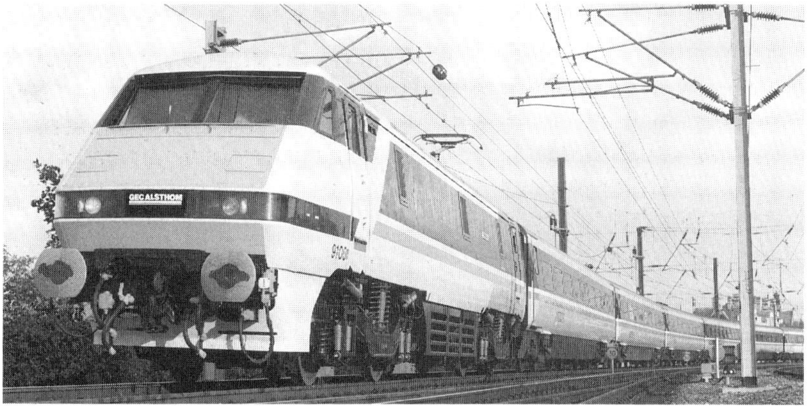

Electric train

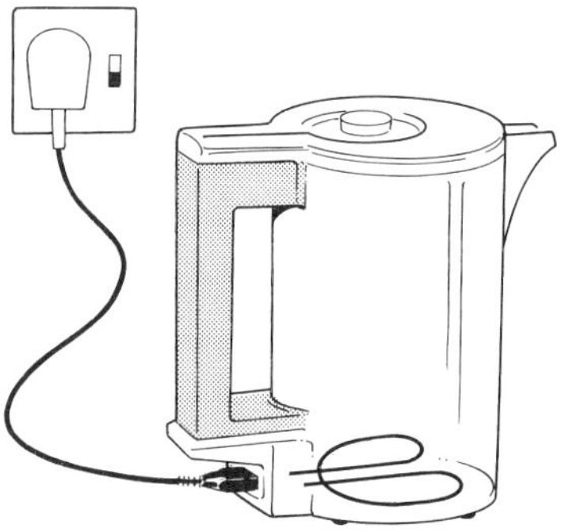

the electricity supply. Fuel burned in the power station produces electricity which passes through the motors of the locomotive, which pulls the train. We shall call this **electrical energy**.

Producing electrical energy: the dynamo or generator

You have probably seen a bicycle dynamo like the one shown in photograph (a) on the next page. The lamp connected to it does not light when the bicycle is at rest. But when the bicycle is moving the bicycle wheel turns the dynamo wheel, and that causes a current to flow which lights the lamps of the bicycle. The energy flow diagram is shown below.

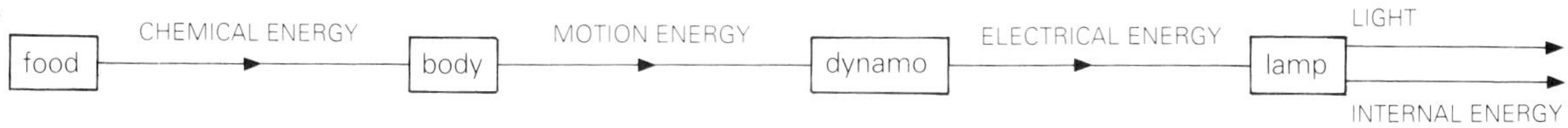

(a) Bicycle dynamo

(b) Mounted dynamo

In photograph (b), a similar dynamo has been mounted so that the lamp lights when the handle is turned.

Electrical generators in power stations work in a similar way (although their design is obviously very different from the bicycle dynamo). The generator has to be turned in order to generate electricity (see the photograph on the lower righthand side of page 140). The problem for the engineer is to find some way of producing the turning effect.

The water turbine

The photograph shows the wheel of a water mill in which moving water from a stream falls on to the paddles of the wheel and causes them to turn, so providing energy for the machinery in the mill.

The water turbine is a machine in which blades are turned when water passes through it (see Fig 11.17). The water enters at a high pressure and leaves at a low pressure and the strain energy in the water gives energy to whatever machinery the turbine is attached. Such turbines can then be used to turn generators to produce electrical energy (see also page 142).

Wheel of a water mill

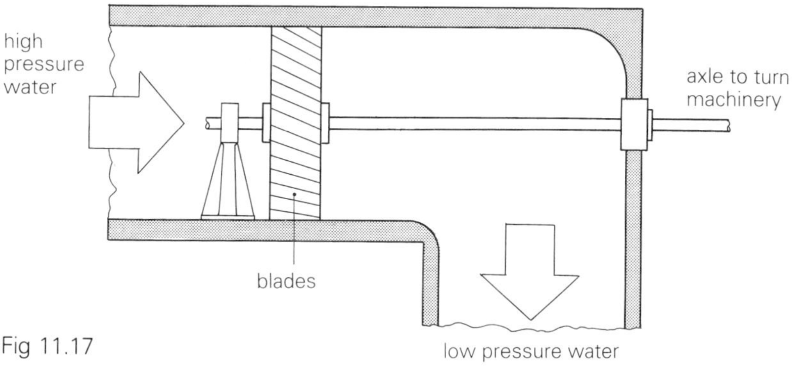

Fig 11.17

Dams and hydroelectric power stations: Hoover Dam, USA (left) and Maraetai, New Zealand (right)

Often large dams are built to create artifical lakes, and hydroelectric power stations are put at the foot of such dams (see photographs). The height of the reservoir provides sufficient uphill energy to drive the turbines in the power station. Here is an energy flow diagram for a hydroelectric power station.

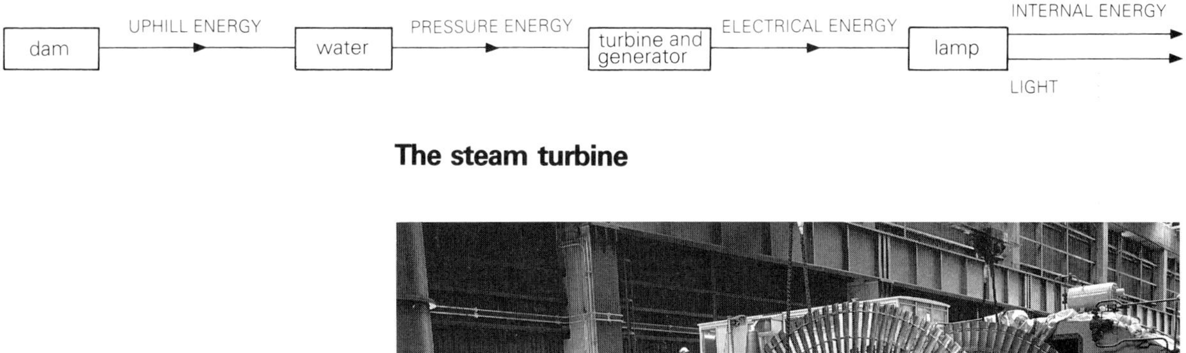

The steam turbine

Steam turbines at West Burton power station

The type of turbine used in most power stations has the blades turned by steam. The steam is produced by heating water either by burning oil, gas or coal or by using the energy generated in a nuclear reactor. The energy flow diagram for a fuel burning power station is shown below.

Experiment 11.1
Looking at energy transfer

These experiments are about ways of changing energy from one form into another. Often a 'device' is needed to do this. The solar cell is an example of such a device; it transfers light into electrical energy. The energy change can be shown in a flow diagram.

The device or **transducer**, as it is usually called, is named in the box. The forms of the input energy and of the output energy are named along the arrows.

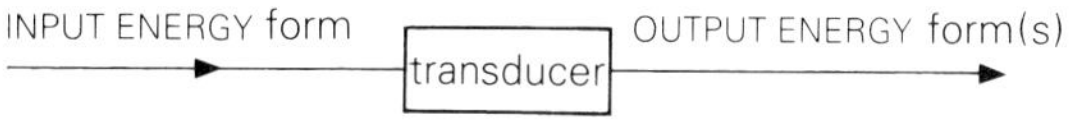

a Lighting a match.

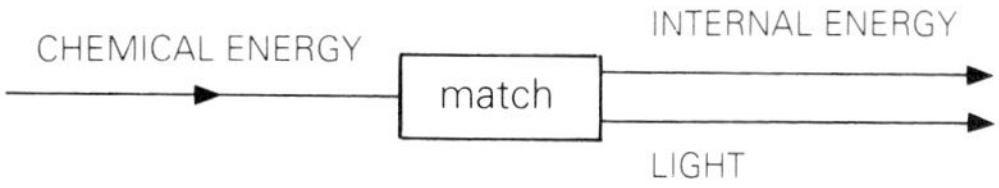

b Using a Bunsen burner.

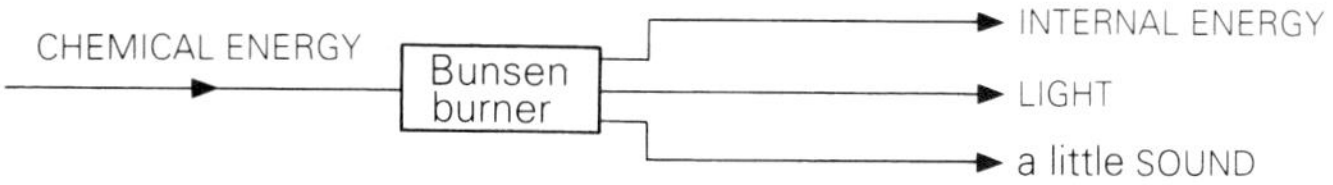

c Hammering a thin strip of lead.

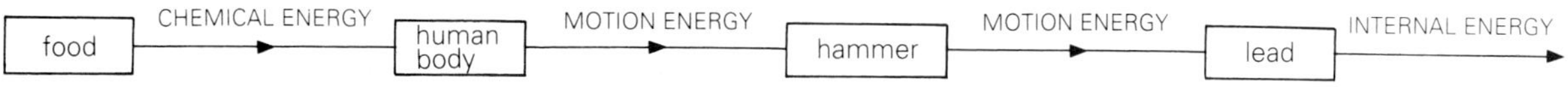

d Using a battery to light a lamp.

e Using a model steam engine to raise a load (Fig 11.18).

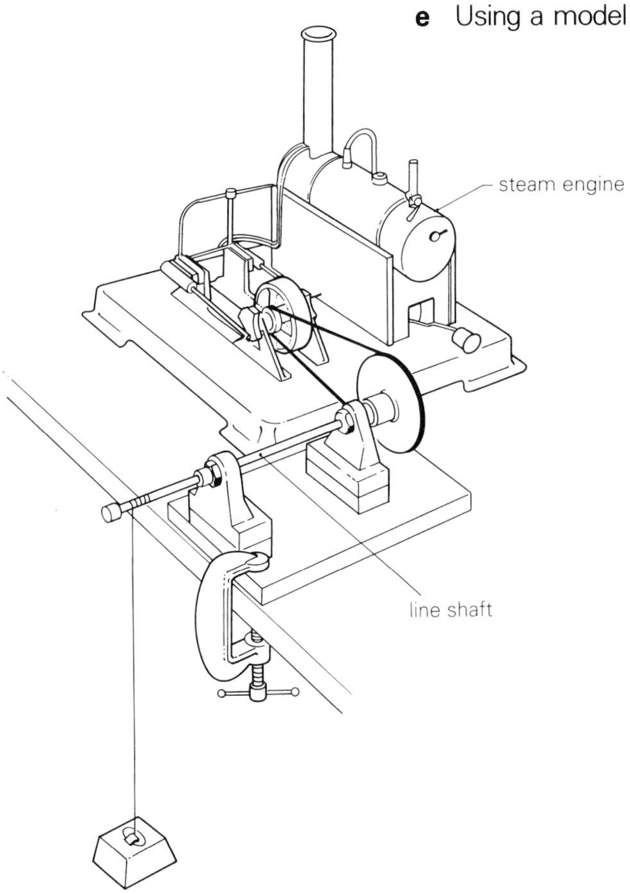

Fig 11.18

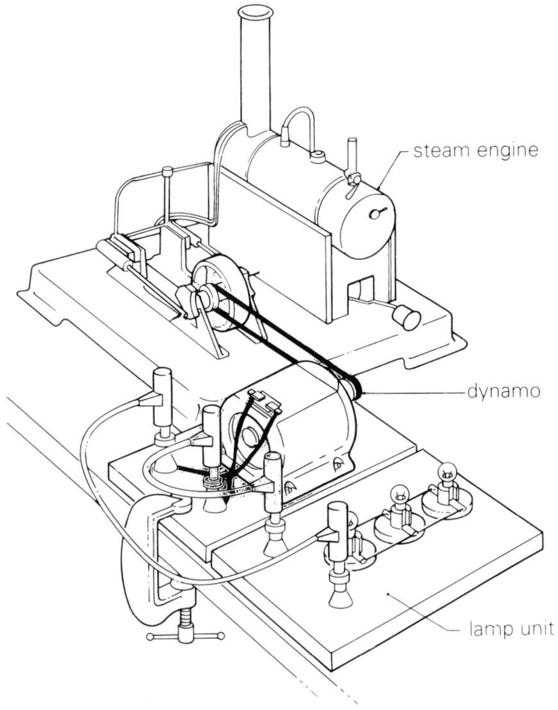

Fig 11.19

f Using a model steam engine to drive a dynamo to light a lamp (Fig 11.19).

g Using a battery to drive a motor, which drives a dynamo, which lights a lamp (Fig 11.20).

Safety note

Before using a steam engine, always check that the safety valve moves freely, fill the boiler with distilled (or at least boiled) water. Use solid fuel in a properly-designed burner.

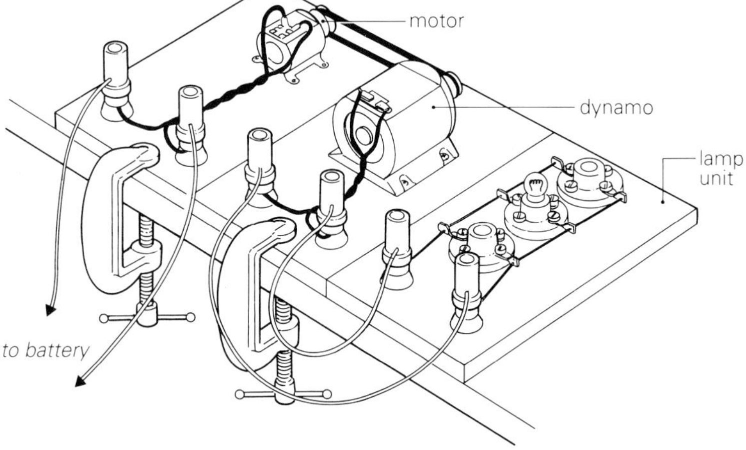

Fig 11.20

h Using a battery to drive a motor to lift a load (Fig 11.21).
i Using a storage cell to light a lamp.

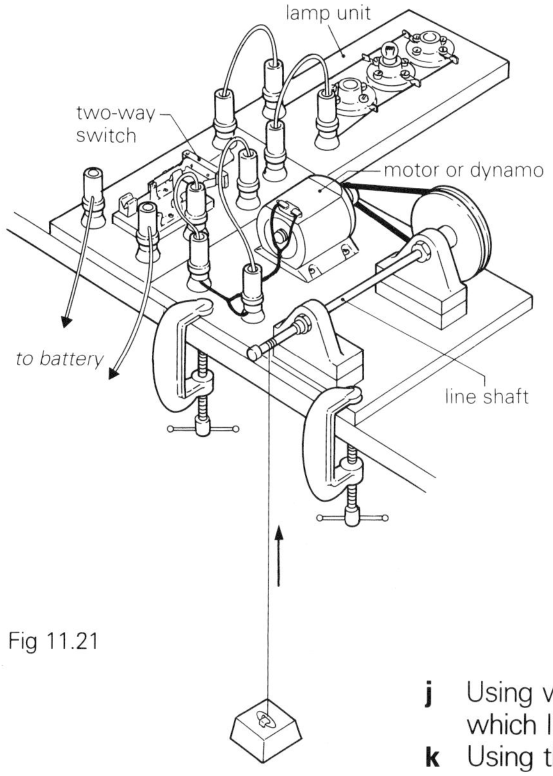

Fig 11.21

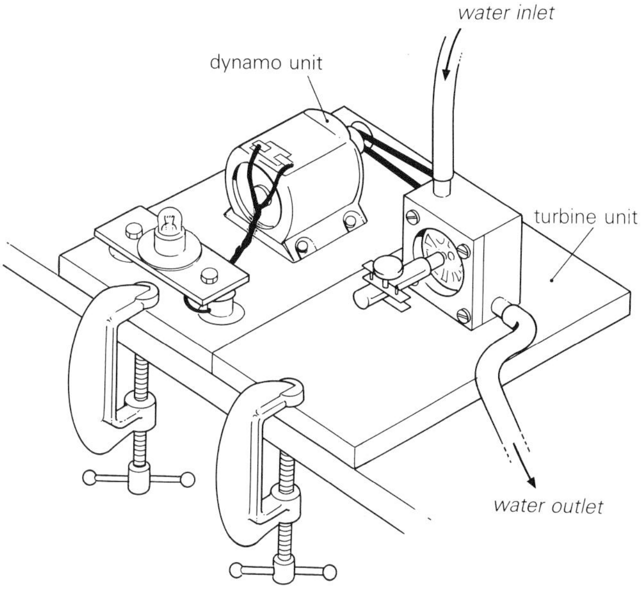

Fig 11.22

j Using water to drive a turbine, which in turn drives a dynamo, which lights a lamp (Fig 11.22).
k Using the turbine in reverse as a pump (Fig 11.23).
In this experiment, a battery drives a motor, which turns the turbine unit as a pump so that water is taken from a lower level to a higher one. The energy flow diagram is:

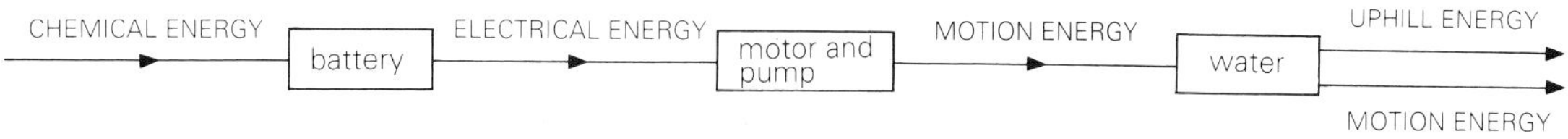

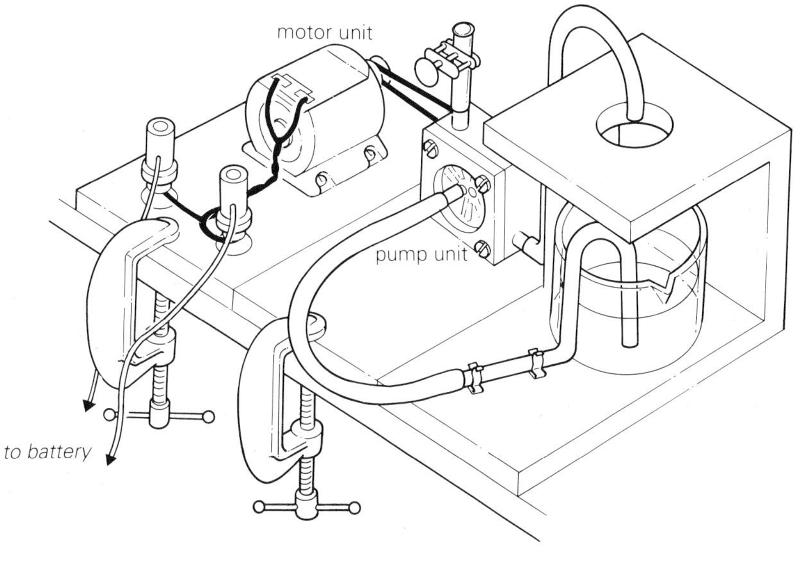

Fig 11.23

l Using coupled pendulums (Fig 11.24).
This experiment works well if the stands are rigidly clamped to the bench and the lengths of the two pendulums are exactly the same. Set one of the pendulums swinging and watch the energy being transferred to the other and then back again to the first.

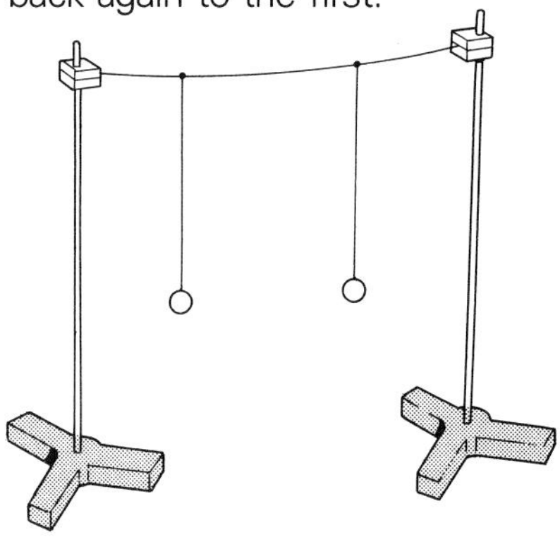

Fig 11.24

m Using a solar cell and motor to lift a tiny weight (Fig 11.25).

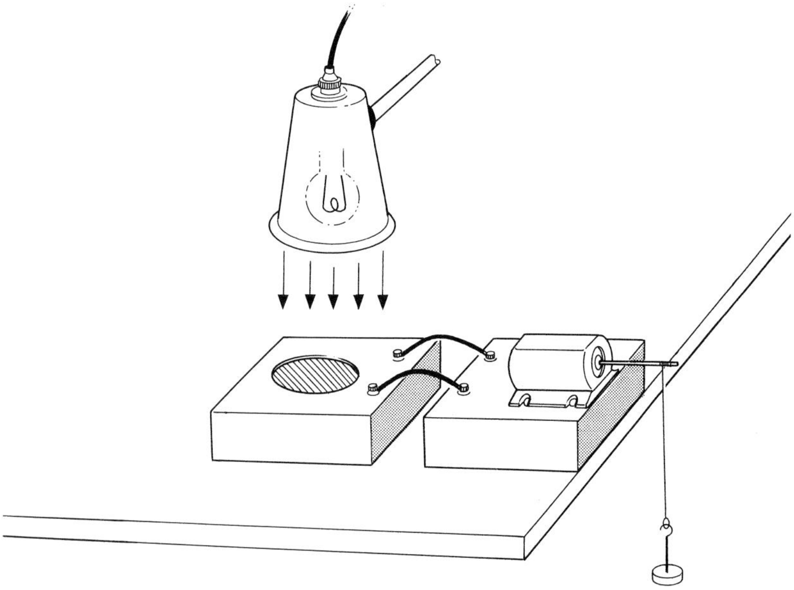

Fig 11.25

Light from a bright lamp is shone on to a solar cell connected to a motor, which is able to lift a tiny weight.

n Using a motor to drive a flywheel (Fig 11.26).
Energy from the battery is transferred to the motor, which drives the flywheel. If the switch is changed, the flywheel will drive the motor as a generator, which will light a lamp.

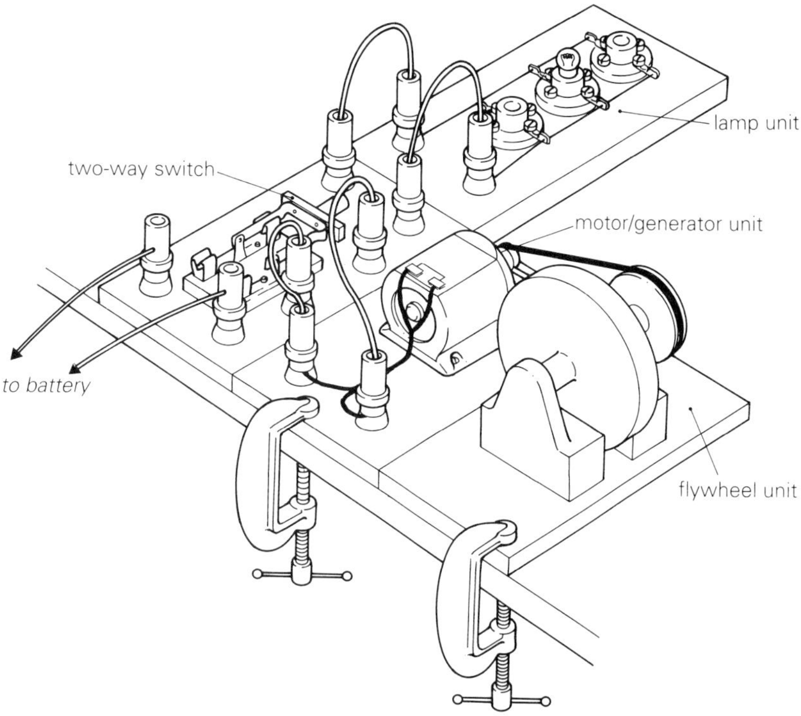

Fig 11.26

Summary: different forms of energy

1 Uphill (or potential) energy
2 Motion (or kinetic) energy
3 Strain (or spring) energy
4 Strain energy of a liquid or gas ('pressure' energy)
5 Electrical energy
6 Chemical energy
7 Internal energy
8 Sound energy
9 Nuclear energy
10 Radiation energy (including light and radiowaves)

Questions for homework

1 Draw a flow diagram for the energy transfers in the following experiments.
  a A model steam engine raising a load.
  b A model steam engine driving a dynamo, which lights a lamp.
  c A battery which drives a motor, which drives a dynamo, which lights a lamp.
  d A water supply which produces a jet of water, which drives a turbine, which turns a dynamo, which lights a lamp.
  e A motor which drives a flywheel, which later drives a dynamo, which lights a lamp.

2 A girl makes a catapult with a piece of elastic. She stretches it and then fires a piece of folded paper across the room so that it hits a boy. Write a short account of this explaining the energy changes at each stage. You should mention each of the following: motion energy, strain energy, chemical energy, internal energy.

3 **a** Coal is burned at a power station to make steam. This steam drives a steam turbine which is coupled to an electric generator. Describe the energy transformations which take place.
b A company uses electric power to drive an electric motor coupled to a pump pumping water from a deep well to a hilltop reservoir. Describe the energy transformations in this process.
c How could the store of water in the reservoir be used to produce electrical energy?
d How could the energy stored in the reservoir be increased?

4 A toy 'tank' can be made from a cotton reel, an elastic band, a pencil, a match stick, a drawing pin and a thin slice cut from a wax candle. Such a tank is illustrated in Fig 11.27.

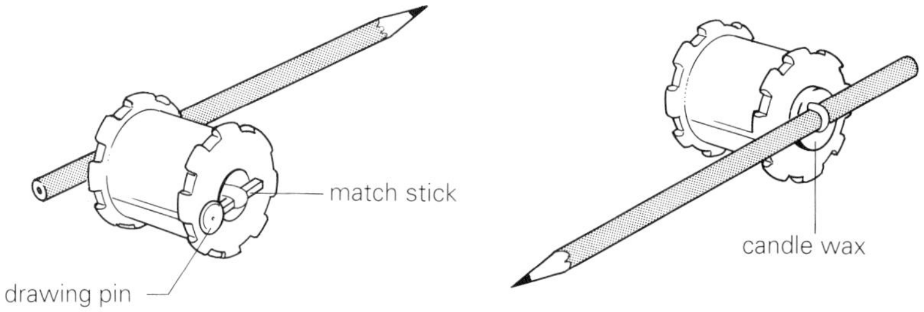

Fig 11.27

If you wind up the elastic band using the pencil and put the tank on the floor it will creep along.
a Where does the tank obtain the energy to move?
b What is the purpose of the matchstick and the drawing pin?
c Is the pencil necessary? Would the tank work as well if the pencil were replaced by another matchstick?
d Why do you think the thin slice of candle is used?
e Do you think that such a tank, placed on a very smooth surface which is slightly tilted, would travel the same distance uphill as it would downhill before coming to rest? Give a reason for your answer.

5 Complete the following table.

Input energy form	Transducer	Output energy form
motion energy	dynamo	electrical energy
electrical energy	motor	. . .
light	. . .	electrical energy
electrical energy	. . .	light
sound	. . .	electrical energy
. . .	loudspeaker	. . .

Can you think of any more examples?

Background Reading

Sources of energy

Primitive people had only one source of energy: the food which they gathered or hunted. By 20 000 BC, they had learned to control fire: this allowed them to eat a wider range of foods and gave them warmth, whilst the light from the flames extended the active part of their day. The domestication of animals made it possible for oxen to pull loads (about 3000 BC) and a donkey mill existed in 500 BC. In the Middle Ages wheels and windmills took energy from rivers and the wind, although wood continued to be the main domestic source of energy.

The Chinese were the first to use coal as a fuel. The Romans used it in Britain two thousand years ago, and it was rediscovered in England in the twelfth century when 'black rocks' near the North-East coast were found to burn. This early coal, coming from near the surface, gave out such unpleasant fumes and smoke that it was banned in London by Edward I. But when supplies of wood began to dwindle in the 1500s, coal was used to fill the gap and it was coal which fuelled the Industrial Revolution in the 1800s. Coal powered the steam engines on land and on the seas.

Coal mining has always been dangerous and miners live close to disease and death. These were disadvantages not shared by oil which proved safer to reach and easier to transport. Oil was first used about 1880 and since then its use has grown at a rapid rate.

Natural gas was discovered near a coal mine in Lancashire in 1664, but the quantities found in other parts of the country were small and of little importance. Coal gas, made from coal, was first used for lighting in a Birmingham factory in 1798, but because of the danger of explosions and fire, its use for this purpose grew slowly. Westminster Bridge was illuminated by gas in 1813 and by the end of the nineteenth century many streets were lit by gas. Gas was also found to be convenient for cooking and heating and became popular in many homes.

In 1956, a huge natural gas deposit was found in the Sahara Desert. It was piped to the Mediterranean coast where its temperature was lowered in order to liquefy it, so that it could be transported to the UK in special refrigerated tankers, LNGs or Liquid Natural Gas tankers. One of these could carry enough to fill 70 gasometers.

In 1959 gas fields were discovered in the Netherlands and exploration for similar fields began in the North Sea. The discovery of large quantities of natural gas in 1965 led the Gas Board to change from coal gas to natural gas, which can be transported very cheaply by high-pressure pipeline, and is easy to control. The transporting of oil or gas is substantially cheaper and more efficient than the transporting of solid fuels. All three sources, coal, oil and natural gas, can be used in power stations to generate electricity.

Electricity is much easier to transport and its use has far less effect on the environment than the burning of coal. Furthermore, electrical energy is particularly convenient both in the home and the factory since it can be made available at the flick of a switch. Such is its convenience that the demand for it almost doubles every ten years.

Electricity however is a **secondary** source since it is produced in power stations using one of the **primary** sources – coal, oil or natural gas. Now another source for generating electricity is being used – nuclear energy.

Fig 11.28

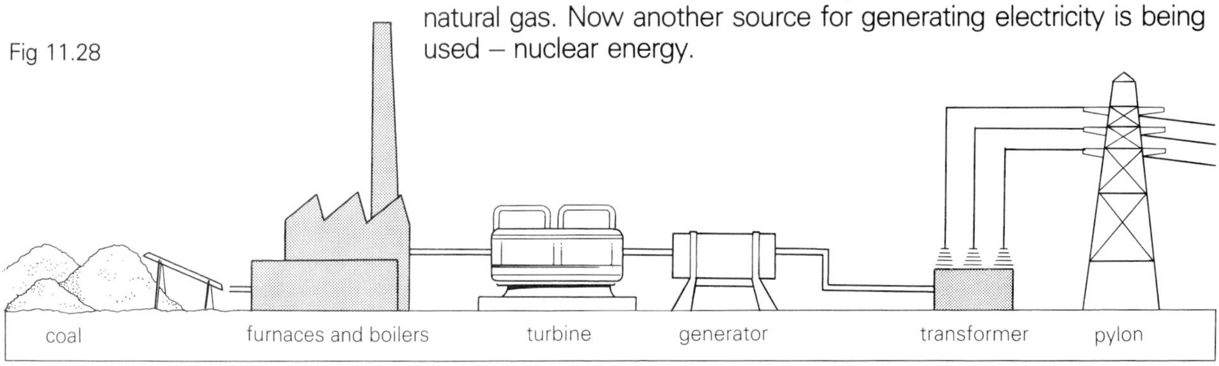

a) Eggborough coal-fired power station

b) Turbine blades

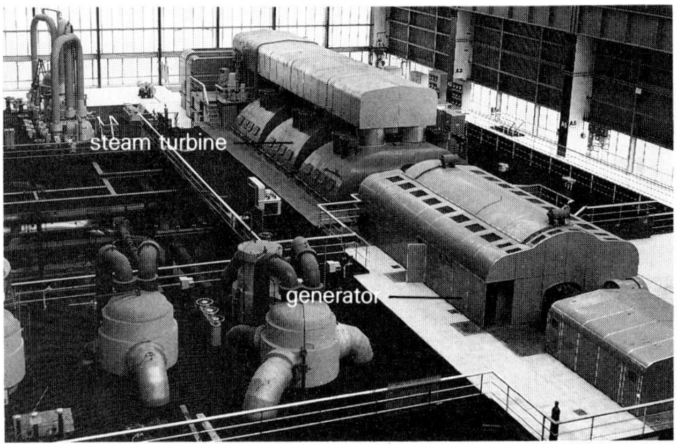

c) Turbine and generator

The basic principles behind a power station are the same whatever the initial source of energy. It is only in the first stage that coal-fired, oil-fired and nuclear power stations differ. The coal or the oil is burned to produce steam by boiling water. Likewise a nuclear reactor also produces steam by boiling water. In all cases high-pressure steam is produced which drives turbines, which drive generators, which produce the electricity (Fig 11.28).

Photograph (a) shows a coal-fired power station: you can see the dump of coal near the station. Photograph (b) shows parts of the turbine (the turbine blades) during construction and photograph (c) shows the finished turbine and the generator.

Hydroelectric power and pumped storage schemes

Hydroelectric power is widely used in many parts of the world, but the power stations can be built only where the land is suitable. When water falls from a high level to a lower one it can be made to drive turbines, which drive generators to produce the electricity (Fig 11.29). Natural sites like the Niagara Falls are ideal, but more often it is necessary to build a reservoir or dam, as for example the Aswam Dam in Egypt.

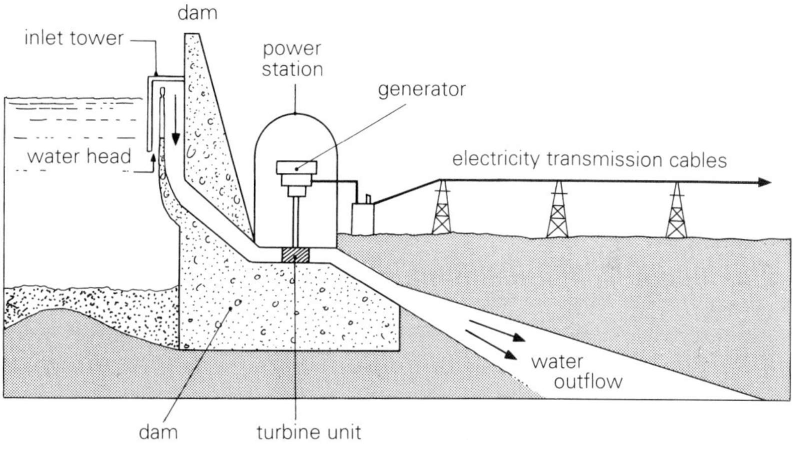

Fig 11.29

The British Isles have few natural hydroelectric resources, but there are several hydroelectric stations in Scotland. A hydroelectric scheme in Wales on the River Rheidol generates sufficient electricity to meet the peak demand of the whole of the Colwyn Bay and Llandudno areas. Important as hydroelectric power is in Britain, it only contributes about 4% to our electricity supply and it is unlikely that it will be able to contribute much more in the future as there are few further sites that can be used.

The power station at Rheidol required a dam to be built high in the Plynlimon mountains. A $2\frac{1}{2}$-mile pressure tunnel conveys 4000 gallons per second to the Dinas power station, where there is a turbine. A further $2\frac{1}{2}$-mile pipeline carries water from Dinas to Cwm Rheidol, where there are two more turbines. Water then flows down the river towards the sea at Aberystwyth.

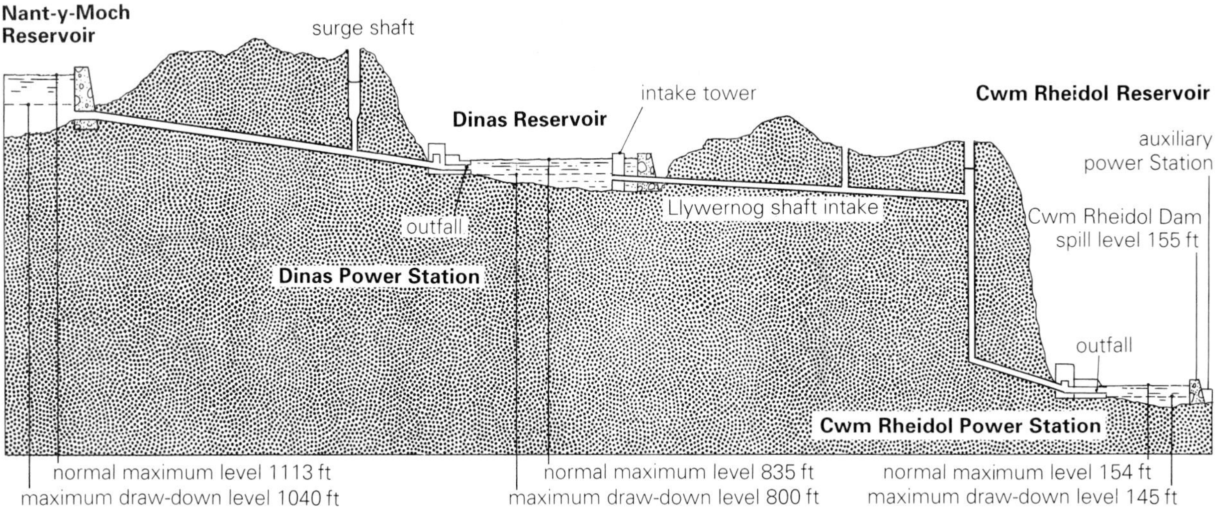

Fig 11.30

Pumped storage schemes are making an important contribution to power supply in Britain. A pumped storage scheme can be linked with a hydroelectric station. The power station operates as normal when there is demand for the electricity it generates, and

Fig 11.31

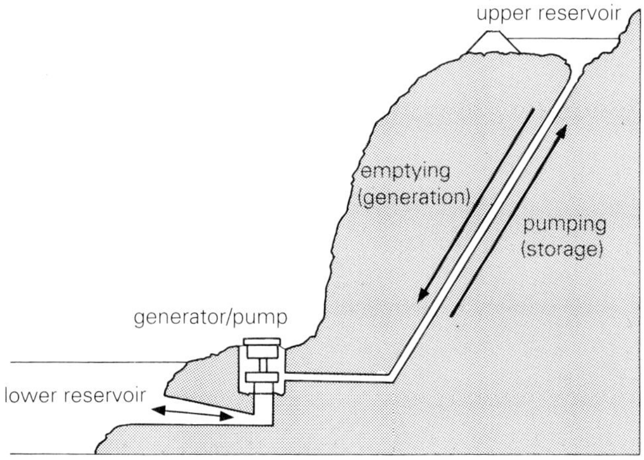

Blaenau Ffestiniog pumped storage station

water flows from a high-level reservoir to a low-level one. But at times of low demand, these stations pump water from the lower reservoir back up to the higher reservoir, using surplus power from continuously operated power stations.

Wales has two pumped storage stations, at Blaenau Ffestiniog and Dinorwig. Scotland also has two, one at Cruachan on Loch Awe, the other at Foyes on Loch Ness.

The great advantage of the scheme is that such stations can be brought to full output in a very short time (less than a minute). So when, on a Sunday evening, the film on television comes to an end and everyone gets up to put on the kettle for a cup of tea, the extra electricity can immediately be provided by 'switching on' the Blaenau or Dinorwig stations!

Explorations 2

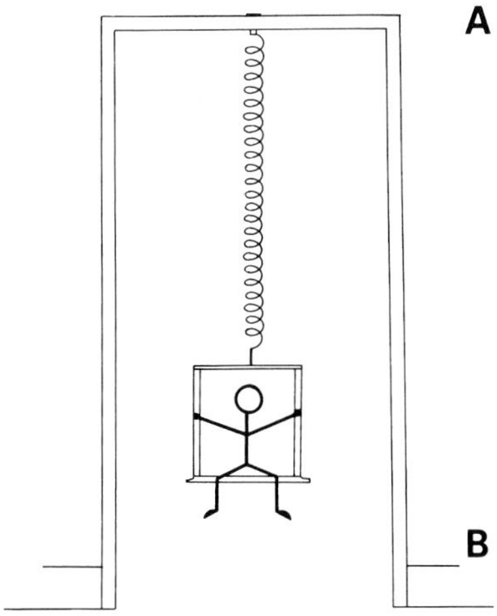

A A baby-bouncer consists of a secure harness for the baby on the end of a long spring. It is usually suspended in a doorway so that the baby can bounce up and down.

Use the apparatus provided to explore what effect the mass of the baby has on the vibration.

What effect would you expect if there were two similar springs side by side?

Why is a baby-bouncer useless as an adult-bouncer?

B People say that you can tell which of two eggs is a hard-boiled egg and which is uncooked by letting them roll down a slope, stopping them suddenly, and quickly placing them on a flat surface. The uncooked egg starts to roll, the hard-boiled egg does not.

Explore whether you can tell the difference between a can full of sand and one full of water in the same sort of way.

Does a can full of water differ in its behaviour from a can full of oil?

C When you blow up a balloon, it seems to be hard to do at first and then much easier as the balloon grows.

Explore how the pressure in a balloon changes with its size.

The diagram will help you to see how it can be done.

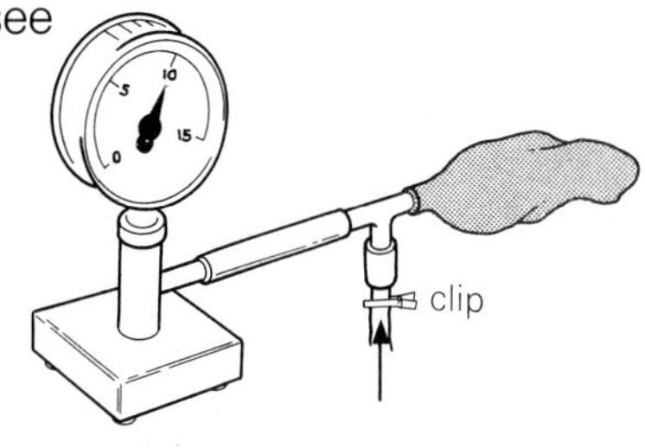

What is the best way to record the results of your exploration in order to tell other people about them?

D The first figure shows a jib crane. The second figure shows an inn sign, the mass of which is 10 kg. This exploration is to find out what force the steel wire AB exerts on the part CB in each case.

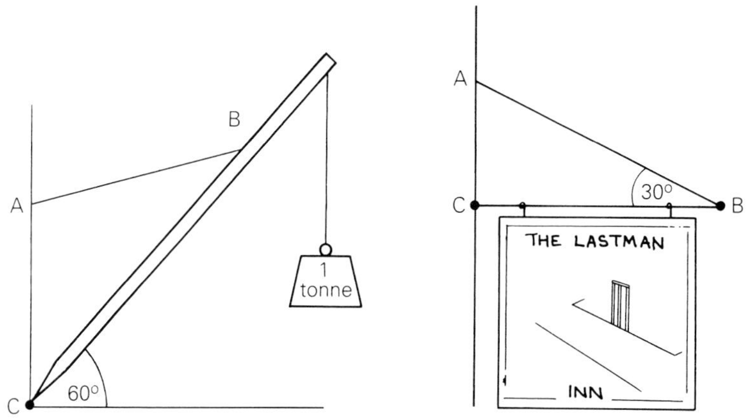

Of course you will have to make a model of these situations. Use a dowel rod for the part CB with cotton thread for AB, A being attached to a spring balance.

From your measurements calculate what the force in the steel wire (the tension) would be in the real case.

How does the force change if the angle marked is changed?

Can you see how it should be possible to *calculate* the tensions in the real case simply by thinking about the turning effect of the forces involved?

E A 'density bottle' is a device which can be used to compare the density of liquids. It consists of a small bottle and stopper. A fine hole runs down the length of the stopper.

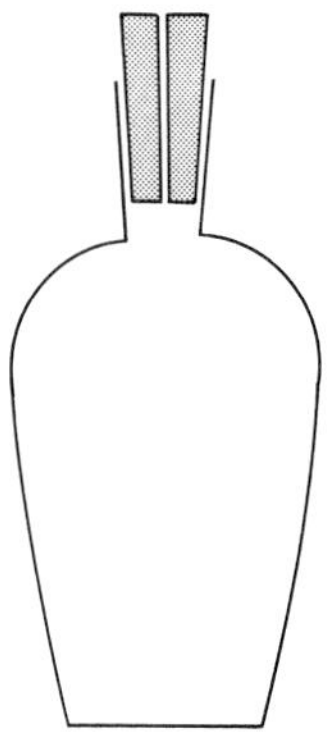

Explore how this can be used to compare the densities of brine and paraffin with the density of water.

What do you think is the reason for the fine hole through the stopper?

Assuming that the density of water is 1 000 kg/m^3, calculate the densities of brine and paraffin. How 'accurate' do you think the result is? What decides whether it is accurate or not?

With the bottle full of water and the stopper in place, hold the bottle clasped in your hands. What do you notice happening? What does this tell you about densities?

F The density of liquids is often measured using an instrument called a hydrometer. The hydrometer floats in the liquid and the depth at which it floats is a measure of the liquid's density.

Make a hydrometer using a test-tube, weighted with sand so that it floats upright with a strip of graph paper inside it to act as a scale.

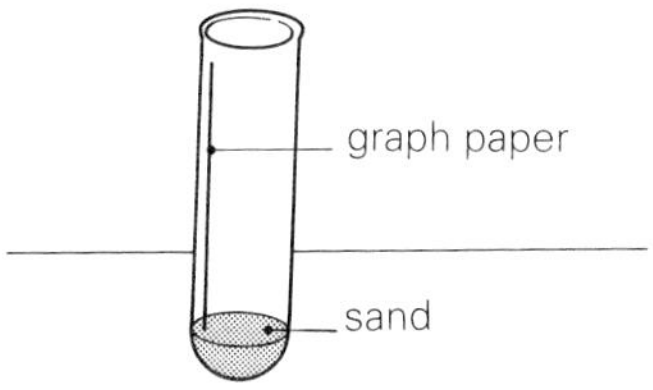

Use brine, water and paraffin to make a rough calibration of your scale.

Find out about the Plimsoll line on the side of a ship.

Chapter 12 Electric circuits

Electricity plays a very large part in the lives of us all, so much so that we tend to take it very much for granted. We are used to switching on the electric light, an electric heater, a record player, a radio, or a spin-drier. Cars would not work without electricity, and aeroplanes would not fly. The telephone would not exist without it and there would be no television. In our homes we would once more have to use oil lamps or gas lighting. We owe a great debt to those people whose investigations made all modern electrical engineering possible.

There is another reason for studying electricity. It has been found that the study of matter is closely related to the study of electricity. If we are to learn more about atoms, it is essential to have a knowledge and understanding of electricity.

In this chapter we will be looking at various electric circuits, and in the following chapters at some electrical devices and some of the effects of an electric current. A lot of experiments will be discussed and of course it is most important that you do these experiments yourself. It is through them that you will come to understand electric circuits.

You may have circuit boards or separate modules to do the experimental work that follows. In the first three experiments of this chapter the drawings will show both methods in use. In later experiments, circuits will be shown by means of circuit diagrams and, where necessary, using drawings of modules.

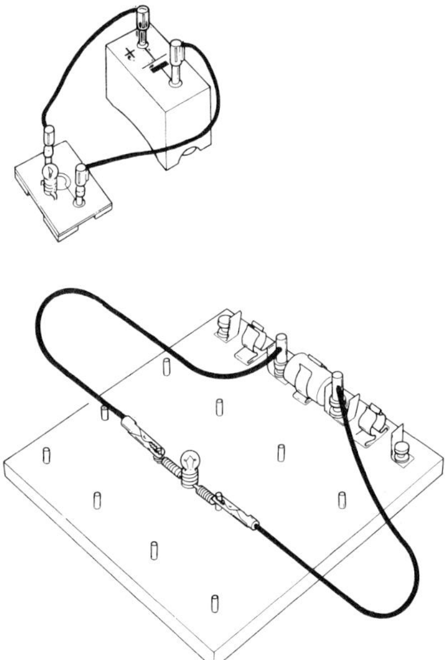

Fig 12.1

Experiment 12.1
Using a cell* and a lamp

a Use one cell and one lamp in its holder. Use two leads to make the lamp light (Fig 12.1).

b Find out what happens if you put the cell in its holder the other way round. Does it make any difference?

c Does it make any difference if you put the lamp holder the other way round?

d Does it make any difference to the brightness of the lamp whether the leads are long or short? Does the shape of the circuit matter?

* In everyday use, the word *battery* is often used instead of the word *cell*. Strictly speaking, a battery consists of a number of cells, in the same way that a battery of guns consists of a number of guns. We shall use the scientific word *cell* throughout this book and only use the word battery when we mean a number of cells.

The electric circuit

These experiments show that a complete path is necessary for the electricity to flow. We refer to this complete path as an electric circuit. The lamp does not light if there are any gaps in the circuit. It is usual to speak about an electric current flowing round an electric circuit, though what exactly we mean by a current will not be clear until later.

Experiment 12.2
Using several lamps and several cells

a Put two lamps in line and connect the cell to the ends as shown in Fig 12.2. How does the brightness compare with the brightness when the cell was connected to only one lamp?

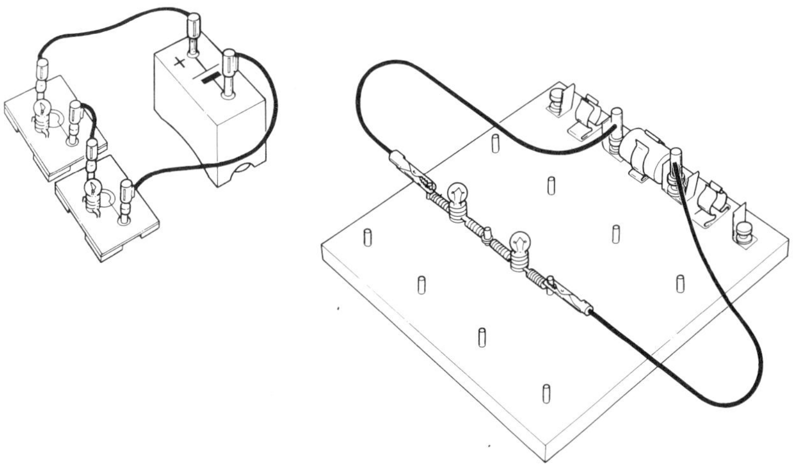

Fig 12.2

b Put three lamps in line and connect one cell across them (in other words, connect one side of the cell to one end of the line and the other side of the cell to the other end of the line). What happens to the brightness? (If by any chance you cannot see the lamp glowing, shield it with your hand and look closely for a faint glow).

c Now try two cells and two lamps. How does the brightness compare with the brightness when one cell was connected to only one lamp?

d Leave one of the cells alone, but turn the other round. Does it make any difference to the brightness?

e Fix three cells in line, all of them the same way round. Put three lamps in line and connect the cells across them. What happens to the brightness?

f Now connect the three cells across two of the lamps. What happens to the third lamp? What happens to the brightness of the other two?

g Then connect the three cells across one of the lamps. What happens to this lamp?

Normal brightness of a lamp

We will say that the lamp is glowing with *normal brightness* when a single cell is connected across it.

When two cells are across one lamp, the brightness is greater than normal brightness. And when three cells were across one lamp, the brightness was very much greater than normal brightness; in fact you had a miniature 'photoflood' or it may have been so bright that the lamp filament melted.

When one cell was put across two lamps, they glowed only dimly. The brightness was less than normal brightness. With one cell across three lamps, the brightness was even less and the lamps glowed very dimly indeed.

One lamp with one cell across it glowed with normal brightness. But so did two lamps with two cells. What happens with three cells with three lamps? Try it and see. They all glow with normal brightness.

We can summarise the results in the following table.

Number of cells	Number of lamps	Brightness
1	1	normal
1	2	dim
1	3	very dim
2	1	bright
2	2	normal
2	3	dim
3	1	very bright
3	2	bright
3	3	normal

You will notice that if the number of cells equals the number of lamps, the lamps all glow with normal brightness.

If the number of cells is greater than the number of lamps, the lamps will be brighter than normal. If the number of cells is less than the number of lamps, the lamps will be dimmer than normal.

Those results are the results which the authors of this book, and probably your teacher, would expect you to give and it is really the 'correct' answer. But it may happen that someone in your class will put three cells across three lamps and they may not all glow with normal brightness, in fact one may be slightly brighter, one might be slightly less bright. Is this 'wrong'? Is there some 'fault'? Certainly not; he or she has done the experiment and that is what happened. There are often slight variations between one lamp and another, even though they are made by the same manufacturer. Lamps from different manufacturers can differ quite a lot even though they are supposed to be the same. Cells can also differ from one another (especially when one has been used a lot and another has not). This is the real, everyday, practical world in which we live. Teachers will probably do their best to see that your lamps are as nearly the same as possible to make matters easier, but we must take a sensible view when there are slight variations. On the whole it is true to say that three lamps across three cells will glow with the same normal brightness as one lamp across one cell.

Experiment 12.3
Lamps in series and parallel

When lamps are arranged in line, as in Fig 12.3, they are said to be in series. Another arrangement is shown in Fig 12.4 in which the lamps are placed side by side. These lamps are in parallel.

Fig 12.3

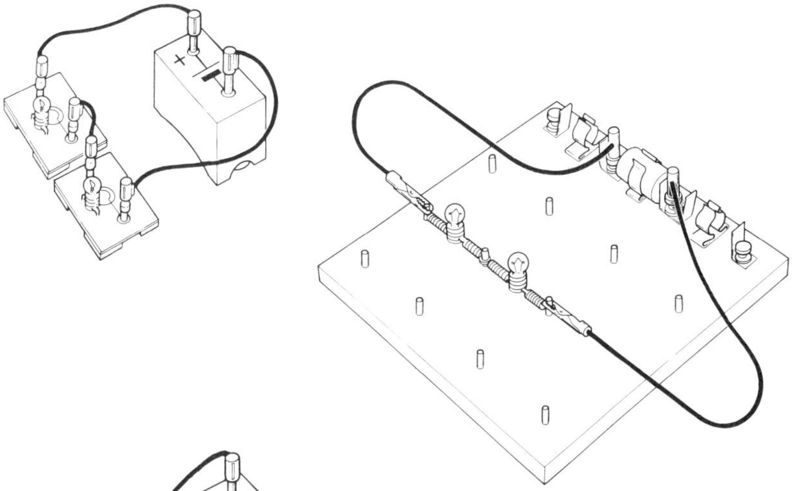

Fig 12.4

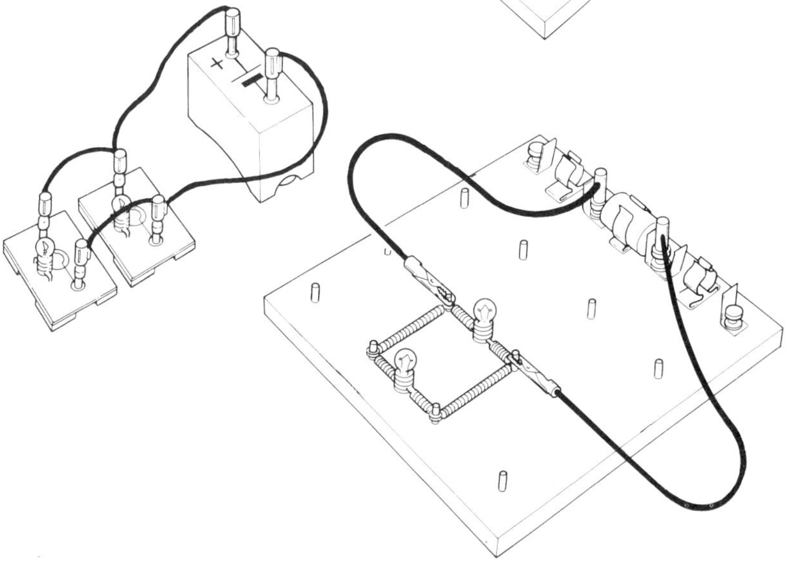

a Put two lamps in series with two cells across the ends. The lamps should glow with normal brightness.

b Now connect the two lamps in series with only one cell across the ends. What happens to the brightness?

c Disconnect the two lamps and reconnect them in parallel. Put one cell across the ends. What happens to the brightness?

d What do you think would be the difference in brightness if you put one cell across three lamps in series and then put it across three lamps in parallel? When you have decided, try the arrangements and see if you were right.

Circuit diagrams

To draw a picture of three lamps in parallel across two cells would be an awkward business if it had to be drawn like the drawings on the previous page. For this reason electric circuits are drawn using special signs or symbols.

The symbol for a cell is two parallel lines as shown below, one longer and thinner than the other. We have already seen that it matters which way round a cell is used in a circuit. You will notice that some cells are labelled + and – . With your cells, the central 'button' is the positive terminal and the metal base at the other end is the negative terminal. In the symbol for the cell, the long thin line represents the positive terminal, the short fat line the negative one.

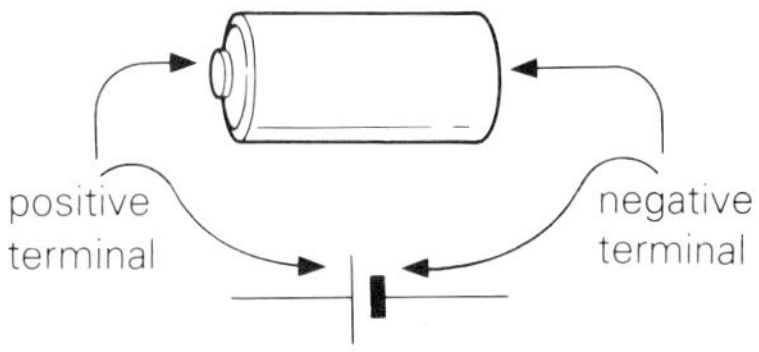

In your circuits you sometimes had three cells in series (connected + to – in a line). This can be shown on a circuit diagram in either of the following ways.

Several cells in series might be shown like this.

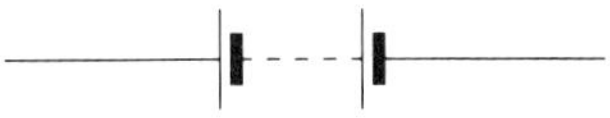

You found in your experiments above that it was necessary to have a complete circuit in order to get a lamp to light. The complete circuit must be shown in your diagram with lines. Where lines meet the junction is marked with a dot as shown below.

The circuit for one lamp and one cell would be drawn like this.

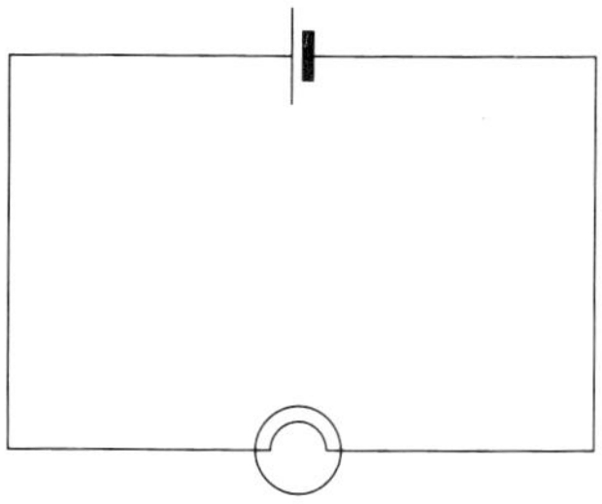

And the circuit mentioned on the previous page, with three lamps in parallel across two cells might be drawn in either of these ways (they are different ways of drawing the same thing).

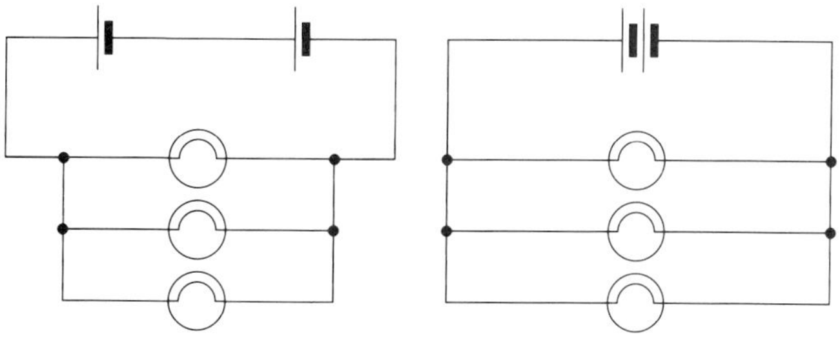

Questions on circuit diagrams

1 Draw a circuit diagram for each of the following
   a one cell across two lamps in series
   b one cell across two lamps in parallel
   c three cells across three lamps in series.

2 What is wrong with these circuit diagrams if you want all the lamps to light? Copy the diagrams and put them right.

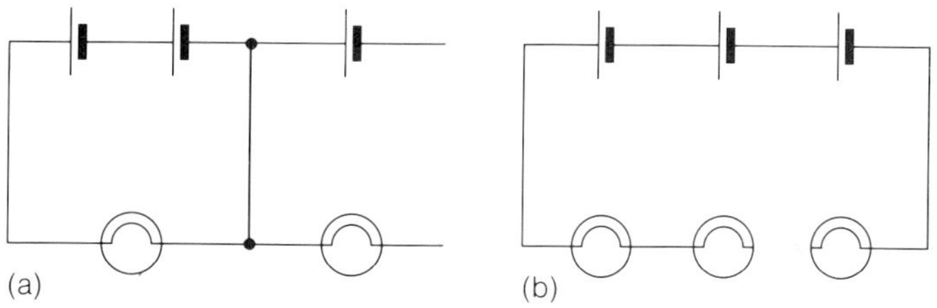

3 Look carefully at this circuit. Do you think the lamp will glow brighter than normal, with normal brightness, dimly or not at all? Give a reason for your answer.

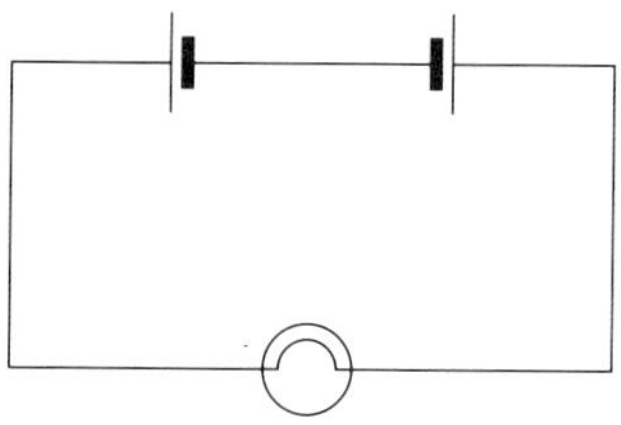

4 These two circuit diagrams are drawn differently. Are they electrically the same or different?

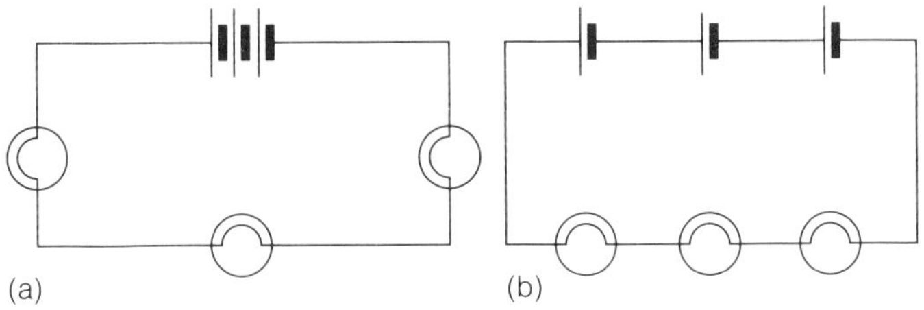

(a) (b)

5 These two circuit diagrams are drawn differently. Are they electrically the same or different? Give a reason.

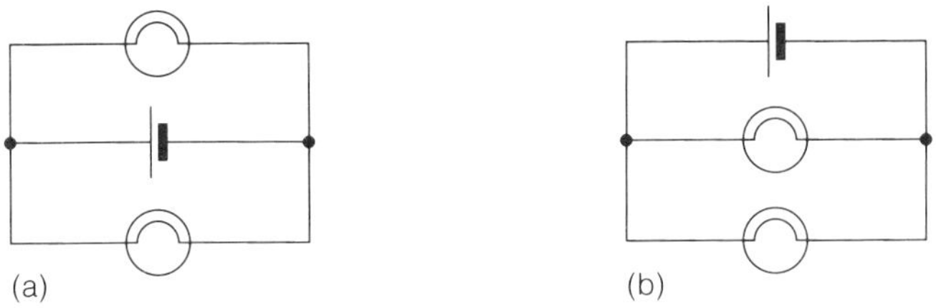

(a) (b)

6 Set up the circuit shown below. Which lamps glow brighter than normal and which glow less brightly than normal? Explain what you see.

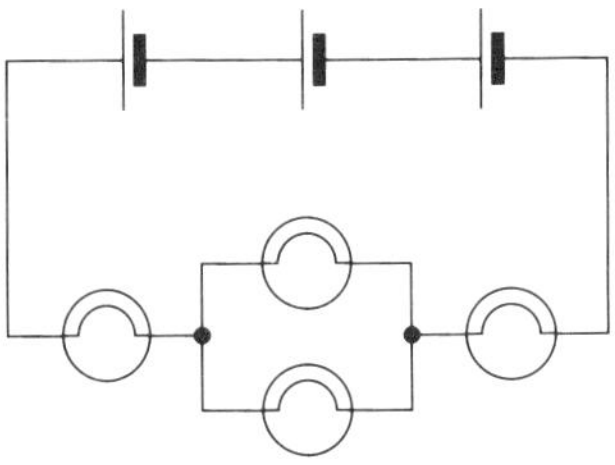

Experiment 12.4
Circuits with switches

You have already found that you must have a complete circuit; there must not be a gap in the circuit if the lamp is to light. So it is possible for you to use a very simple switch in your circuits. With the switch in the open position there is a gap in the circuit; but when the switch is pressed, the gap is closed and the circuit is completed.

a Connect a cell, a lamp and a switch in series so that you can switch the lamp on or off.

b Now put two cells, two lamps and a switch in series so that you can switch both lamps on or off at the same time.

c Put two lamps in parallel across a cell, so that they both glow with normal brightness.

d Add two switches to the circuit so that one switch operates one of the lamps and the other operates the other lamp.

e When you have solved it, draw a circuit diagram of your arrangement. Use this symbol for a switch.

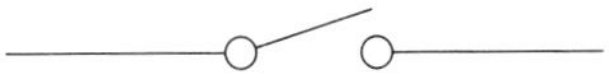

Experiment 12.5
Conductors and insulators

You know that an air gap prevents a current from flowing in a circuit. We say that air is an **insulator**. A strip of metal and the leads you have been using allow a current to flow: these are therefore called **conductors**.

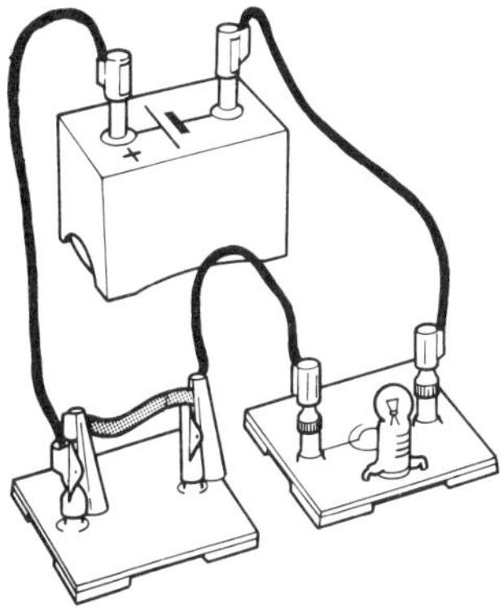

Fig 12.5

Set up the circuit shown in Fig 12.5. The clips are a convenient way of fixing various objects in the circuit to find out if they are conductors or insulators. If they are good conductors, the lamp will light; if they are insulators, it will not. You may use two cells in series if you wish.
Collect as many different objects as you can and test each of them. Make a list showing which are conductors and which are insulators. Your investigation might include the following, though it is hoped you will try other things as well:

strip of paper	a wooden pencil	tin can
piece of wood	pencil lead	protractor
strip of copper	aluminium foil	plastic cup
thread of nylon	comb	saucer
piece of material	handkerchief	hair
piece of lead	knitting needle	expanded polystyrene

Experiment 12.6
Resistance wire

For this experiment you are provided with some special wire: eureka wire is the name for it, but there is no need to remember that name.

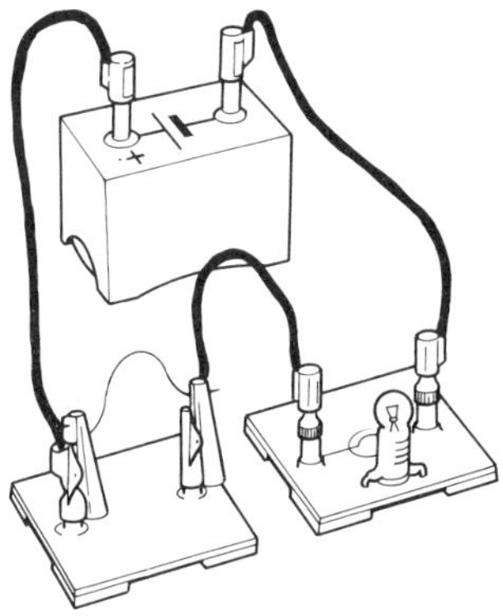

Fig 12.6

a Use the same apparatus as in Experiment 12.5, but only one cell is necessary for this experiment. Put a very short length of the wire between the clips (Fig 12.6). Does the lamp light or not? Is the wire a conductor or an insulator?
b Then try a longer piece of eureka wire between the clips. Is there any change in the brightness of the lamp?
c Now try a very much longer piece of the wire. What happens to the lamp now? If your piece of wire is very long, see it does not touch part of itself. What happens if it does?

The dimmer

The previous experiment has shown that when a short length of eureka wire is used, it seems to be a good conductor and the lamp glows brightly. But as a longer and longer piece is used, it is harder for current to flow and the lamp gets less bright (see the circuits below). It appears to discourage the current and for that reason such wire is called **resistance** wire.

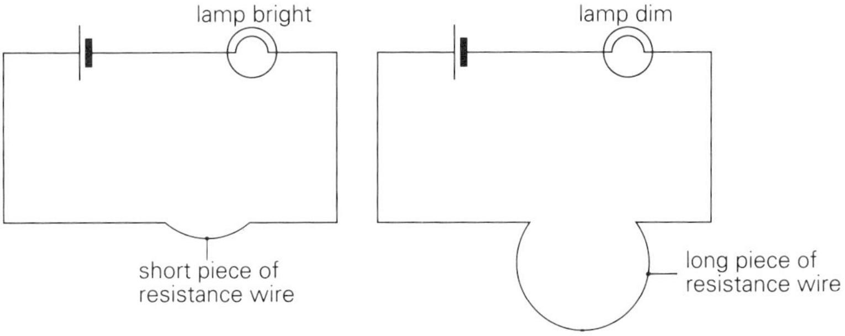

With a short length of resistance wire in the circuit, the lamp was bright. With a long piece the lamp was dim. This suggests that such a length of wire included in a circuit would make a very good dimmer for changing the brightness of lamps.

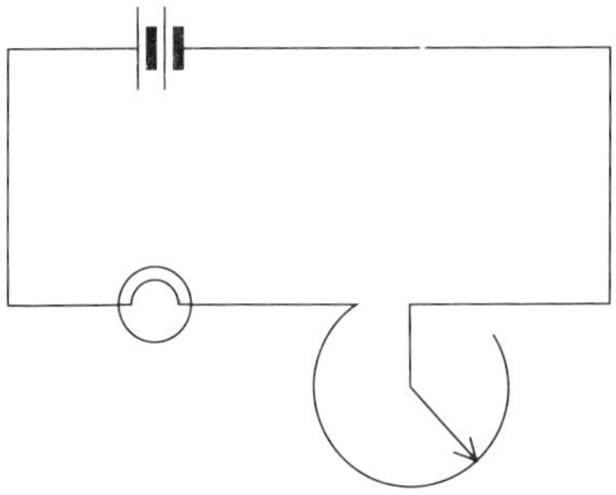

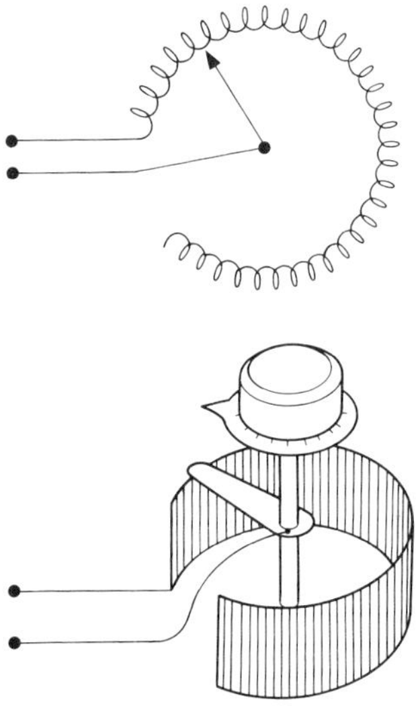

Fig 12.7

But a long piece of wire like that would be very inconvenient. So manufacturers wind it up into a convenient coil with a control knob on top which allows the length of wire through which the current flows to be varied (Fig 12.7).

A piece of wire or other substance which offers some resistance to a current is called a **resistor**. A dimmer is sometimes called a **variable resistor** and another name sometimes used for it is a **rheostat**. In a circuit diagram, the old symbol for a resistor was a zigzag line, but it is now usual to use the rectangular symbol shown below.

The symbol for a variable resistor is the same, but with an arrow through it.

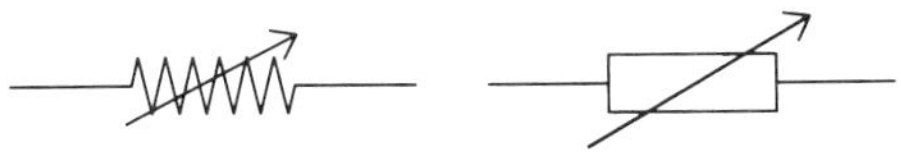

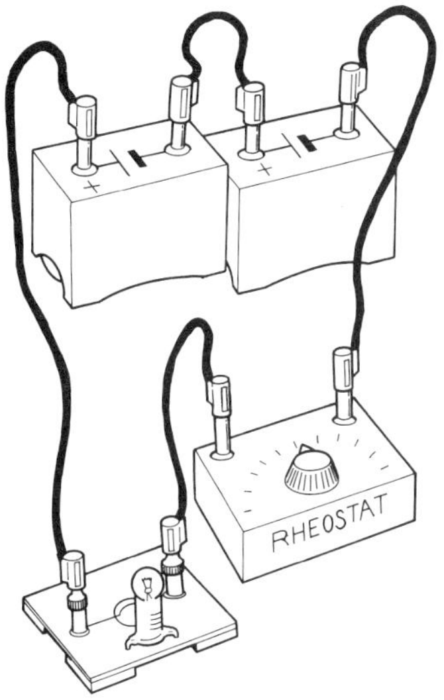

Fig 12.8

Experiment 12.7
The dimmer

a Set up a circuit with two cells across one lamp so that it glows more brightly than normal.

b Insert a dimmer in the circuit so that you can control the brightness of the lamp (Fig 12.8). Does it make any difference on which side of the lamp you put the dimmer?

c Now set up the circuit shown here with two lamps in parallel.

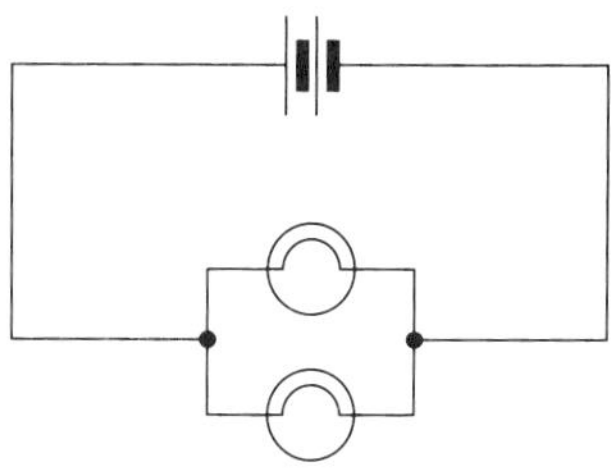

d Insert two dimmers in the circuit so that you can control the brightness of the two lamps independently.

Experiment 12.8
Magnetic effect of a current

In this experiment you will find that a current passing through a coil of wire can influence a compass needle; the current in the coil has a magnetic effect.

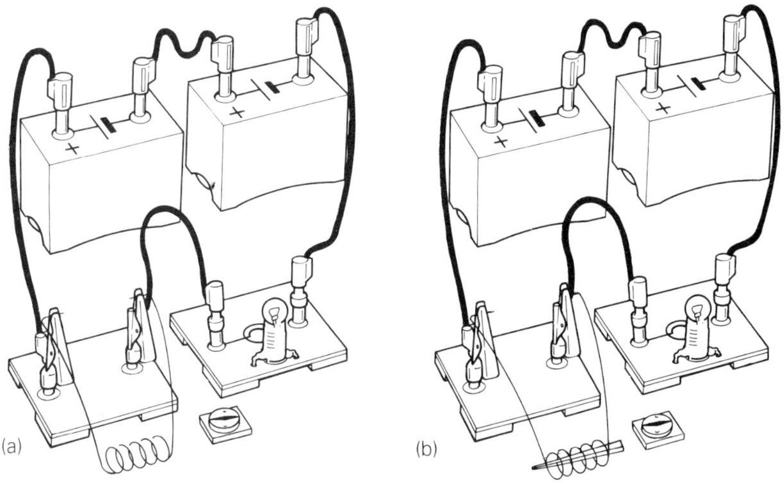

Fig 12.9 (a) (b)

Set up a circuit with two cells and one lamp as shown in Fig 12.9. Wind a coil of wire loosely round a pencil and then fix the bared ends of the coil to the clips (Fig 12.9a). Put a small compass next to the coil. It will point north and south, and the coil should be arranged so that it is east or west of it.

Switch the current on and off. Does the coil have any effect on the compass needle?

Take an iron nail. Put it in the coil (Fig 12.9b). Switch on the current through the coil. Is the effect on the compass needle now much greater?

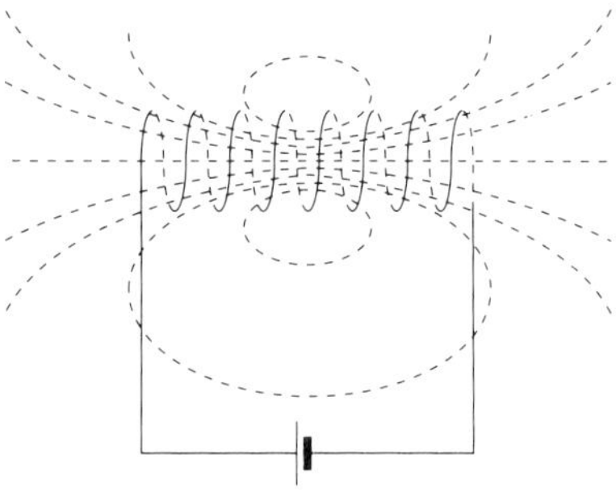

Fig 12.10

The electromagnet

The current passing through the coil produces a magnetic field. You can map the field with a plotting compass. It has the shape shown in Fig 12.10 and it is strongest inside the coil.

With an iron nail inside the coil, the field magnetises the iron and the magnetic effects become much stronger. The most easily magnetisable iron is called *soft iron*, and it loses practically all its magnetism when the current is switched off. A piece of steel is also magnetised by the field of the current in the coil but not as strongly as soft iron, and it retains more of the magnetism when the current is switched off. Indeed, a coil carrying a current can be used to magnetise bars of steel alloys for use as permanent magnets.

When soft iron is used inside the coil, the arrangement is called an **electromagnet**. Electromagnets can be made extremely strong by wrapping a coil of many turns round a piece of soft iron and passing a large current through it. Fig 12.11 shows a section through an electromagnet designed for lifting heavy masses of magnetic metals, such as in scrap metal yards.

Electromagnets have the advantage that the magnetic effect disappears when the current is switched off so that the load drops off it.

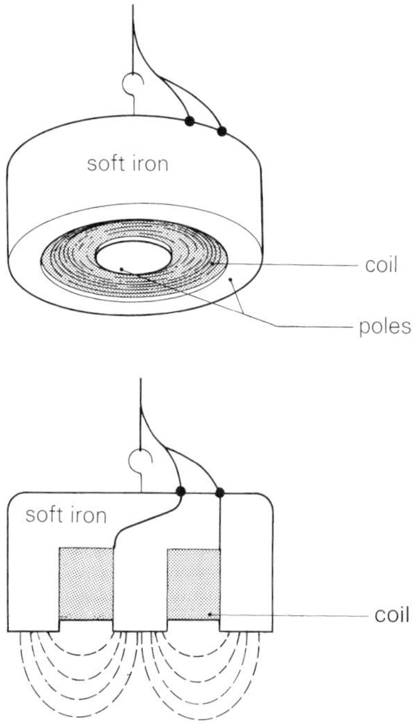

Fig 12.11

A use for an electromagnet – the electric bell

Fig 12.12 shows how a battery-operated bell can be made. At the heart of the bell is a small electromagnet and a piece of soft iron (the armature) fixed to a short length of springy steel. When a current flows through the coil of the electromagnet, the armature is attracted to the poles of the magnet and the hammer strikes the gong. When the current stops, the magnetism disappears and the springy steel moves the armature away from the electromagnet.

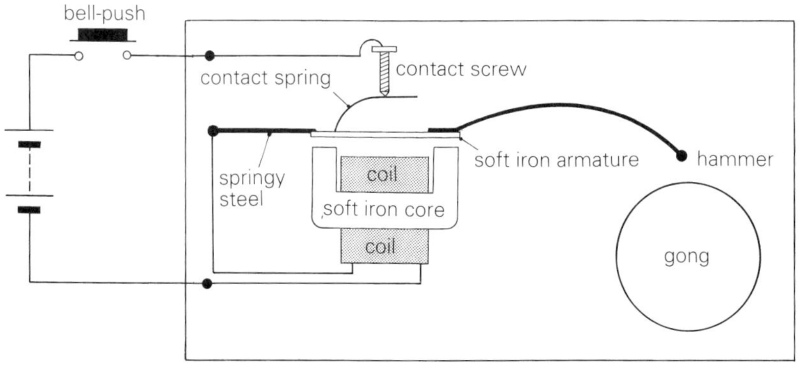

Fig 12.12

In order to make the bell ring continuously, a contact screw and spring are included in the circuit. When the armature is attracted to the electromagnet and the hammer hits the gong, the circuit is broken at the contact screw. When the current stops, the magnetism disappears and the steel spring pulls the armature away from the electromagnet.

This remakes contact at the screw, the armature is attracted again and the hammer hits the gong again, and so on. These processes continue for as long as the bell-push is pressed.

Questions for homework or class discussion

1 When a lamp is lit by a single cell, we shall say that it has *normal brightness*.

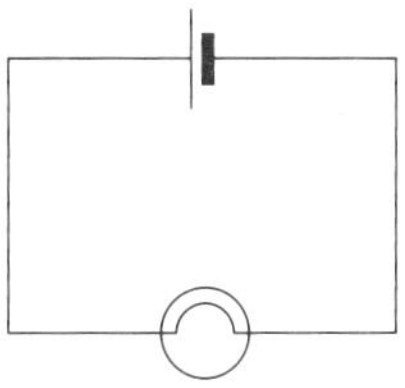

All the cells in the circuits below are similar, and all the lamps are similar. State whether each of the lamps will be extra bright, at normal brightness, dim or out.

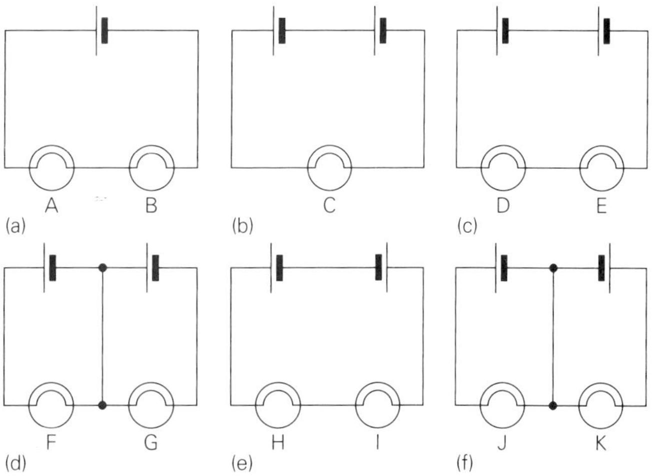

2 You are given three similar lamps and one cell. Draw the circuit diagrams that would give each of the following results:

a all three lamps glow dimly with the same brightness,

b all three lamps glow with normal brightness,

c two lamps glow equally dimly and one with normal brightness.

What will happen if one lamp is unscrewed from its holder in **b**? In **c**, what will happen if one of the dimly lit lamps is unscrewed?

3 The circuit here shows three cells lighting three lamps.

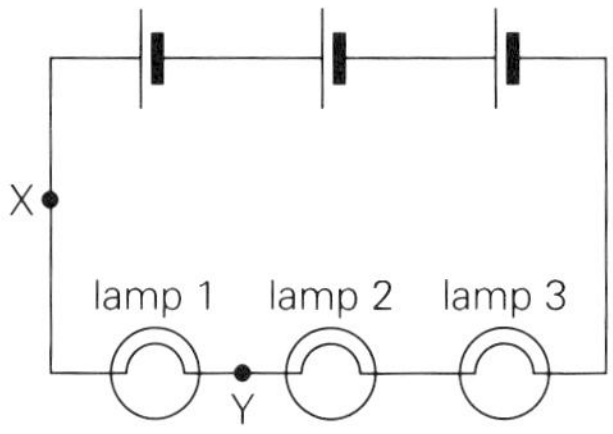

Points X and Y are joined by a piece of copper wire (a good conductor). What happens to the brightness of lamp 1? What happens to the brightness of lamps 2 and 3?

4 What would you expect to happen to the lamp in this circuit

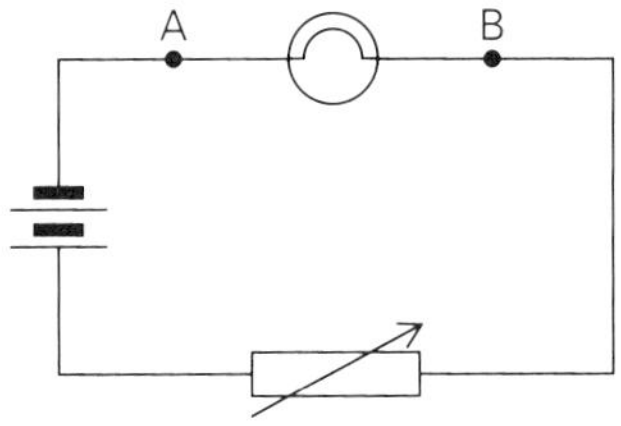

a if the resistance of the variable resistor is reduced,
b if a piece of copper wire (a good conductor) is connected between the points A and B,
c if a second, similar lamp is connected to the points A and B in parallel with the first lamp,
d if one of the cells is turned round?

5 In each of the circuits shown below, describe what changes will happen to each lamp when the switches are closed (they are drawn in the open position).

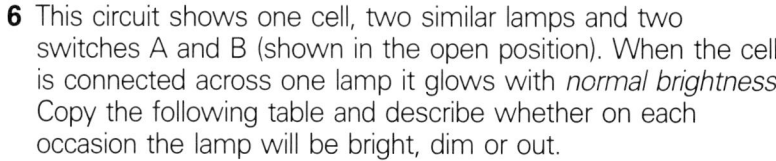

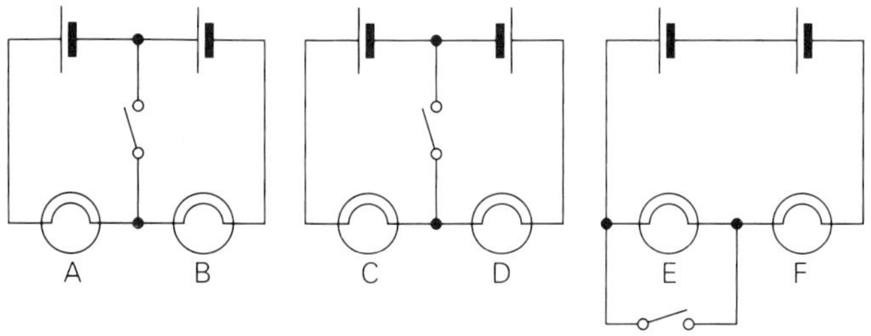

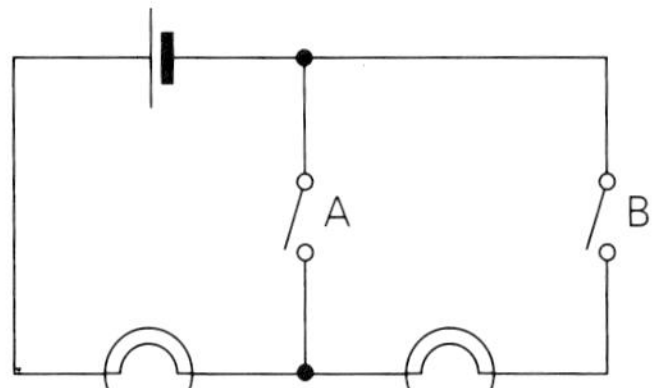

6 This circuit shows one cell, two similar lamps and two switches A and B (shown in the open position). When the cell is connected across one lamp it glows with *normal brightness*. Copy the following table and describe whether on each occasion the lamp will be bright, dim or out.

Switch position	Lamp 1	Lamp 2
A open, B open A open, B closed A closed, B open A closed, B closed		

Chapter 13 Electrical measurements

Ammeters

As a way of measuring a current, the brightness of a lamp is not very satisfactory since there are many different kinds of lamp and even lamps which are supposed to be the same may differ from one another. It is more usual to measure current with an instrument called an *ammeter*, in a unit called an **ampere** (often abbreviated to A). Mass-produced commercial ammeters are easy to use. The photographs show two kinds of ammeter.

Analogue ammeter

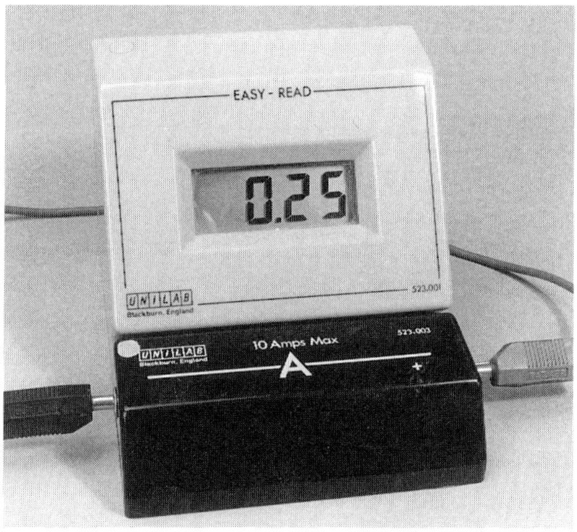

Digital ammeter

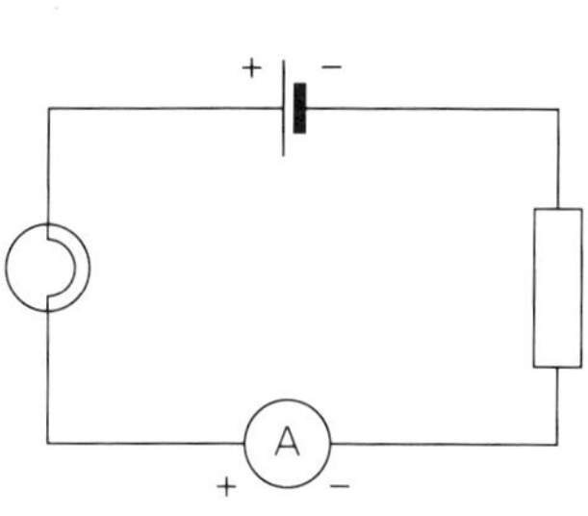

In order to avoid damaging an ammeter, it is important to connect it in a circuit the right way round. One terminal of an ammeter is often coloured red or it may be marked +. This terminal of the meter must be connected to the wire which eventually goes to the + terminal of the battery or cell. The black or − terminal is connected to the wire which eventually goes to the − terminal of the battery or cell (see the circuit on the left). Whatever other items there are in a circuit, the ammeter must always be connected in this way.

The symbol used in circuit diagrams for an ammeter is an A with a circle round it, as shown in the circuit on the left.

Experiment 13.1
Currents in circuits

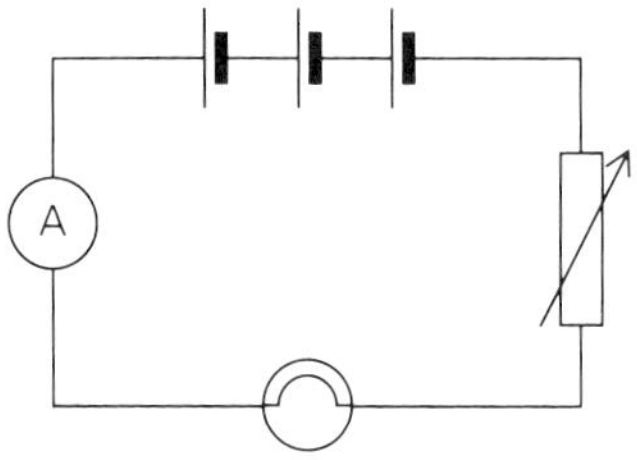

a Set up the circuit shown here. The current can be altered by changing the variable resistor. Notice the reading of the ammeter when the lamp is very dim. Adjust the resistor until the lamp is at normal brightness. Then adjust until the lamp is very bright. How much current flows each time?

b Set up the circuit shown, using two cells and two lamps. What does the ammeter read when inserted in the circuit at A?

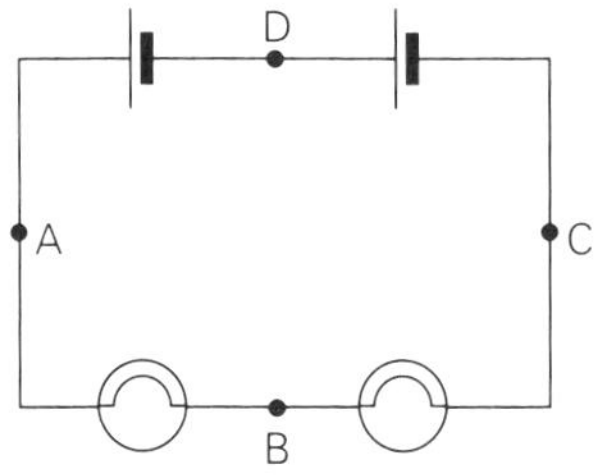

Put the ammeter into the circuit at B instead of A. Notice the ammeter reading. Then put it at C and again notice the reading.

Finally insert the ammeter at D between the two cells. What is the ammeter reading this time?

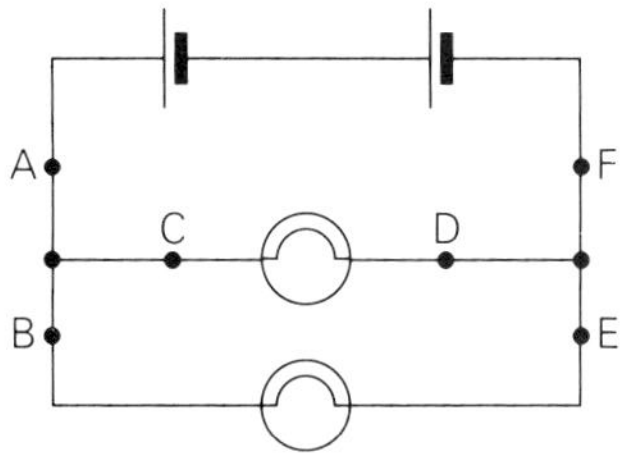

c Set up the circuit shown and insert the ammeter at each of the points A, B, C, D, E, F in turn.

From Experiment **b** we would expect the readings at C and D to be the same. Are they? How do the readings at B and E compare? And those at A and F?

How does the reading at A compare with the reading at B and that at C?

d Look at the circuit shown below. Will an ammeter at B read more than, the same as or less than an ammeter at C? Then decide how the reading of an ammeter at A will compare with the readings at B and C. Having forecast what will happen, set up the circuit and see if you were right.

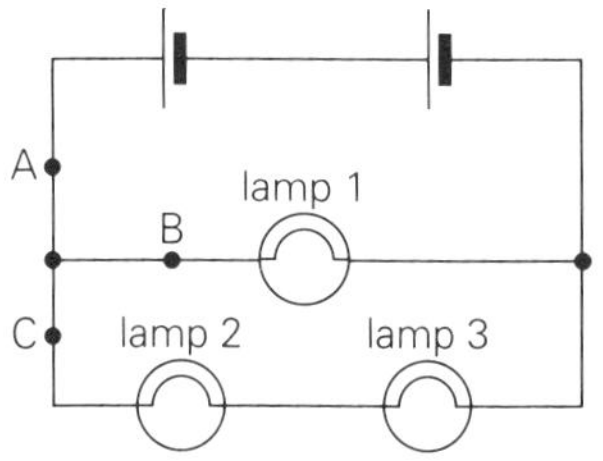

Currents in electric circuits

Some important facts about electric currents follow from the experiments which you have done.

In a simple series circuit the current is the same all the way round the circuit

It is untrue to say that 'the current gets used up going round the circuit': it is the same current everywhere, even between the cells of the battery. It is the same current through each component in the circuit, including the cells themselves.

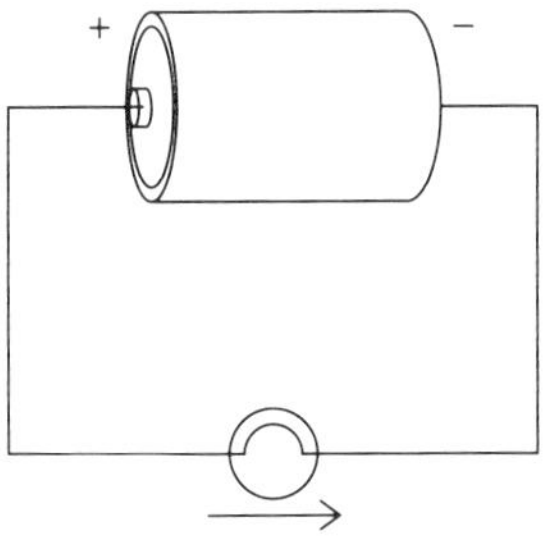

Currents have direction

In fact, none of our experiments have shown which way the current is flowing, but it is a convention to assume that the current flows from the positive (+) terminal, through the circuit and back to the negative (−) terminal of the cell, as shown.

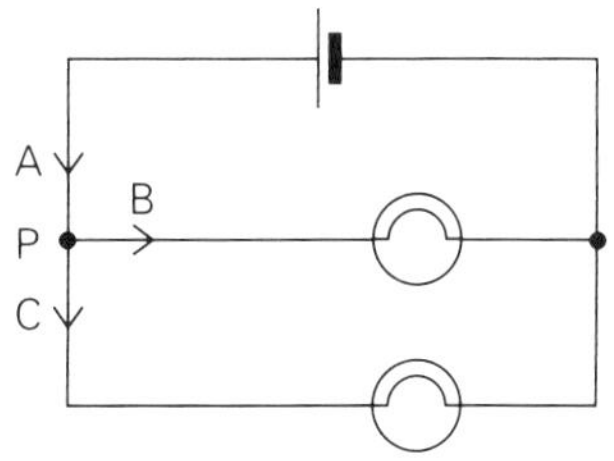

As much current leaves a junction as reaches it

In the circuit on the left, the current leaving the point P is equal to the current at B + the current at C. The current arriving at junction P is the current at A. It therefore follows that

current A = current B + current C.

This result is not really very surprising as the following may show.

Suppose two roads P and Q merge into a wide road R, and that they are one-way streets (Fig 13.1). If there is one car travelling along road P every minute and if there are two cars per minute along road Q, how many cars per minute does common sense tell us must be travelling along road R?

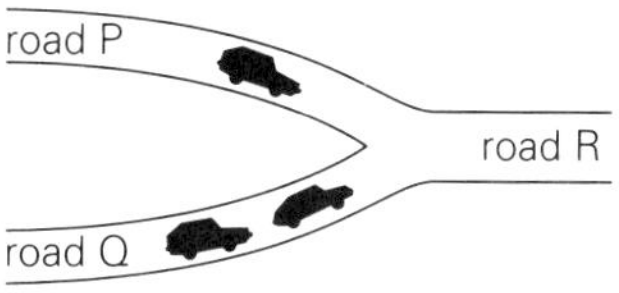

Fig 13.1

Imagine also three pipes through which water is flowing at a rate of 1 litre per second, 2 litres per second and 3 litres per second, as shown in Fig 13.2.

The rate of flow of water through the pipe after the smaller ones have joined is 6 litres per second. Similarly, currents of 1 ampere, 2 amperes and 3 amperes reach the point P and the current leaving is therefore 6 amperes.

We have mentioned cars moving along roads and water flowing through pipes. It is therefore an obvious question to ask what is flowing when there is an electric current in a wire. You cannot see an electric current. Our experiments have in fact merely shown us how an electric current behaves. We will see later that what is flowing is electric charge, but for the present we will concentrate on what an electric current does.

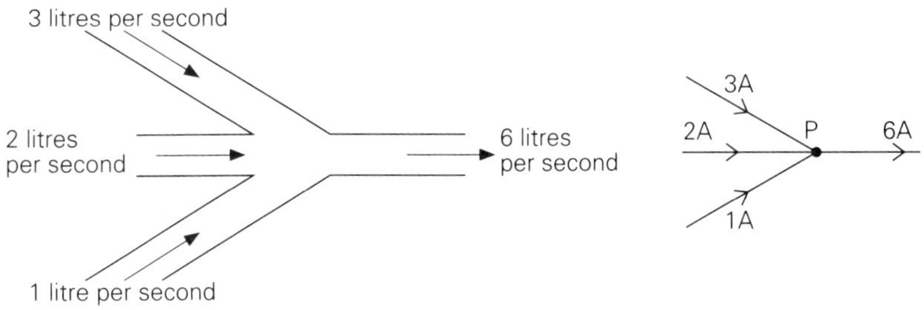

Fig 13.2

Background Reading

Electricity in the home – 1

In the circuits you have been using, the current was driven round the circuit by cells. It flowed steadily from the positive terminal of the cell round the circuit to the negative terminal. Steady currents flowing in one direction only are called *direct currents* (d.c.).

The current which flows from the mains electricity supply is not like that. The two wires which carry the electricity to the lamp or fire are called the **live** wire and the **neutral** wire. At one moment the current flows out of the live wire, through the lamp or fire, and back to the generator through the neutral wire. One hundredth of a second later, it is flowing out of the neutral wire and back through the live wire. One hundredth of a second later, it flows out of the live wire and back by the neutral wire again... and so on. The changes do not take place suddenly; the current grows to a peak value and then reduces to nothing before growing to the peak value again but flowing the other way. The graph in Fig 13.3 shows the changes that take place. We call it **alternating current** (a.c.).

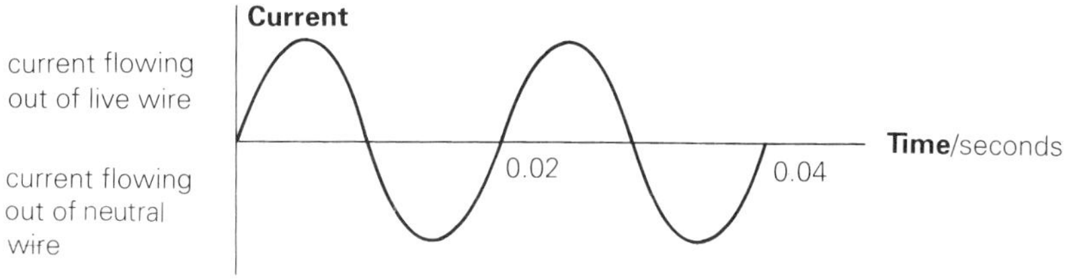

Fig 13.3

It takes *two* hundredths of a second for a complete variation – called a **cycle** – so that there are 50 cycles every second. How many cycles there are in one second is called the **frequency**. The mains supply to the homes and factories in the UK has a frequency of 50 cycles/second. Scientists have agreed to call the unit the *hertz* (Hz) – so the frequency is 50 Hz. (Your teacher may show you these changes taking place by using an oscilloscope.)

You may wonder why the electricity is supplied in this way. Later you will find that it is easier to distribute electricity efficiently in this form – that is, with less waste of energy. And, of course, the lamp will light and the fire will get hot whichever way the electricity flows.

Why does the lamp not flicker when the current is alternating? It does, very slightly, but the changes take place so quickly that the filament has not got time to cool very much as the current changes direction and the remaining flicker is too rapid for our eyes to detect anyway.

What about those things which need direct current in order to work, radios and television sets for example? The alternating current supply has to be converted into a supply of direct current. This is not difficult to do using diodes (you will meet the diode in Chapter 14), and every radio and TV set has the necessary circuit built into it.

Questions for homework or class discussion

1 In each of the circuits below, the cells are all similar and the lamps are all similar. When one of the cells is put across one of the lamps, the lamp glows with normal brightness and the current that flows through the lamp is 0.2 A.

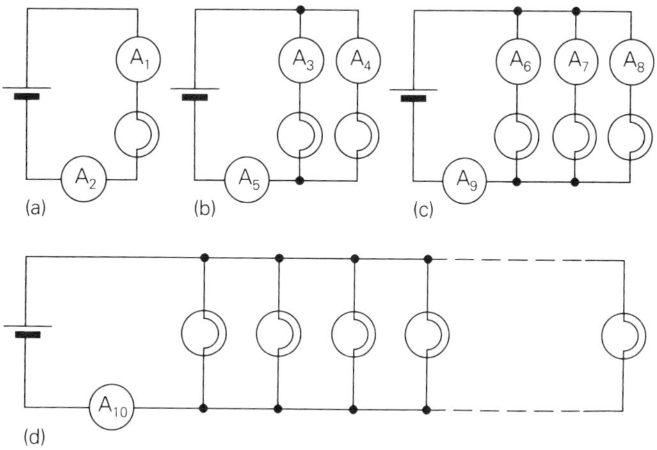

a What is the reading of ammeters A_1 and A_2?

b In circuit (b), the cell is across two lamps in parallel. What is the current through ammeters A_3 and A_4? What is the current through A_5?

c In circuit (c), the cell is across three lamps in parallel. What is the current through A_6, A_7, A_8 and A_9?

d Circuit (d) is a way of showing that the cell is across a large number of lamps, though it is not clear from the diagram exactly how many. If there were ten lamps, what would you expect the current to be through A_{10}? Do you think this is what the reading would be if you did this experiment? To find out, the best thing is to do Experiment 13.2 to see for yourself.

2 In the circuit shown below, what is represented by the symbols labelled a, b, c and d?

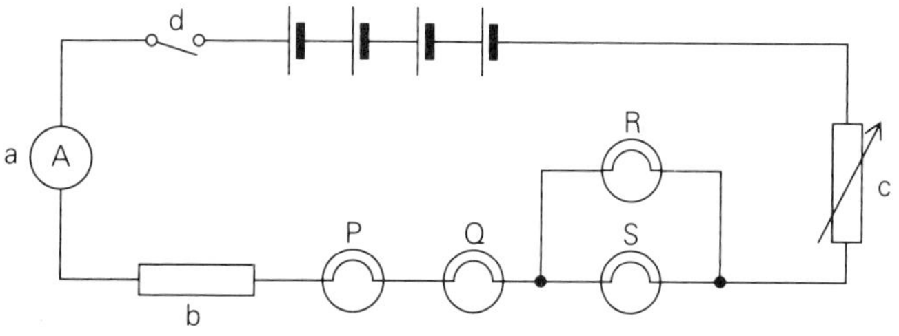

Which pair of lamps P and Q, or R and S, are in series? Which pair are in parallel?

If the lamps P, Q, R and S are all similar, would P be more bright, equally bright or less bright than Q? Would P be more bright, equally bright or less bright than R?

3 All these circuits use similar cells and similar lamps. In circuit (a) the current is 0.2 amperes. For each of the other circuits, say whether the current will be zero, between 0 and 0.2 A, 0.2 A, or greater than 0.2 A. Give a reason for each answer.

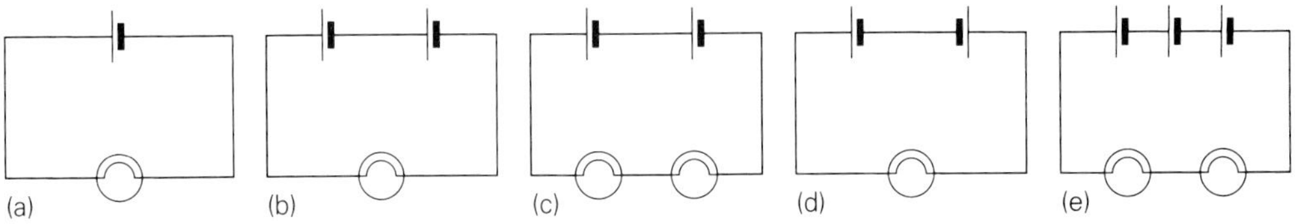

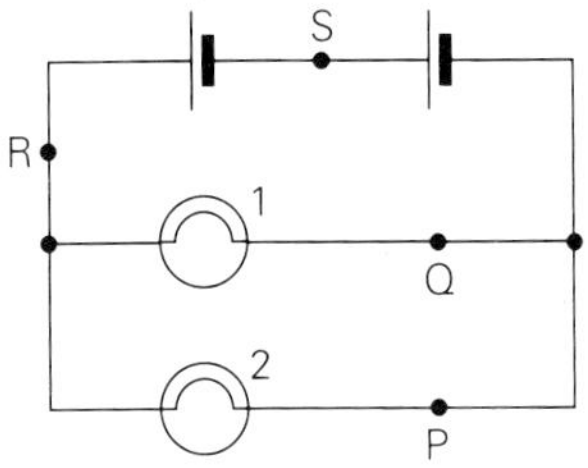

4 Two cells are connected across two similar lamps as shown on the left. The current passing through an ammeter at P is 0.2 A.

a What is the current at Q?
b What is the current at R?
c What is the current at S?
d Redraw the circuit and insert a switch (which you should mark with the letter A) which will switch both lamps at once, and a switch (marked with the letter B) which will switch off only the lower lamp.

5 The three lamps in the circuit below are all similar. The three ammeters A_1, A_2 and A_3 are also similar.

Which ammeter will show the largest reading? Which ammeter will show the smallest reading?

If another ammeter was inserted in the circuit at the point marked P would it read more, less or the same as ammeter A_1?

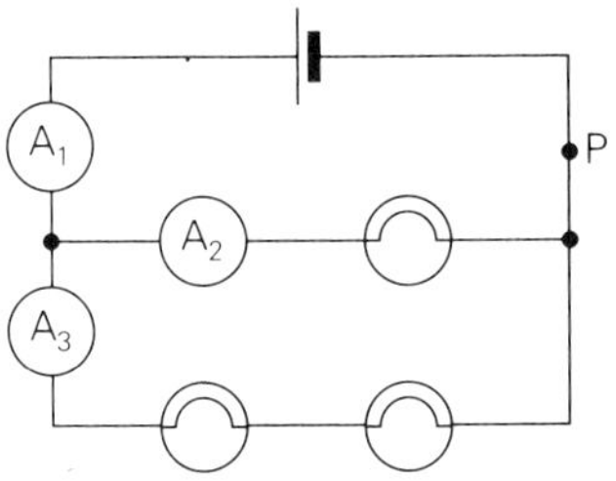

Experiment 13.2
Lamps in parallel

a Set up the circuit shown in Fig 13.4. What is the reading of the ammeter?
b Put two lamps in parallel as shown in circuit (b) on the next page. What is now the reading of the ammeter? Is it exactly double the reading when there was only one lamp?
c Then put three lamps in parallel (circuit (c) on the next page). Finally try four lamps in parallel. Each lamp has the cell connected across it, but they do not glow with the full normal brightness, and the current is not quite four times what it was originally.

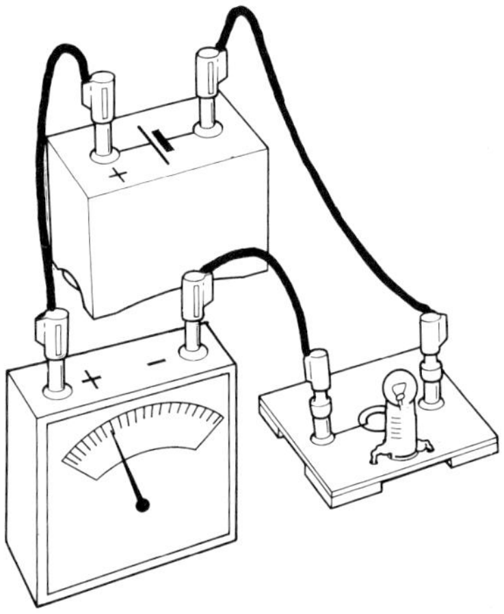

Fig 13.4

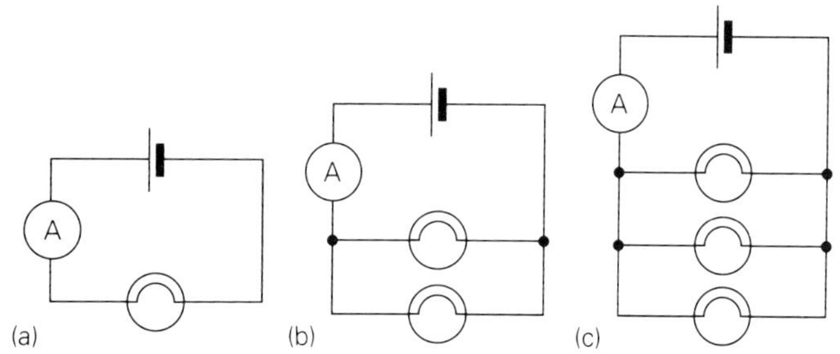

d If you like, you can try more lamps in parallel. You will find you cannot go on getting more and more current from your cell: there seems to be a maximum current it will give.

If an ammeter is connected straight across a cell without any lamps in the circuit, it will be found that the current is about 1.2 A. This is therefore the maximum current the cell will give. There is of course some resistance in the connecting leads and in the ammeter itself, but the main resistance which restricts the current to a maximum value is the resistance inside the cell itself. This is usually called the **internal resistance** of the cell.

How long will a cell last?

Consider the three circuits below. The lamp will glow brightest in circuit (a), but there are two cells. The lamp will glow with normal brightness in circuit (b). In circuit (c) there are two cells, but three lamps each of which will glow with less than normal brightness. In which case will the cell or cells run down most quickly?

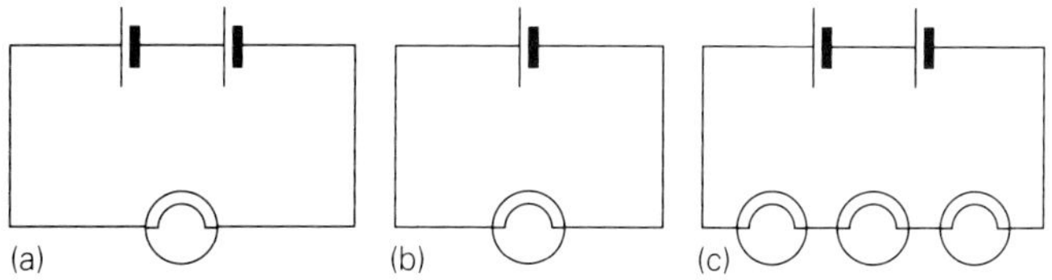

In circuit (b) the current may be about 0.2 A, but in circuit (a) the current is greater, probably over 0.3 A. It is important to remember that this current flows all round the circuit, which includes the two cells. Thus over 0.3 A flows through each cell in circuit (a) and only 0.2 A through the cell in circuit (b). For this reason both the cells in circuit (a) will run down faster than the cell in circuit (b). The current is least in circuit (c), so those cells will last longest.

In other words, to decide which cell will last longest, you merely have to find which cell has the smallest current flowing through it.

Of course, a cell contains a fixed amount of chemical energy. There is not an unlimited supply, and with more current you transfer the chemical energy at a greater rate, so the cell 'wears out' more quickly.

Cells in parallel

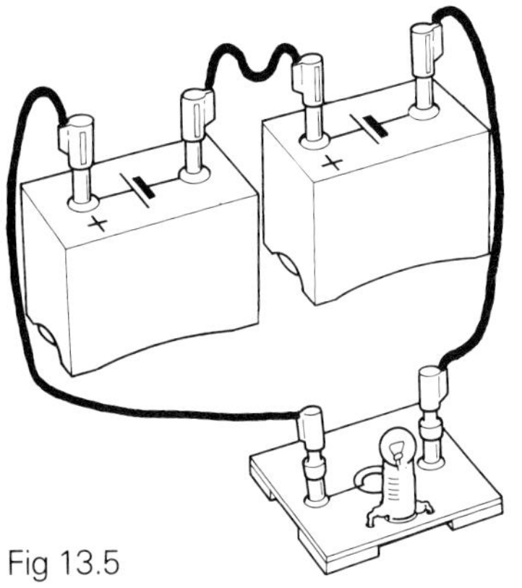
Fig 13.5

Until now we have considered cells only in series, as shown in Fig 13.5 and the circuits below.

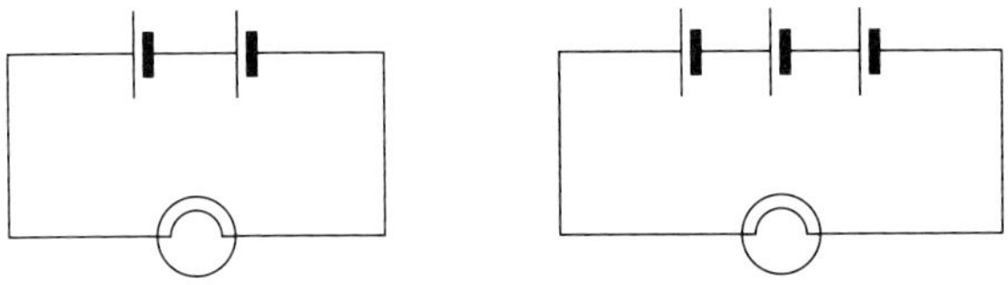

In the first circuit, the lamp glowed brighter than normal and in the second circuit very much brighter.

What would be the effect of putting similar cells in parallel, as shown in Fig 13.6 and the circuits below?

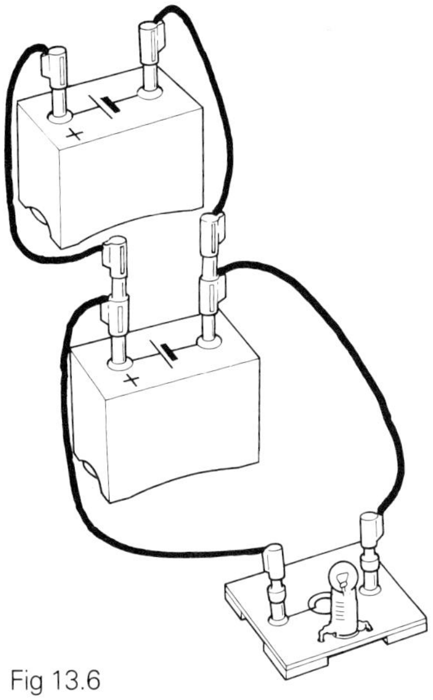
Fig 13.6

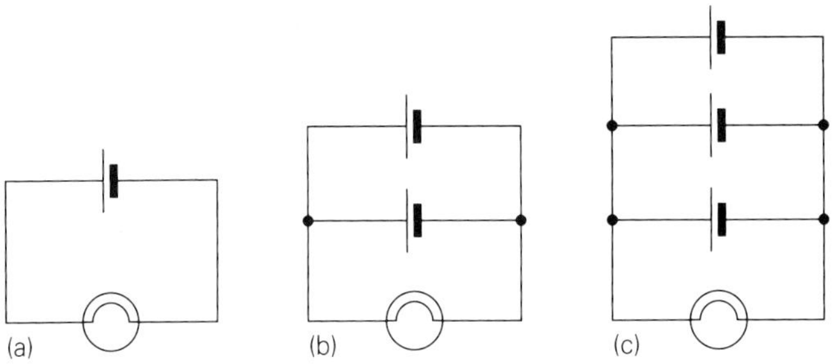

As far as the lamp is concerned, it would be just the same as putting one cell across it. In each case, it would merely glow with normal brightness. The difference would be the amount of current taken from each cell. In circuit (a), the current through the lamp might be 0.2 A, and 0.2 A would flow through the cell. In circuit (b), the current through the lamp would still be 0.2 A, but only 0.1 A would flow through each cell. What would be the current through each cell in circuit (c)?

Experiment 13.3 Resistance

Set up the circuit in Fig 13.7 so that different things can be put into the circuit between the clips X and Y. Try each of the items in the list below. First guess what will happen to the reading of the ammeter and then see if your guess is correct.

a A length of copper wire and then the same length of very much thinner copper wire.
b A strip of copper and then a strip of paper.
c A short length of eureka wire (resistance wire) and then a longer piece of the same wire.
d A length of eureka wire and then eureka wire of the same length, but larger in diameter.

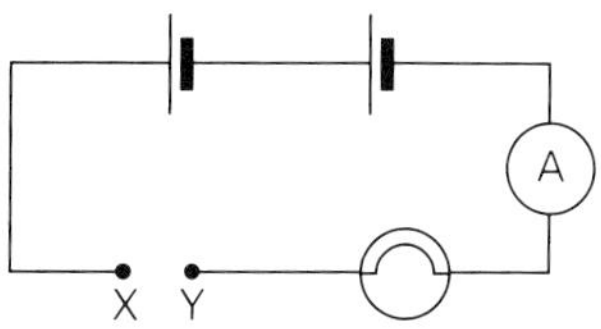

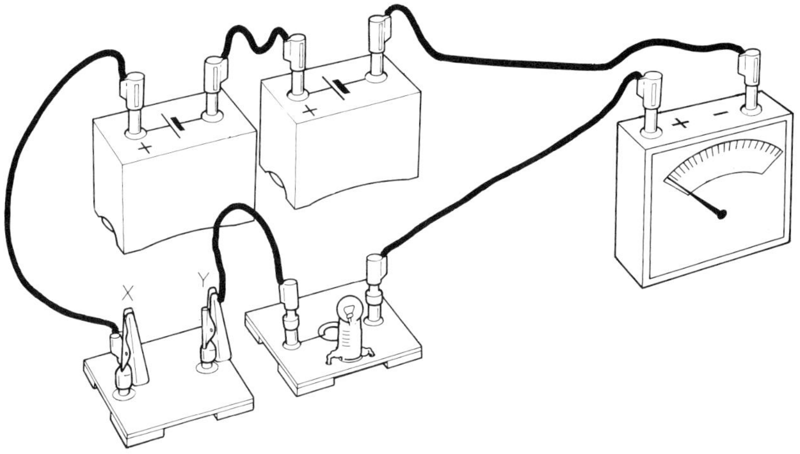

Fig 13.7

Paper has a very high resistance. For metal wires, a long length has more resistance than a short piece; and for the same length of the same material, thick wires have less resistance than thin wires. (You get the same thing when water flows through pipes: it is easier for water to flow – there is less resistance – when the pipe is wider than when it is narrower.) Copper is a better conductor than eureka.

The experiments will have shown of course that the higher the resistance the smaller the current.

Experiment 13.4
The heating effect of a current

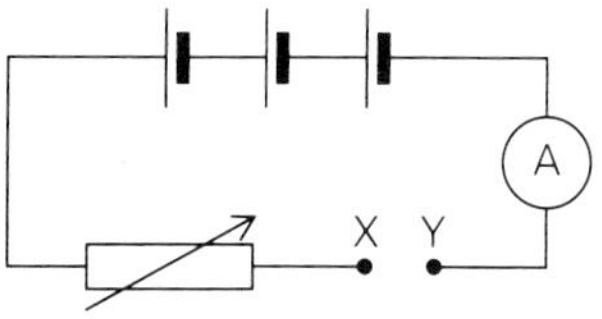

a Set up the circuit on the left using three cells, an ammeter, a variable resistor and clips at the points X and Y. Between the two clips stretch a few strands of steel wool. Start with a small current. Gradually increase the current by reducing the resistance. The steel wool will be found to get hot. The greater the current the hotter it will get.

b Try blowing hard on the wire so that it is cooled. You should notice that the ammeter reading increases. In other words the resistance depends considerably on temperature: the greater the temperature of the steel wool the greater the resistance.

c If only one strand of steel wool is used, it may get so hot that the circuit is broken and no current flows.

This is the principle on which a fuse is made. In a fuse, a very fine wire melts if it gets too hot. A fuse is included in a circuit to prevent too great a current causing damage. When the current is too great, the wire melts, the circuit is broken and current can no longer flow.

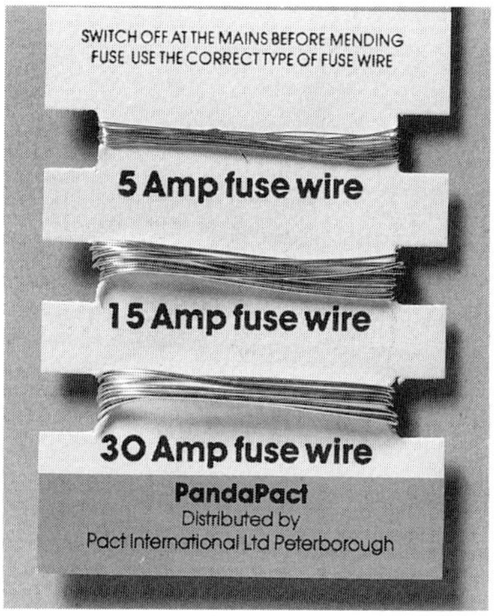

Fuse wire

The symbols used for a fuse in circuit diagrams include those shown above. The photograph shows typical fuse wire and

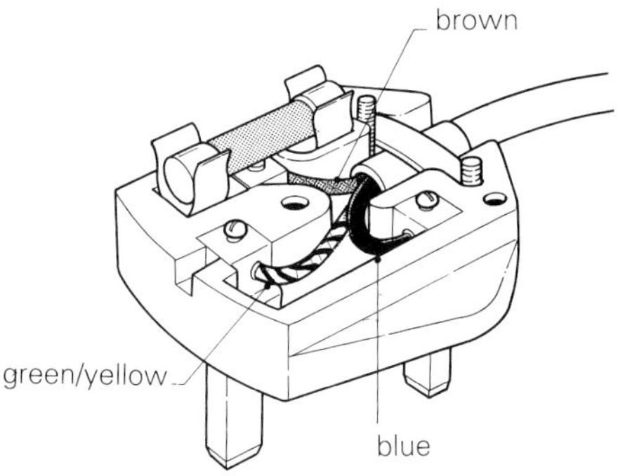

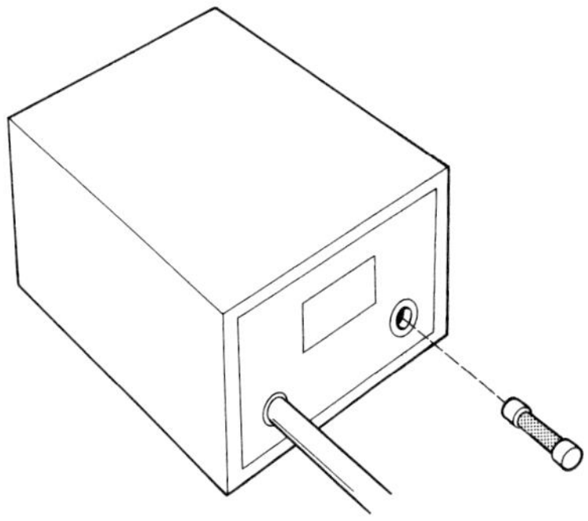

Fig 13.8

Fig 13.8 shows the small cartridge fuses used in plugs and apparatus. The cartridge fuse has a fine wire inside which melts as soon as the current exceeds the stated value.

Experiment 13.5 Testing fuses

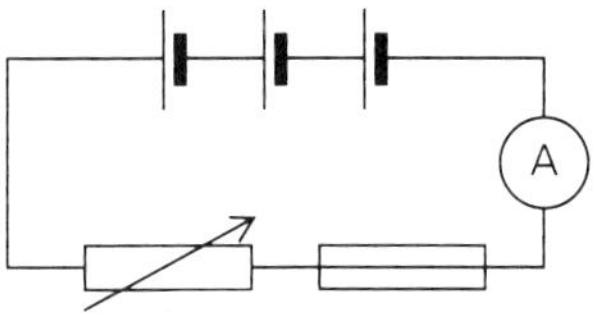

To test a cartridge fuse, set up a circuit with three cells, an ammeter, a variable resistor and a $\frac{1}{4}$-ampere fuse held between two clips.

Start with a small current. Then decrease the resistance, watching the ammeter reading until the fuse 'blows'.

Electrical pressure

Suppose you have two syringes, A and B, with a tube fixed between them, and that the space inside is full of water (Fig 13.9).

Fig 13.9

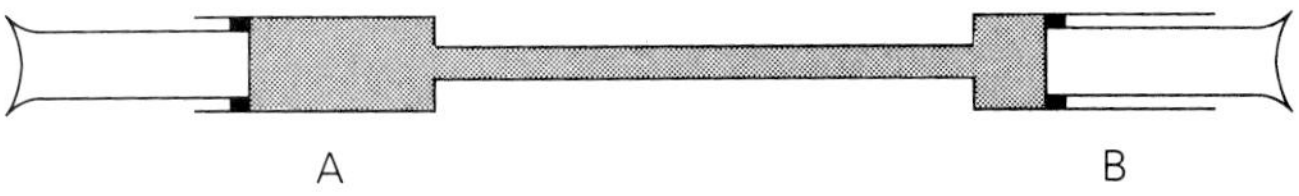

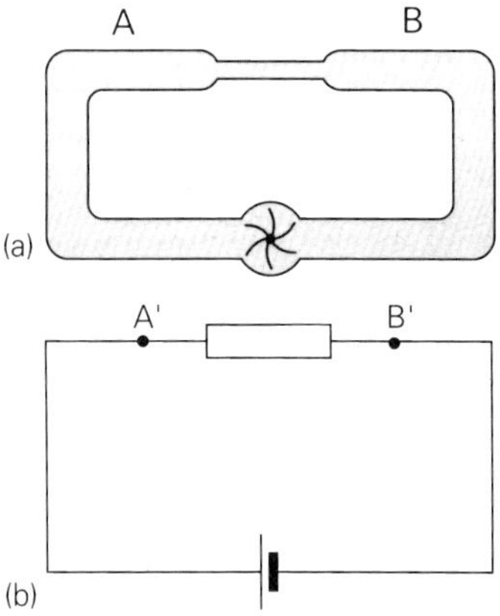

Fig 13.10

You can make water pass from A to B through the tube by pressing the piston of syringe A. Doing this makes the pressure in A greater than the pressure in B and this *difference* in pressure causes water to flow through the tube. The harder you push on the piston of A, the greater is the pressure difference and the faster the water flows so that there is a bigger current. Of course, the current will stop when you stop pushing or when the piston cannot move any further.

If a current is to flow all the time, then a water pump has to be used to maintain a difference in pressure between A and B.

This is just like an electrical circuit in which there is a cell (an electricity 'pump') causing electric current to flow through a resistance wire (Fig 13.10). The cell pumps electric charge through the wire, and we call that movement of charge an electric current. The cell causes a difference of electrical 'pressure' between A′ and B′, and the effect of this is an electric current flowing from A′ to B′ through the resistance wire.

If the pipe between A and B in the water circuit is shorter, then the same difference of pressure between A and B causes a larger current, because it is easier to push water through a short pipe than through a long one. And the electrical circuit is like that, too – a shorter piece of resistance wire allows a larger current to flow. What do you think would be the effect of using a fatter wire?

There are other similarities between water circuits and electrical circuits.

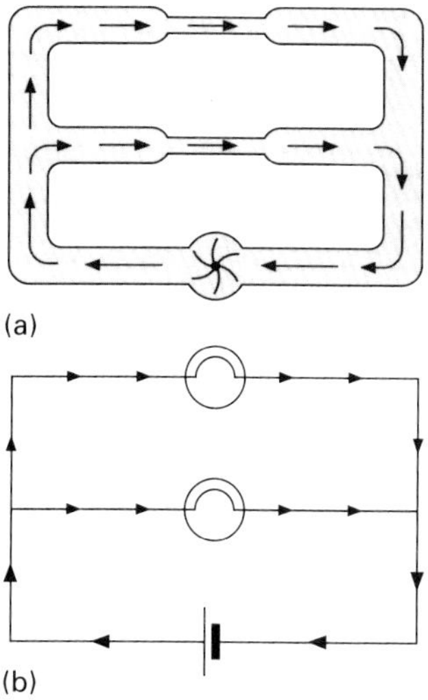
(a) (b)

Fig 13.11

1 If the circuits have parallel branches, some current flows through each branch. For both circuits in Fig 13.11, the sum of the currents flowing in each branch equals the current flowing from the pump.

2 If one of the pipes had a stop-tap in it, we could use the tap to stop the water current even with the pump still working. The water pressure on one side of the tap would then be bigger than on the other side, but water could not pass through the tap.

Electrical circuits can be like that if we put a switch in the circuit: there is no current when there is a gap in the circuit, but there is a pressure difference across the switch.

3 Notice, too, that the water pump does not create water! The water is there all the time, and all the pump does is to move it through the pipes. So it is with the electrical circuit. The battery does not manufacture electric charge. The electric charge is in the wires all the time, and all the cell does is to move the charge through the wires.

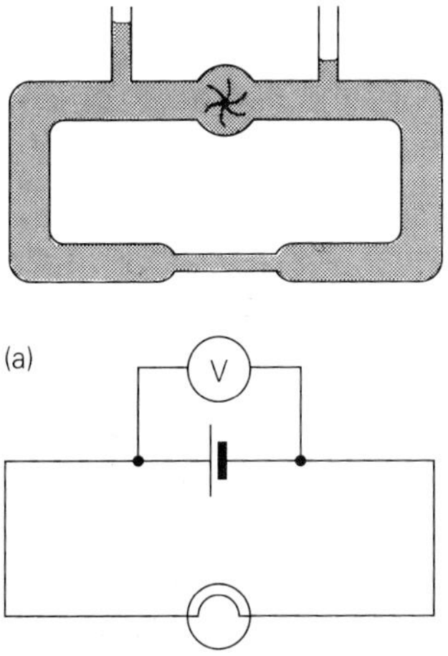

(a) (b)

Fig 13.12

Measuring electrical pressure difference

In a water circuit we could measure the difference in pressure between the two sides of the pump by inserting a long tube on each side of it (Fig 13.12a). On one side of the pump, the pressure would be 'high' and on the other side 'low', and the difference in pressure would be measured by the difference in the levels of the water in the two tubes.

Electrical pressure difference is measured with an instrument called a **voltmeter** (and in a unit called a volt, often abbreviated to V). The circuit diagram symbol for it is shown overleaf. As with the water circuit, the voltmeter has to be connected between the points whose pressure difference we require; that is to say, it is connected in parallel with the cell, or *across* it (Fig 13.12b). Notice that this is different from the ammeter which is connected in a circuit in series.

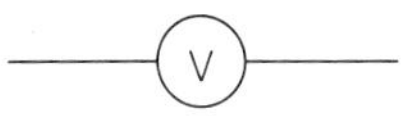

Like an ammeter, the voltmeter must be connected the right way round. One terminal is usually coloured red (or marked +) and this should be connected to the positive side of the cell, or the point where the pressure is higher. The other terminal is usually black (or marked −), and this should be connected to the negative side of the cell, or the point where the pressure is lower.

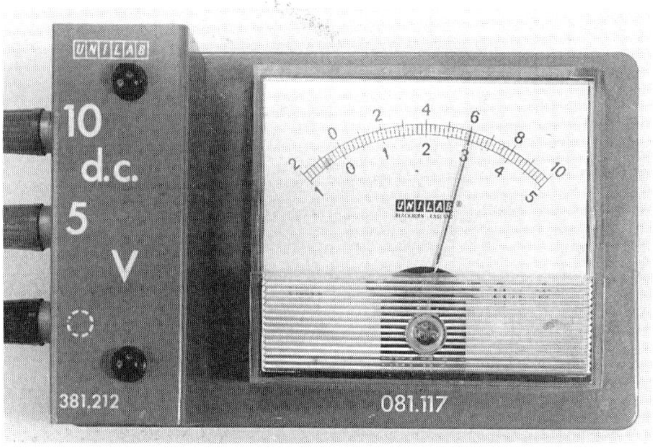

Analogue voltmeter

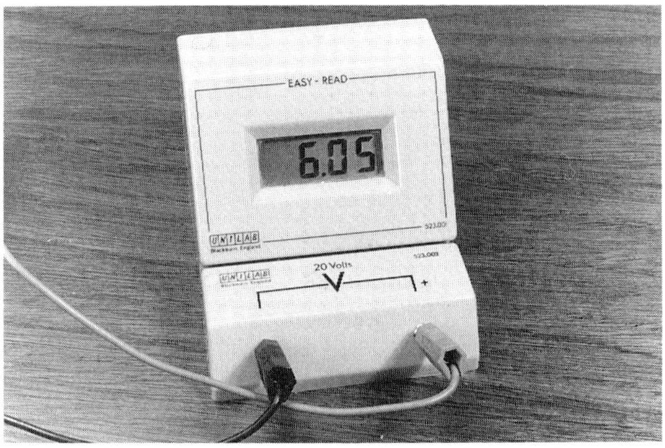

Digital voltmeter

Experiment 13.6 Using a voltmeter

a Connect three cells in series as shown on the left. Measure the pressure difference or voltage between P and Q with a voltmeter. Note its reading.

b Now measure the voltage between Q and R, and then between R and S. The readings should all be close to 1.5 volts.

c Then measure the voltage of two cells (between P and R, or between Q and S). Finally, measure the voltage of three cells in series (between P and S).

Each of the cells you have used produces a pressure difference of about 1.5 volts, or 1.5 V. A battery of two cells in series produces twice as much, about 3.0 V, and a battery of three cells gives 4.5 V. Batteries using this type of cell and giving 3, 4.5, 6 or 9 V can be bought. Different kinds of cell can also be obtained; one type produces 1.2 V, another 2.0 V, for example. See if you can find out what voltage a car battery gives, and how many cells there are in it.

Putting cells in series to form a battery produces a more powerful 'pump', and more powerful pumps cause more current to pass through a circuit. The voltage from a cell or

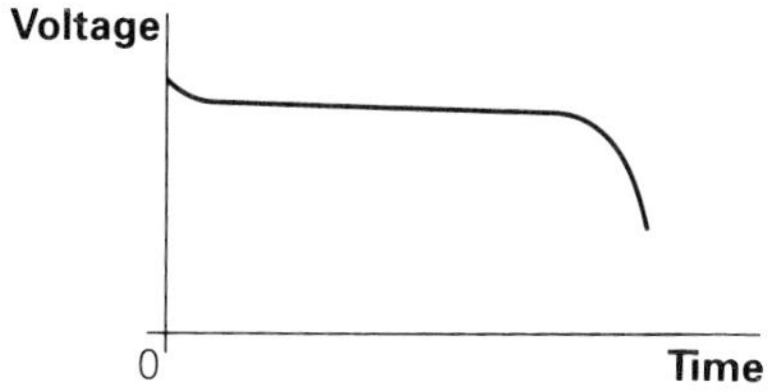

Fig 13.13

battery remains fairly steady over most of its life and then falls towards the end of it. A graph of its voltage against time might look like the one in Fig 13.13, though how rapidly it falls depends on the current which flows through it.

Experiment 13.7
Investigating how the current through a lamp changes with the voltage across it

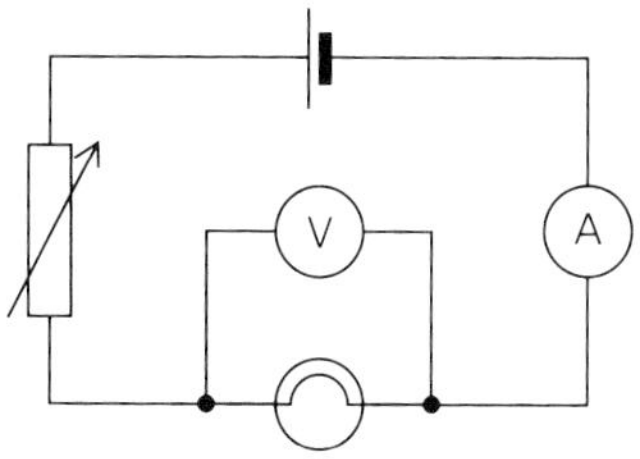

Current in amperes	Voltage in volts

a Connect the circuit shown so that the voltmeter measures the pressure difference or voltage across the lamp.
Change the current in the circuit and notice what happens to the voltage as the current changes.

Larger currents require larger pressure differences across the lamp, but you may have noticed that, if the voltage is doubled, the current increases by *less* than a factor of two. You may be interested to find out how the voltage varies as the current changes.

b Copy the table. Set the current at a low value and measure the voltage. Enter these corresponding values in your table. Now increase the current slightly. Measure the voltage again and record those values. Do this about five times to obtain different sets of readings and then draw a graph to display those readings.

Doubling the voltage across the lamp does not cause the current to double because the lamp filament gets hotter as the current increases. And a hotter filament has a bigger resistance. Doubling the voltage only causes the current to double if the resistance of the conductor does not change.

Questions for homework or class discussion

1 A lamp connected to a cell as in circuit (a) glows with normal brightness.

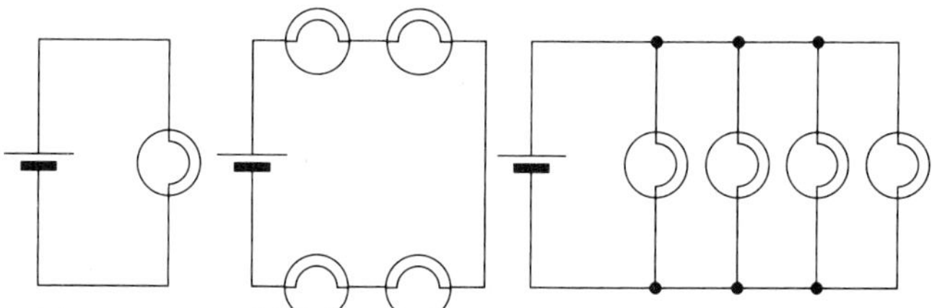
(a) (b) (c)

What would happen to the brightness if four similar lamps were connected to the cell as in circuit (b)?

What would happen to the brightness if the four lamps were connected as in circuit (c)?

In which of the three cases would the cell run down most quickly? Give the reason for your answer.

2 In these circuits, all the cells are similar and all the lamps are similar. Assume in each case that all connections are made with very good conducting wire.

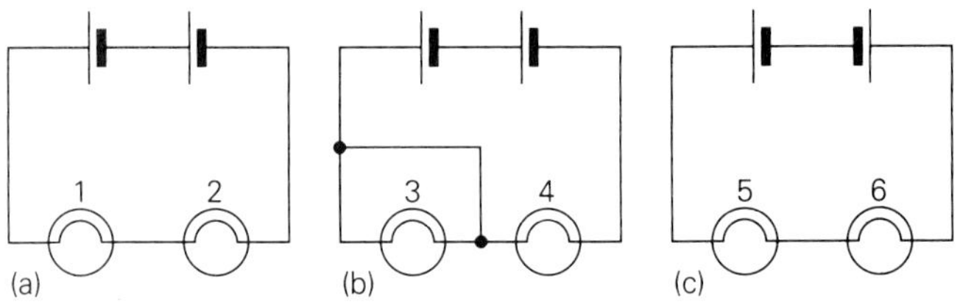

In the first circuit the lamps 1 and 2 glow with normal brightness. State whether the lamps 3, 4, 5 and 6 will be brighter than normal, normal brightness, less than normal brightness or will not glow at all.

In which case will the cells run down most quickly? Give a reason for your answer.

3 Two similar model boats A and B are fitted with similar electric motors to drive the propellers. Similar cells are used to drive the motors, but the circuits used are as shown below.

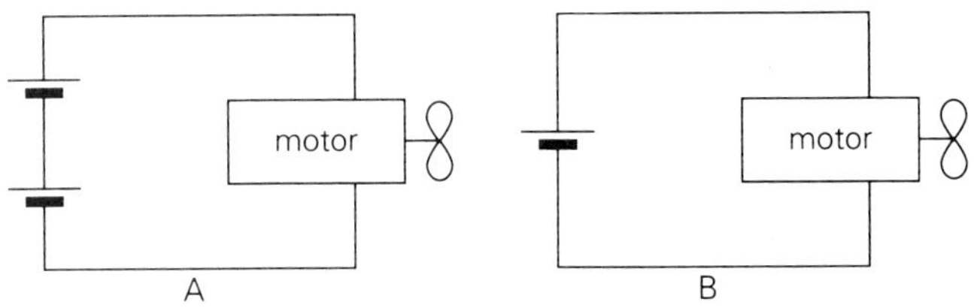

Which of the two boats is likely to go faster? Give a reason for your answer.

In which of the two boats will the cell or cells run down faster? Give a reason.

4 In this circuit, the battery X is used to light the two lamps P and Q. It is found that P glows more brightly than Q.

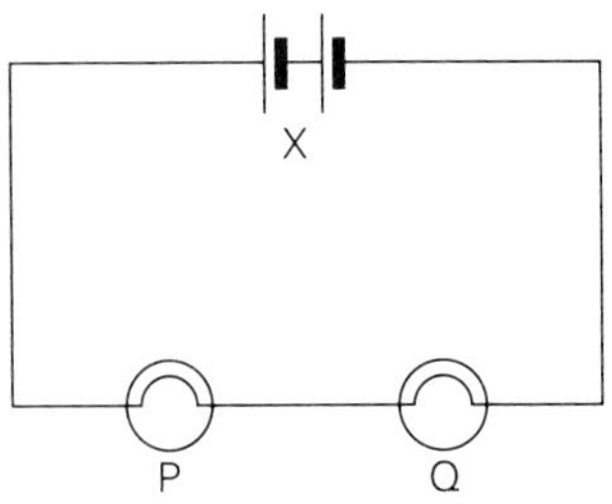

Someone suggests that this is 'because the current decreases as it moves round the circuit'. How would you show this to be either correct or incorrect

a if you had a suitable ammeter,

b if you had no other apparatus?

If you think the suggestion was incorrect, give a more likely explanation.

5 Describe what will be the effect of decreasing the resistance of the variable resistor in each of the following circuits.

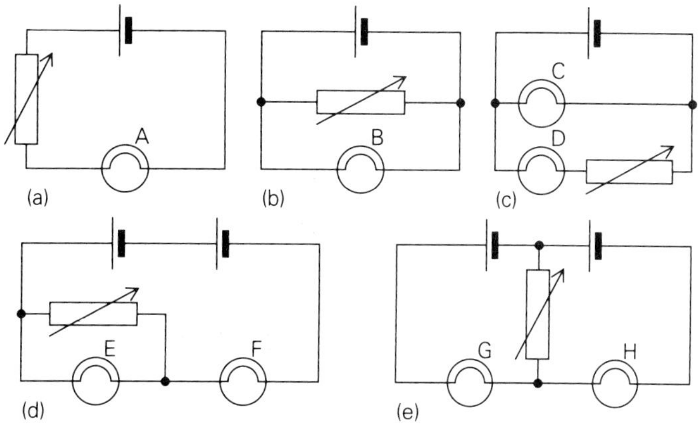

6 Explain why the circuit below is a 'fool's circuit'.

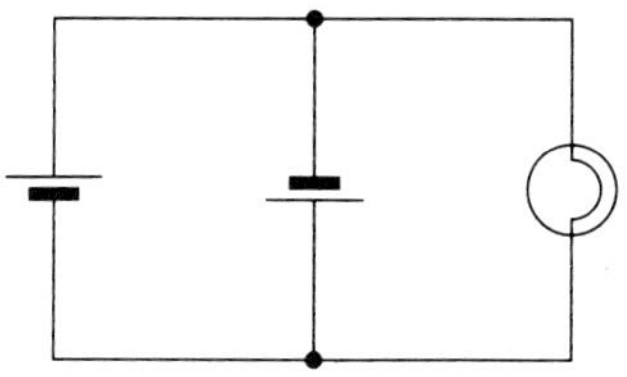

7 Describe what will be the effect of closing the switch in each of the following circuits.

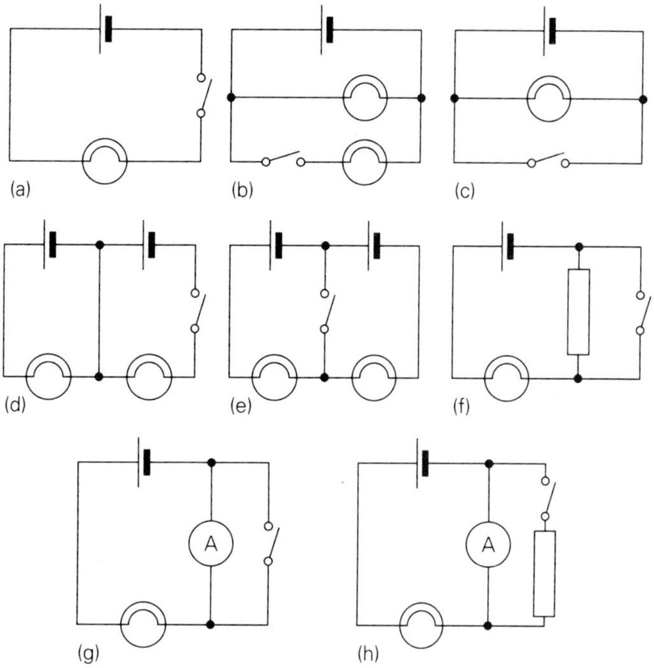

Background Reading

Electricity in the home – 2

The current which operates all the electrical devices in our homes and factories flows via the live and the neutral wires. This is because the generators in the power stations provide a varying voltage, and the electricity boards supply that to us using the live and neutral wires at a voltage of 240 V. This is a special kind of average value – the actual voltage varies between + 340 V and – 340 V (Fig 13.14). This is a high voltage and could be dangerous if safety measures were not taken.

Fig 13.14

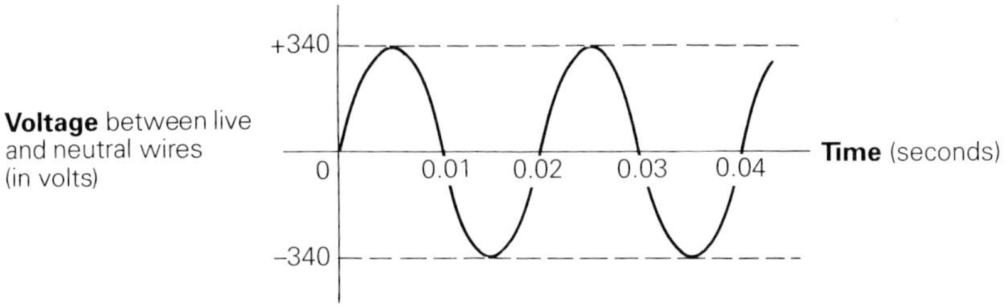

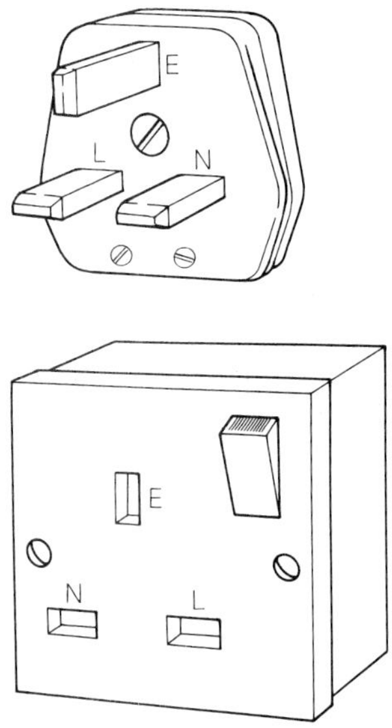

Fig 13.15

One of the dangers is that of fire. If the current through a cable is too high, the live and neutral wires in the cable become hot. They could become hot enough to melt the plastic insulation round them and set it on fire. Since the cables in a home often run under the floors and along roof rafters, it is very important to prevent the cables getting too hot. The safety device in this case is a fuse which is placed in the live lead and which melts when the current exceeds a certain value, so breaking the circuit.

Another danger is that of electric shock. To provide some protection against this danger, the electric sockets and plugs have a third wire called the **earth** wire (Fig 13.15). How that is used and how it works will be described later (Chapter 15, page 205). In what follows the earth wire will be omitted but it is, of course, an essential part of the installation.

The Electricity Board's installation

The cable bringing the supply to your home usually enters from underground, and goes to the Electricity Board's fuse box. This has a 100-amp fuse in it and is meant to protect the underground cables so that a serious fault in your home does not result in the supply to your neighbours being interrupted. The Board's fuse box is sealed and must not be opened by the householder. If the fuse in this box blows, a worker from the Electricity Board has to be called out to repair it.

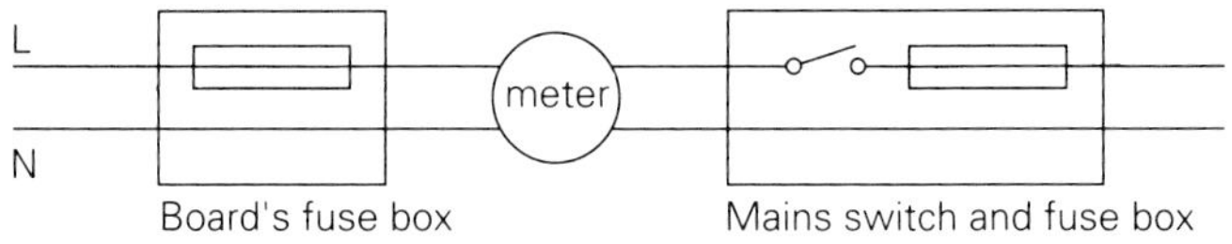

Fig 13.16

The wires then pass to the meter which records how much electrical energy is supplied, and from the meter the wires go to the mains switch and householder's fuse box (Fig 13.16). The mains switch controls the supply to the whole house – when it is 'off', no electricity is supplied to any lights or sockets.

The electricity is distributed to the rooms in the house from the fuse box. The leads from the mains switch are connected to two thick copper bars known as *bus bars*, from which leads go off to various circuits in the house (see Fig 13.17).

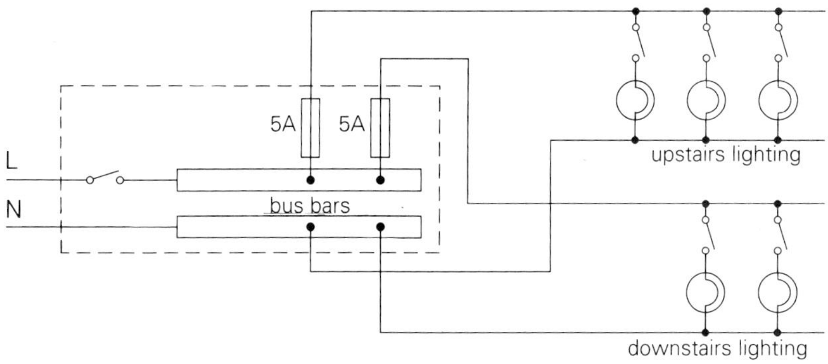

Fig 13.17

The lighting circuits are usually wired with 6-amp cable, (cable capable of carrying up to 6 A safely), and the fuses protecting these cables will be rated at 5 A. A 100-watt lamp will require a current of about 0.4 A, so that the number of lamps that can be used in parallel on a particular fuse is limited.

The ring main system

Power to the wall sockets in the house are supplied nowadays by means of a ring main. The sockets are meant to supply electricity to more powerful devices than lamps: things like electric fires, toasters and irons. The cable to carry the circuit is usually rated at 15 A, and it is wired in a ring as shown in Fig 13.18.

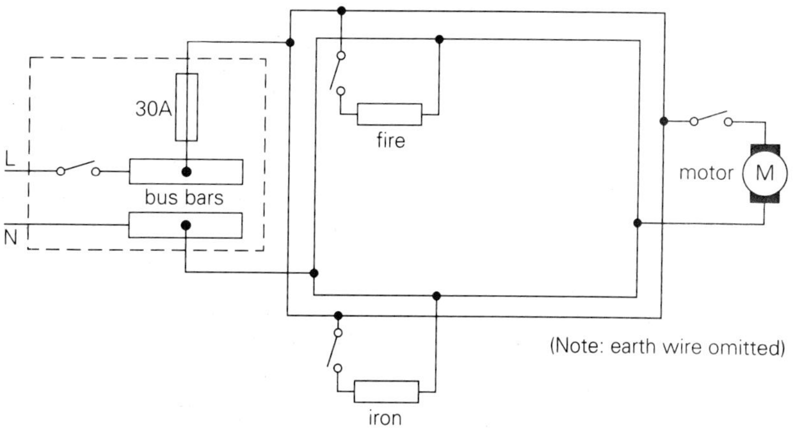

Fig 13.18

The fuse in the fuse box is then rated at 30 A because there are two routes to any one appliance, and when the cable is carrying its limit of 15 A, the total current will be 30 A (see Fig 13.19).

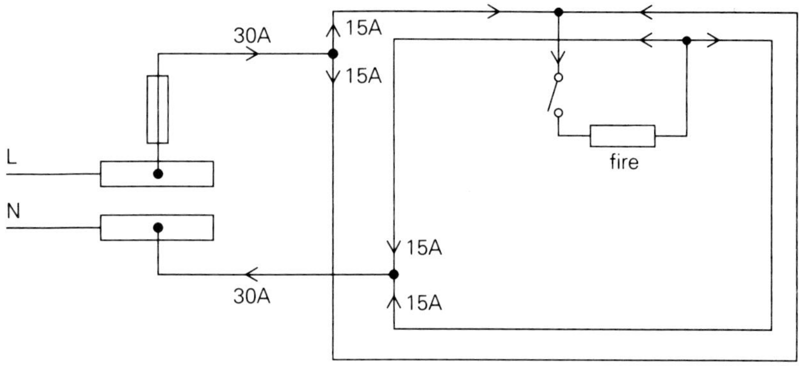

Fig 13.19

A house may well have two or three ring main systems to supply the sockets in the living rooms, kitchen, bedrooms, garage, *etc.* There will also be special circuits for the devices which take a lot of current – the immersion heater and the electric cooker – and you must not forget that all these circuits have an earth wire as well.

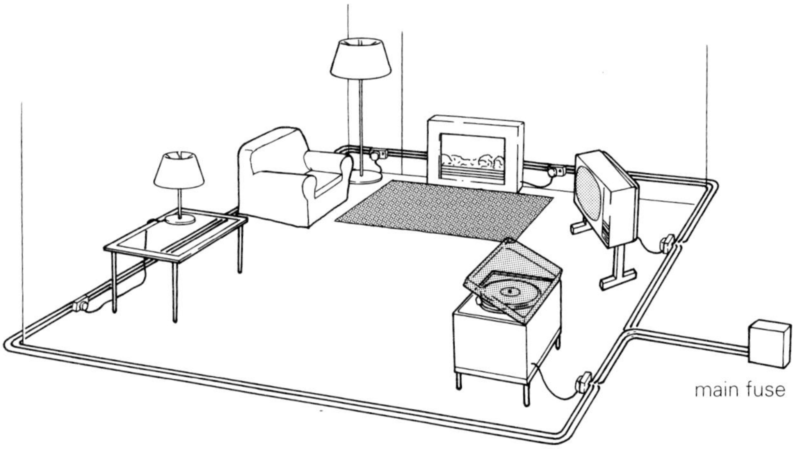

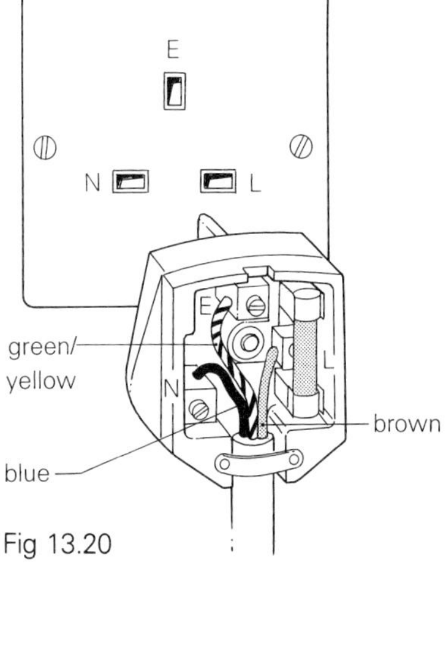

Fig 13.20

The fused plug and socket

An appliance is connected to the ring main by means of a three-pin plug with a fuse in it. The fuse, which is in the live lead, is meant to protect the cable from damage due to too much current and also to isolate the appliance should a certain type of fault develop in it (see Chapter 15 page 206).

Notice how the plug is wired (Fig 13.20). The insulation of the wire connected to the live pin of the plug via the fuse is coloured brown, the insulation of the wire to the neutral pin is coloured blue and that of the earth lead is coloured green and yellow. Of course, the wires must be bared of insulation where they are screwed to the pins. The strands of each lead should be twisted together so there are no stray ends and carefully trimmed to the correct length so that there is a minimum of bare wire showing. When the wires have been firmly screwed to the pins, the cable should be securely clamped by the cable clamp.

The rating of the fuse to be used in such a plug depends on the appliance being supplied. Mains fuses are made with ratings of 1 A, 2 A, 3 A, 5 A, 7 A, 10 A or 13 A, and for the greatest safety, the rating should be just above the current required by the appliance, always assuming that the cable is capable of carrying that current. So an electric fire which requires 4 A should have a fuse rated at 5 A, and a hair-drier requiring 2.5 A should be fused at 3 A.

Chapter 14 Some electrical components

In this chapter we shall consider a number of electrical components, many of which play an important part in modern electronics.

The diode

The diode is sometimes called a **rectifier**, but in this book it will always be called a **diode**.

Experiment 14.1
The diode

a Set up a circuit with two cells, a lamp and two clips as shown in Fig 14.1. Complete the circuit by connecting the diode between the two clips. What happens to the lamp?

b Now turn the diode round. What happens to the lamp this time?

c You might try the following circuits. Each time when you have set them up, turn the cells round to see what difference it makes. Can you explain what happens?

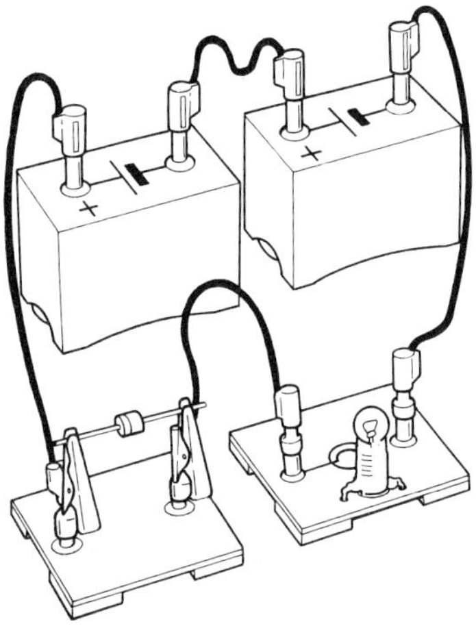

Fig 14.1

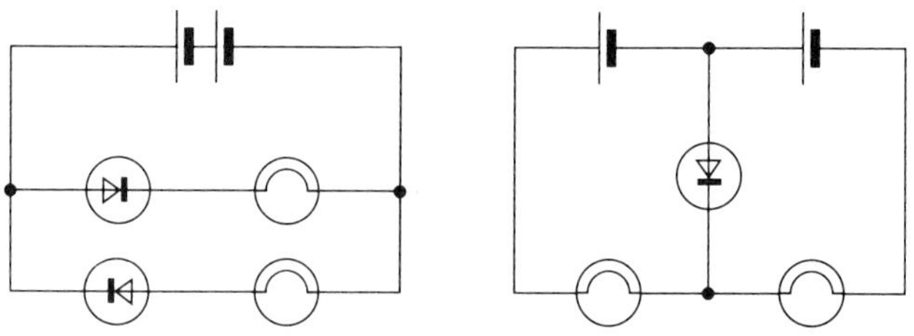

The diode has a very high resistance one way round (it is virtually an insulator) and a low resistance the other way round. In effect it allows current to travel one way through it, but not the other way.

In circuit diagrams a diode is shown by the symbol below.

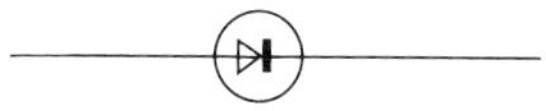

The current flows in the direction of the arrowhead and it will not flow the other way.

The light emitting diode (LED)

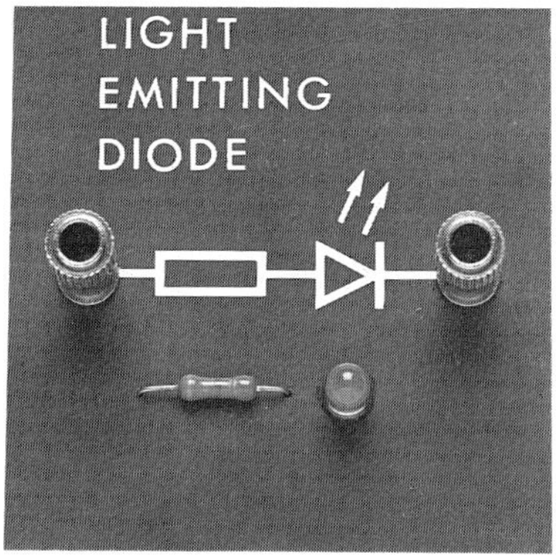

The **L**ight **E**mitting **D**iode, or 'LED' as it is usually called for short, is an inexpensive device widely used in electronic circuits in order to show that a current is flowing. The circuit diagram symbol for an LED is shown below.

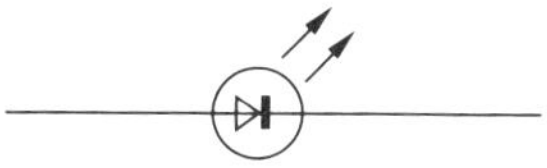

As an LED can be damaged if too big a current flows through it, the LED module has a resistor in series with the LED and this prevents the current from getting too large (see the photograph). In all the circuit diagrams which follow, that resistor is always shown.

Experiment 14.2
The light emitting diode (LED)

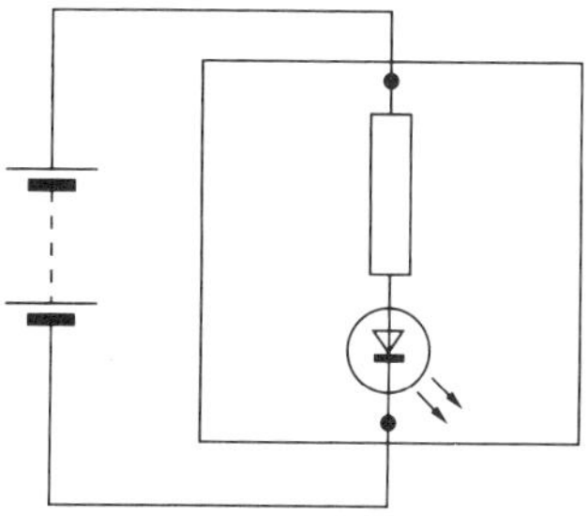

a Connect an LED module and a battery as shown on the left.
b What happens to the LED when the circuit is connected?
c Change round the connections to the LED module. What happens this time?

An LED will allow current to pass through it in only one direction. As with an ordinary diode, the arrowhead in the circuit diagram symbol points the way in which current can flow. But the LED differs from the ordinary diode since the passage of current through it causes light to be given out.

Experiment 14.3
Brightness and current

a Connect the circuit shown here using the LED module, the resistor module (see the photograph) and a battery. The resistor module has separate connections so that the resistance can be high, medium or low. Use the low value first. Notice how brightly the LED glows.

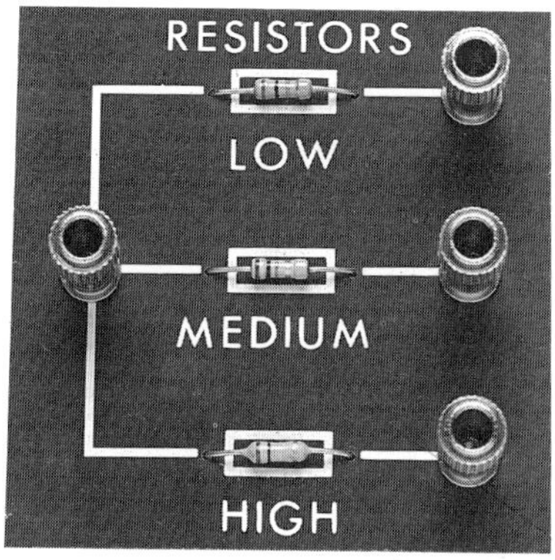

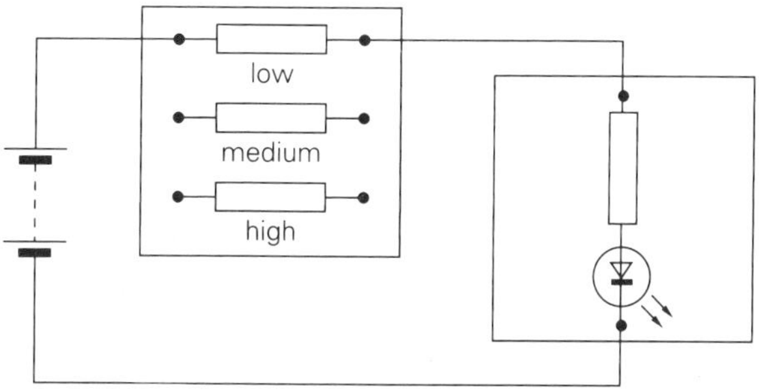

b Now use the medium resistance in place of the low one. What happens to the brightness of the LED?

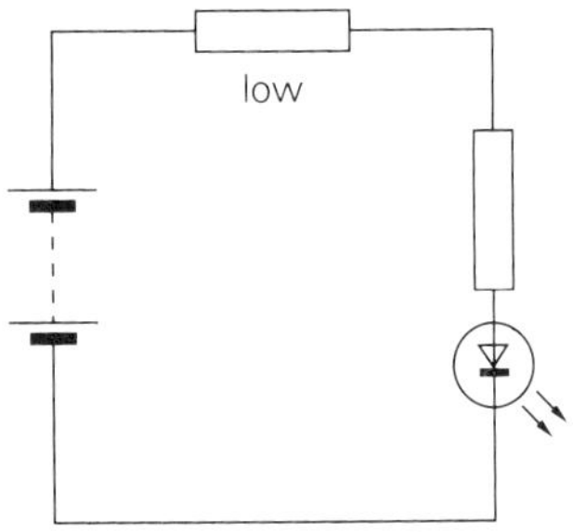

c Finally use the high resistance. What happens to the brightness this time?

d As the resistance gets greater, so the current in the circuit gets less. How does the brightness of the LED depend on the current passing through it?

The circuit diagram on the left shows how the arrangement in the experiment on the previous page would normally be drawn.

All you need to know about resistors at this stage is that they can have different resistances. Resistance is measured in units called **ohms.** The values of the resistances in the resistor module are about 27 thousand ohms ('high'), about 2.7 thousand ohms ('medium') and about 270 ohms ('low'), but it is not necessary for you to remember this.

Experiment 14.4
LEDs in parallel

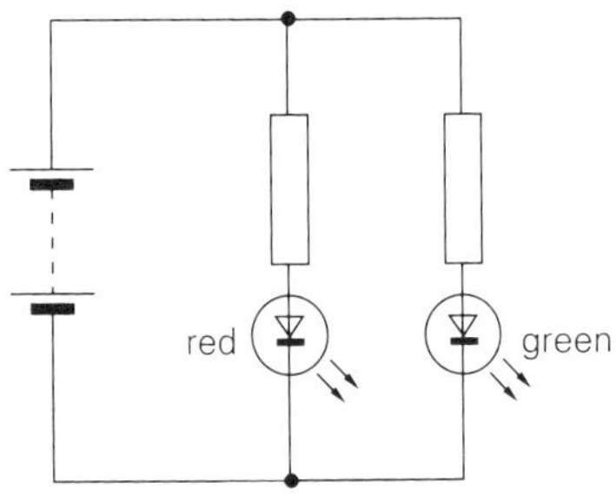

a Set up the circuit shown here using one red LED module, one green LED module and a battery. Why do both LEDs glow?

b If another LED were connected in parallel with the two on the left, how brightly would you expect it to glow? Test your answer with another LED.

Positive and negative supply rails

Parallel connections are often used in electronic circuits. The last experiment is a useful introduction to the idea of positive and negative supply rails, as shown in the circuit below.

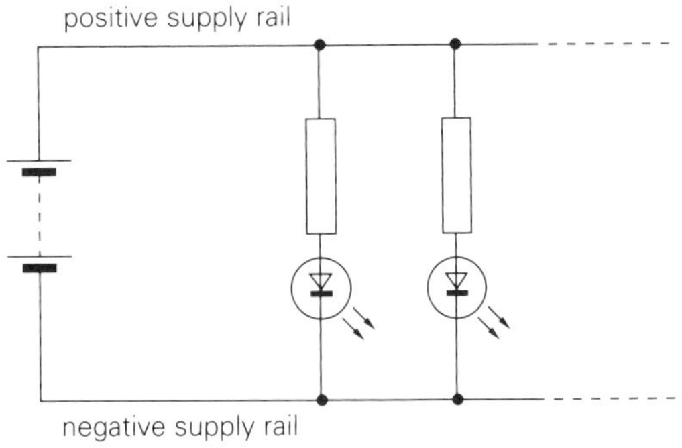

If an LED is connected between the positive supply rail and the negative supply rail, a current will flow through it (assuming, of course, that it is connected the right way round).

Project
A current direction indicator

In this chapter there are a number of projects which you may like to try to solve for yourself, though of course you can get advice from your teacher if necessary. However it is much more fun to do it on your own.

The problem here is to use a red LED module and a green LED module to construct a current direction indicator. It should be such that when the battery is connected one way round the green LED glows; and when the battery connections are the other way round the red LED glows.

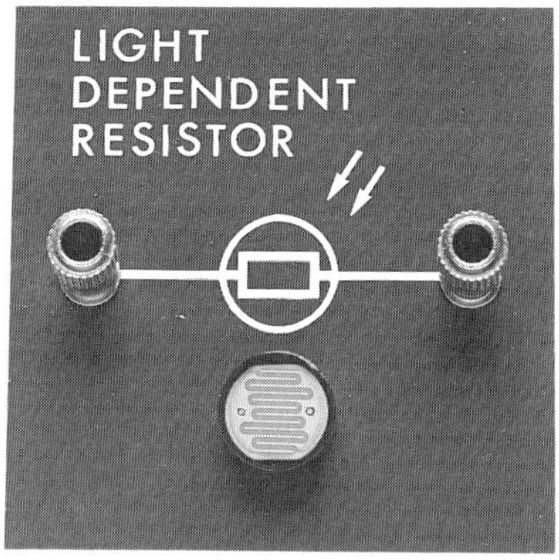

The light dependent resistor (LDR)

In circuit diagrams, the **L**ight **D**ependent **R**esistor, or LDR, as shown in the photograph, is represented by the symbol below.

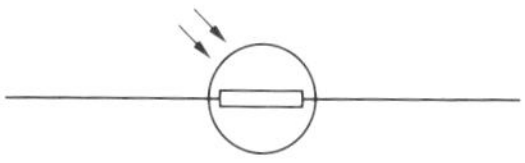

Experiment 14.5
The light dependent resistor (LDR)

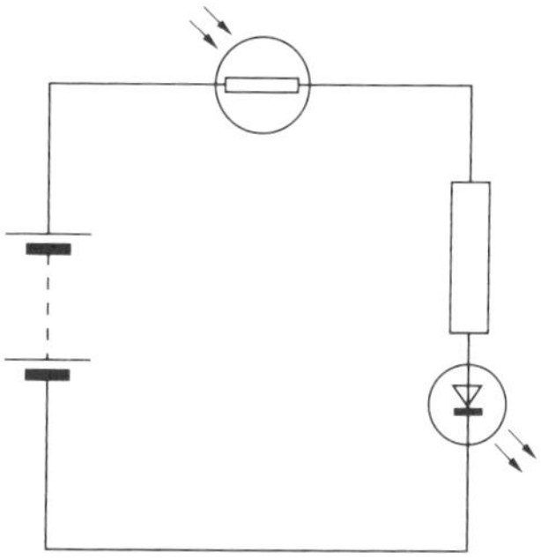

a Set up the circuit shown on the left using a light dependent resistor module in series with an LED module and a battery.

b What happens to the brightness of the LED when light is shone on the LDR? What happens when the LDR is covered up?

c What does this tell you about the resistance of the LDR in the light and in the dark?

d Does the circuit behave any differently if the connections to the LDR are changed round?

The experiment shows that current will flow in either direction through the LDR, as with an ordinary resistor. However, when it is dark, the LDR has a high resistance and therefore allows little current to pass. In bright light, the LDR's resistance falls to a low value and a much bigger current can flow. The resistance of the LDR is perhaps a million ohms in the dark, falling to about 100 ohms in a bright light.

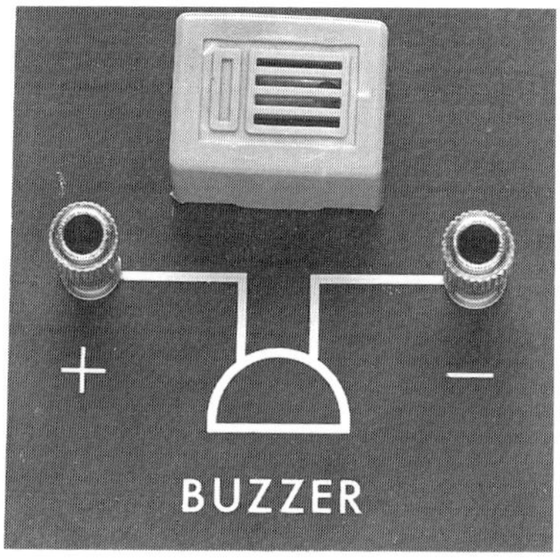

The buzzer

When an electric current passes through the buzzer (see the photograph), it makes a sound. In circuit diagrams it is represented by the symbol below.

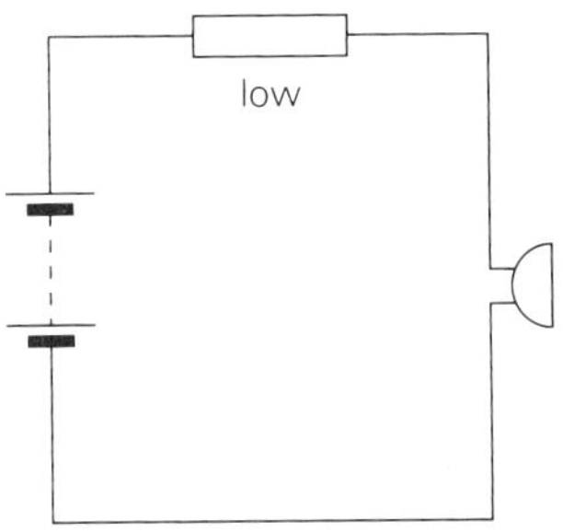

Experiment 14.6
The buzzer

a Set up the circuit shown here, using the buzzer module, the resistor module and the battery.
b Investigate whether it makes any difference which way round the buzzer is connected in the circuit.
c Replace the low value resistor with the medium and then the high value resistor. What effect does this have on the operation of the buzzer?

Project
A very simple burglar alarm

The problem is to use an LDR module, a buzzer module and a battery to make a circuit which will sound an alarm when a light is switched on. If burglars were foolish enough to turn on the light in a room they had entered, or if the light from their torch fell on the LDR, the circuit could be used to warn the householder.

The motor

The symbol used in this book for the motor in a circuit diagram is shown below.

Experiment 14.7
The motor module

a Connect the motor module to the battery. What happens when the current flows?
b Reverse the battery connections to the motor. What difference does it make?

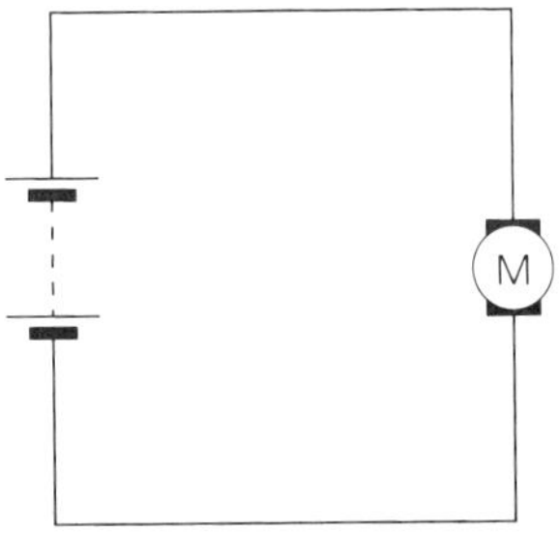

This experiment shows that the motor can be driven in either direction depending on the direction in which the current flows through it. It is not always easy to see which way the motor in the module is rotating. You could add a small propeller made from thin card to help. An interesting extension to this experiment would be to add a green and a red LED to the circuit in such a way that the red LED lights when the motor rotates one way and the green LED lights when it rotates the other way.

Switches

Switching plays a very large and important part in modern electronics. Computers, for example, use electronic switches of

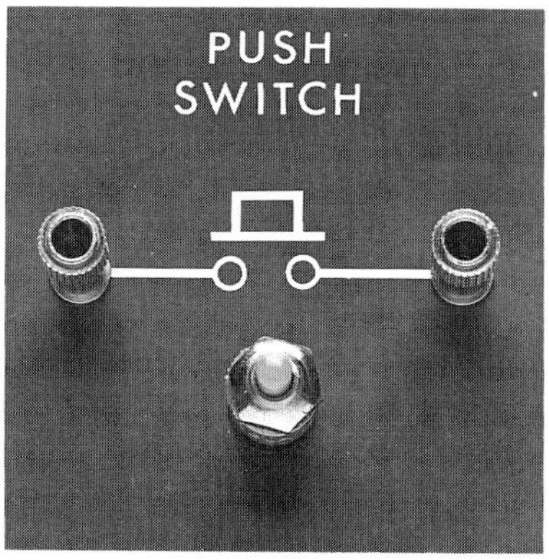

various kinds, and the fact that computers can be used to control robots or aeroplanes is a result of their ability to switch things on and off.

The simplest switch is an on/off switch. The symbol for such a switch in a circuit diagram is shown below on the left.

We know that a current will not flow in a circuit which has a gap in it: when the switch is open, there is a gap, but when it is closed the circuit is completed and a current flows.

The push-button switch is a type of on/off switch that makes contact when the button is pressed. The symbol used for it in this book is shown below on the right.

(a) (b)

Experiment 14.8
Circuits with switches

a Connect a battery, an LED module and a push-button switch in series so that you can switch the LED on or off.

b Now connect the battery, two LED modules in parallel and one switch so that you can switch both LEDs on or off at the same time.

c Finally connect the battery, two LED modules and two switches so that one switch operates one of the LEDs and the other switch operates the other LED.

d When you have arranged your apparatus to work as described in **c**, draw a circuit diagram of it.

The pressure pad as a switch

The pressure pad shown in the photograph is another form of on/off switch, in which the contacts are normally open. In other words, the two leads coming from it are not connected. As soon as any pressure is applied to the pad (by stepping or sitting on it), the switch is closed and the two leads are connected.

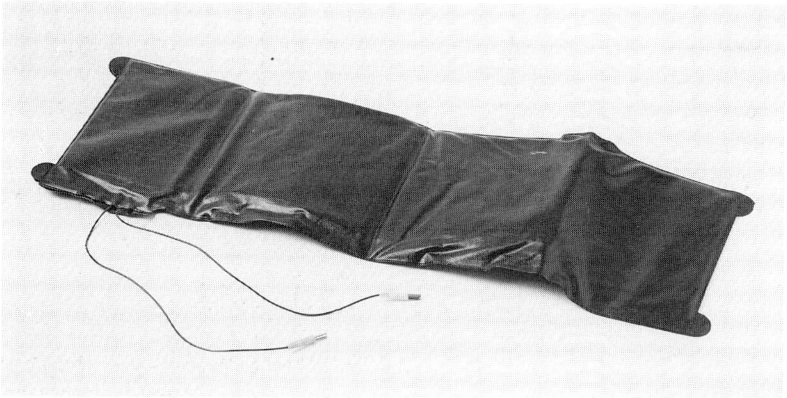

Projects with a pressure pad

1 Arrange a pressure pad, buzzer and battery so that the buzzer sounds whenever someone sits on a chair (it would be best to hide the pressure pad under a cushion).

2 Arrange a pressure pad under a mat outside the door of a classroom so that a warning is given when the teacher is approaching.

Experiment 14.9
Simple AND circuit

a Connect two push-button switch modules in series with an LED module and a battery as shown below.

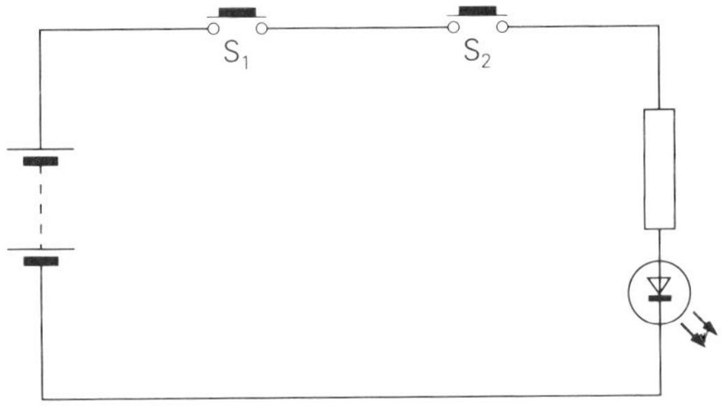

b What happens when S_1 alone is pressed?
c What happens when S_2 alone is pressed?
d What happens when both switches are pressed at the same time?
e Why do you think this is called an AND circuit?

Experiment 14.10
Simple OR circuit

a Using the same modules as in the last experiment, set up the circuit shown here.

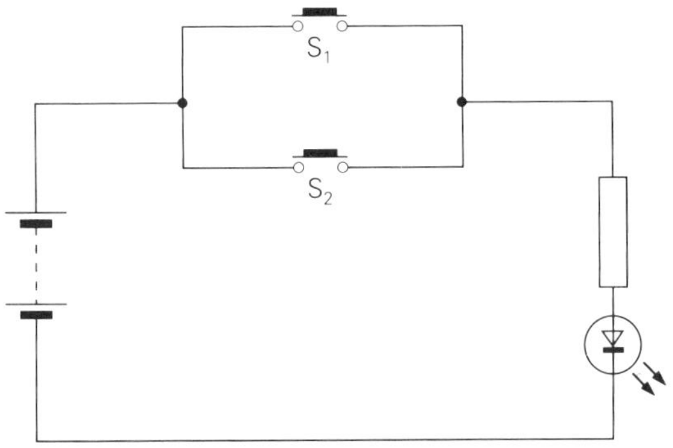

b What happens when S_1 alone is pressed?
c What happens when S_2 alone is pressed?
d Why do you think this is called an OR circuit?

Uses of AND and OR circuits

There are many situations in which an AND circuit might be useful. For example, in a motor car we might want a 'ready to start' light to come on only when the driver had fixed the seat-belt AND closed the door. In a bank, it would be a useful precaution if the door to the strong room could be opened only when a switch beside the door AND a switch on the manager's desk were pressed at the same time. Can you think of any more?

An obvious use for an OR circuit is a simple burglar alarm. A pressure pad is placed under a carpet near a door so that the switch is closed by the pressure of the intruder's foot. The OR circuit could be used to protect two doors. An alarm would sound if entry were through either door 1 OR door 2. Can you think of any more uses of an OR circuit?

The double throw switch

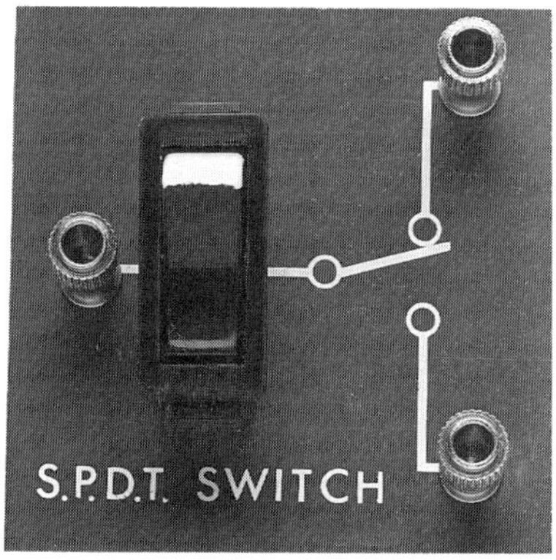

A simple on/off switch, whose symbol is shown below, is sometimes called an SPST switch, which stands for a '**S**ingle **P**ole **S**ingle **T**hrow' switch. There is one moving contact (the pole) and there is one position where it makes contact (the throw).

The photograph shows an SPDT switch module. This switch is a '**S**ingle **P**ole **D**ouble **T**hrow' switch and the symbol for it is shown below. When the switch is in one position, the lead A is connected to B; when it is in its other position, lead A is connected to C. It is sometimes called a **change-over** switch.

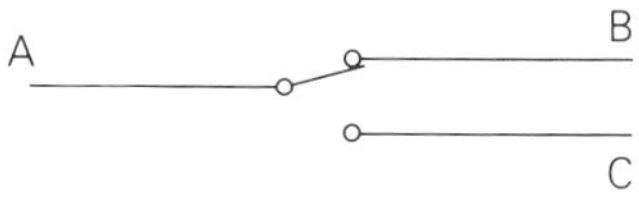

Project
Manual control of 'stop–go' traffic lights

A *single* set of 'stop–go' traffic lights is to be used to control cars entering a car park through a single lane. The operator sits at one end of the lane to control the lights so that, when a car is leaving, other cars are prevented from entering.

The operator could use the circuit of Experiment 14.8c on page 183 to control the lights, but this would not be very satisfactory. Why not?

Instead, use a red and a green LED module together with an SPDT switch and a battery to show how a more satisfactory system could be constructed.

In practice, it would be better if the operator had *two* sets of lights, each having a green and a red lamp. Devise a circuit which will allow cars to pass through the lane safely from either end. You will need two red LED modules, two green LED modules, an SPDT switch and a battery.

Project
Staircase lighting

The problem is to use two SPDT switch modules together with an LED module and a battery to construct a simple staircase lighting system. Suppose one switch is at the top of the stairs and the other at the bottom. It must be possible to turn the light on or off using *either* switch.

Experiment 14.11
Experiment with a motor

The circuit shown, involves two SPDT switch modules, a battery and the motor module. This is a problem experiment in that you should look carefully at the circuit diagram to decide for yourself what it will do. When you have decided, set up the apparatus to find out if you were correct.

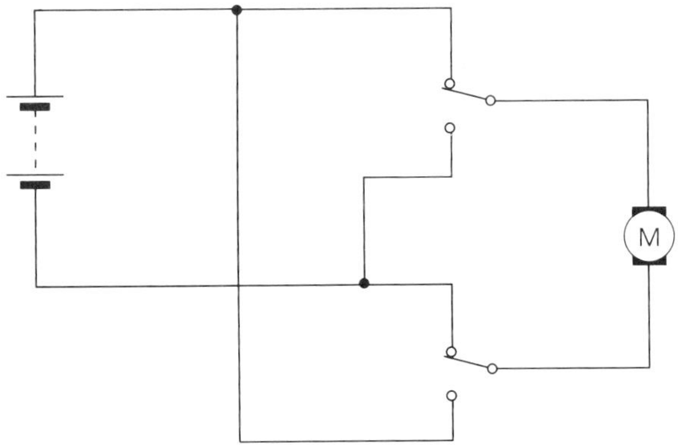

The reed switch

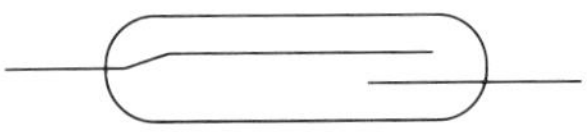

The symbol used in this book for a reed switch is shown on the left.

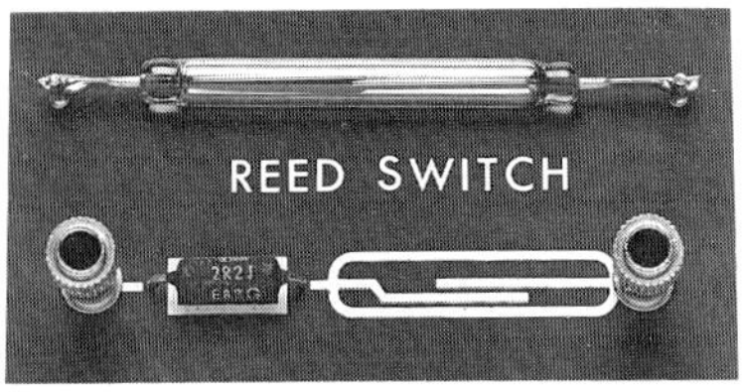

Experiment 14.12
Reed switch and magnet

a Take the reed switch module (shown in the photograph) and examine the two metal contacts inside the glass envelope with a magnifying glass. These contacts are normally open. Bring a magnet to the side of the glass envelope, as shown in the

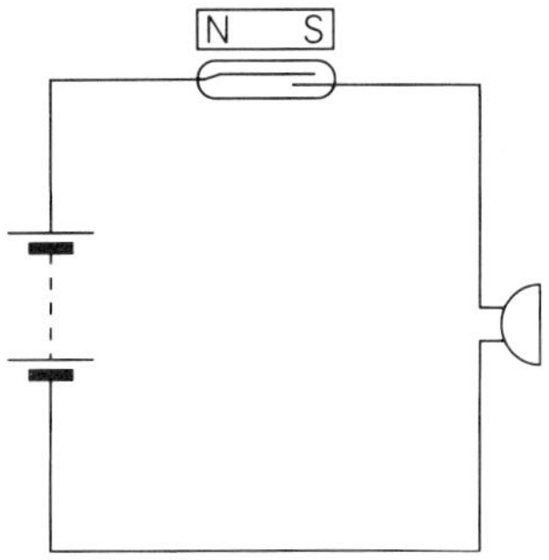

circuit on the left, and listen as you do so to see if you can hear a click as the contacts come together.

b Then connect the switch in series with a buzzer and a battery. Bring a small bar magnet close to the reed switch. What happens? What must have happened inside the glass envelope? Use your magnifying glass to see if you are right.

The reed switch consists of two metal contacts (called *reeds*) inside a glass envelope filled with an inactive (inert) gas to prevent corrosion. As the reeds are made of a metal containing iron, they can be magnetised by a magnet. If the magnet is brought close to the switch (see Fig 14.2), the metal strips are magnetised, one end being a North pole and the other a South. They are therefore attracted to each other, so that contact is made.

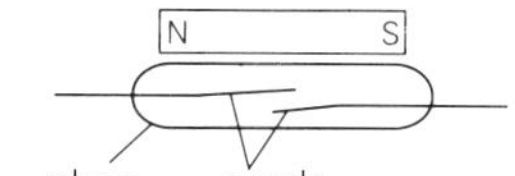

Fig 14.2

As the magnet is taken away, the strength of the magnetism in the reeds decreases and the springiness of the metal is able to pull the contacts apart again.

In the reed switch module, there is either a resistor or a fuse in series with the switch to make sure that the current is never too big to damage the switch. In the circuit diagrams in this book we will not draw this safety device.

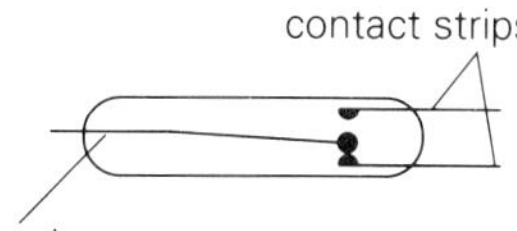

Fig 14.3

The switch described above is an SPST switch: it is either open or closed. We shall later make use of a reed switch which behaves as an SPDT switch. In this switch there is one long reed which acts as the pole of the switch, and two short contact strips (Fig 14.3). The reed is at first in contact with the lower contact strip, which is made from a non-magnetic material. But when the magnet is brought near, the reed and the top contact strip are magnetised and they come together.

The reed relay

When you first learned about electric current, you found that it had both a heating and a magnetic effect. If a current is passed through a small coil, the coil behaves like a magnet.

The reed relay consists of a reed switch like the one you have used, but it is not operated by a magnet. Instead, it has a coil around it (Fig 14.4). If a current passes, the coil acts like a magnet and, if the current is large enough, the reed switch closes.

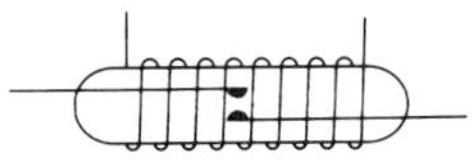

Fig 14.4

In circuit diagrams it would be confusing to draw the coil round the reed switch, so it is usual to draw it at one side, as shown.

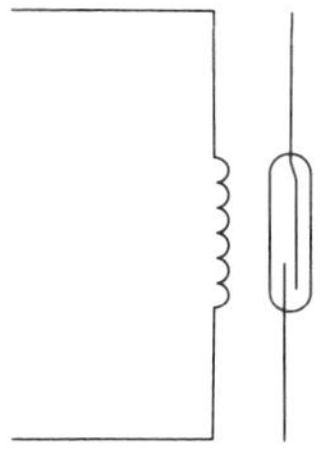

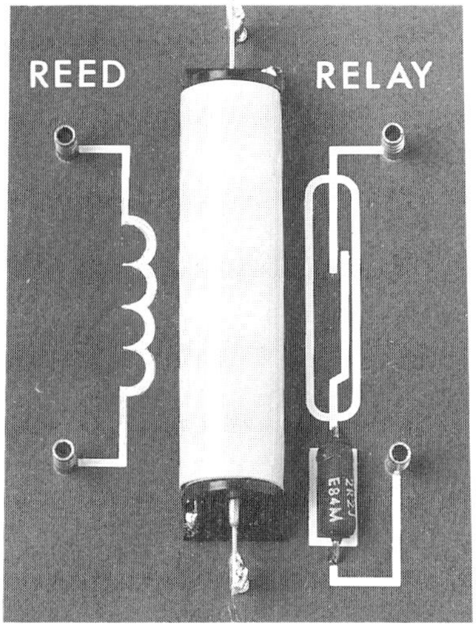

Experiment 14.13
The reed relay

a Using the reed relay module (see the photograph), a push-button switch, an LED module and two batteries, set up the circuit below.

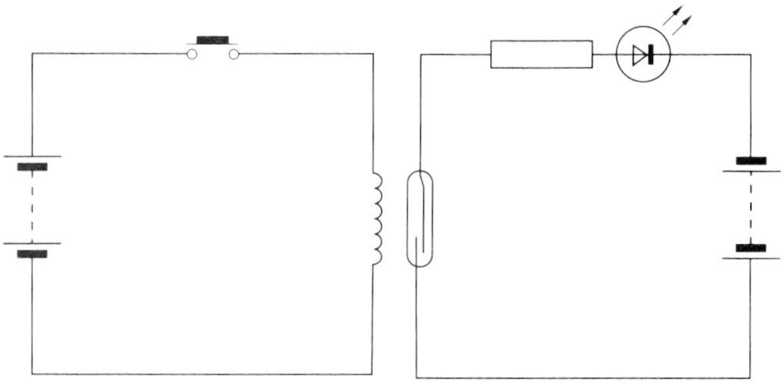

b What happens to the LED when the push-button switch is closed?

This is an important experiment because this time the switching is produced by the flow of an electric current. The two circuits in the experiment are quite separate from each other, but what happens in one is controlled by what happens in the other.

In practice the current through the coil, which is needed to operate the reed switch, is usually much smaller than the current through the switch contacts. If the LED and resistor were replaced by an electric motor needing a much larger current to make it work, it could be controlled by a much smaller current passing through the coil circuit.

A good example of this is the starter motor of a car. The starter motor may require a very large current of about 50 amperes. This means that the leads from the battery to the motor need to be as short as possible, and, in any case, you do not want large currents of that size going to the dashboard of the car. So a much smaller current is switched on at the dashboard by the ignition switch, and this operates a relay which switches on the much larger current for the starter motor.

Using one circuit to control another plays an important part in electronics. The next experiment is another illustration of this.

Experiment 14.14
The reed relay used to control a motor

a Use the LDR module, the reed relay module, the motor module and two batteries to set up the circuit on the next page.
b What happens to the motor when the LDR is covered up? What happens when a light is shone on to the LDR?
c If possible, put an ammeter in each circuit. Is the current in the first circuit very much less than that in the second?

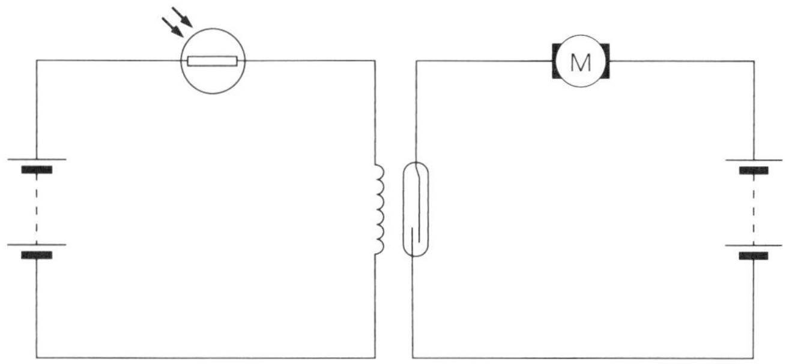

d You may wonder why it is necessary to use the reed relay at all in this experiment. Why not use the circuit below where the LDR operates the motor directly? Try it and see.

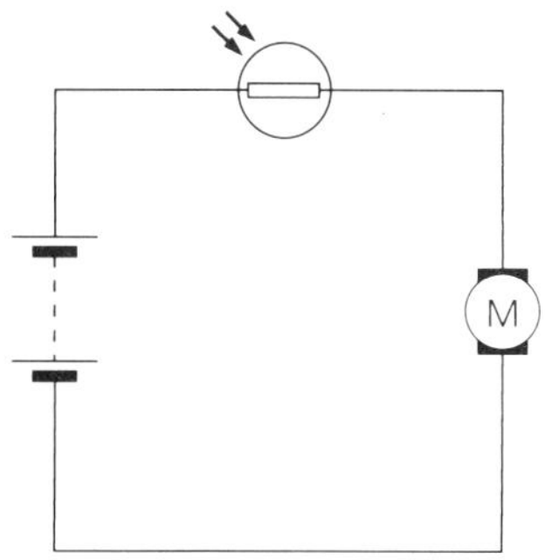

It does not work because the motor needs a large current to operate it and the resistance of the LDR does not decrease enough to make it possible. Even if the resistance of the LDR did fall to a low enough value, it would probably be damaged by the very large current needed to operate the motor.

Experiment 14.15
Controlling a motor with a single power supply

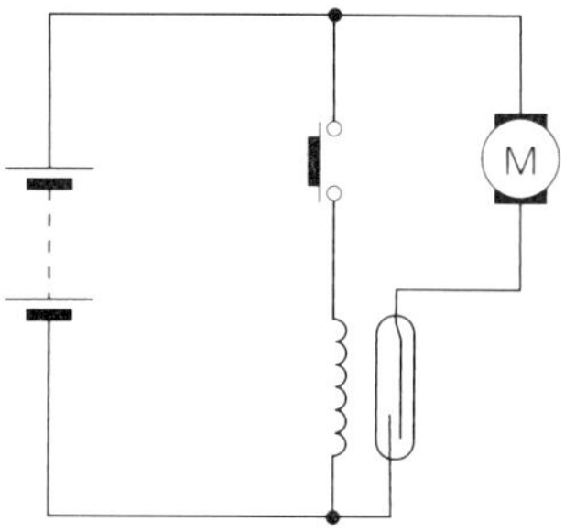

a In Experiment 14.14 two batteries were used. In this experiment, only one battery will be used. Set up the circuit on the left with one battery, a reed relay module, a push-button switch and the motor module.

b What happens to the motor when the switch is closed?

c In this circuit the relay coil and the relay contacts are connected in parallel. Copy the diagram and mark it with arrows to show the direction of the currents from the battery through each branch of the circuit and back to the battery again.

Project
To make a light-controlled motor

Alter the circuit in Experiment 14.15 so that the motor is controlled by a light beam. When the light shines, the motor should work.

Another kind of relay

A relay which uses a small electromagnet to operate a switch is shown in Fig 14.5. When a current flows through the coil, it

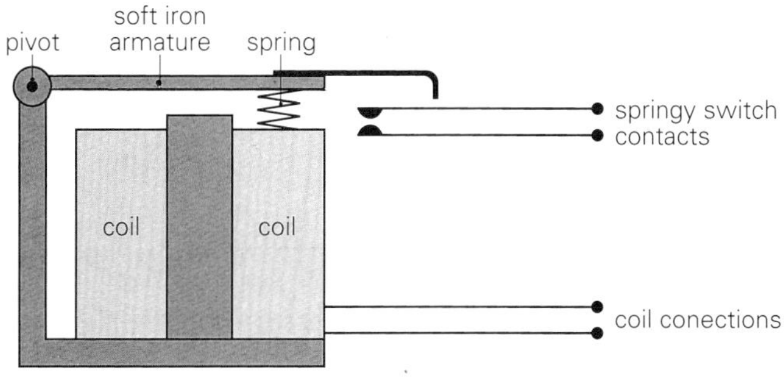

Fig 14.5

magnetises the electromagnet and the soft iron armature is attracted towards it. This causes the switch contacts to touch, in other words the switch is closed. When the current to the coil is switched off, the magnetism disappears and the spring pushes the armature away from the electromagnet, allowing the switch contacts to spring open.

This kind of relay makes it possible for a small current to the coil to switch on a large current which can flow through the spring contacts. These contacts can be made large enough to carry big currents such as those required by the starter motor of a car.

Questions for homework or class discussion

1 Describe the effect of closing the switches in each of the following circuits. The switches are all shown in the open position.

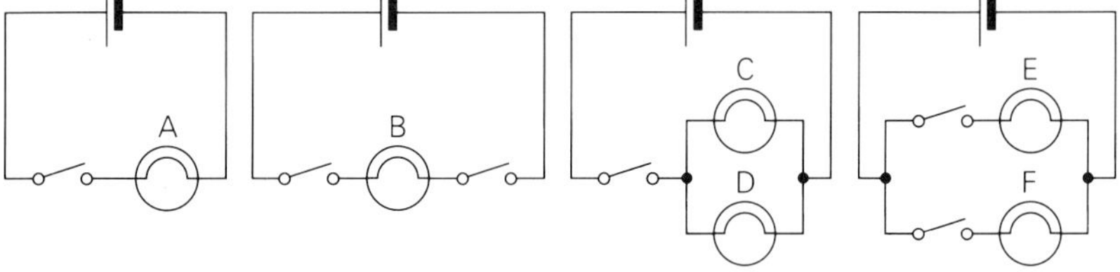

2 The following circuits include diodes. Which of the lamps (A, B, C, . . .) will light?

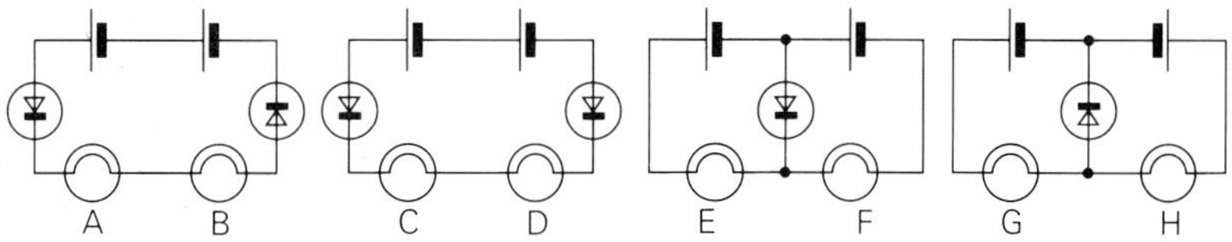

3 The table shows the four possible arrangements for the SPDT switches, A and B, in the circuit below. Copy the table and complete it to show whether each of the lamps, X, Y and Z is on or off.

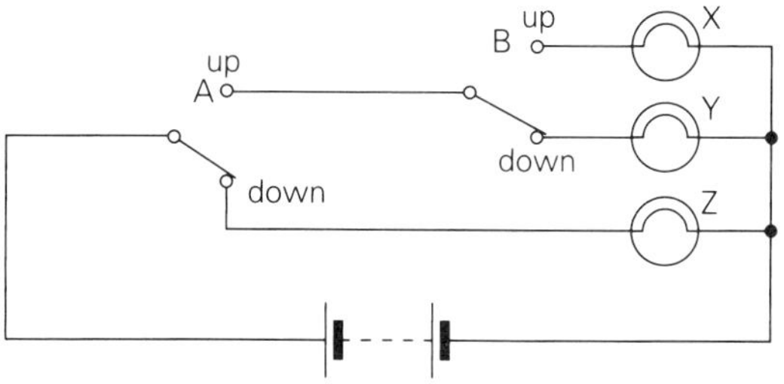

Position of A	Position of B	X	Y	Z
down	down			
down	up			
up	down			
up	up			

4 The circuit shown below has two SPDT switches, A and B, in it.

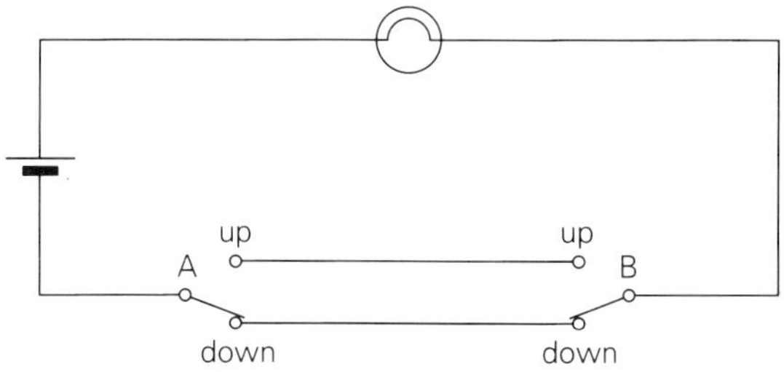

Position of A	Position of B	Lamp
down	down	
down	up	
up	down	
up	up	

a The table shows the four possible arrangements for switches A and B, but the column showing whether the lamp is on or off has not been completed. Copy the table and complete it.

b Where are you likely to find a circuit similar to this one being used?

5 In most cars, the interior light comes on when either of the two front doors is opened. Each door operates an SPST switch which is in the open position when the door is closed.

a Is an OR circuit or an AND circuit needed to do this? Explain your answer.

b Draw a diagram of a circuit which would work in this way.

c Add to your diagram another switch which would allow the driver to switch the light on if both doors were closed.

6 A spin-drier has a start–stop switch, but the motor will only spin the drum if the lid is closed. Closing the lid closes an SPST switch inside the machine, and the motor can then be started with the start–stop switch.

a Is an OR circuit or an AND circuit needed? Explain your answer.

b Draw a diagram of a simple circuit which could be used.

c Add a buzzer to your circuit diagram so that the buzzer will sound if the lid is not closed and the start–stop switch is operated. (Remember that a buzzer requires much less current than a motor does.)

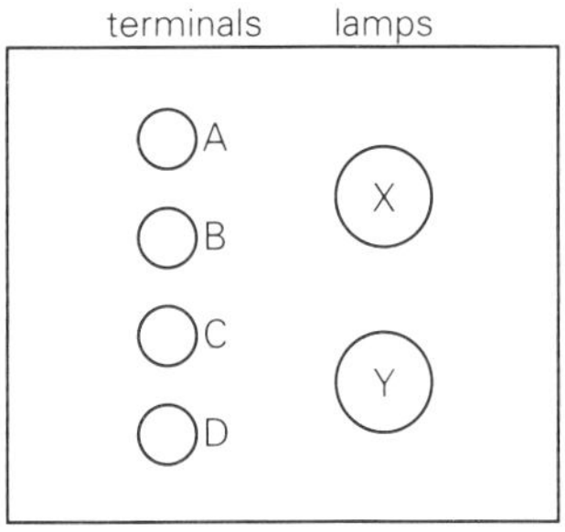

7 A closed box has four terminals, A, B, C and D, and two similar lamps, X and Y, screwed into sockets in it, as shown.

When a cell is connected to terminals A and D, lamp X lights normally. When the cell is connected to A and C, nothing happens until B and D are connected by a piece of copper wire when both lamps light dimly.

a Draw a diagram to show the connections inside the box.

b How would you connect the cell to make lamp Y light normally without lamp X lighting?

c How would you get both lamps to light brightly?

Background Reading

Electronic control systems

In this chapter, you have been concerned with devices which can be used to switch motors, LEDs and buzzers on and off. Circuits have been used to control these devices. When electronic circuits are joined together to control things, the whole arrangement is called a control **system.** Systems are used, for example, to control what happens in aircraft, in power stations, in household appliances such as washing machines and microwave ovens, and for road traffic control and so on.

Robots operated by electronic control systems are used nowadays, for example, in car factories. Electronic control systems are of ever increasing importance in the world today, and much of the art—and the fun—of electronics lies in designing such systems.

You have heard about microelectronics and the so-called 'chip'. Microelectronics is about very small electronic circuits. The devices in this chapter were quite large, but scientists have found

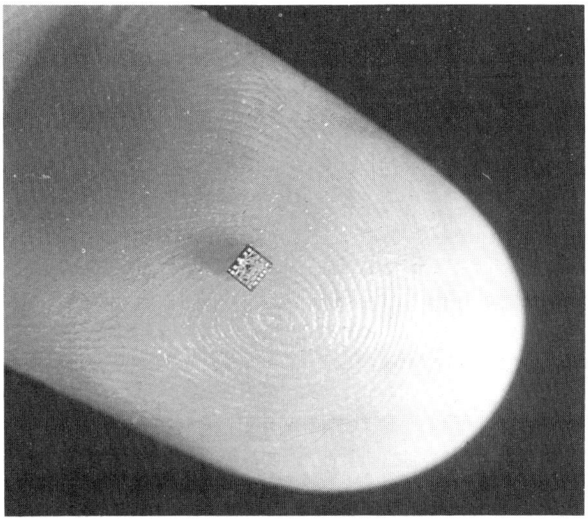

Microchip on finger

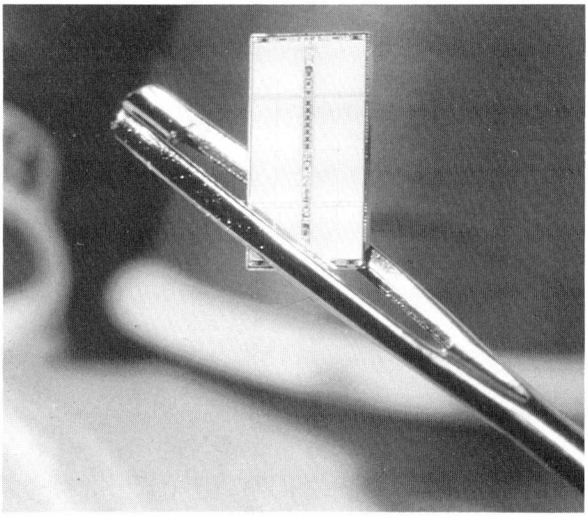

Microchip in the eye of a needle

ways of making switching circuits very small in thin layers of silicon. It is this piece of silicon which is known as a chip. The photograph shows a chip resting on a finger.

The important thing is that many switching circuits can now be built into the same chip, as ways have been found to make the circuits smaller and smaller. Today more than 100 000 can be put on a single chip of silicon only 5 mm square! These switching circuits are the basis of all computers.

The other photograph shows a chip in the eye of a needle. Chips like these form the basis of small computers equivalent to the whole processing power of a 1960 computer.

(Based on *Electronics* by G E Foxcroft, J L Lewis and M K Summers, published by Longman.)

Chapter 15 Electric charge and electric current

Fig 15.1

You have probably done experiments at home using a comb to lift pieces of paper. You might also try the following. They will work best under very dry conditions.

a Try combing your hair with a nylon comb in front of a mirror in a darkened room. What can you hear? Can you see anything?
b Pull off a nylon garment in the dark. Do you hear strange crackles or see flashes of light?
c Comb your hair or rub the comb on a woollen sleeve. Hold the comb over some small pieces of paper on a table (Fig 15.2).
d Attach a grain of puffed rice to a fine nylon thread. Support one end of the thread under a book and let the puffed rice hang freely. Comb your hair and bring the comb near the puffed rice (Fig 15.3).

Fig 15.2

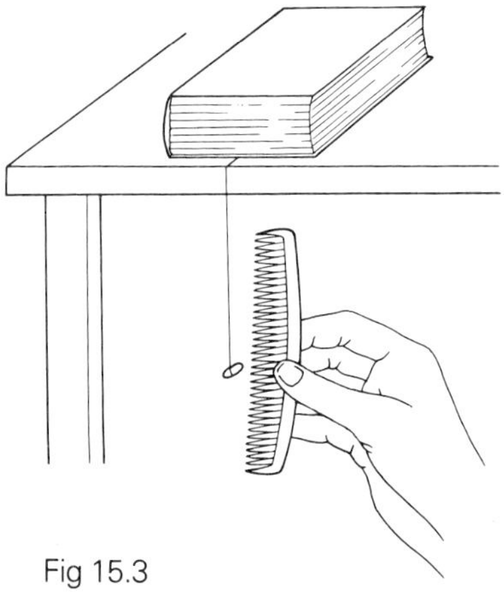

Fig 15.3

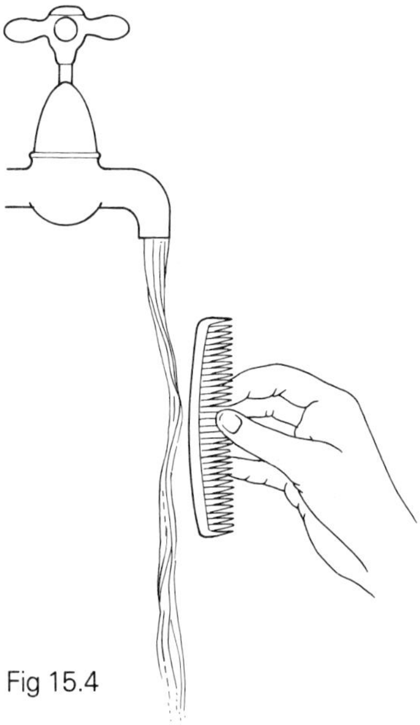

Fig 15.4

e Turn on a water tap to get a fine jet of water and hold the comb near the jet (Fig 15.4).
f Place a piece of plastic on top of some books (Fig 15.5). Put some shapes made from tissue paper under the plastic. Rub the plastic with a duster. What happens to the tissue paper shapes?

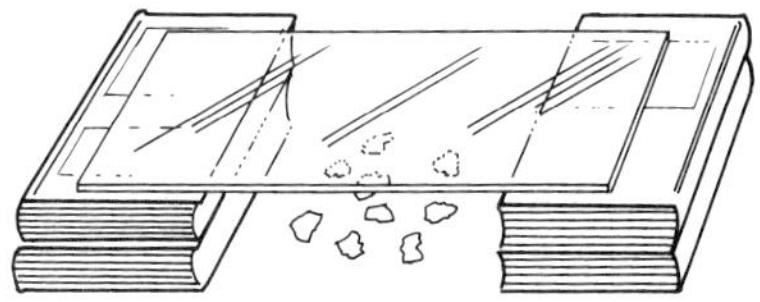

Fig 15.5

g Blow up a balloon and rub it on a carpet. Then put the balloon against the wall of a room or touch the ceiling with it (Fig 15.6). Does it stay there when you let go?

Fig 15.6

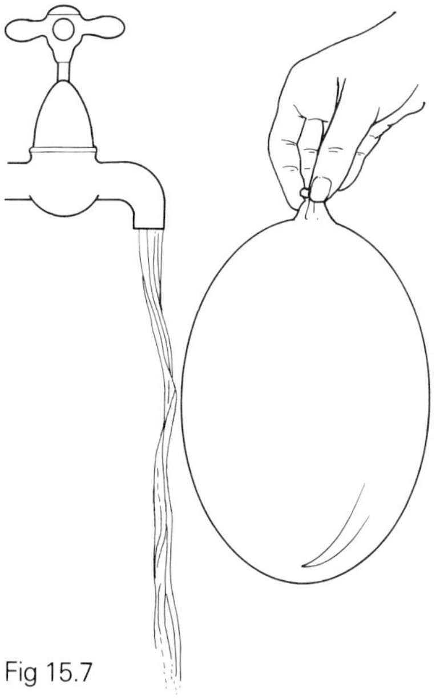

Fig 15.7

h Rub a balloon on a carpet and hold it near to a jet of water (Fig 15.7).

These are all strange effects, and scientists find the best way to explain them is by introducing the idea of **electric charge**. We cannot see electric charge, any more than we can see an electric current, but we believe it to exist because of its effects. We say that when a comb, a balloon or a plastic sheet is rubbed, then it has become **charged**.

Experiment 15.1 Evidence for two kinds of charge

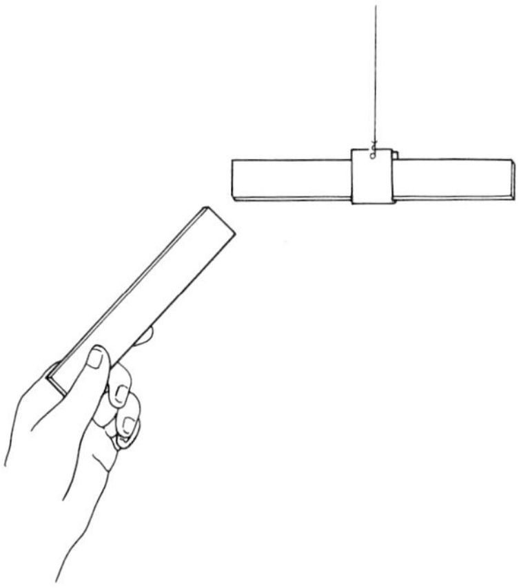

Fig 15.8

For this investigation you will need strips of polythene and cellulose acetate, together with a duster or large handkerchief. Try the following experiments.

a Rub both a polythene strip and a cellulose acetate strip to show they can both pick up small pieces of paper. This shows that both strips become charged when rubbed.

b Use a wire stirrup (or one made from a strip of paper) to hang up a polythene strip so that it hangs freely from a length of nylon thread. Rub both ends of the strip with the duster.

Also rub another polythene strip. Bring *one* of the ends close to *one* end of the suspended strip (Fig 15.8). What happens? Try the other end. What happens?

As both rods are of the same material and were rubbed with the same duster, they must be charged the same. You notice that, whichever end is used, the charged rods repel each other.

c Now repeat all of Experiment **b**, but use instead two cellulose acetate strips. Once again whichever ends you use, the charged rods repel each other.

d Finally hang up a polythene strip. Rub it with the duster. Then rub a cellulose acetate strip with the duster and bring it near one end of the polythene strip. This time they attract each other.

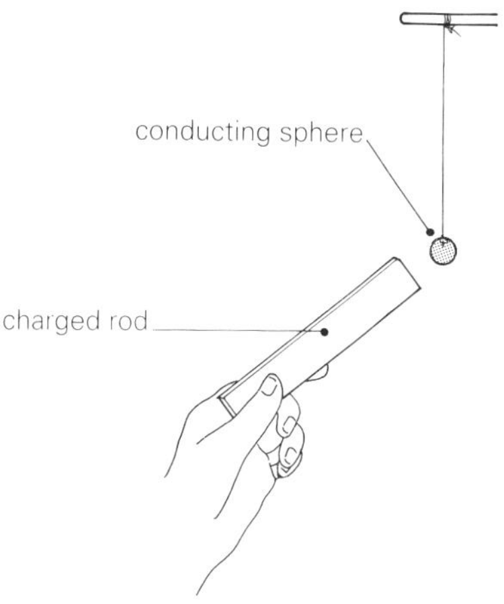

Fig 15.9

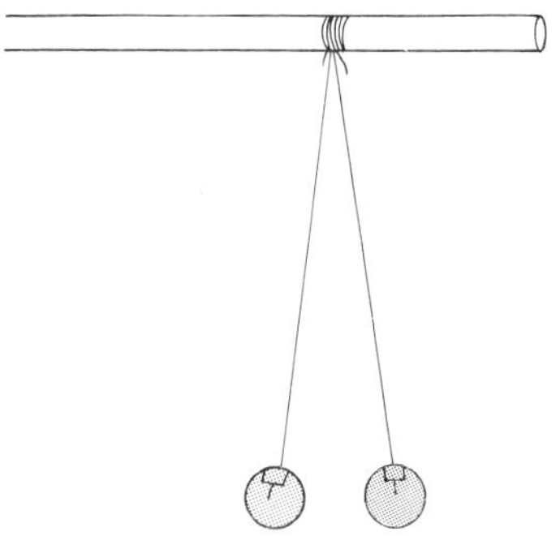

Fig 15.10

Thus it appears that there are two kinds of charge: one is found on a rubbed polythene rod, the other on a cellulose acetate strip. It also appears that like charges repel each other, and unlike charges attract. For convenience it is usual to refer to these two kinds of charge as **positive** and **negative**.
The charge on the cellulose acetate strip is positive charge and that on the polythene strip is negative.

Experiment 15.2
Investigation of charges using light spheres

As well as the polythene and cellulose acetate strips, you will need two light, conducting spheres suspended from fine nylon thread: these can be very light expanded polystyrene spheres coated with aluminium paint to make them conducting. They can be fixed to the nylon thread with a small piece of sticky tape.

a Hang up one of the conducting spheres. Bring a charged polythene strip near to it (Fig 15.9). What happens?
b Let the sphere touch the strip so that it is charged by it. Remove the strip and then bring it slowly towards the sphere. What happens this time? Can you explain this?
c Bring up a charged cellulose acetate strip near to the charged sphere. What happens this time?
d Now hang two uncharged spheres side by side (Fig 15.10). Wipe one sphere with the edge of a charged polythene strip and the other with the edge of a cellulose acetate strip. Wiping like this will transfer charge to the spheres. What happens?
e Repeat **d**, this time wiping both spheres with the same charged rod. What happens?

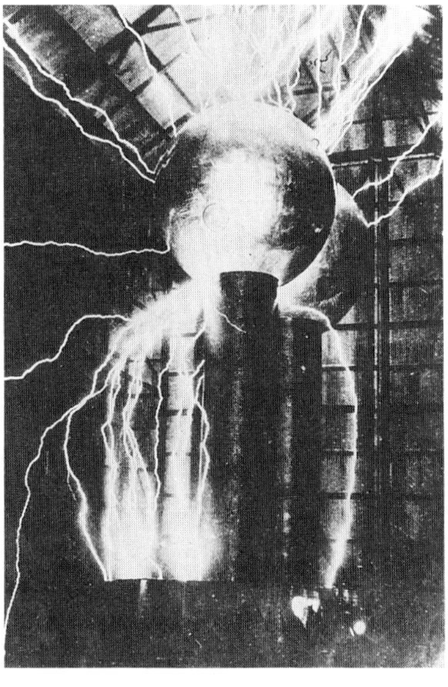

One of the original generators made by van de Graaff in 1931

The van de Graaff generator

We have been producing charges by rubbing. An ingenious way to produce large charges was invented in 1931 by van de Graaff. A photograph of a large van de Graaff generator is shown on the left. We use a small one in school laboratories: it can build up very large charges on the big collecting sphere at the top when the machine is driven either by hand or by an electric motor.

Demonstration experiment 15.3
Experiments with a van de Graaff generator

a Hang a small conducting sphere near to the big collecting sphere of the generator. Let the small sphere touch it so that the small sphere becomes charged and then, as like charges repel, the small sphere will be pushed away (Fig 15.11).

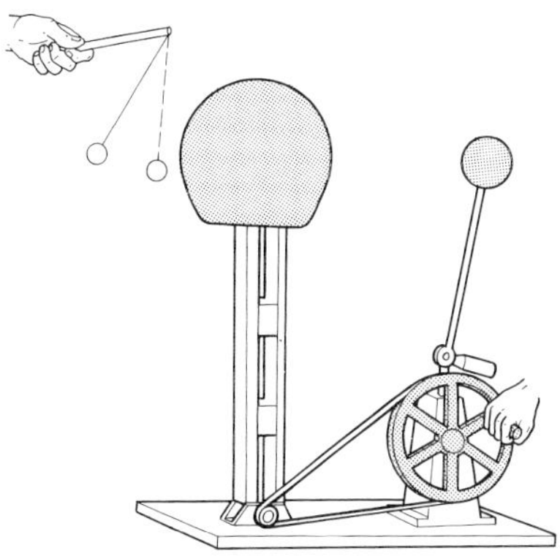

Fig 15.11

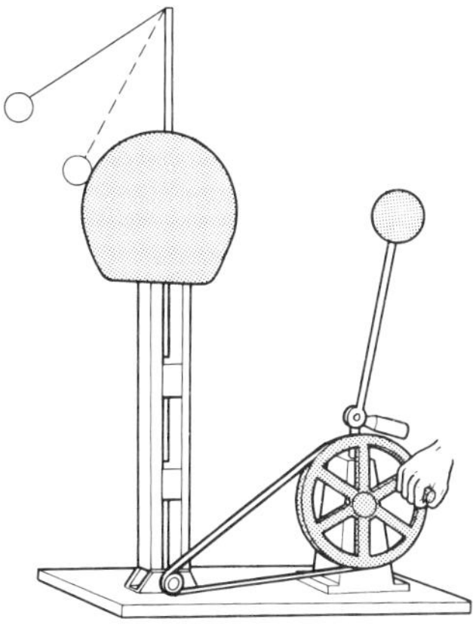

Fig 15.12

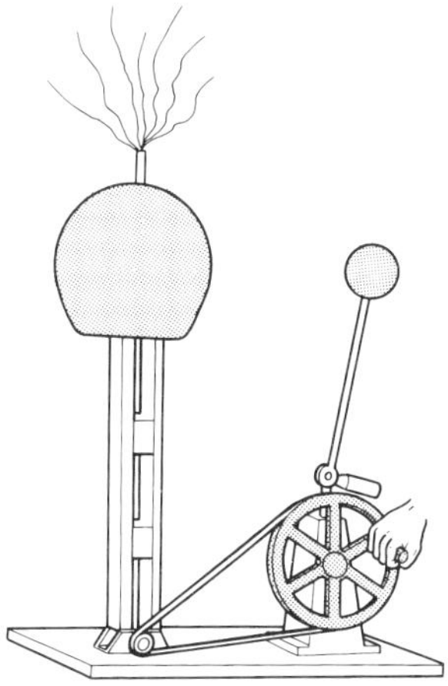

Fig 15.13

b Fix a rod with a small conducting sphere on the top of the generator (Fig 15.12). Again the small sphere will be repelled.
c Fix a 'head of hair' on the top of the generator. Charge it up and watch the repulsion (Fig 15.13).
d Hold your head near the generator and feel the effect on your hair.
e If someone in rubber boots stands with a hand on the conducting sphere while it is being charged up, he or she will also be charged up. Charge can then be dramatically transferred to someone else who touches the other hand!
f Finally, bring the small sphere near to the collecting sphere. When they are close together, sparks will pass between them as the handle is turned.

Moving charge and current

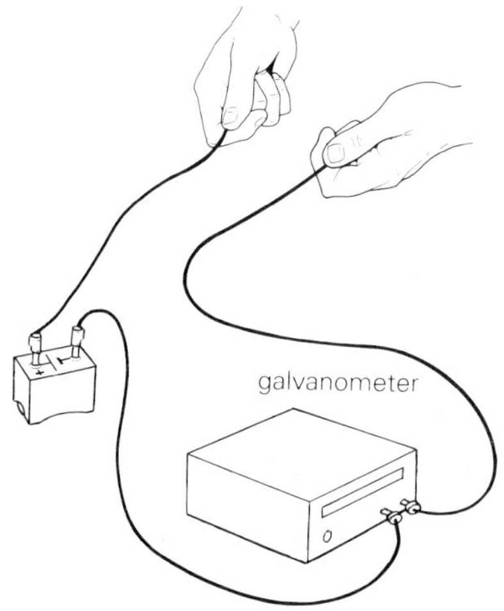

Fig 15.14

We have learned in this chapter about electric charge and we already know about electric currents, but we have not yet shown a relationship between them. For that experiment we need to use a new instrument called a **galvanometer**. This is a detector of very small electric currents. (It is named after an Italian professor of anatomy, Luigi Galvani, who in 1780 used the legs of dead frogs to detect electric currents: the leg twitched when a current passed through it.)

You can show how sensitive a galvanometer is by putting it in series with a cell and a person (Fig 15.14). With an ordinary ammeter you would not be able to detect any current. (Try it and see.) The resistance of a person is so high that with only one cell the current is very small indeed, but it can be detected with a galvanometer. This shows that the galvanometer gives a reading when a very small current passes through it.

Demonstration experiment 15.4
A current and moving charge

A table-tennis ball is coated with graphite (Aquadag) to make it conducting, and it is hung by a very long thread between two metal plates as shown in Fig 15.15. One plate is connected to the collecting sphere of a van de Graaff generator. The other plate is connected to one terminal of the galvanometer. Finally, the other terminal of the galvanometer is connected to the terminal near the handle of the van de Graaff generator.

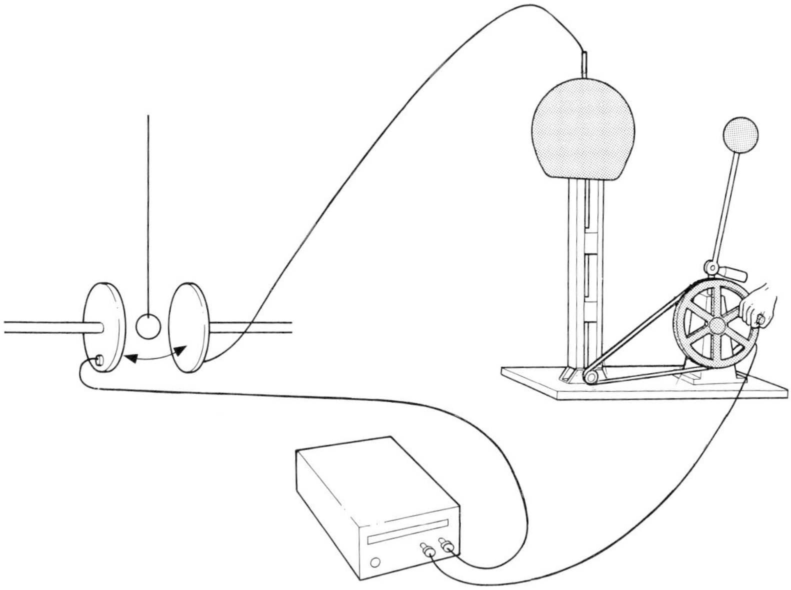

Fig 15.15

The van de Graaff generator is set going so that the large sphere is charged. The table-tennis ball oscillates between the two plates carrying charge across from one to the other, and the galvanometer records a current.

This experiment shows us that moving charge causes a current. From now onwards we can assume that an electric current is the movement of electric charge. Of course the experiment has not shown whether it is positive charge moving one way or negative charge moving the other. A decision on that will have to wait until later in your physics course.

Finally, remove the oscillating ball and the two plates and connect the galvanometer directly across the van de Graaff generator (initially uncharged, of course). Then turn the handle and see what happens.

Conduction of electricity in liquids

In previous chapters, we found that certain solids were conductors of electricity, whereas other substances were insulators. We shall now investigate whether liquids are conductors or insulators.

Experiment 15.5
Investigation of liquids as conductors

Connect three cells in series with an ammeter and two bare copper leads as shown in Fig 15.16. The bare ends of the copper wires should hang down into the beaker, but they must not touch each other. The beaker will be filled in turn with the various liquids to be investigated.

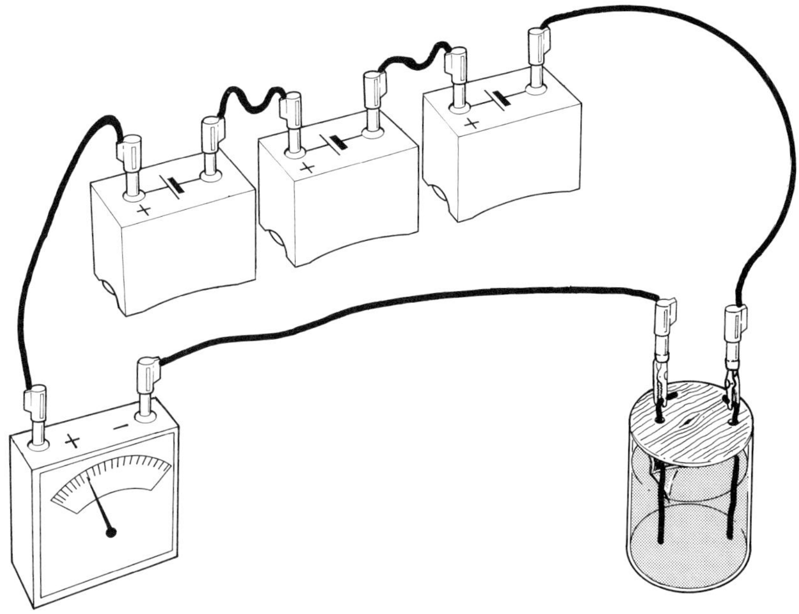

Fig 15.16

a First try distilled water. Does any current flow?
b Add a little common salt to the water. Stir the solution and see what happens. If you see anything happening at the copper wires, break the circuit temporarily to see whether it is due to the current or not.
c Empty the beaker and wash it out. Refill it two-thirds full with distilled water and this time add a little dilute sulphuric acid and see what happens. Switch off the current. Does the effect stop? This shows that it is due to the current and not to some chemical reaction in the solution.
d Repeat with fresh distilled water to which sugar has been added. Alternatively, try paraffin on its own.
e Finally, try tap water.

Safety note

Dilute sulphuric acid is irritant. It must be kept off the skin.

Electrolysis

You found in the last experiment that very pure water (distilled water) did not conduct electricity, but very small amounts of certain chemicals made it conduct. You have doubtless heard how dangerous it is to have any mains-operated electrical appliance in a bathroom: your dirty bath water is a long way from being pure water and is quite a good conductor of electricity.

You will have noticed that when a current passed through the slightly acidified water it produced certain chemical changes. In all scientific work it makes it easier to discuss details if we give names to parts of the apparatus. A solution through which the current passes is called an **electrolyte**. The two electrical contacts immersed in the electrolyte are called **electrodes**: the one connected to the positive terminal of the battery is called the **anode** and the one connected to the negative terminal the **cathode**. The whole process is called **electrolysis**.

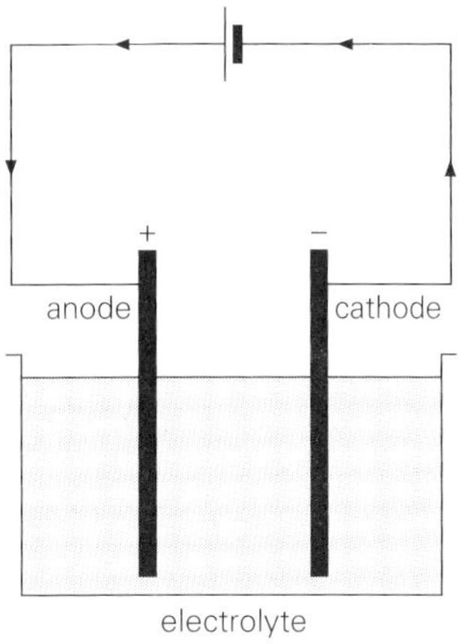

Fig 15.17

The chemical changes which occur during electrolysis are often complicated. They depend both on the electrolyte and on the nature of the electrodes. We shall not consider electrolysis in detail here, except for the interesting case shown in the following experiments.

Experiment 15.6
Electrolysis of copper sulphate solution

For this experiment use the same circuit as in Experiment 15.5, but, for electrodes, use two strips of copper foil slipped down the inside of the beaker and bent over at the top (Fig 15.18). They can be connected into the circuit with crocodile clips. Make certain they do not touch each other.

Safety note

Copper sulphate is harmful and must not be taken internally.

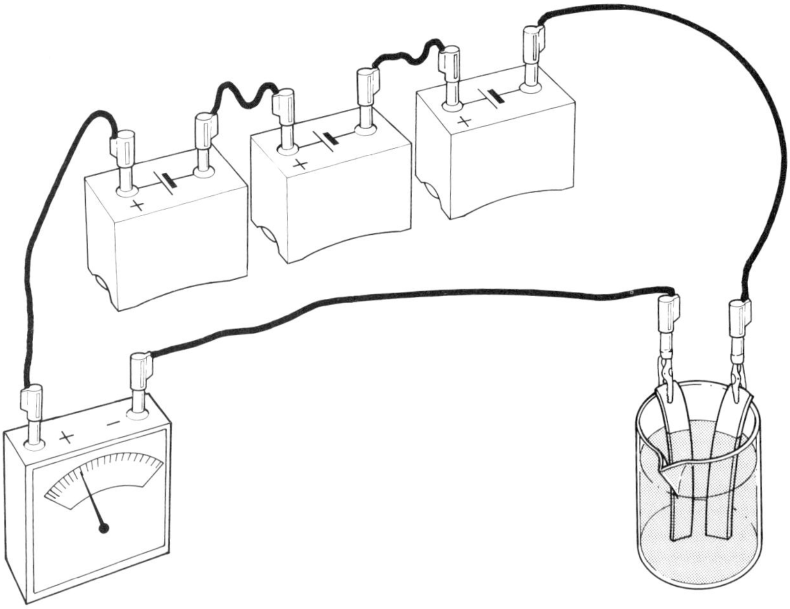

Fig 15.18

Fill the beaker with some strong copper sulphate solution. Switch on the current. Let it flow for several minutes and then look at the electrodes to see if there is any difference. You should see that clean, pure copper has been deposited on the cathode, and you could confirm this by measuring its mass with a sensitive balance both before and after the current flowed. Similarly you would find that the anode has lost mass.

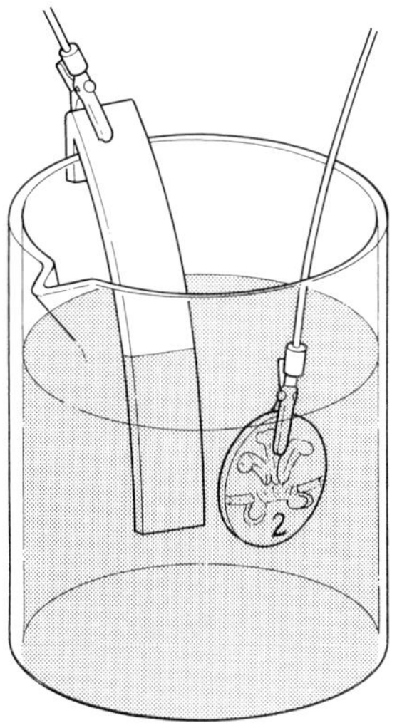

Fig 15.19

Experiment 15.7
Copper plating

Use the same circuit as in the previous experiment, but instead of a cathode made of copper foil try other articles, like a brass screw, a paper clip or a coin. (It is better to avoid objects made of zinc or iron for these will react of their own accord with the solution.) Fill the beaker with strong copper sulphate solution and pass the current through it.

Safety note

Copper sulphate is harmful and must not be taken internally.

How does a current flow in a liquid?

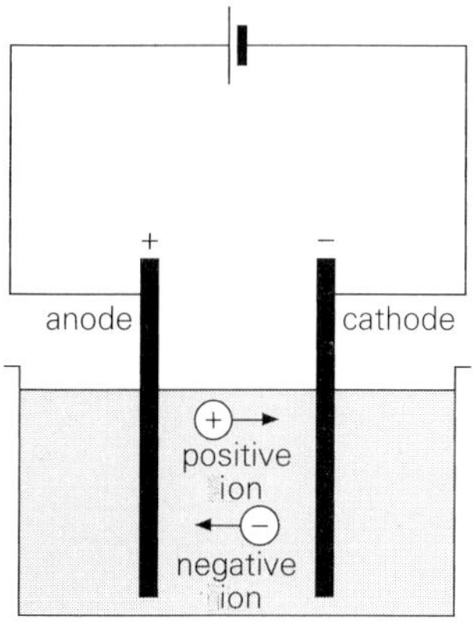

Fig 15.20

How can we explain electrolysis? It is outside the scope of this book to discuss it in detail and you will learn about it in your chemistry course, but one thing seems clear from these experiments. Something is moving through the electrolyte.

A possible explanation might be that something charged moves through the liquid. Of course this is only a guess, but you will find that other experiments, particularly in chemistry, support this idea. We know already that opposite charges attract each other. These **ions**, as they are usually called, have charge: the negative ions move to the positively charged anode and the positive ions to the negatively charged cathode. It is these moving ions that are responsible for the current through a liquid.

Background Reading

Electroplating

The process by which a metal is deposited on another is called **electroplating**. It is done for a variety of reasons. Steel is strong and comparatively cheap but it rusts very easily. Electroplating the steel protects it from rust and gives it a more attractive finish.

At one time, the steel was electroplated with nickel, but nickel goes dull and soon appears rather drab. Furthermore, it is not very hard and scratches easily: deep scratches, through to the steel, lead to rusting. Chromium is a much harder metal, and it stays bright and shiny. Unfortunately if it is of a certain thickness it is liable to crack, leading to rusting underneath. What is the solution to all this?

Fig 15.21

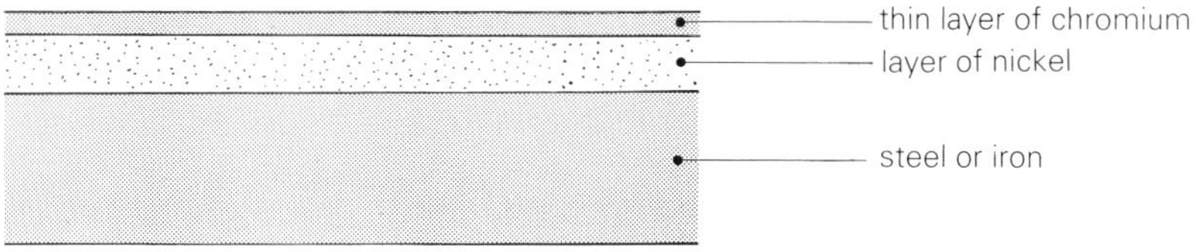

The usual process now is to nickel-chromium-plate the steel. Using the correct electrolyte, a layer of nickel is first deposited to protect the steel from rusting. (This may be about 0.000 4 cm thick, though the actual thickness depends on whether the appliance is to be used outside in severe weather conditions; it can be thinner if the plated appliance is used indoors.) Then, with a different electrolyte, a thin layer of chromium (usually about 0.000 004 cm thick) is deposited on top of the nickel. This protects the nickel from scratches, as it is so much harder, and also gives a shiny finish.

Very few people can afford to eat with solid silver spoons and forks. Furthermore, they are weak when used. A strong spoon or fork can, however, be made from a nickel alloy and then electroplated with silver, giving it the pleasant, shiny appearance of a solid silver spoon. You will often find EPNS stamped on the back of spoons and forks: this stands for Electro-Plated Nickel Silver (see the photograph).

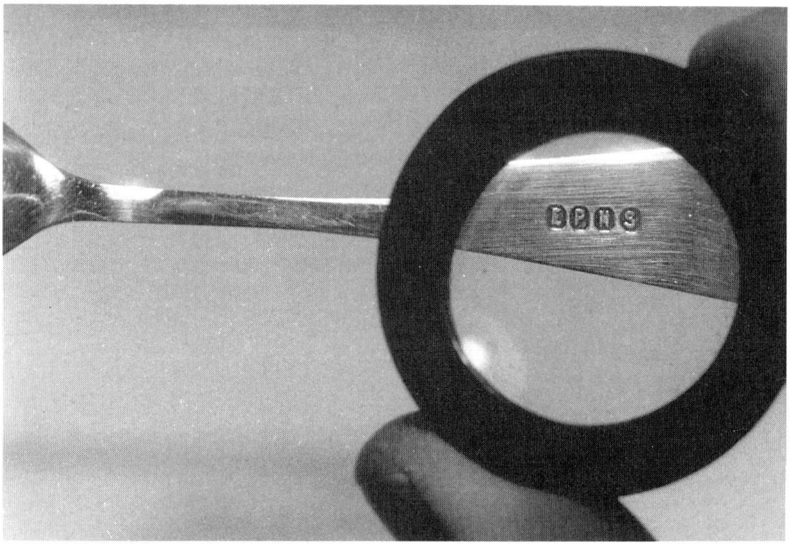

EPNS stamped on a spoon

Large numbers of articles can be electroplated automatically. They are usually hung on metal frames which are carried slowly through an electrolytic tank by a moving belt. They pass right through in about an hour; in successive stages they are cleaned with weak acid, rinsed, nickel-plated, washed and finally dried.

(Adapted from Longman Physics Topic *Electric Currents* by J L Lewis and P E Heafford.)

Conduction of electricity in gases

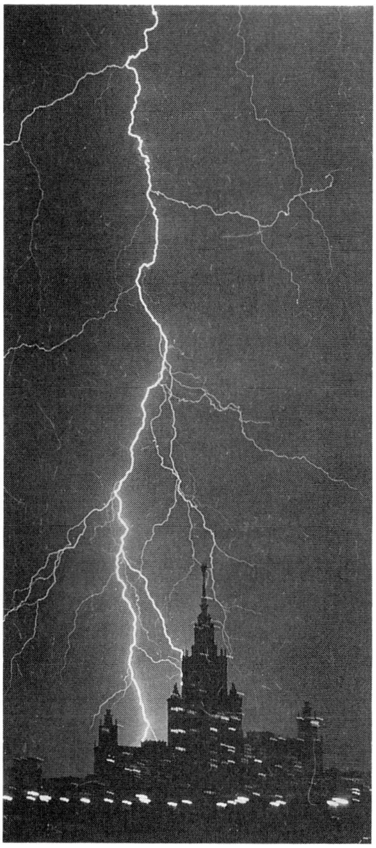

We have seen already that some solids are conductors and others are insulators. We found the same with liquids: some conduct, others do not. What about gases?

In your experiments with circuits, you found that any gap in the circuit stopped the current from flowing: air does not conduct electricity. Think how awkward it would be if it did conduct: current would flow from an electric socket at the wall even if there were nothing plugged into it and even if the switch were off. And batteries would not last very long!

However, you have now seen sparks produced by a van de Graaff generator. In the same way, when a cloud becomes highly charged, a lightning flash may occur: a huge spark passes through the air either from one cloud to another, or from the cloud to the Earth (see the photograph).

What makes it possible for a current to flow through the air? Perhaps **ions** are present, in the same way that we suggested positive and negative ions might carry the current through a liquid. The next experiment gives some evidence for ions in the air.

Experiment 15.8
Ions in the air produced by a candle

Two metal plates, about 5 cm apart, are set up above a lighted candle. One of the plates is connected to the large sphere of the van de Graaff generator, the other to the terminal near the handle. This means that when the generator is operating there will be several thousand volts between the plates. A strong light source casts a shadow of the flame on a screen (Fig 15.22).

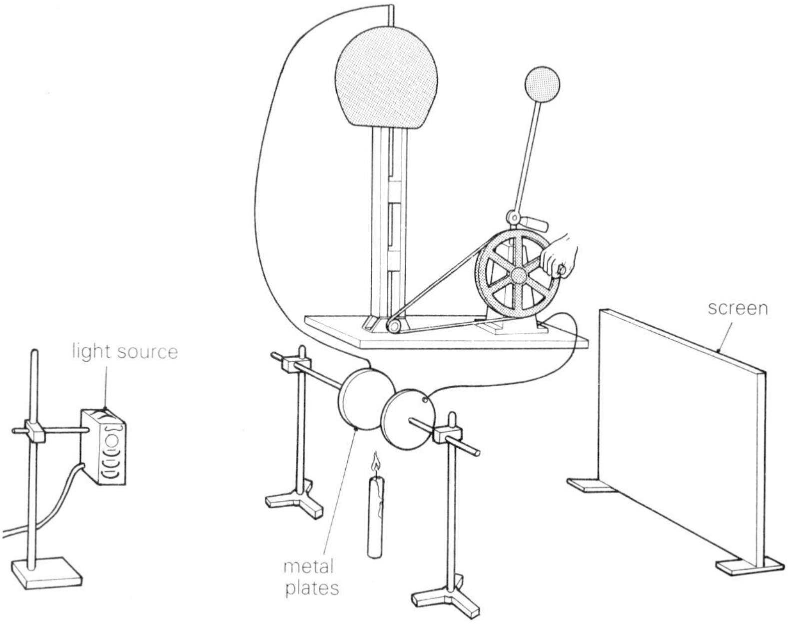

Fig 15.22

What happens to the shadow when the voltage is applied? It should be seen that the flame divides into two parts, one towards the positive plate and one to the negative plate. It looks as though the candle produces ions in the air, negative ions which move one way and positive ions the other.

Questions for class discussion

1 A thin piece of copper wire can safely be used to connect a lamp to a cell. Yet the same wire when connected across a mains socket would immediately produce a disaster. Explain what happens in both cases and account for the difference.

2 If a wire is connected straight across the terminals of a 12-volt car battery, a current of 50 amperes might flow. A current of 1 ampere is sufficient to kill a man. If you touch the two terminals of a car battery you will not feel even the slightest shock. Explain this.

3 You can get rid of the charge on a polythene or cellulose acetate strip by pulling the strip through the flame of a Bunsen burner or a candle. Can you explain why it loses its charge?

4 Six conducting balls are hung from insulating threads. Ball 1 is positively charged.

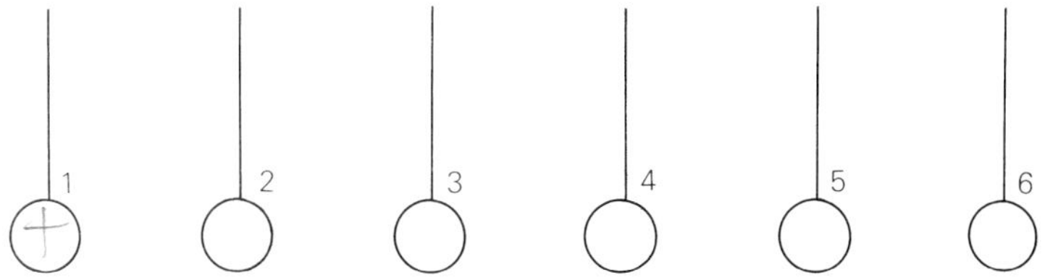

Fig 15.23

In succession, different balls are brought near to each other and various effects are observed.

a When ball 2 is brought near ball 1, it repels it.
b When ball 3 is brought near ball 2, it repels it.
c When ball 3 is brought near ball 4, it attracts it.
d When ball 4 is brought near ball 5, it repels it.
e Ball 6 is found to be attracted to both ball 3 and ball 4 when brought near to them.

Decide whether the balls 2, 3, 4, 5 and 6 are positively charged, uncharged or negatively charged. Give your reasons.

5 You have seen a van de Graaff generator in operation.

a Why does it not work well when the weather is damp?
b Why is an electric shock received from a van de Graaff generator (about 50 000 volts) less dangerous than one received from the domestic mains supply (240 volts)?
c Why do you not receive a shock if you stand on an insulating platform and charge yourself up by holding the large sphere?

d What do other people in the room notice about your appearance as you are being charged up?

e The machine is now switched off and you shake hands with your friend who has been standing on the floor beside you. Why do you both receive a shock?

Background Reading

Earthing

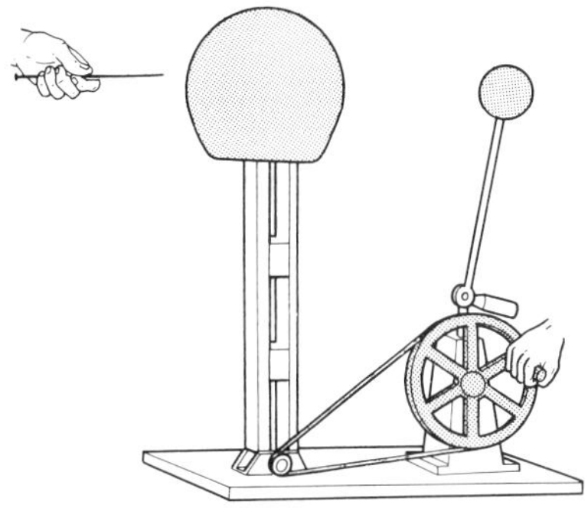

Fig 15.24

If you have used a van de Graaff generator, you may have found out that an easy way to discharge the dome is to hold a pointed conductor near it. Indeed, if you hold a large pin near the dome while the handle is turned, and the room is dark, you may see sparks at the point of the pin. Electricity seems to travel more easily through the air near a point.

In a thunderstorm, electric charge builds up in the clouds and on the Earth beneath. The lightning is the electric current (or spark) which passes from one part of the cloud to another part, or from the cloud to the ground. If the lightning strikes a building or a tree, the large current can cause severe damage. In the case of a tree, the heat produced by the current boils the sap and the tree explodes.

Tree damaged by lightning

To prevent damage to a building, a lightning conductor is usually fixed at its highest point so that the lightning strikes the conductor rather than the building. The lightning conductor is a thick piece of metal, with one or more large spikes on it, connected to a thick strip of copper. The copper strip runs down the outside of the building to a large metal plate buried in the earth (see the drawing on the next page). The effect of the

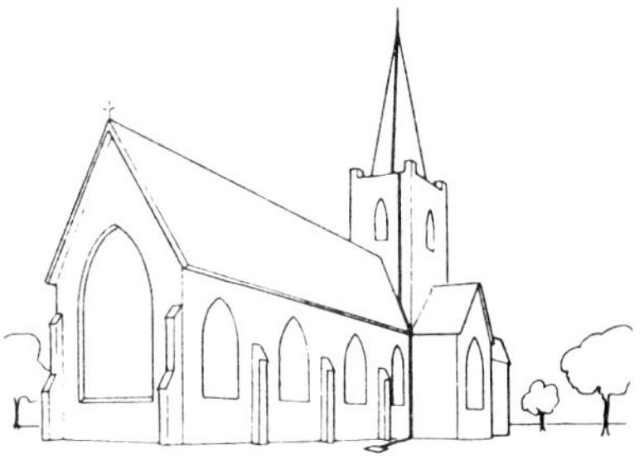

spikes is to help discharge any thundercloud overhead (just like the pointed conductor near the van de Graaff generator) and to provide an easy route for the electric current if the lightning strikes.

'Earth' connections are provided with the electrical wiring in houses, also to ensure safety. The largest pin of the three-pin plug is to enable you to make a connection to earth for safety purposes.

There are three wires to the *socket*. One of these, either a bare copper wire or one with a green and yellow insulation, is connected directly to a metal plate buried in the ground or to a metal (not plastic) water pipe. The electricity for any apparatus connected to the socket flows along the 'live' (L) and 'neutral' (N) wires. The neutral wire (blue insulation) is connected to 'earth' at the substation. The voltage present on the 'live' wire (brown insulation) is high and it is dangerous. If you were to touch it, then a current would flow through you to earth, and if that current were more than a few milliamps, you might be killed. How much current flows will depend on your resistance. If you were wearing rubber or plastic gloves and standing in shoes with rubber or plastic soles, then your resistance would be very high and, though you might feel a shock, the electricity would be unlikely to kill you. However, if your shoes were wet or some other part of your body were connected to earth, then your resistance might be quite low and the larger electric current could easily cause death.

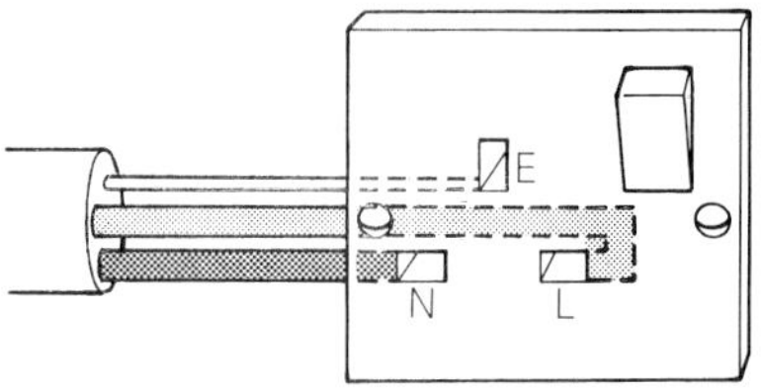

Fig 15.25

Obviously, people have to be protected from the danger of electrocution if a fault develops in, say, an electric fire. If only the live and neutral wires were used and the live wire happened to touch the metal case, then, if you touched the metal case, you would get an electric shock. So the earth wire is connected to the case to protect you. Then, if the live wire touches the case, a large current will flow through the very low resistance to earth, and the fuse in the circuit will melt and disconnect the fire from the live lead.

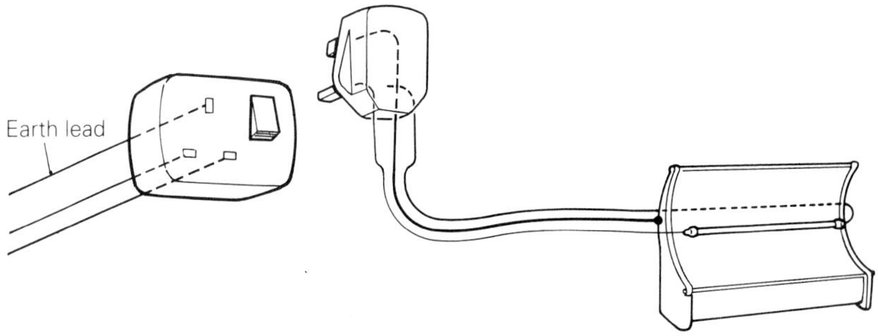

Fig 15.26

This earth lead is therefore most important when connecting appliances, even though the appliance could work without it. And, of course, the fuse must be in the live lead to ensure protection.

Chapter 16 Motors and dynamos

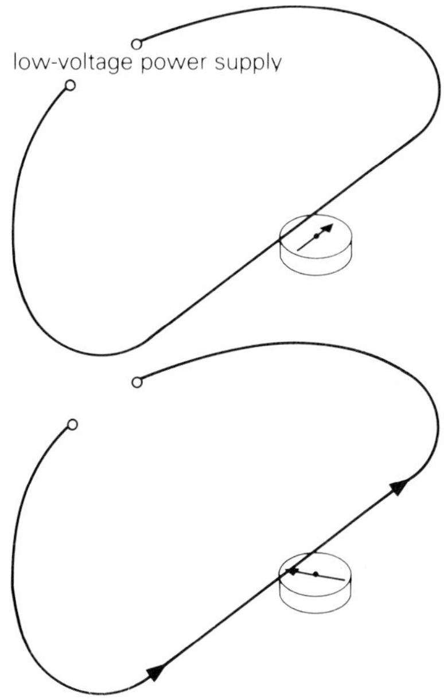

Fig 16.1

You have seen that an electric current has a magnetic effect which can be used to magnetise iron and so make an electromagnet. It was Oersted who first discovered that there was a magnetic field near a wire with a current flowing through it.

Experiment 16.1
Oersted's experiment

Allow a compass needle to come to rest in a north—south direction. Connect a length of wire (1 m long) to a low-voltage power supply capable of delivering large currents. Hold the wire just over and parallel to the compass needle and then switch on the current. Note what happens and then switch the current off. What happens if the current flows in the opposite direction over the compass?

When the current flows, the needle deflects and points at right angles to the direction of the wire. The current creates a magnetic field round the wire. The direction of the magnetic field depends on which way the current is flowing.

Experiment 16.2
The magnetic field due to a current in a wire

The field can be plotted using the apparatus shown in Fig 16.2. The wire passes through a hole in a piece of card supported on wooden blocks. To make the field strong enough to be shown with iron filings, it will probably be necessary to loop the wire through the hole several times.

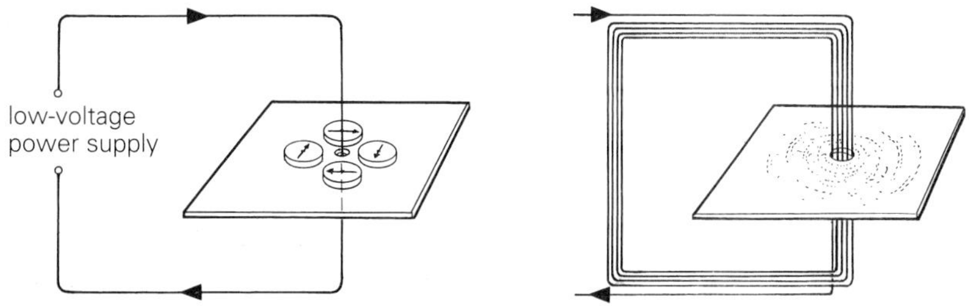

Fig 16.2

The magnetic field lines are circular round the wire. They are not like those of a magnet which start at a N pole and finish at a S pole. These are continuous with no start or finish.

Experiment 16.3
The magnetic field due to current in a coil

The coil which you used to magnetise an iron nail (page 155) also produces a field pattern. If you did not plot the field then, you might do it now. Use a coil with about 6 turns, spaced so that you can plot the field inside the coil, and use the low-voltage power supply to provide a large current. The coil may be slotted on to a piece of card supported on wooden blocks. Try using iron filings too.

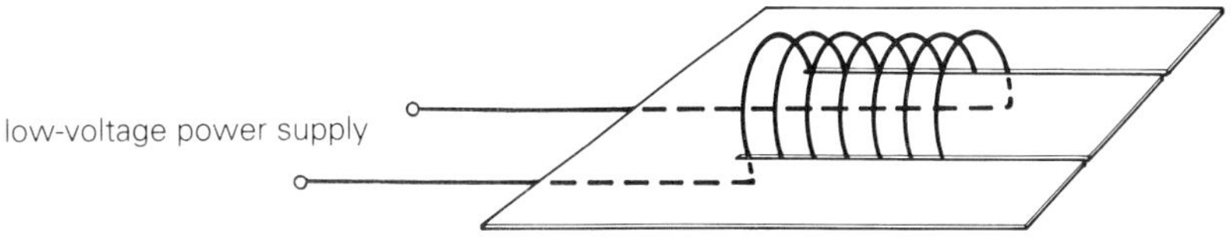

Fig 16.3

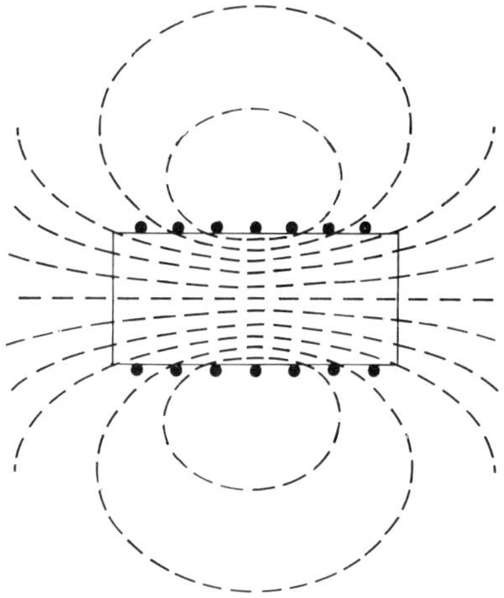

Fig 16.4

Outside the coil, the field pattern is just like that of a bar magnet, but the lines do not start and stop anywhere; they continue through the coil (where the field is strongest) and again form closed loops round the wires.

The force between a current and magnets

When a wire carrying a current is in a magnetic field, produced by magnets for example, there is a force on it. This is sometimes called the *motor force* because it is the force which makes an electric motor turn.

Experiment 16.4
The motor force

To show the motor force, you need to have a large current and a strong magnetic field. Use the field formed between two slab magnets on a U-shaped piece of steel. The slab magnets have their poles on the flat faces and it is important to make sure that opposite poles face each other when you place the magnets on

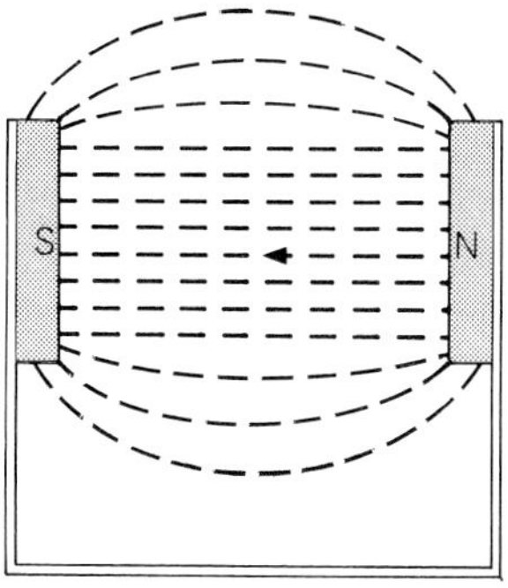

Fig 16.5

the steel. The piece of steel is now like a horse-shoe magnet with a field of the shape shown in Fig 16.5 between the poles.

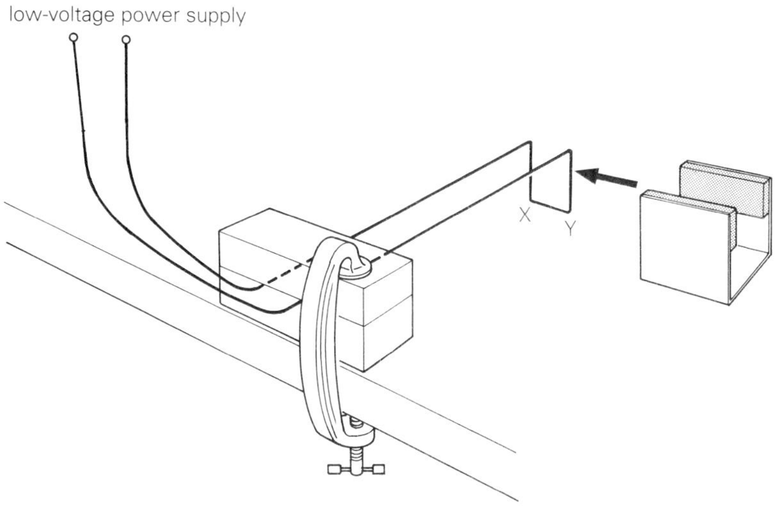

Fig 16.6

Make a long narrow loop of thin, bare, copper wire as shown in Fig 16.6, and clamp it between two small blocks of wood. Connect the loop to a low-voltage supply (switched off) and move the magnets round the wire, as indicated in the diagram. Now switch on the current. What happens? What happens if the direction of the current is reversed? What happens if the direction of the U-shaped magnet is turned round? What happens if the magnet's field is vertical rather than horizontal? What happens if the field is parallel to the piece of wire XY?

You should have found that the effect is most marked when the magnets are around the wire as in Fig 16.6, with the magnetic field horizontal and at right angles to XY. The force is then up or down depending on the direction of the current. The motor force is greatest when the magnetic field and the current are at right angles to one another, and the force is then at right angles to both of them (see Fig 16.7). The direction of the force depends on the directions of the field and the current.

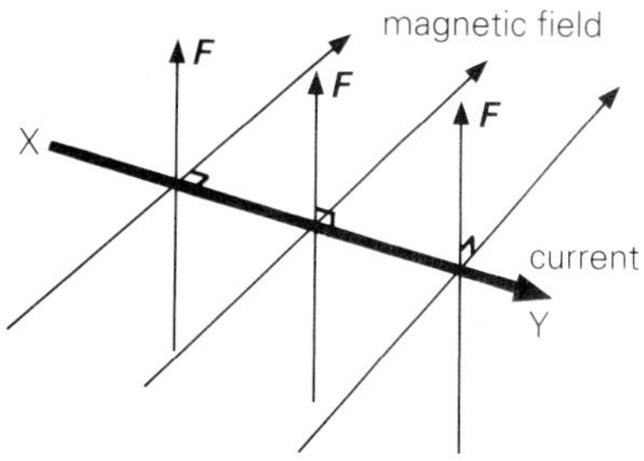

Fig 16.7

The loudspeaker works because of the motor force. You will find the loudspeaker described on pages 294–295.

The electric motor

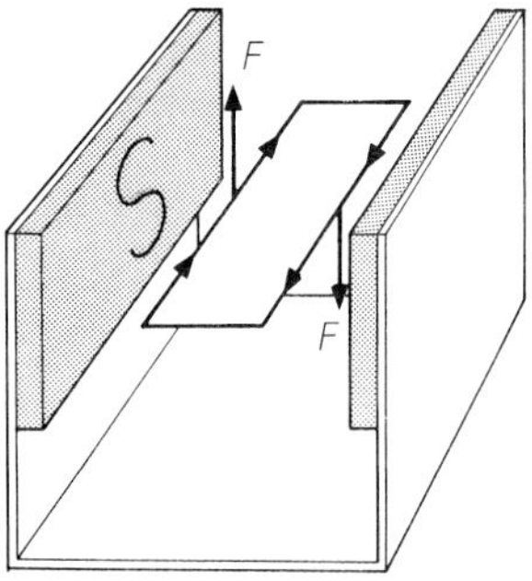

Fig 16.8

The simplest motor has a flat coil which rotates in the field of a strong magnet (Fig 16.8). If a current is passed round the coil, there will be forces, *F*, on the sides of the coil. (Can you see why these forces have opposite directions?) These forces have a turning effect and, if the coil is pivoted on a horizontal axis, it will turn (see Fig 16.9). But, when it is vertical, the forces are in line and have no turning effect. Even if the coil overshoots the vertical position, the forces now slow it down and eventually bring it back to the vertical.

To be a motor, the coil must go on turning. Some sort of automatic switch is needed which turns the current off momentarily as the coil reaches the vertical position and then causes the current to flow round the coil in the opposite direction so that the forces are reversed (compare Fig 16.9 with Fig 16.10). This automatic switch is called the *commutator*.

Fig 16.9

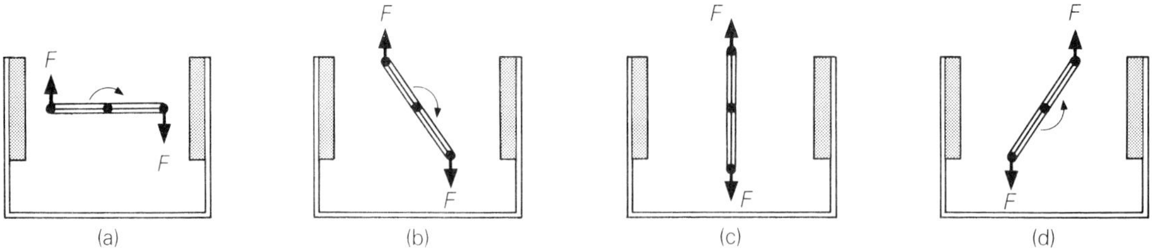

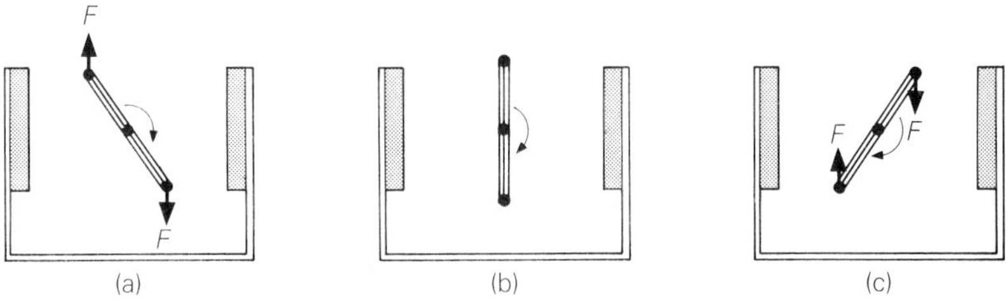

Fig 16.10

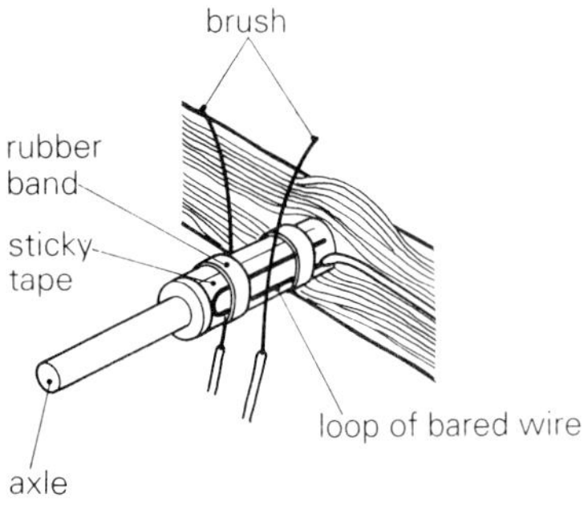

Fig 16.11 Details of commutator and brushes

To make a simple commutator, the ends of the wire forming the coil are bared and fixed to the axle as shown. Of course, the axle must be insulated so that it does not short-circuit the coil. Then, two other bare wires are arranged so that they press against the axle, as shown in Fig 16.11. These wires are called the brushes.

The brushes are connected to a battery or a low-voltage supply. Fig 16.12 shows how the current flows through the brushes to the coil as the coil turns. The brushes and the commutator act as an automatic reversing switch.

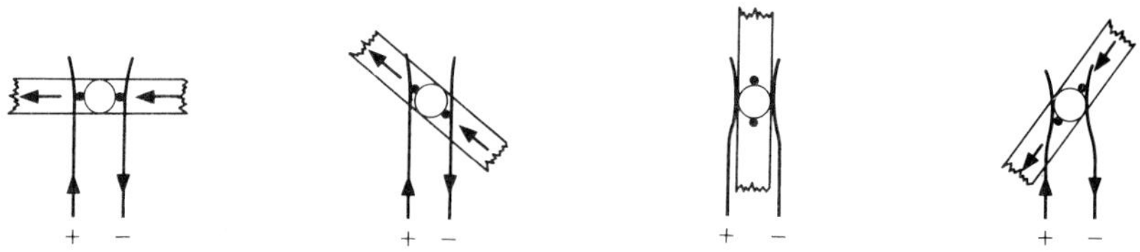

Fig 16.12

Experiment 16.5
Making the model motor

You should now be able to make the model motor. The diagrams which follow should be sufficient to show how it is done using a simple kit of parts.

Fig 16.13

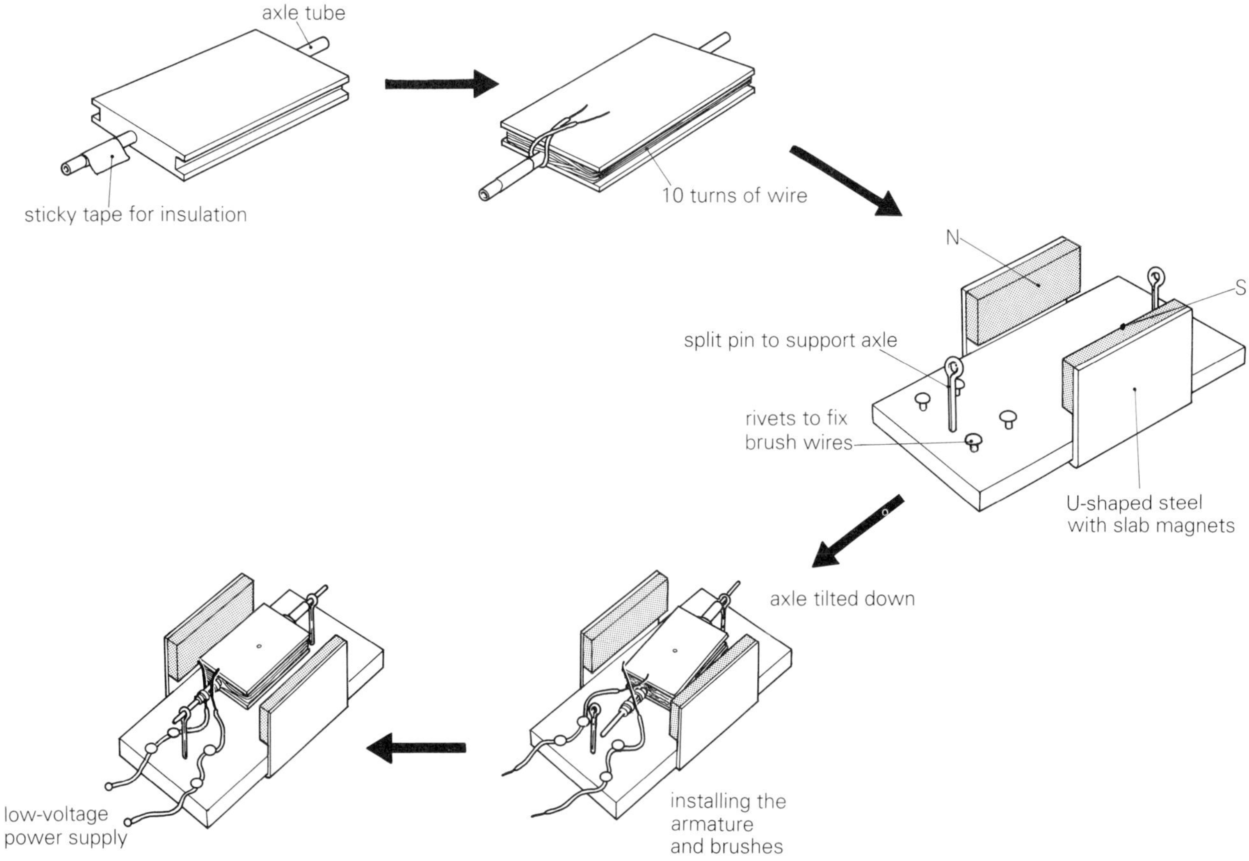

When you have constructed your model motor, connect it to a cell or to a low-voltage supply. If it does not work, the fault will most likely be in the brushes and commutator. Make sure the commutator is in the right position for the current to be switched when the coil is vertical, and there is enough 'bend' in the brush wires for them to press against the commutator as the coil turns.

Do not dismantle your motor until you have done Experiment 16.6.

Real motors

The model motor is not very powerful and runs somewhat jerkily. It would run more smoothly if there were two coils at right angles to one another and the commutator had 4 contacts, 2 for

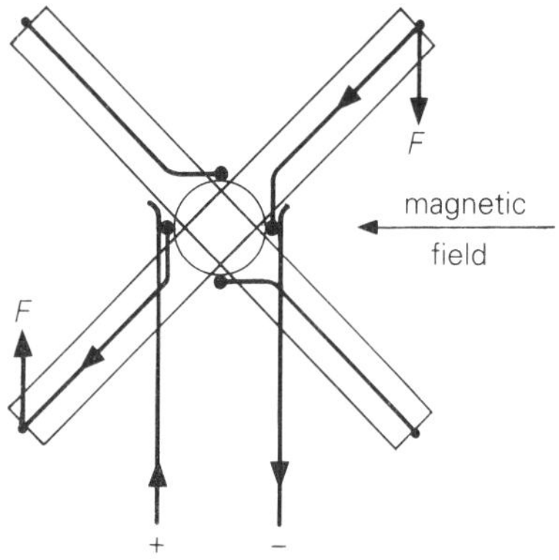

Fig 16.14

each coil. They could be arranged so that each coil was used in turn and switched into operation at the right moment to provide the greatest turning effect. Fig 16.14 shows this arrangement.

In a powerful motor, such as the motor in an electric drill, there are many coils, equally spaced, and a commutator with many segments. The coils are wound round a cylinder of soft iron in order to make the magnetic field as strong as possible. The field is provided by electromagnets rather than permanent magnets. (Permanent magnets are used only in small motors such as those used for driving model trains.) The brushes are two small blocks of carbon which are kept pressed against the commutator by springs. These brushes gradually wear away with use and must be replaced occasionally.

Fig 16.15 is a drawing of the inside of an electric drill. Notice the brushes and the commutator, and the coils wound in grooves in the soft iron cylinder.

Fig 16.15

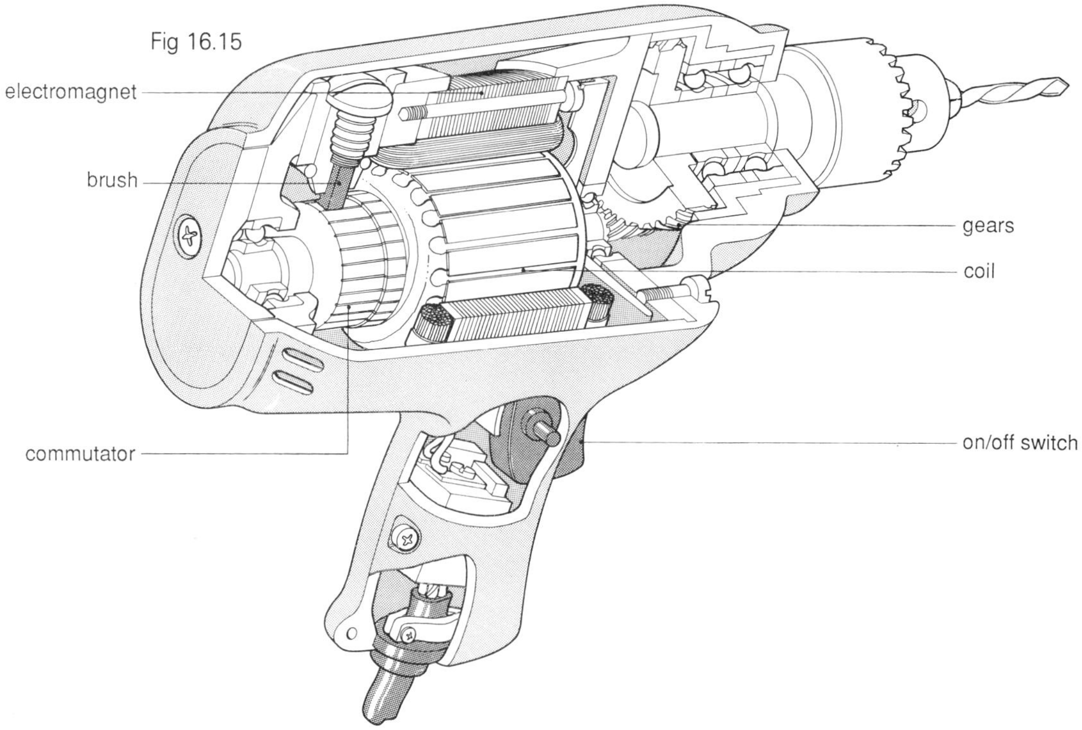

Experiment 16.6
The model motor as a dynamo

Connect the brush wires of your model motor directly to a galvanometer (without a power supply, of course) and spin the coil with your fingers. What happens?

As the coil moves in the magnetic field, the galvanometer needle should deflect. A voltage is created which causes a current to flow as the coil moves. This effect is known as **electromagnetic induction**. It was discovered by Michael Faraday in 1831.

Background Reading

Michael Faraday and electromagnetic induction

'Michael Faraday, English chemist and physicist, was born at Newington in Surrey on 22nd September 1791. His parents had migrated from Yorkshire to London, where his father worked as a blacksmith. Faraday himself, at the age of 14, was apprenticed to a bookbinder. He continued to work as a bookbinder until March 1813 when he was appointed an assistant in the laboratory of the Royal Institution on the recommendation of Sir Humphry Davy, whose lectures on chemistry had inspired him to enter the service of science. Faraday carefully wrote up the notes on those lectures, bound the sheets, drew neat illustrations and sent the volume to Davy with the request that he be given a place as an assistant in the Institution!

Faraday was largely self-educated. He accompanied Davy on a tour through France, Italy and Switzerland from October 1813 to April 1815. He was appointed Director of the laboratory in 1825; and in 1833, he became the Professor of Chemistry in the Institution for life without the obligation to deliver lectures. He died at Hampton Court on 25th August 1867 and lies buried in Highgate cemetery.'

(Adapted from an article in the *Encyclopaedia Britannica*, 1951 edition)

Michael Faraday was a great experimenter and lived during the times when many exciting things were being discovered about electricity. Faraday turned his attention to this part of physics and soon began making discoveries. Indeed, he is now far better known for his work as a physicist than as a chemist.

Why a motor can be used as a dynamo

Michael Faraday knew that electric current caused resistance wire to get hot (the heating effect). He also knew that if the junctions of a thermocouple were at different temperatures, a small electric current would flow (the **thermoelectric** effect). Fig 16.16 shows a thermocouple, connected to a galvanometer, generating a current.

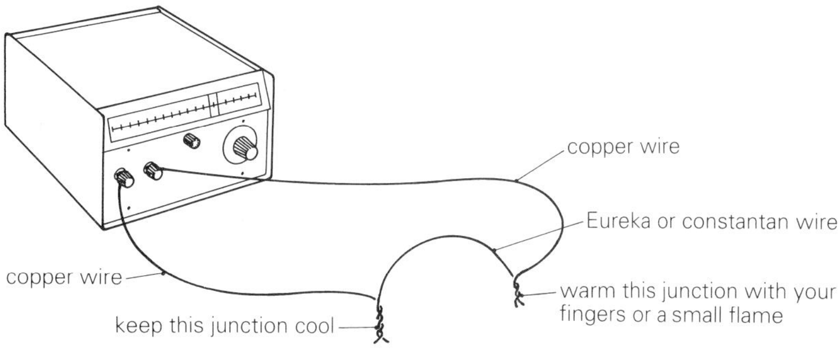

Fig 16.16

Faraday also knew that electric currents could be produced from chemical changes (a cell) and that chemical changes could result from the passage of an electric current (electrolysis). No doubt he thought that, because electric currents produced magnetic effects, (see Expt 16.2), it ought to be possible to produce electric currents from magnetic fields.

In trying to do this, he used two coils of many turns wound on a wooden cylinder. He made a magnetic field by passing currents from a battery through one of the coils (A in Fig 16.17). The other coil, B, was connected to a sensitive galvanometer to see if a current flowed. Of course, great care was taken to make sure there were no electrical connections between the two coils. Faraday noticed that the galvanometer showed a sudden slight deflection in one direction when switch S was closed, and another slight deflection in the opposite direction when switch S was opened. But, there was no deflection (that is, no current through B) when the current through A was steady. Faraday had discovered that a current could be *induced* to flow in one circuit (coil B) when there were *changes of current* in another circuit (coil A) nearby.

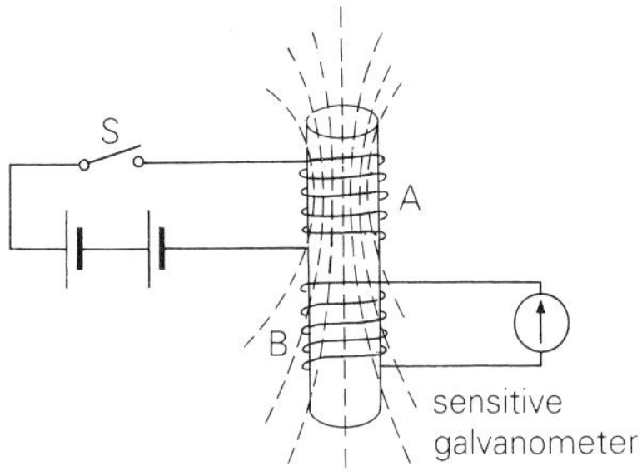

Fig 16.17

Faraday believed that this was due to the magnetic effect of the current flowing in coil A. This magnetic field passed through coil B and current was induced in B when the field changed. The simplest way of exploring the effect is to use a magnet to produce the magnetic field.

Experiment 16.7
Induced voltage due to a moving magnet

Connect a coil to a galvanometer as shown below.

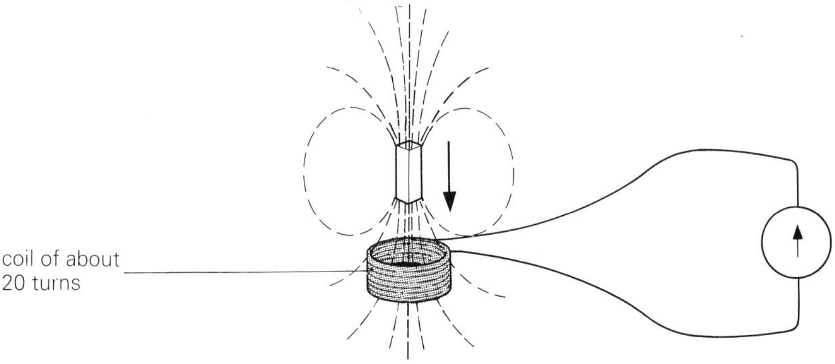

Fig 16.18

Push the magnet slowly into the coil and then pull it out again. The galvanometer flicks one way as the magnet moves in, and the other as it moves out. The galvanometer needle only moves whilst the magnet is moving. It is the change of the magnetic field passing through the coil which causes a voltage to appear (an induced voltage) and that voltage causes a current to flow.

Even with this simple experiment there are many things to try.

What happens if you poke the other pole of the magnet into the coil?

What happens if you move the coil towards and away from a stationary magnet?

What happens if you move the magnet faster or slower?
What happens if you use two magnets held side by side?
What happens if you have more or fewer turns on the coil?
What happens if you have a coil of larger diameter?
What happens if you move the pole of the magnet sideways across the top of the coil?

To obtain large induced currents, a coil of many turns is needed with strong magnets. The movement should be rapid but it does not matter whether it is the coil or the magnet which moves, or whether the magnet moves towards and away from the coil or across the top of it.

Now you can begin to understand why a motor can also be used as a dynamo. Inside the motor there are magnets and coils. When you use the motor as a dynamo, you have to rotate the moving parts by hand or by another motor. Now the coils rotate past the magnets and an electric voltage is induced in the coils.

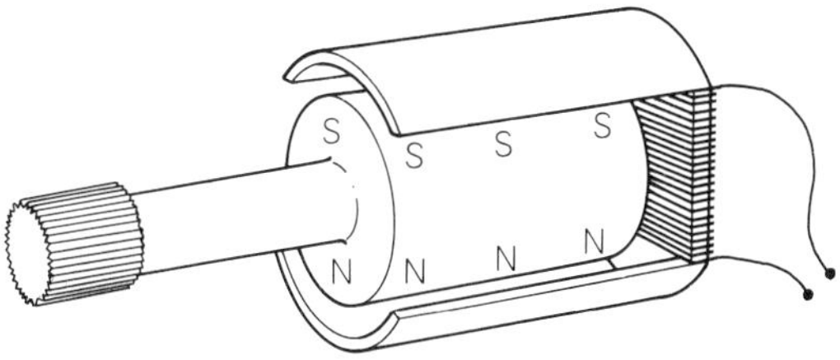

Fig 16.19

Fig 16.19 shows the fixed coil and the moving magnet of a simple bicycle dynamo. The coil used is wound on a piece of soft iron. This is easy to magnetise and makes the magnetic field which passes through the coil much stronger. The magnet rotates between the ends of the iron. If you can find an old bicycle dynamo, take it to pieces to see how it is made.

Experiment 16.8
The effect of iron

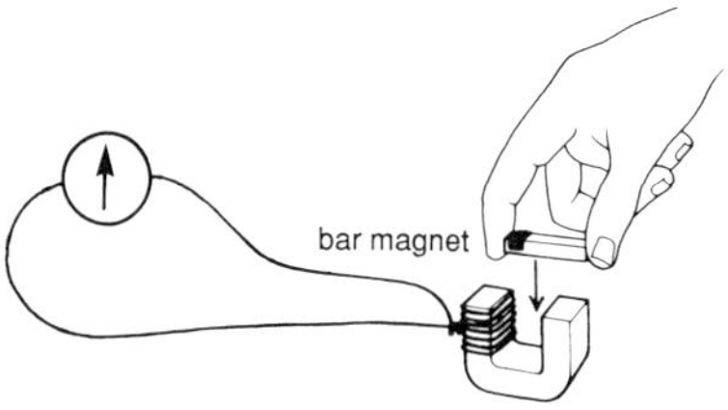

Fig 16.20

In the bicycle dynamo, a larger voltage is obtained by winding the coil on a piece of soft iron. To show this, wind a coil of 20 turns on an iron C-core and connect it to your galvanometer. Now bring a bar magnet up to the C-core (Fig 16.20) and notice the size of the kick. Take the coil off the C-core and try the experiment again. Is there any difference?

The effect is much greater with the iron there because the magnet magnetises the iron and the change of magnetic field in the coil is much greater. Even larger effects can be achieved using an electromagnet in place of the bar magnet.

Experiment 16.9
Induced effects using an electromagnet

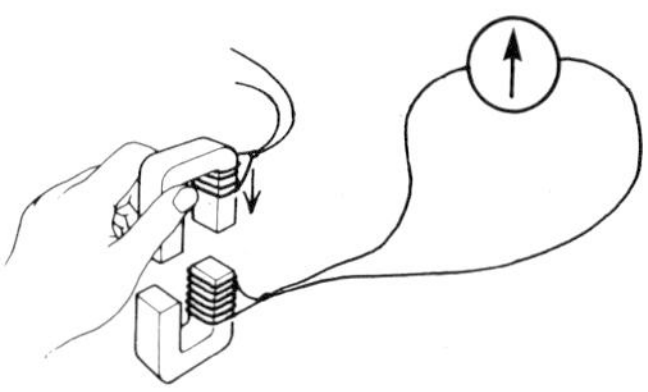
Fig 16.21

Wind a 10-turn coil on each of two C-cores. Connect one of the coils to a galvanometer and the other to a cell in order to make it into an electromagnet. Bring the two C-cores together and then separate them. The induced effect is large.

Summary

Faraday showed that, whenever a change in the magnetic field passing through a coil takes place, an induced voltage appears in the coil. The size of this voltage depends on the rate at which the field changes: that is to say, a large field changing quickly results in a bigger voltage than a small field changing slowly. If the coil is part of a complete circuit, the induced voltage causes a current to flow just like a battery does.

Background Reading

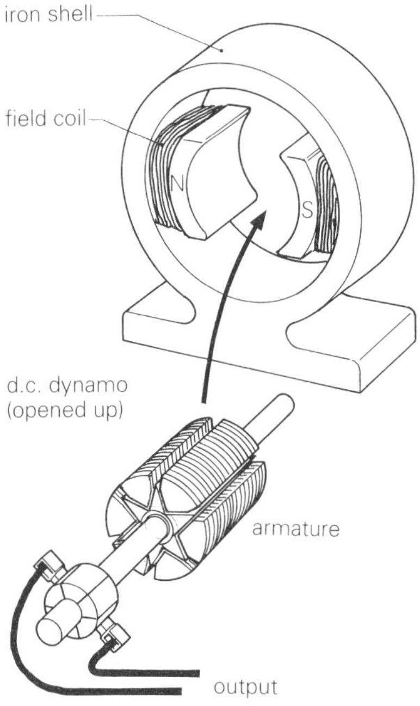

Fig 16.22

Dynamos

A dynamo spinning at constant speed with its magnet kept at constant strength produces a constant voltage — like a well-behaved battery of cells. Even if there is no output current, a dynamo still produces that voltage. It is *ready* to drive a current. When we let it drive a current, by connecting something to its output terminals, the amount of current depends on the resistance of the thing we connect.

Only small dynamos have permanent magnets to make the magnetic field. Large dynamos have electromagnets whose coils are fed by a little of the dynamo's own output current.

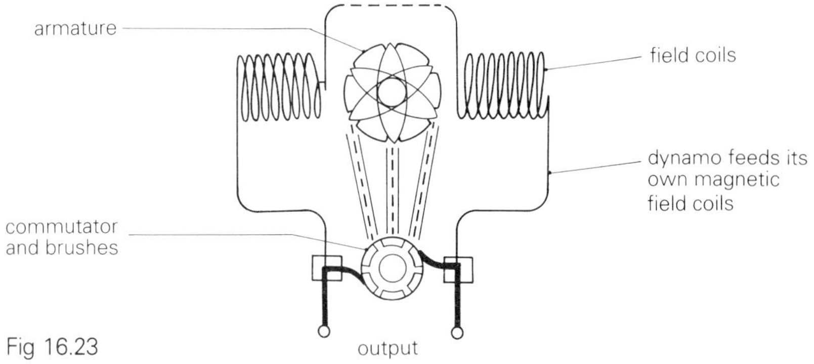

Fig 16.23

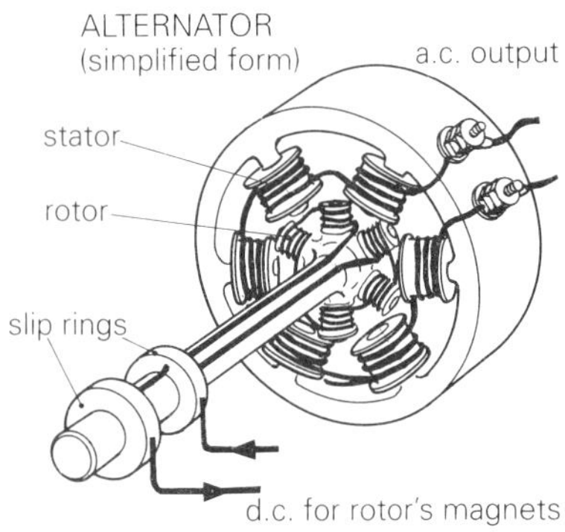

Fig 16.24

The very large a.c. generators in power stations are called *alternators*. In them, it is the assembly of electromagnets which rotates, driven by a turbine; so that is called the *rotor*. The coils in which the output voltage is generated are held in a frame outside the rotor and remain stationary. So they are called the *stator*.

This arrangement is more convenient for big machines because it does not need brushes and commutator. The spinning rotor's electromagnets are supplied with the small direct current they need through brushes and slip-rings. That direct current usually comes from a small d.c. dynamo on the same spinning shaft as the big generator.

(Adapted from Longman Revised Nuffield Physics *Pupils' Text Year*

The transformer

Experiment 16.10
Induced effects using an electromagnet

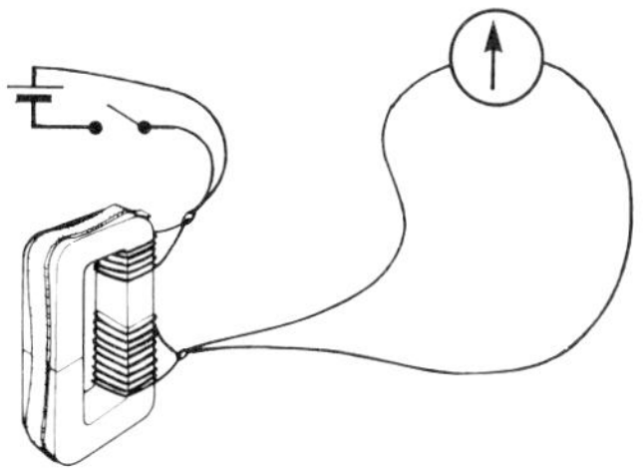
Fig 16.25

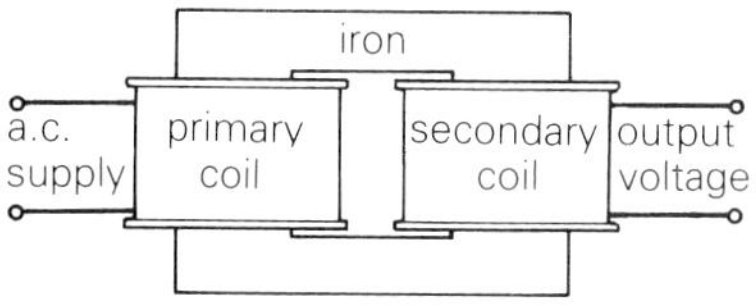

Fig 16.26

This experiment uses the same apparatus as Experiment 16.9. Clip the two C-cores together and then switch the current on and off. This 'starts' and 'stops' the magnetic field in the C-cores and there is a large induced effect because now the changes are very rapid indeed.

This experiment suggests that, if the electromagnet coil (called the *primary* coil) is supplied with an alternating current, the changing current will induce an alternating voltage in the second coil (called the *secondary* coil). This is because the changing current in the primary coil causes the magnetic field in the iron core to change and these changes induce a voltage in the secondary coil. If there is a circuit connected to the secondary coil, then a current will flow. The largest voltage is induced in the secondary coil if both coils are wound round a continuous 'ring' of soft iron (called the *core*) (Fig 16.26).

Thus, a transformer is a device which has to be supplied with electrical energy and then gives out electrical energy at a different voltage. Very little energy is 'wasted' in this process; the output energy is only a little less than the input energy. We say that the transformer has a high efficiency. If a transformer is 95% efficient, that means that 95 units of electrical energy are delivered at the output for every 100 units supplied to the input. What do you think happens to the other 5 units?

For a transformer which is 100% efficient,

$$\frac{\text{voltage output across the secondary coil}}{\text{voltage applied to the primary coil}} = \frac{\text{number of turns on the secondary coil}}{\text{number of turns on the primary coil}}$$

Using the C-cores as in Experiment 16.10, see if you can make a transformer which will light a 2.5 V lamp correctly using a 1 V a.c. supply. You should use a primary coil with 10 turns on it.

Transformers are used in a vast number of different circumstances. For instance, electrical energy is distributed over Great Britain at a very high voltage and this has to be reduced to 240 V for use in our homes. Transformers are used to do that. Transistor radios can often be run from the mains supply using adaptors. The adaptor includes a small transformer so that the 240 V of the mains can be reduced to about 6 V a.c. (which has to be changed into d.c. using diodes) from which the circuits operate. In a TV set, the high voltage needed to work the picture tube is obtained from the mains supply voltage by using a transformer to step up the voltage, whilst the low voltage required for the electronic circuits is from a transformer which steps down the voltage. Indeed nearly all low-voltage equipment which works from the mains supply includes a transformer – just like the low-voltage power supply used in your laboratory. And all this – dynamos, alternators, transformers – stems from Faraday's discovery of how to generate an electric current using a magnetic field!

Questions for homework

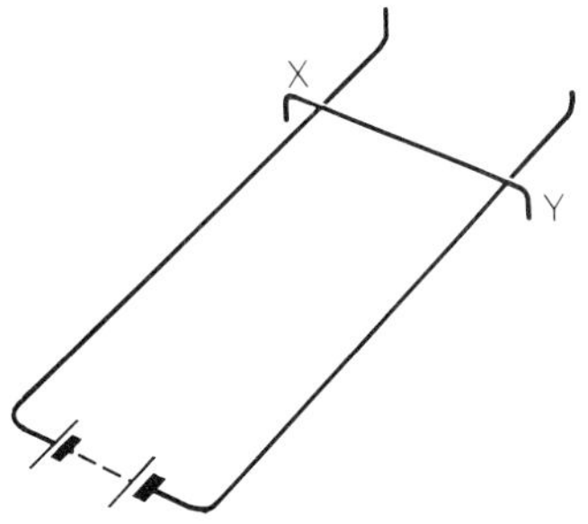

Fig 16.27

1 The diagram shows two bare, parallel wires held horizontally and connected to a battery. Another piece of bare wire, XY, is laid across them so completing a circuit. Explain with the help of a diagram how you would hold a U-shaped magnet so that XY slides along the parallel wires.

2 Say what you can about the forces acting in each of the following cases.

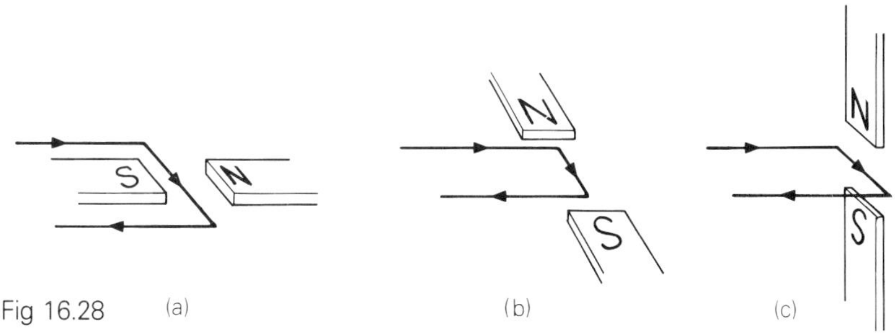

Fig 16.28 (a) (b) (c)

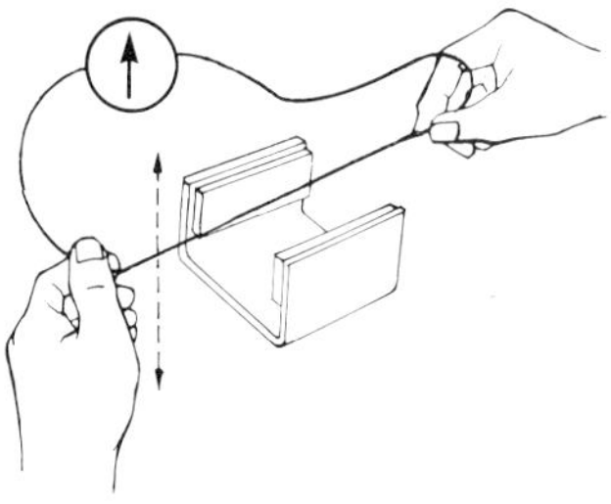

Fig 16.29

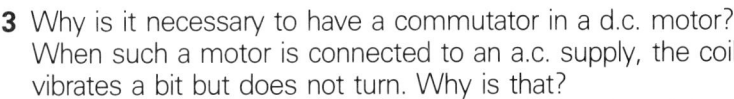

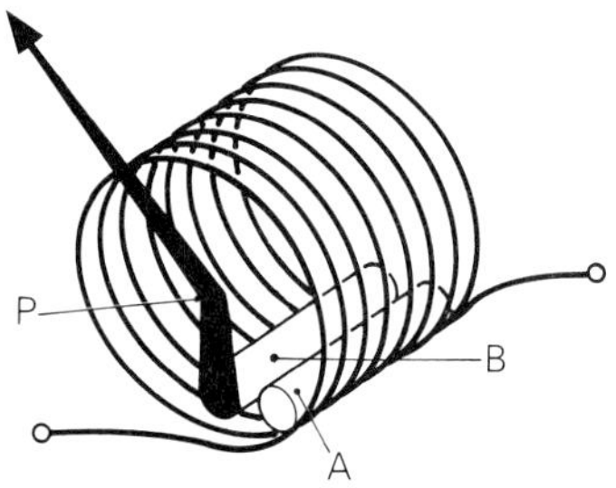

Fig 16.30

3 Why is it necessary to have a commutator in a d.c. motor? When such a motor is connected to an a.c. supply, the coil vibrates a bit but does not turn. Why is that?

4 Using the apparatus illustrated in Fig 16.29, it is found that the galvanometer kicks two divisions to the left when the wire is moved downward through the magnetic field. How would you make the galvanometer kick
- **a** two divisions to the right,
- **b** four divisions to the left,
- **c** one division to the right?

How could you move the wire so that the galvanometer does not deflect at all? (There is more than one correct answer.)

5 A transformer used for driving a model train works from the mains supply, 240 V, and has an output voltage of 12 V. If the transformer is very efficient and the primary coil has 1000 turns, how many turns are there on the secondary coil?

Explain what is meant by saying that the transformer is 93% efficient.

6 Fig 16.30 shows the working parts of one kind of ammeter. How do you think it works? (A and B are two soft iron rods inside a coil of wire, A being fixed to the coil. B is fixed to the end of the pointer which is pivoted at P.)

How do you think an ammeter designed to measure a small current will differ from one meant to measure a large current?

Explorations 3

A For an athlete to jump as far as possible in a long jump, it is important that he or she takes off at the best angle.

Use a jet of water from a tap to explore what is the best angle to project the water so that it goes as far as possible.

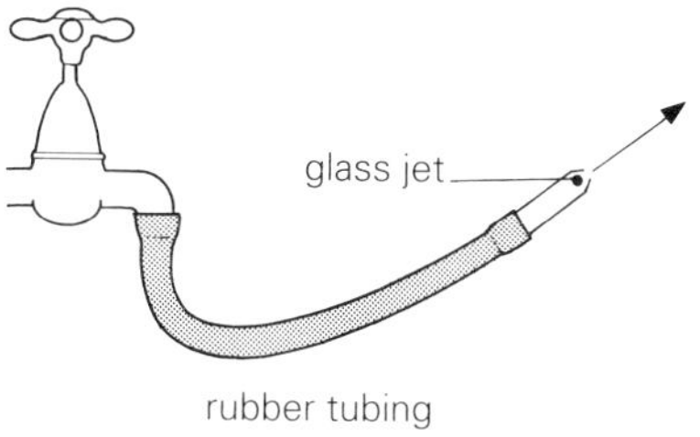

How does the distance travelled change with the angle?

B Explore how drops are formed. This is a test of observation.

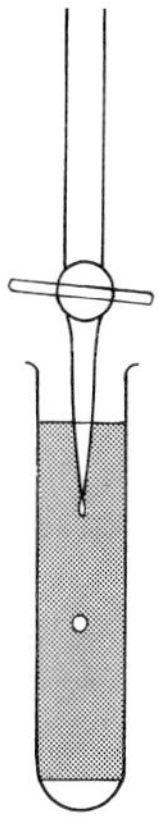

Do it by allowing drops of water to fall from a burette into a test tube of olive oil (or drops of glycerol into brine).

C A manufacturer of copper-plated paper clips needs to plate them as quickly as possible, but not so quickly that the copper does not adhere to the steel properly.

Do some experiments which will help him to know what current to use.

D An electromagnet can be made from a small piece of iron rod with turns of insulated wire round it. Explore if the strength of the magnet depends on the current flowing or on how many turns of wire there are.

You will need to use the circuit shown and you will have to decide on some reliable way of showing the strength of the magnet.

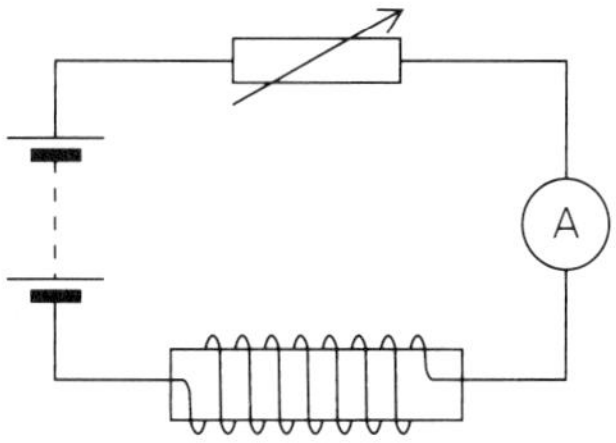

E The object of this exploration is to find out how cells run down and how long they last.

Using three new cells (of the same make and type) set up the three circuits shown. Adjust the variable resistor so that the readings of the three ammeters are as different as possible.

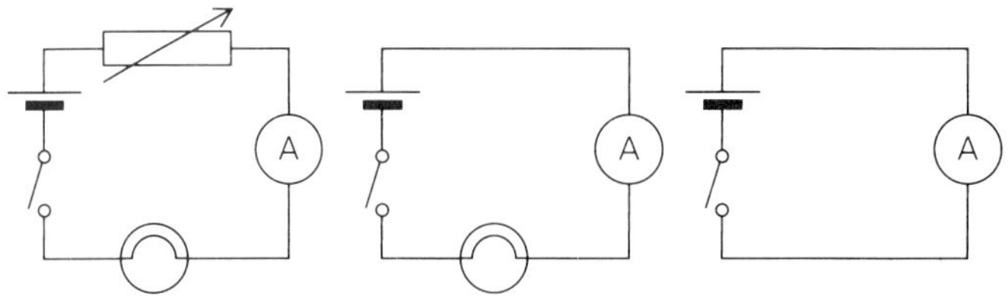

Find how the currents change with time after the currents are switched on.

F This is an exploration of LEDs. Use the circuit shown in the diagram below.

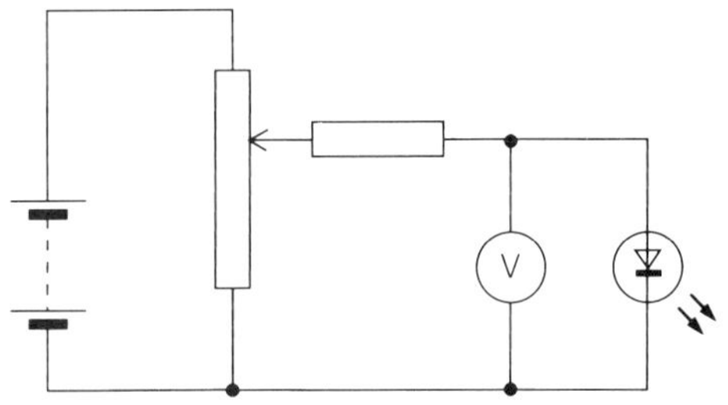

As you apply more voltage across an LED, more current flows through it and the brighter it glows. But does there have to be a certain voltage across the LED before it emits any light?

And if that is so, is the voltage for a red LED always the same?

And is it different for a green LED?

(To do this, you will have to shield the LED from external light in order to find out when it starts glowing.)

Chapter 17 A further look at energy

Measuring energy transferred

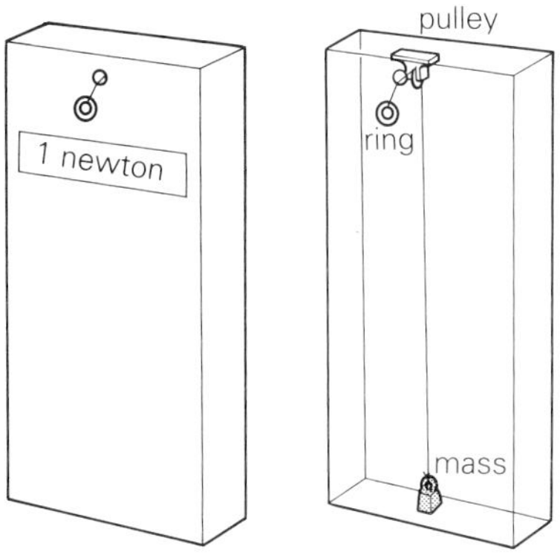

Fig 17.1

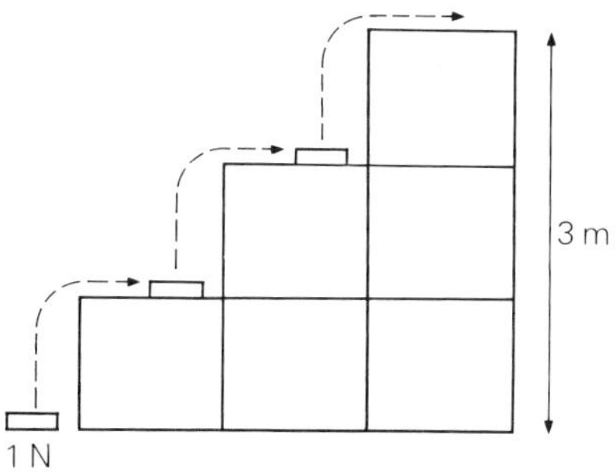

Fig 17.2

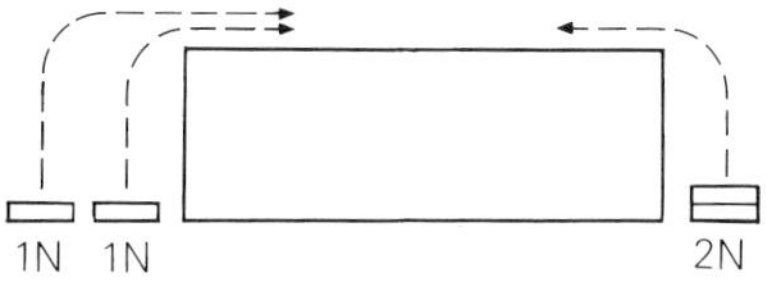

Fig 17.3

In Chapter 11 you were introduced to energy in several different forms. There were many examples of energy being transferred from one form to another. In this chapter we will discuss how the energy transferred is usually measured.

You are certainly familiar with the word *work*. We speak about doing work digging a garden or lifting sacks of potatoes; we may say we are doing some hard work reading a history book. But this everyday use of the word *work* is a bit vague and in science a precise meaning has been given to it. **Work is a measure of the amount of energy transferred.**

Energy is transferred whenever a job of work is done. When that happens *forces* are used to lift things or to move things. How much energy is transferred has to do with the force used and with the distance moved.

You are already familiar with the forces demonstration box. If you pull on the ring, you can feel a force of 1 newton. You will also know how the box is made: when you pull with a force of 1 newton, a mass inside the box is raised (Fig 17.1). If you pull the force through a distance of 1 metre, the mass is raised 1 metre. In other words chemical energy in your body has been transferred into uphill energy of the mass. Another name for uphill energy is **gravitational potential energy**, or more shortly **potential energy**, and we will use this in this chapter. The more usual scientific name for motion energy is **kinetic energy** and we will also use that in future.

If an object weighing 1 newton is lifted a height of 1 metre, it gains a certain amount of potential energy. Each time it is lifted this distance, its potential energy changes by the same amount. So, if the 1 newton weight is lifted 3 metres, its potential energy changes by three times as much as when it is lifted only 1 metre (Fig 17.2).

How much energy is gained also depends on the force needed to lift the object. Lifting a 2 newton weight is the same as lifting two separate 1 newton weights (Fig 17.3). So a 2 N weight gains twice as much energy as a 1 N weight would if lifted the same distance.

Scientists have therefore decided to measure the work done by multiplying the force used and the distance moved.

WORK = FORCE × DISTANCE MOVED

The work done (and the energy transferred) is measured in a unit called the **joule** (J). If a force of 1 newton is used to lift an object 1 metre, then the object has gained 1 J of potential energy. (This has come from the chemical energy in your body, but more than 1 J of chemical energy will have been transferred as you will also be a little hotter. So, chemical energy will have been transferred to internal energy as well.)

We have defined work as the energy transferred and have said that it is calculated from the force applied multiplied by the distance moved. But we must be careful about what we mean by the distance moved.

Imagine a large block of ice with a mass of 5 kg resting on a flat glass surface (Fig 17.4a). It will be very easy to push the block sideways through a distance of 1 metre since there will be only a small frictional force of a few newtons (Fig 17.4b).

Fig 17.4

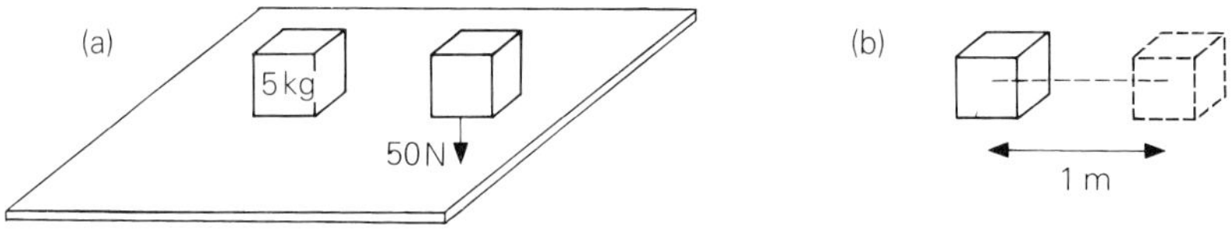

But the downward force due to gravity is 50 newtons and a lot of energy has to be transferred to lift the block upwards a distance of 1 metre.

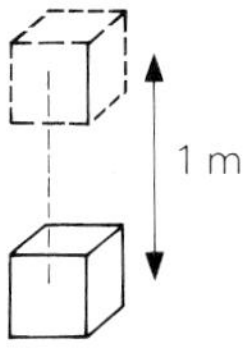

Fig 17.5

Each time, the weight of the block was 50 N, but when the block was moved to one side, it was not moved in the direction of the 50 N force and so the work done in sliding the block 1 m is considerably less than the work needed to lift the block upwards through 1 m.

So when we say the work equals the force × the distance, it must be the distance moved in *the direction of the force.*

Questions for class discussion

1 If one joule of energy is transferred when raising a mass through 1 metre, how much energy is transferred raising it through 2 metres? Through 3 metres? Through 10 metres?

2a If the strength of the gravitational field is 10 N/kg, what is the force exerted by gravity on a mass of 10 kg?

b What is the work done, measured in joules, when the 10 kg mass is raised through 10 metres?

c How much energy, measured in joules, has been transferred from chemical energy in your body to potential energy of the mass?

d How much energy would be transferred if the mass were 20 kg instead of 10 kg?

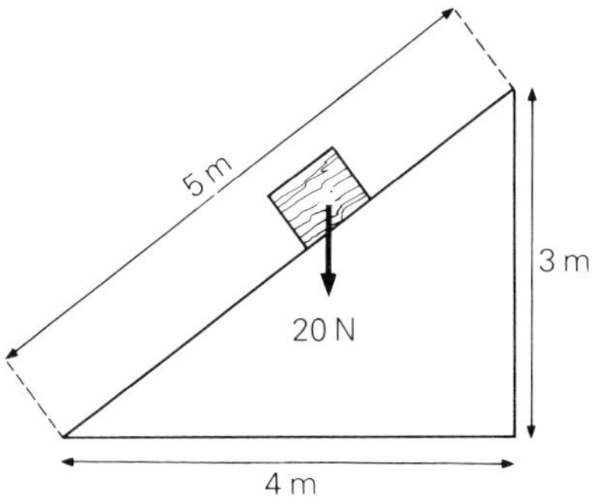

Fig 17.6

3 There are two shelves on a wall, one is 1 metre and the other is 2 metres above the ground. Pete lifts a 1 kg mass from the ground onto the lower shelf, he rests a moment and then lifts it on to the second shelf. Clare lifts another 1 kg mass from the ground and puts it straight on to the top shelf. Did Pete do more work, the same work or less work than Clare?

4 A block has a downward force on it due to gravity equal to 20 newtons. It slides down a *smooth* slope as shown in Fig 17.6. The potential energy at the top turns to kinetic energy near the bottom. In calculating the energy transferred, one student says it is the force (20 N) times the distance (5 m), in other words, 100 J. Another student says it is 20 N × 3 m, that is 60 J. Which statement is correct?

5a A crate has a weight of 100 N. A man lifts it on to the back of the lorry which is 1 m above the ground (Fig 17.7a). What is the potential energy gained by the crate? Where does this energy come from?

Fig 17.7

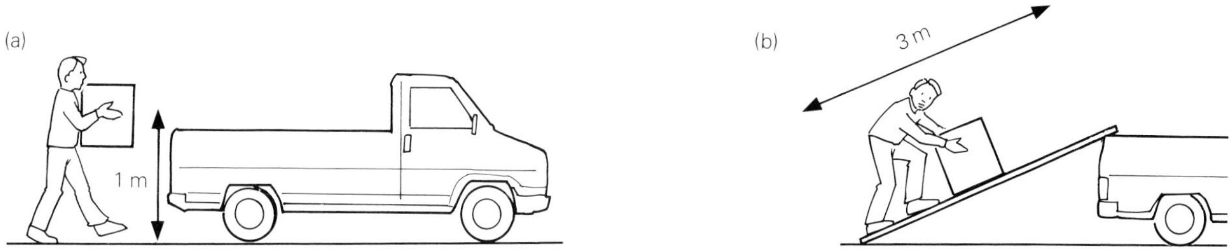

b He finds it easier to get the crate on to the lorry if he slides it up an inclined plank of wood 3 m long (Fig 17.7b). What is the potential energy gained by the crate this time? Where does this energy come from?

c In fact the man will have to transfer rather more energy pushing the crate up the rough plank than lifting it straight on to the lorry. Where has this extra energy gone?

d Why does he find it 'easier' to get the crate on to the lorry using the inclined plank?

Conservation of energy

It has been found that, whenever there is a change and energy is transferred, no energy is lost, nor is any created. The *total* amount of energy remains the same. Scientists refer to this as the **Conservation of Energy**. In some ways, energy is like money: there is a transfer of money from you to the shopkeeper and you get some sweets. You have less money, the shopkeeper has more money, but the total amount of money has not changed. And, as a result of the transfer of money, you have some sweets. When there is a transfer of energy, the amount of the initial form of energy goes down, and the amount of the final form goes up, but the total amount of energy does not change. As a result of the transfer of energy, some work has been done.

Energy is like money in another way too. Think about the explosion of a firework. Chemical energy is transferred into kinetic energy, potential energy, internal energy, strain energy, radiation, sound; it is transferred into many different forms of energy. When you spend your pocket money, you will probably spend it on many different things. The amount you spend on some things will be so small that you will hardly notice the expense, and you will wonder, when all your pocket money has gone, where you spent it. When energy is transferred, it is usually changed into several different forms, and some transfers are so small that they are hardly noticeable, so that energy *seems* to have disappeared. *Very* often, energy transfers result in an increase of the internal energy of the surroundings – in the walls of a room, for example – but the rise in temperature, as a result, is too small to notice. Nevertheless, careful experiments have shown that energy is *never* lost.

Human energy

How much energy do we need? This depends very much on our age and what we spend our time doing, but the following figures give some idea of what energy is needed per day. The energies are measured in kilojoules (kJ), and one kilojoule equals 1000 joules.

Child (1 year old)	4 000 kJ
Child (5 years old)	6 000 kJ
Child (10 years old)	8 000 kJ
Boy (12 years old)	11 000 kJ
Girl (12 years old)	11 000 kJ
Boy (18 years old)	14 000 kJ
Girl (18 years old)	10 000 kJ
Adult (light work)	11 000 kJ
Adult (very heavy work)	20 000 kJ

The energy required for a man working ranges from 10 kJ to 50 kJ every minute depending on whether the work is heavy or light. Just resting in bed requires nearly 4 kJ per minute.

This energy comes from the food we eat, and some idea of what is available from different foods is given below. In each case, the value given is the energy available from 1 kg.

Butter	32 000 kJ	Fried fish	9 000 kJ
Cheese	18 000 kJ	Eggs	7 000 kJ
Sugar	16 000 kJ	Potatoes	3 000 kJ
Beef	14 000 kJ	Oranges	1 500 kJ
Bread	10 000 kJ		

Questions for class discussion

1 How much energy would a man get from a plate of fried fish and chips? (This is an estimate question. First you must decide on the mass of the fish, then the mass of the potatoes. After that you can calculate the number of joules obtained from them.)

2 A man eats a breakfast of a glass of orange juice, two boiled eggs, two slices of bread and butter. How many joules does this give him?

If all this energy is used in lifting sacks of potatoes a distance of 10 metres, how many sacks could he lift? Would he, in fact, be able to lift that number on such a breakfast?

Kinetic and potential energy

Whenever anything is moving, it has kinetic energy. A falling raindrop, an aeroplane in flight, a crawling ant – all these have kinetic energy. Experiment 17.1 will show that rotating objects also have kinetic energy. Internal energy is at least partly due to the kinetic energy of moving or vibrating particles. The hotter the body is, the greater is the speed of movement; the internal energy is greater partly because of the greater kinetic energy. Sound energy too is in part due to the kinetic energy of the particles of the substance through which the sound is passing.

Gravitational potential energy is a kind of stored energy. Strain energy is another form of stored or potential energy. Experiment 17.2 will show the potential energy or strain energy stored in a clock spring which has been wound up.

Since energy is so often being transferred from one form to another, it is wise to measure it in the same units, namely joules, whatever kind of energy it is.

Experiment 17.1
Rotational kinetic energy

Set up the apparatus shown in Fig 17.8. Energy from the battery is transferred to the motor, which drives the flywheel. The flywheel has gained kinetic energy, but it is rotational kinetic energy, not quite the same as that of a ball thrown across the room, for the flywheel stays where it is!

If the switch is changed, the flywheel will drive the motor as a generator, which will light a lamp. It is interesting to see what happens if the flywheel is used to light first one lamp, then two lamps and finally three lamps. Make sure the flywheel is going about the same speed to begin with for each experiment. You

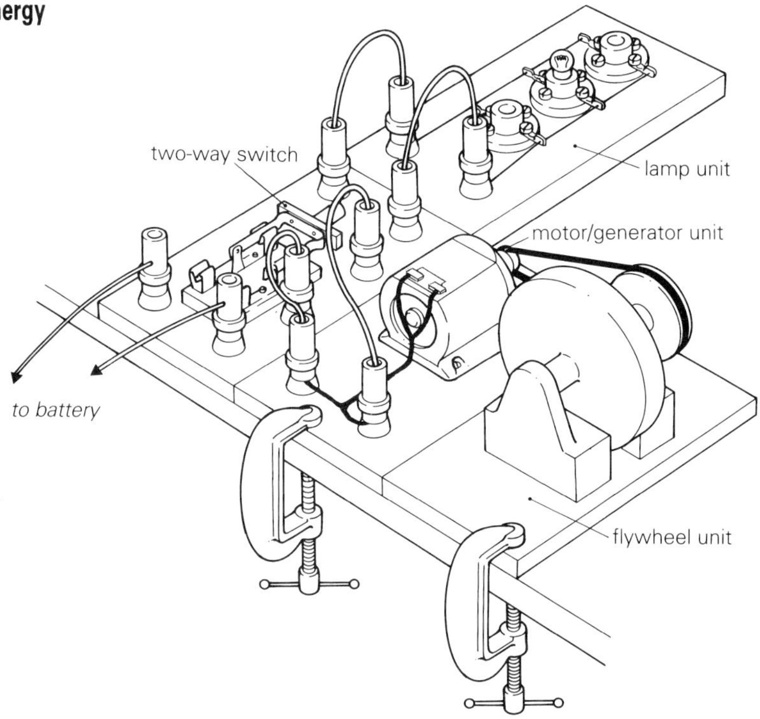

Fig 17.8

should find it slows down more quickly the greater the number of lamps. Can you explain why, in terms of energy?

Experiment 17.2
Potential energy stored in a clock spring

Set up the apparatus shown in Fig 17.9. Energy from the battery drives the motor, which winds up the clock spring. The energy is stored as potential energy in the spring.

When the energy is released, it can be used to drive the motor as a generator and the lamp will flash momentarily.

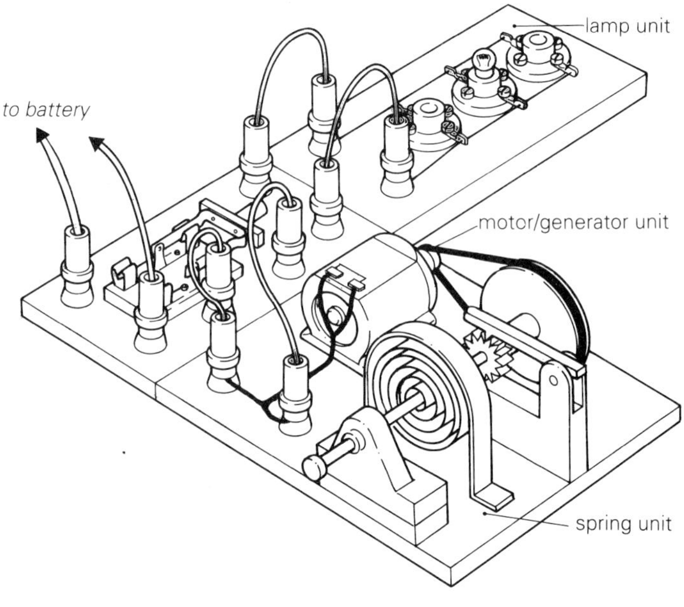

Fig 17.9

Stopping cars: forces which change kinetic energy

A force is required to change the speed of an object. For example, a forward force on a car causes it to speed up, and we say the car is **accelerating**. The greater the forward force, the more rapidly does the speed change; in other words, the greater is the acceleration. Similarly, a force in the backward direction will cause a car to slow down. This is called **deceleration**. The greater the backward force, the more rapidly the car slows down; in other words, the greater is its deceleration.

Larger forces cause the speed of an object to change more rapidly. Any size of retarding force will cause a moving car to stop, but a small force will only cause the speed to fall slowly (a small deceleration) and it will take a long time before the car stops. And it will travel a long way in slowing down to a halt. A large force causes the speed to change more rapidly and the car comes to a halt sooner, and in a shorter distance.

Fig 17.10

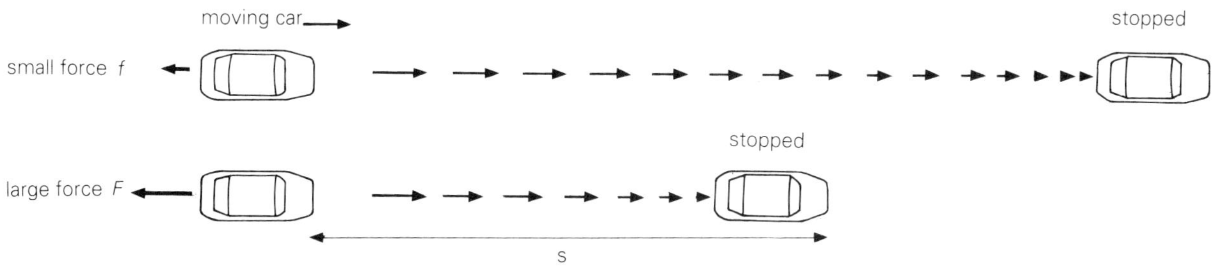

Of course, when the speed changes so does the kinetic energy. The amount of kinetic energy transferred is measured by the product of the average force applied, F, and the distance moved in the direction of the force, s.

$$\text{Energy change} = \text{Work done} = F \times s$$

Thus, if a car is brought to a stop, $F \times s$ must equal the original amount of kinetic energy that the car had. The larger the braking force is, the smaller is the stopping distance s.

A family car travelling at 50 kilometres per hour (about 30 miles per hour) has a kinetic energy of about 100 000 J. If it is stopped in a distance of 100 m, then the total retarding force, F, on the car is given by

$$100\,000 = F \times 100$$

so that $$F = \frac{100\,000}{100} = 1000\ \text{N}.$$

Most of this force comes from friction between the tyres and the road, and, if the road is icy so that the frictional force cannot be as large as that, the car will skid and travel further than 100 m before stopping.

To stop the car in a distance of 10 m instead of 100 m would require a force ten times as great, namely 10 000 N (which is a force equal to the weight of 1 tonne). The likelihood of skidding would be very much greater.

Passengers in the car also have to stop as the car stops. A passenger has less kinetic energy than the car because the passenger is less massive than the car. For example, a 50 kg passenger travelling at 50 kilometres per hour has about 5000 J of kinetic energy. If the car stops in a distance of 100 m, then so does the passenger. For the passenger

$$\text{average force} = \frac{\text{kinetic energy}}{\text{distance to stop}} = \frac{5000}{100} = 50\text{ N}.$$

That force is usually provided by the seat and from the feet pressing on the floor. A greater deceleration would require a larger retarding force. If the seat cannot provide that force, the passenger would not decelerate as quickly as the car and he or she would move forward in the car. If this happens quickly, the passenger appears to be thrown forward.

Forces involved in collisions can be very large when the distance travelled by the car in stopping is very small (or the time taken to stop is very small). If the car were to hit a wall and come to a halt in a distance of 10 cm, then the force involved would be about 1 000 000 N (the weight of 100 tonnes) and that would easily crush the car. For extra safety, some manufacturers make their cars so that the bonnet compartment crumples in a collision. The passenger compartment can then travel about 1 m during the deceleration so reducing the forces on that part of the car. The kinetic energy of the car is largely 'used up' in crumpling the front compartment.

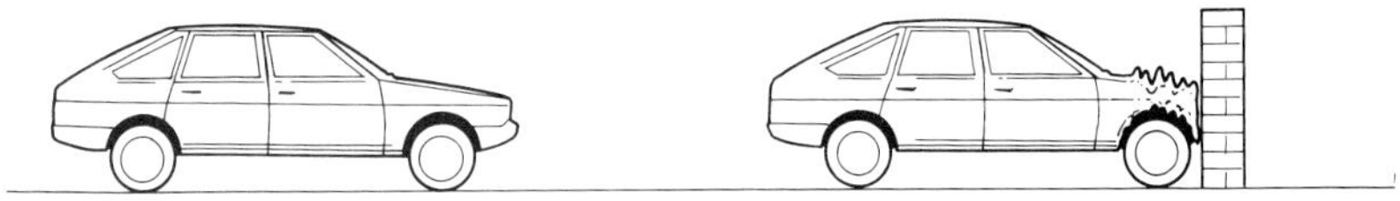

Fig 17.11

Seat belts provide further protection for passengers. In a collision, passengers not using seat belts are 'thrown forward' – particularly the upper parts of their bodies not in contact with the seat. They are then stopped suddenly in a very short distance by the part of the car immediately in front of them. The force exerted can be very large, resulting in very serious injuries. Seat belts restrain the passenger from being 'thrown forward'. The retarding forces are applied to the body over a larger distance (and longer time) and the body is brought to rest before it hits the part of the car in front. Thus the forces on the body are smaller than they would be without a seat belt, and the risk of serious injury is reduced.

Similar considerations apply when a parachutist lands. If the parachutist were to land on hard ground with legs held straight and rigid, deceleration would take place in a small distance and the force of impact would be large enough to shatter the parachutist's legs. Parachutists land with their knees bent so that the knees 'give' on landing and the body is decelerated over a larger distance, that is, more slowly. The forces required are smaller and there is less likelihood of fractures. (*Question*: Why does a good fielder draw the hands backwards when catching a fast ball at cricket?)

Machines and energy

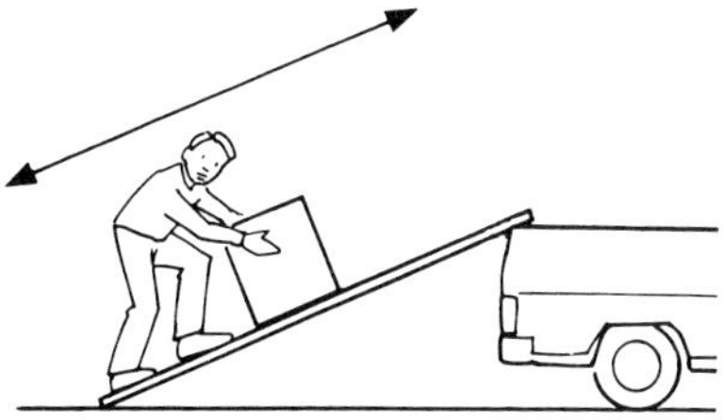

Fig 17.12

To put a crate on the back of a lorry, the crate could be lifted on or it could be pushed up an inclined plank. In both cases the same energy is transferred to the crate, but the plank enables us to do it more easily.

A machine is a device which enables us to do things more easily, that is, with smaller forces than would otherwise be necessary. A simple example of a machine is a lever.

A mass of 8 kg is placed 2 units to the right of the fulcrum. The downward force on it due to gravity is 80 N. A force of 20 N, applied 8 units to the left, will balance this. A very slightly greater force will lift the load. Using the lever, it becomes possible to move the mass with a force which is slightly greater than 20 N. But 80 N would be necessary without the lever. It thus makes things easier. Is there any gain in energy?

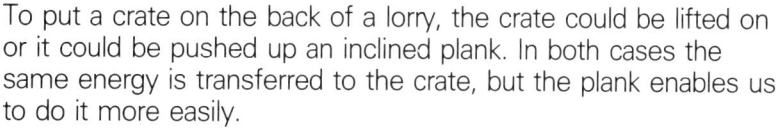

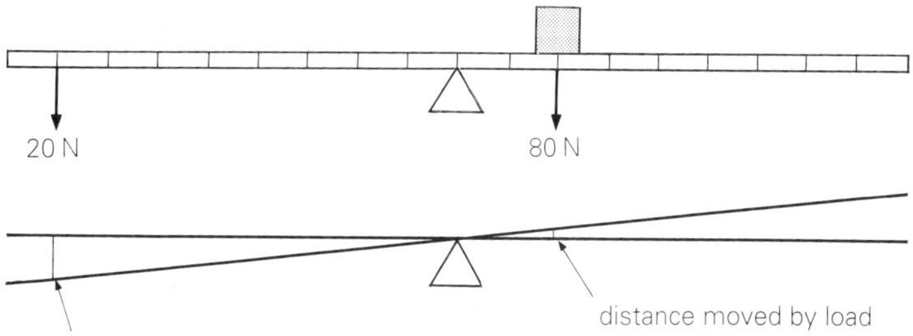

Fig 17.13

Suppose the load is moved through 1 cm. How far will the 20 N force move? From the second drawing, you can see that it is four times as far. The potential energy given to the load is

$(80\ \text{N}) \times (\tfrac{1}{100}\ \text{m})$ or 0.8 J.

The work done moving the lever is slightly greater than

$(20\ \text{N}) \times (\tfrac{4}{100}\ \text{m})$ or 0.8 J

because the force has to be slightly greater than 20 N.

In other words, the lever makes it easier to move the load, but there is no gain in energy. The smaller force moved through the larger distance. A pulley system is yet another device designed to make things easier.

Experiment 17.3
Investigation of a pulley system

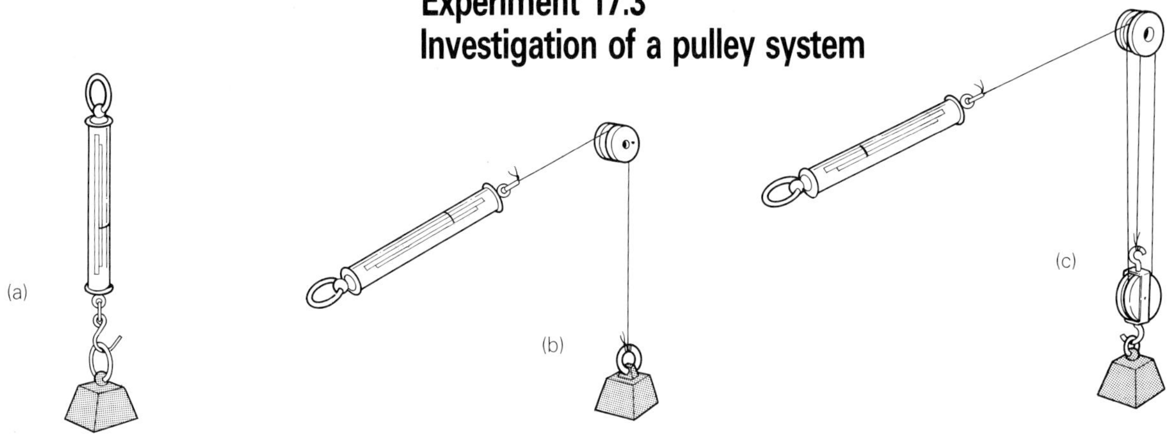

Fig 17.14

Attach a spring balance to a 1 kg mass (Fig 17.14a). Because the gravitational field is 10 N/kg, the reading will be 10 N.

Set up a pulley as in Fig 17.14b. The spring balance should still read 10N assuming the frictional forces are negligible. *The effect of the pulley is merely to change the direction of the force.*

Then set up the pulley as shown in Fig 17.14c. The reading on the spring balance is much less than 10 N, and by pulling on the string you can get the mass to rise with a smaller force than was necessary without the pulley system. Is there an increase in energy this time?

Find out how far you have to pull the string in order to raise the load 10 cm. Calculate how much energy is transferred to the load and how much energy is supplied by the effort pulling on the string. Once again you should find that there is no gain in energy. Indeed, this time you may well have to use an appreciably larger amount of energy because of the weight of the lower pulley block and the friction in the pulleys. A pulley system merely makes things easier.

Safety note

Take care not to drop weights on your toes!

Gears

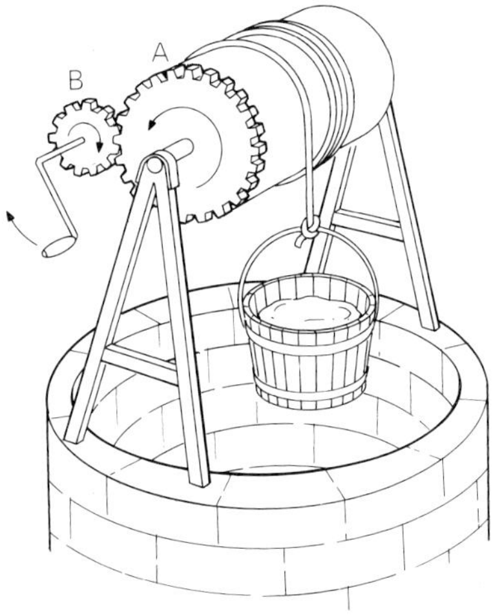

Fig 17.15

Another sort of machine uses gears. Gear wheels usually have teeth on them which mesh with one another as shown in Fig 17.15.

Rotating one wheel causes the other to rotate. Since all the teeth have to be the same size if they are to mesh correctly, wheels of different diameters have different numbers of teeth. If wheel A in the diagram has twice as many teeth as wheel B, then B will turn round completely twice for every complete turn that A makes.

If a person turns wheel B, energy is transferred to wheel A and its load. The operator has to make twice as many turns to lift the load than if the handle were fixed to A directly, but the turning force will be smaller (about half as big). (Will the energy transferred by the person turning the wheel be different?)

Questions for homework or class discussion

1a 'Water from a reservoir runs down pipes to drive turbines. These turbines produce electricity which is used to drive a railway train.'

b 'At night, power stations supply electricity to pumps. These pump the water back up to the higher reservoir.'

For each of the above, list the energy changes involved at each stage, stating what form the energy is in.

Suggest the reason for using electricity to pump water to the reservoir when it might be used for heating our homes.

2 Give one example of each of the following energy transfers.
- **a** Kinetic energy to gravitational potential energy.
- **b** Potential energy to kinetic energy.
- **c** Kinetic energy to internal energy.
- **d** Chemical energy to internal energy.
- **e** Electrical energy to radiation (light) energy.
- **f** Potential energy to electrical energy.

3 A man has a mass of 70 kg. He walks up a hill 200 metres high.
- **a** If the strength of the gravitational field is 10 N/kg, what is his weight?
- **b** What is the gravitational potential energy gained in climbing the hill?
- **c** This energy comes from the chemical energy stored in the man's body through the food he has eaten. The actual chemical energy used in climbing the hill is about three times as great as the gravitational potential energy calculated in **b**. What has happened to the rest of the energy?
- **d** When he walks down the hill, what happens to the potential energy which was gained on the way up?

4 Use the details given on page 224 to work out how much butter, cheese, sugar, beef and eggs a heavy manual worker might need to eat in a day.

5 In a bicycle, the 'gear' wheels do not mesh. Instead, they have a linked chain going round the chain wheel and the sprocket wheel.

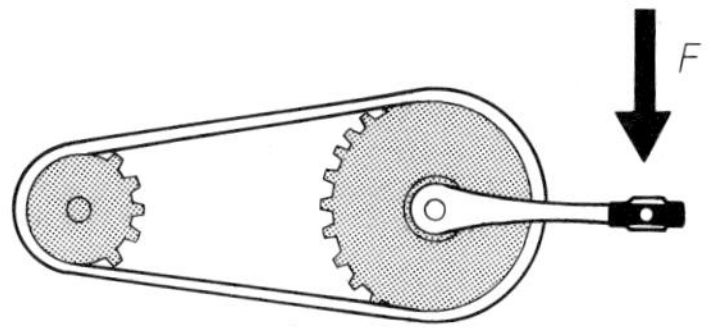

Fig 17.16 sprocket wheel chain wheel

- **a** How does the force *F* compare with what it would be if the pedals drove the rear wheel directly?
- **b** Why is the chain wheel made larger than the sprocket wheel?
- **c** How are different gear ratios obtained?

Chapter 18 **Heating things**

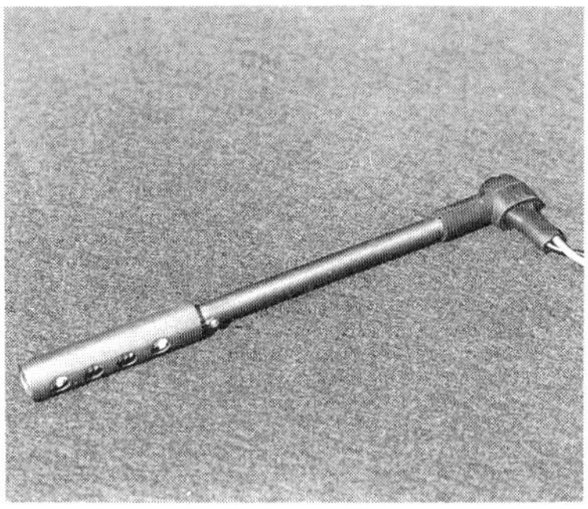

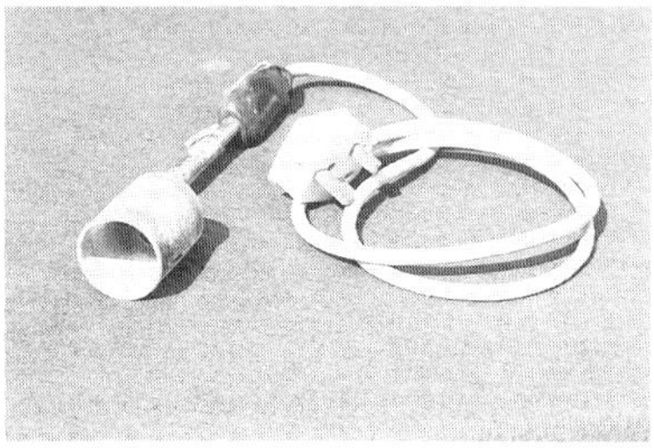

Immersion heaters

When a 12-volt battery is connected to a car headlamp bulb, energy is transferred from chemical energy of the battery to internal energy in the lamp and to light. Because the lamp gets hot, it could be used to warm water if it were immersed in it. In fact that would not be a sensible thing to do as the water might short-circuit the lamp. Indeed, if the lamp had been a mains lamp running from the mains supply, it would be a very dangerous thing to do. Later in this chapter, a heater will be used which can be immersed in water; the two photographs show such immersion heaters, the one below being for the mains supply, and the one above for a 12 V supply.

When the internal energy of a body increases, it gets hotter. Doubtless you know from experience what hotter means. In fact our senses are not always very good at deciding whether something is hot or cold, as the experiment on page 45 showed. It is much more satisfactory to use a thermometer to decide which is the hotter of two objects.

The commonest kind of thermometer which you will use in your laboratory is a mercury thermometer, marked in degrees Celsius (named after the Swedish scientist who suggested the scale). On this scale 0 °C is the temperature at which water freezes. 100 °C is the temperature at which water boils. Normal room temperature might be about 18 °C. A very hot summer day might be 30 °C.

The melting point of water is given the value 0 °C and the boiling point 100 °C in order to mark the scale on the thermometer. For pure water at a pressure of 76 cm of mercury, these temperatures are always the same all over the world and are very convenient for ensuring that we all use the same scale of measurement.

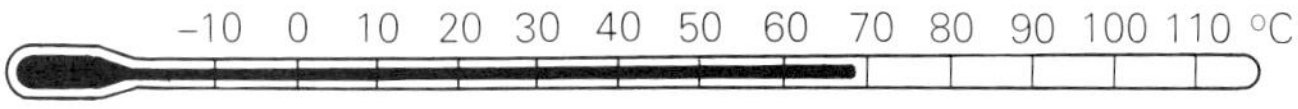

Fig 18.1

You might find it surprising that water continues to boil at 100 °C when you continue heating it. You might expect that if you heat something, it always get hotter, but that is not always true. If you have some ice at 0 °C and warm it, its temperature does not change; it melts and becomes water at 0 °C instead. The energy that is added at 0 °C is used to break the bonds which hold the molecules together as a solid so that it becomes liquid. When water boils, the temperature stays at 100 °C whilst the water is boiling. Water molecules are leaving the water surfaces in the bubbles and becoming steam. As they escape, the molecules still in the liquid attract them back (see page 116) and work is done. The heater which keeps the water boiling supplies the energy for that.

Experiment 18.1
The heating effect of an immersion heater

Fig 18.2

a Put a kilogram of water into an aluminium saucepan. Use a thermometer to measure the temperature of the water. Connect the 12-volt immersion heater to a 12-volt supply and put it in the water. Find out how long it takes for the temperature of the water to rise 5 °C.

When doing this experiment, the water must be kept well stirred. Why is this necessary?

b Pour away the water and do the experiment again, using only $\frac{1}{2}$ kg of water. How does the time compare with that needed for 1 kg?

c How long do you think it would take to heat 1.5 kg of water through 5 °C? Decide what you think is likely to be the answer and then do the experiment and see if the result agrees with your forecast.

d If you have time, it is an interesting experiment to put 1 kg of water in the saucepan, to switch on the immersion heater and to watch how the temperature rises. Keep the water well stirred and read the thermometer every minute. Plot the readings on a graph with temperature along the vertical axis and time along the horizontal axis. The temperature rises steadily at first. Why does it not go on doing so?

Experiment 18.2
Using the heater on a lower voltage

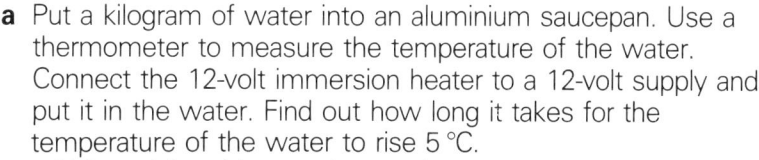

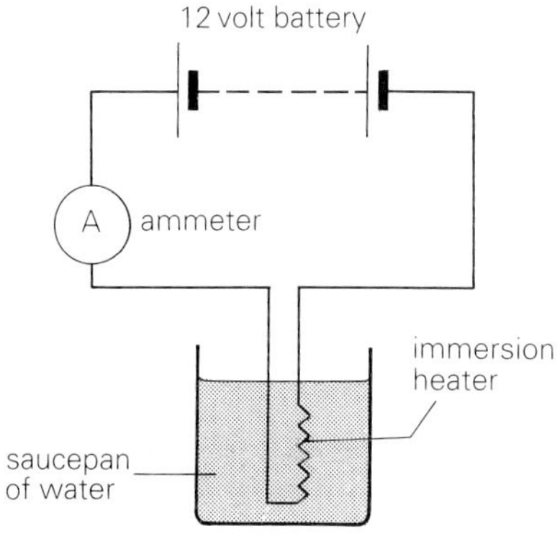

Fig 18.3

Again connect the immersion heater to the 12-volt supply, but this time include an ammeter in series. The current is about 5 A. Use a watch to find how long the heater takes to heat 1 kg of water through 5 °C.

Now connect the heater to a 6 V supply instead of a 12 V supply. Notice what the current is, probably between 2.5 A and 3 A. Again find how long it takes to heat 1 kg of water through 5 °C.

The current is about half what it was originally, but it takes much more than twice as long to heat the water, in fact about four times as long.

Heating effect of an electric current

The amount of energy transferred by a lamp depends on the current through it. The experiment with an immersion heater shows it depends on the voltage as well.

Later in your course, it will be shown that the energy transferred using a voltage, V, and current, I, in time, t, is given by

$$V \times I \times t.$$

For example, if the voltage is 12 V and the current is 5 A, the energy transferred in 1 second is $12 \times 5 \times 1$ or 60 joules.

An immersion heater which transfers 60 joules in 1 second is said to be a 60-watt heater. A 100-watt heater transfers 100 joules in 1 second. In other words, the **wattage** of a heater, lamp or other device gives the number of joules transferred per second. The wattage of a device is usually called its **power**.

Experiment 18.3
Heating water

Put 1 kg of water into a saucepan. Connect the immersion heater to a 12-volt supply with an ammeter in series with it. Put the heater in the water and find how long it takes for the temperature of the water to rise by 5 °C.

Knowing the current, the voltage and the time, calculate the number of joules transferred by the immersion heater. This gives the amount of energy needed to raise the temperature of 1 kg of water through 5 °C. From this, you can calculate the number of joules to raise the temperature of 1 kg of water through 1 °C.

Your value is likely to be a little higher than the usually accepted value, partly because some of the energy will have been transferred to the surrounding air, and partly because some will have raised the temperature of the saucepan as well as the water. More precise measurement shows that 4 200 J are needed to heat 1 kg of water through 1 °C.

Joulemeters

Instruments called *joulemeters* (see the photographs on the next page) are available which will *measure* how much energy is transferred by the electric supply.

If you use a joulemeter in these experiments, then you will simply connect the supply to the 'supply' terminals and the heater to the 'load' terminals of the joulemeter.

The meter provided by your electricity supplier in your home is a joulemeter.

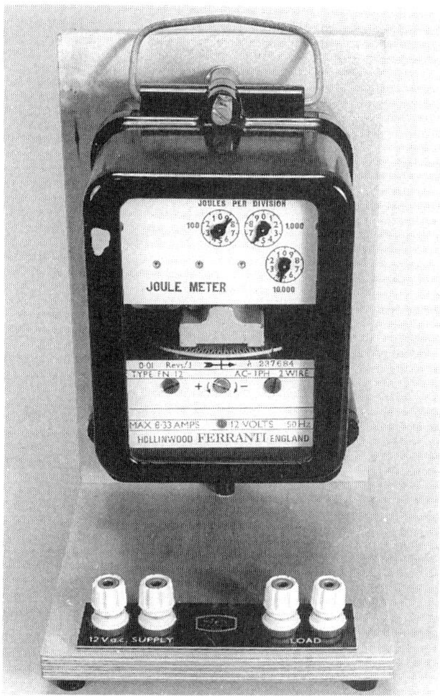

Joulemeter

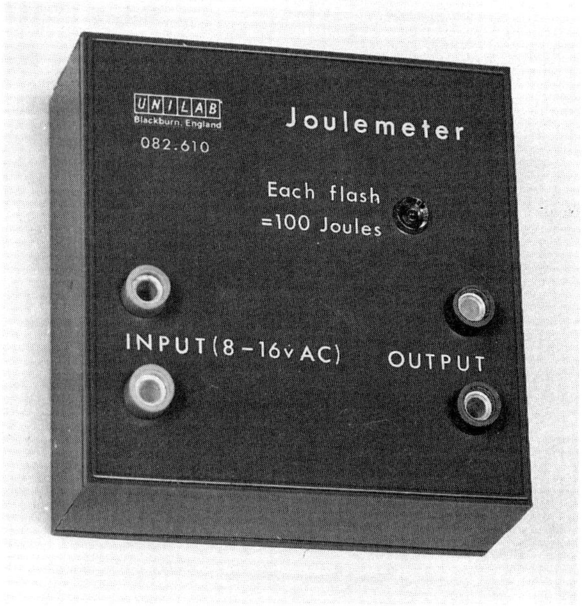

Modern laboratory joulemeter

Questions for class discussion

1 If 4 200 J is needed to heat 1 kg of water through 1 °C, how much energy is needed to heat
a 2 kg of water through 1 °C,
b 2 kg through 5 °C?

2 How many joules of energy must be transferred to heat 5 kg of water through 10 °C?

3 A kettle contains 1 kg of water at 0 °C. Estimate how many joules will be necessary to bring the water to 100 °C.

4 If a lamp has 12 V connected across it and if the current flowing is 2 A, how much energy is transferred
a in 1 second,
b in 1 minute?

5 An electric fire is designed to operate from 200 V and a current of 5 A flows through it. How much energy is transferred
a in 1 second,
b in 1 minute,
c in 1 hour?

6 An electric kettle operating from 200 V takes 5 A. It is filled with 1 kg of water at 0 °C. Estimate how long it will take to bring the water to the boil (at 100 °C).

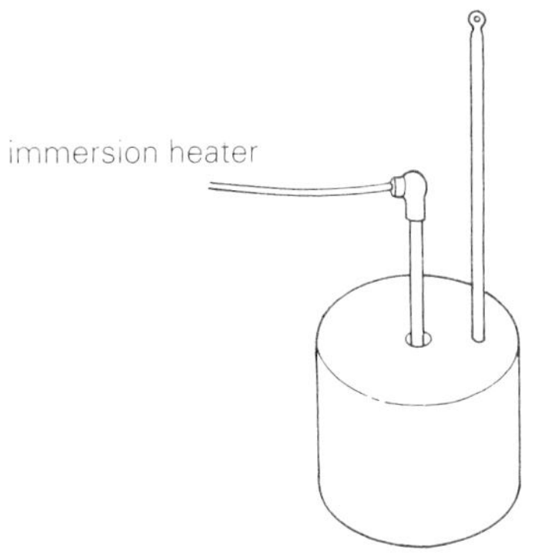

Fig 18.4

Experiment 18.4
Heating aluminium

You are provided with an aluminium block which has a mass of 1 kg. Insert a thermometer into it and measure the temperature. Again connect the immersion heater to a 12-volt supply in series with an ammeter. Put the immersion heater into the block and leave it there for a definite time, say 3 minutes. Measure the current during this time. Remove the heater and measure the highest temperature reached by the aluminium block.

Calculate the energy transferred by the heater to the block. For example, if the voltage is 12 V, the current 5 A and the time 3 minutes (that is, 180 seconds) then

$$\text{energy transferred} = V \times I \times t = 12 \times 5 \times 180 = 10\,800 \text{ J}.$$

Your value will be the amount of energy to raise the temperature of 1 kg of aluminium through the temperature rise you measured. You can then calculate the amount required to raise the temperature of 1 kg of aluminium through 1 °C.

This quantity (the energy to raise the temperature of 1 kg of a substance through 1 °C) is given a special name in physics: it is the **specific heat capacity** of the substance concerned.

Some values for the specific heat capacity

Substance	Specific heat capacity (number of joules to heat 1 kg through 1 °C)
aluminium	900
copper	390
iron	450
lead	130
paraffin oil	2100
mercury	140
water	4200

Background Reading

Paying for electricity

A lamp which is labelled 240 V 100 W is much brighter than a lamp labelled 240 V 40 W.

This is exactly what one would expect. W stands for ***watt*** and a watt is a joule of energy transferred every second. In other words a 100 W lamp is transferring 100 joules of energy every second when it is switched on, but a 40 W lamp is only transferring 40 joules of energy every second and so it is not so bright. As it is energy which we are paying for when we pay our electricity bill, it will cost more to have a 100 W lamp than one of 40 W if they are used for the same amount of time.

The following table gives some typical values of the wattage of several electrical appliances. (Note 1000 watts is sometimes called a kilowatt.)

table lamp	40 W
stronger lamp	100 W
refrigerator	120 W
television set	120 W
electric blanket	120 W
hair-drier	600 W
electric iron	720 W
electric kettle	750 W
electric fire (1 bar)	1000 W (1 kilowatt or 1 kW)
immersion heater	3000 W (3 kilowatts or 3 kW)

The Electricity Board charges people for the energy supplied to them. The energy is measured in what the Board calls a 'unit'. One unit is the energy transferred by a 1 kilowatt device in 1 hour, and another name for it is one kilowatt-hour.

To find out how many kilowatt-hours an appliance has 'used', you have to multiply its power in kilowatts by the number of hours for which it was used. Thus a 1-kilowatt fire used for 1 hour uses 1 kilowatt-hour, as does a 100-watt lamp operating for 10 hours and a 3-kilowatt immersion heater in a hot water tank for 20 minutes. Each of those examples would cost the consumer the same. The cost of a 'unit' was at the time of writing about 7 pence.

How many joules can you get for 7 p? When a 1-kilowatt fire is operated, 1000 joules are transferred to internal energy and light every second. So, if the fire is on for 1 hour, the energy transferred is

$$1000 \times 60 \times 60 = 3\,600\,000 \text{ joules.}$$

That energy could lift a 1 kg mass upwards a distance of 360 km (more than 200 miles). And it is the kinetic energy a car has, approximately, when travelling at 300 kilometres per hour (a little under 200 miles per hour)! In our society, energy is cheap and we have become wasteful, but the world's resources are limited and we have a responsibility to our successors to use energy more prudently.

Effects of heating things

In Chapter 10 we discussed how heating could cause solids to melt, and liquids to turn to gases. We saw how this might be explained on a particle model of matter. The expansion of solids, liquids and gases is another important effect caused by heating. We shall look at this expansion in the following experiments.

Experiment 18.5 Expansion of solids

When a rod is heated, the expansion is very small and we must therefore use a special arrangement to detect it. The rod is fixed at one end using a 1 kg mass as shown in Fig 18.5. The other end of the rod rests on a needle. When the rod expands, the needle rotates and that makes the drinking straw turn. (It helps to put the needle on a piece of glass, perhaps a microscope slide, and a mass hung on the rod can stop slipping.) Heat the rod with a Bunsen burner and watch the expansion.

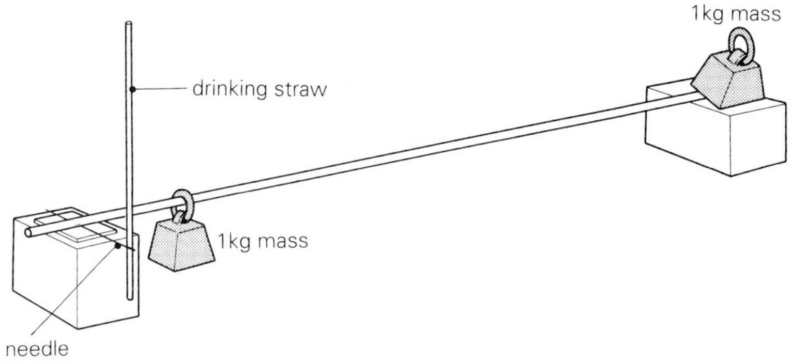

Fig 18.5

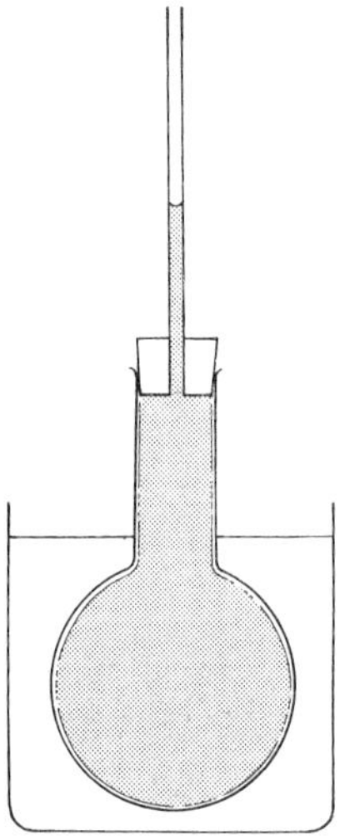
Fig 18.6

Experiment 18.6 Expansion of liquids

Fill the flask with coloured water (a little ink is suitable for the colouring). Insert a bung with a delivery tube through it. Be careful to make certain there are no air bubbles left in the flask.

Carefully hold the flask in a saucepan of warm water and watch what happens to the water in the flask.

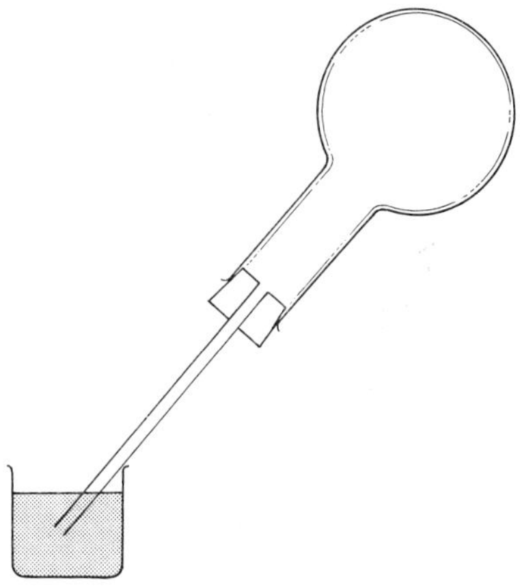
Fig 18.7

Experiment 18.7 Expansion of gases

The flask containing air is held so that the end of the tubing is below the surface of water in a small beaker. Warming the flask with hands causes the air to expand and bubble out through the water. What happens when the air is allowed to cool?

In Chapter 9 you found that you can squash a gas into a smaller volume by increasing the pressure on it, and that reducing the pressure outside a balloon caused the volume of the gas inside to increase. Now you have found that the volume of a gas increases as you warm it. So the volume of a fixed mass of gas depends on both its pressure and its temperature. If you heat a gas and do not allow its volume to change, then its pressure rises as the next experiment shows.

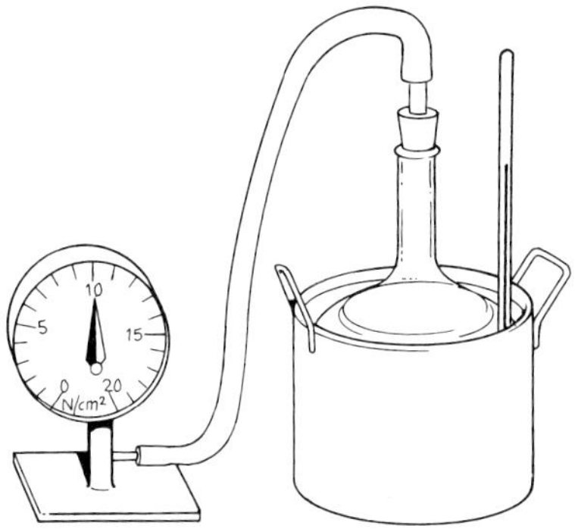

Fig 18.8

Experiment 18.8
Increase in the pressure of a gas when heated at constant volume

In order to keep the volume of the gas (air) constant, it is kept in a flask. The flask is connected to a Bourdon gauge. It is then put in a saucepan of cold water. See what pressure the gauge reads. Then heat up the water and watch how the pressure increases.

This result is to be expected. If the gas had been heated without its pressure changing, it would have expanded, and it would then have been necessary to compress it in order to bring its volume back to the original value. Heating a gas at constant volume causes an increase of pressure.

The kinetic model of a gas also suggests that the pressure should increase. When the gas is warmed, the internal energy increases and the molecules move faster. These molecules will collide with the walls more violently and more frequently so that the pressure will be greater.

Background Reading

Expanding solids

One of the most important consequences of making things hotter is that they expand. This may be a nuisance or it may be useful, but it is never wise to ignore it. This expansion poses problems for the engineer, because, if metals are prevented from expanding, very large forces are created.

The photograph below shows a way of jointing sections of railway lines to allow expansion to take place without causing damage. Lines have been known to buckle on a hot summer's day when allowance for the expansion has been insufficient. Long lengths of welded rails are used today and tapered overlaps between the end of sections, as shown, are essential.

Jointing section of railway line

The next photograph shows how the expansion or contraction of a bridge can occur without damage to its structure.

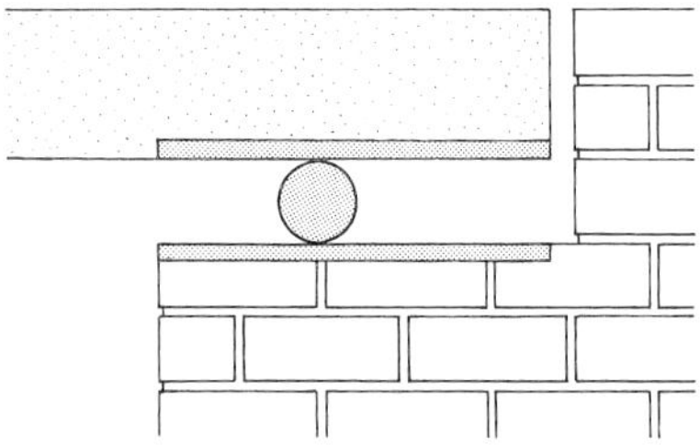

Fig 18.9

Bridge expansion

This photograph shows an oil pipe in the desert. In the desert it is cold during the night and very hot during the day. A two kilometre stretch of oil pipe could expand there by as much as one metre. A loop in the pipe (the pipe itself is hidden behind the bigger pipe near the camera) enables the expansion to take place without buckling the pipe.

Oil pipe loop

(Adapted from Longman Physics Topic *Heat* by A J Parker and P E Heafford.)

Homework assignments

There are many different books which will help you with these assignments. If you do not find what you want in the first book you try, then look in another. Looking in different books becomes a very necessary thing to do as you continue your studies in science.

1 Prepare a short talk on how a mercury thermometer is made.

2 In this book we have talked about the Celsius temperature scale. Another scale is the Fahrenheit scale. Find out how it differs from the Celsius scale.

3 When studying weather conditions, a maximum and minimum thermometer is often used. Find out how it works and prepare a short talk on it.

Background Reading

A clever invention

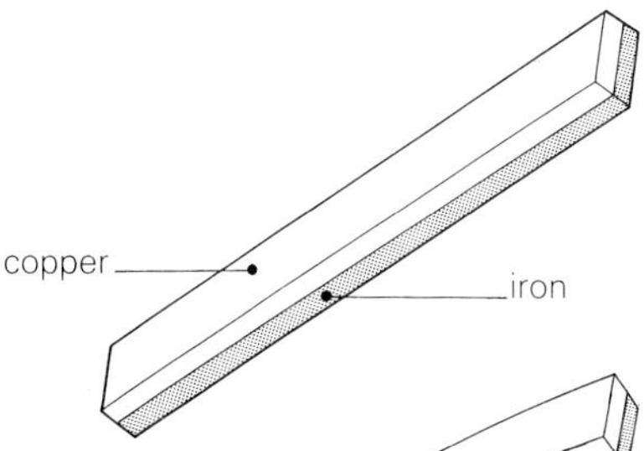

Fig 18.10

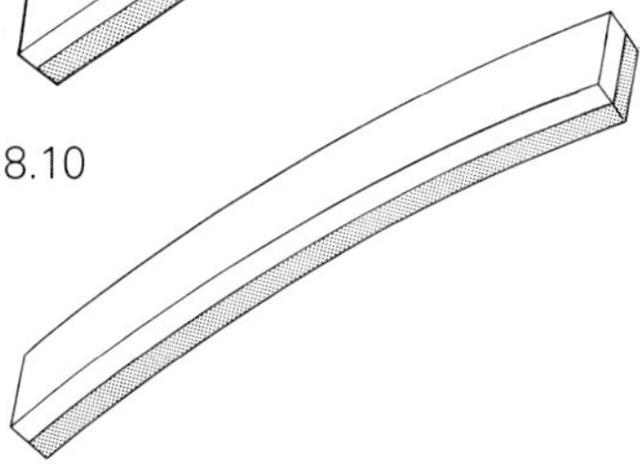

Fig 18.11

Fig 18.10 shows two pieces of different metals fixed together—what is usually called a **bimetallic strip.**

Copper and iron do not expand the same amount when heated through the same change of temperature: copper expands more than iron. So, if the bimetallic strip is heated, it will bend as shown in Fig 18.11. Why must it bend?

This invention has many applications, particularly as a device for controlling temperatures. Fig 18.12 shows an electric iron. The heater is shown on the base of the iron. When the temperature rises, the bimetallic strip, in this case made of brass and iron, bends. As a result it breaks the electrical contact so that the current ceases to flow and the iron does not get any hotter.

There is a knob on the top of the iron which is connected to a screw which can move the lower contact to the bimetallic strip. Can you see how this can be used to control the temperature of the iron?

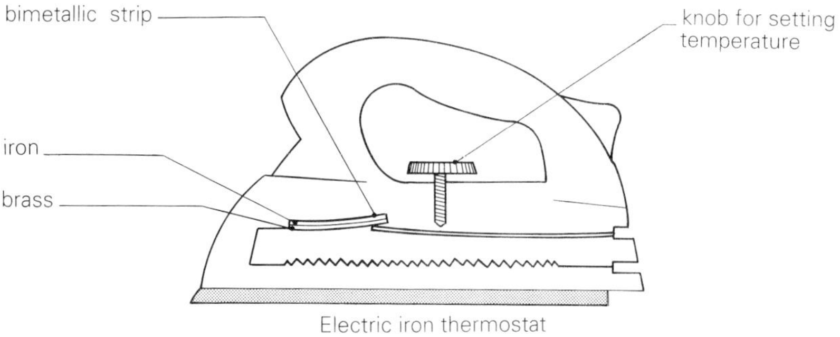

Fig 18.12

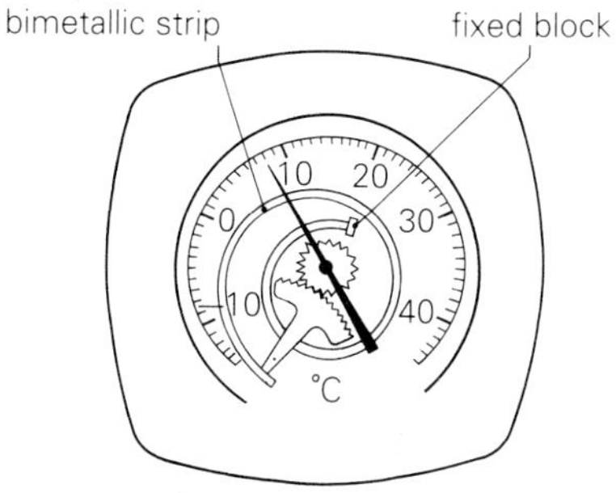

Fig 18.13 Bimetallic thermometer

The bimetallic thermometer is another device using this clever idea. When the temperature rises, the bimetallic strip bends more and more and causes the pointer to move.

(Adapted from Longman Physics Topic *Heat* by A J Parker and P E Heafford.)

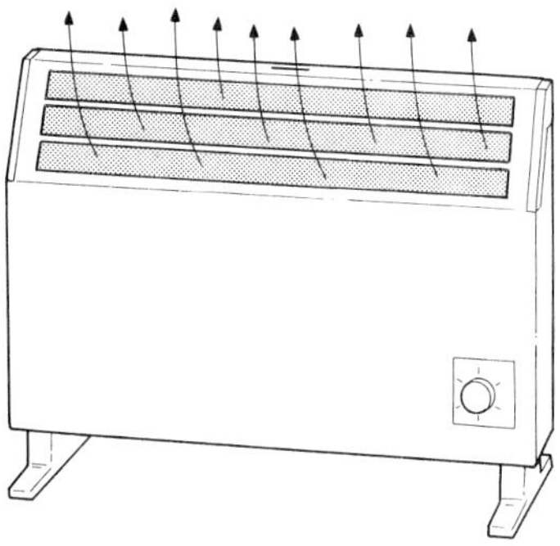

Energy flow

If you want to pass a written message from the back of a crowded hall to the front, you could pass it to someone in front of you, who then passes it to someone in front of them and so on until the message reaches its destination. Another way of sending the message would be to give it to someone at the back of the room who pushes a way through the room until it is delivered to the front. Yet another way would be for you to throw the message from the back of the room so that it reaches the front. There are three ways in which heat energy can be transferred, **conduction, convection** and **radiation**, and these have similarities with the three ways of sending the message.

If one end of a poker is put in a fire, the fire gives energy to the atoms in the poker so that they vibrate more vigorously: in turn they will cause their colder neighbours to vibrate more vigorously and energy is conveyed down the poker. This is conduction, in which energy of vibration is passed from neighbour to neighbour down the poker, from the hot end to the cold end.

An electric convector heater in a room warms the air around it. This air expands and rises upwards because it is less dense, whilst more dense, cooler air comes in to take its place. This in turn gets warmed and rises. This stream of moving air is called a convection current and the process by which energy moves from one part of the room to another is called convection. The molecules move taking the energy with them. The process can only occur in liquids or gases, not in solids.

Both conduction and convection require the presence of matter. As the space between the Sun and the Earth is virtually a perfect vacuum, energy cannot reach the Earth from the Sun by either conduction or convection. The process involved is usually referred to as radiation.

Experiment 18.9
Demonstration experiments on conduction

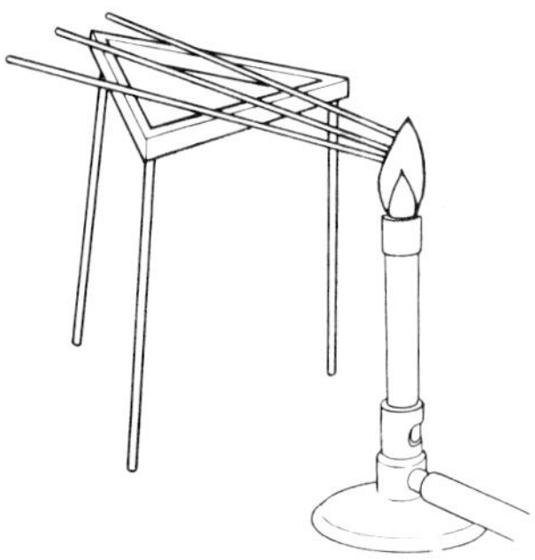

Fig 18.14

a Support rods of the same diameter of various materials (copper, brass, aluminium, iron) on a tripod as shown in Fig 18.14, so that one end of each is in the middle of a Bunsen flame. Heat for a short while and then explore the temperature along each by sliding a finger carefully towards the hot end. Alternatively you can move a match head along the rod instead of a finger and see where it ignites.

Safety note
Take care not to burn your fingers.

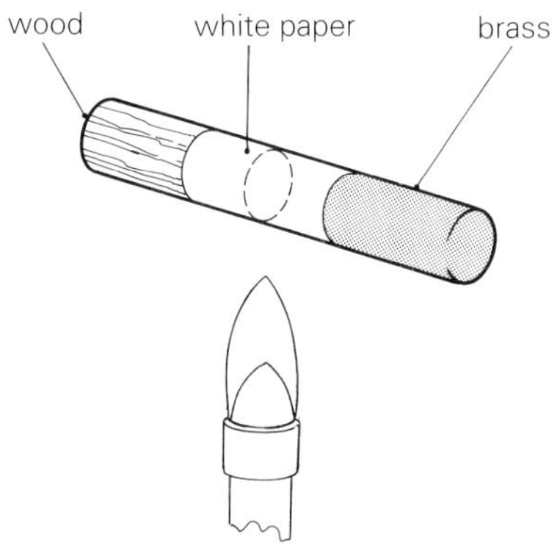

Fig 18.15

b Stick a piece of white paper around the join of a rod, one end of which is wood and the other brass. Hold the rod over a Bunsen flame as shown in Fig 18.15 until the paper chars. Then look at the paper carefully: it will have charred where the paper was over the wood, but not over the brass. Why is this?

c Light a Bunsen burner so that it burns with a blue flame and then bring down a piece of copper gauze from above it into the position shown in Fig 18.16a. Note that the gas does not burn above the gauze, yet there is unburned gas there, for you can light it with a match. You can also light the gas from the burner so that it burns above the gauze but not below (Fig 18.16b). This happens because copper is a good conductor of energy. It conducts energy away from the flame quickly so that the gas on the other side of the gauze does not get hot enough to ignite.

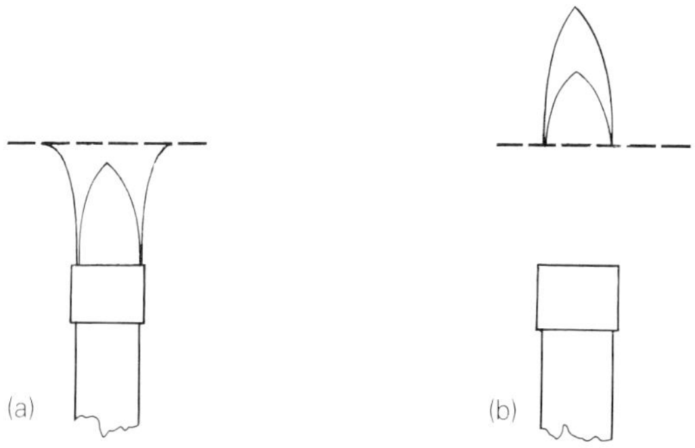

Fig 18.16

The experiments show that in general metals are good conductors of internal energy and much better than non-metals, just as for electricity. However in Experiment 18.9a, care must be taken over the results. The energy that is conducted along the rod is partly lost from the surface and partly used to raise the temperature of the rest of the rod. How much the temperature rises depends on the specific heat capacity as well as how much energy is absorbed. Only when the temperatures have stopped changing can any conclusions be drawn from touching the rods.

Experiment 18.10 Demonstration experiments on convection

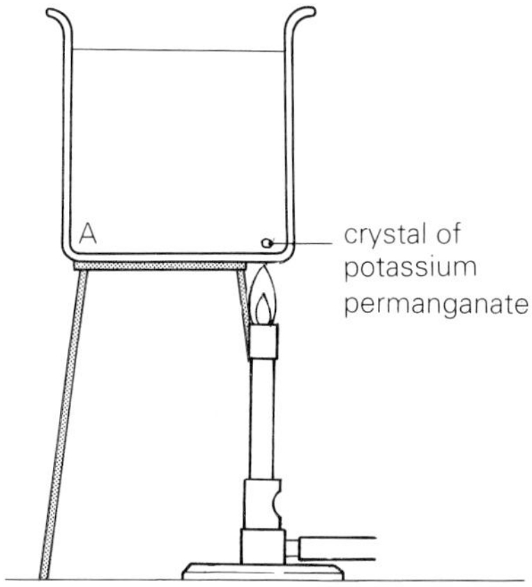

Fig 18.17

a Fill a beaker with cold water. Drop one small crystal of potassium permanganate down a drinking straw to the bottom, towards one side. Put the beaker on a tripod (without gauze — the beaker will not crack if it is Pyrex) and heat it gently with a Bunsen burner. Convection currents will be seen in the liquid.

Repeat the experiment but with the crystal of permanganate placed at A instead of over the Bunsen burner.

b Fill a test-tube with cold water. When the water is still, put a single crystal of potassium permanganate at the bottom. Hold the top of the test-tube in the fingers just above the water level and heat the bottom of the tube in a gentle Bunsen flame (Fig 18.18a). Watch the convection currents and see how long you can hold the tube without discomfort.

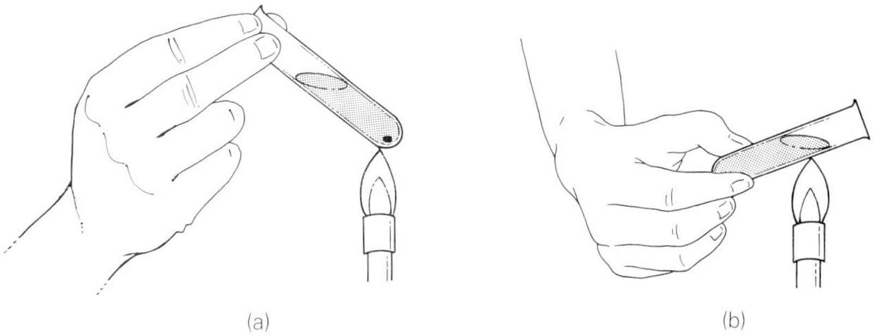

Fig 18.18

Empty the tube, wash it out and refill with cold water. Again put a single crystal of potassium permanganate at the bottom. This time hold the tube at the bottom and heat near the top of the tube, just below the water surface (Fig 18.18b). This time there are no currents and you can hold the tube much longer.

c Convection in air can be shown with the apparatus of Fig 18.19. A lighted candle is placed under one of the chimneys. A piece of burning oily rag or a smouldering drinking straw is held at the top of the other chimney. The smoke will make the convection currents visible.

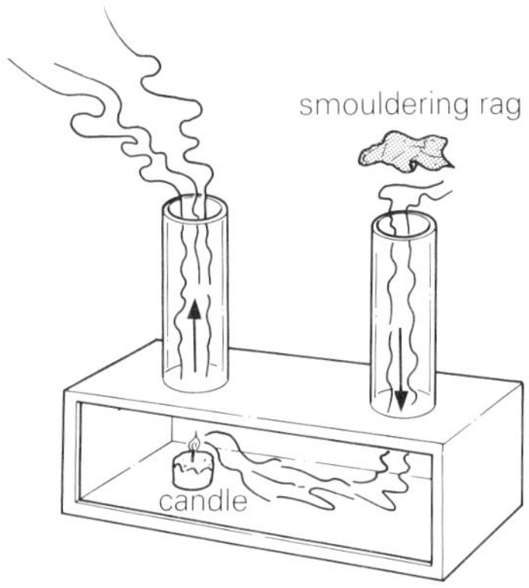

Fig 18.19

Convection occurs because the heated liquid or gas expands and so becomes less dense than the surrounding fluid. The less dense fluid rises vertically carrying its energy with it, whilst its place is taken by colder fluid. Thus a current of moving fluid is formed, so spreading internal energy throughout the fluid.

Background Reading

Domestic hot water system

At the heart of the domestic hot-water system, there is a boiler (B) and a hot-water storage tank (HWT). These are kept full of water from a cistern (C) supplied from the water mains via a tap controlled by a float (see page 87). When water in B is heated, it expands and rises to the hot-water storage tank via pipe X. As it rises, colder water from the tank moves to the boiler B via pipe Y. This convection current, flowing from B via X to HWT and back to B via Y, keeps the tank supplied with hot water.

The expansion pipe is a safety measure to prevent the system bursting should a blockage occur. Can you explain why the level of the water in the expansion pipe is a little higher than the level of the water in the cistern?

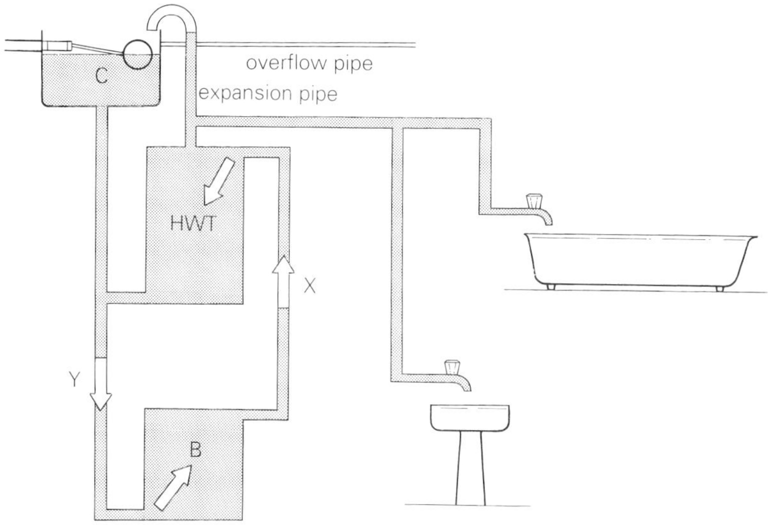

Fig 18.20

The pipe which supplies the taps is connected to the top of the hot water tank. (Why?) Usually the connection is made where the expansion pipe is joined to the tank. The pressure which drives the water out of the taps is due to the difference in levels between the water in the expansion pipe (or cistern) and the taps. To get an adequate flow of water, it is necessary to have the cistern as high as possible and it is usually positioned in the space just under the roof.

Of course, the pipes to and from the cistern have to be lagged so that they do not freeze quickly in cold weather. Lagging is made of a bad conductor so that the internal energy cannot be conducted quickly away from the water in the pipes. The hot-water tank is also lagged to delay the cooling of the hot water. This saves the money spent on the extra boiler fuel which would be burned to replace the energy lost.

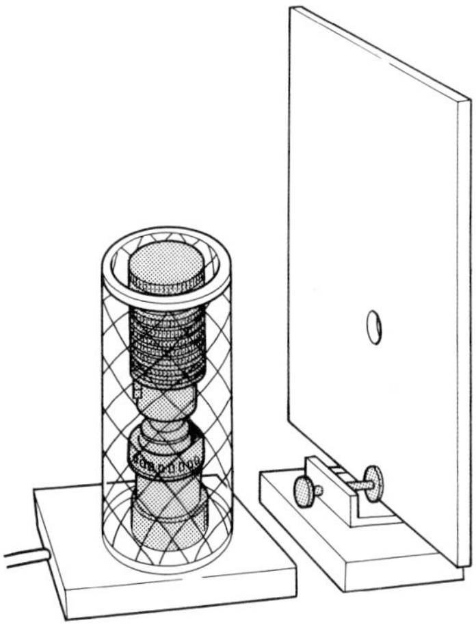

Fig 18.21

Experiment 18.11
Experiments on radiation

a For these experiments the source of radiation is a heating element in front of which is a fire proof screen with a hole in it.

Look at the element through the hole. Put the back of your hand in front of the hole to feel the radiation. Then put your cheek a short distance from the hole. Move your cheek further away and the energy received will be less.

Insert a book between your cheek and the hole. The moment the book is put in the way the radiation is no longer felt on your cheek. Put a thin sheet of glass between the hole and your cheek, then two sheets to see what difference a thick sheet makes.

Safety note

The plate gets very hot.

b Heat a thick copper plate over several Bunsen burners as shown in Fig 18.22. One side of the plate is dull black, the other shiny. Turn the plate sideways and hold your cheek near each side in turn. Much more energy appears to be radiated from the dull black surface than from the shiny side.

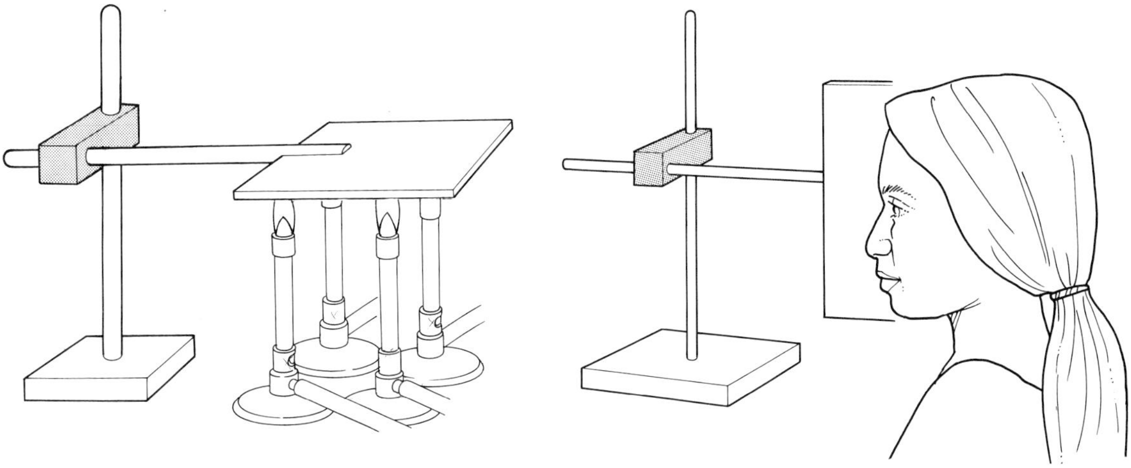

Fig 18.22

Safety note

Take your hand away as soon as your feel it getting hot.

c To investigate how different surfaces absorb radiation, first hold the back of your hand by the hole and feel the radiation for a short while (Fig 18.23a). Then cover the back of the hand with aluminium leaf and hold that by the hole. The hand can be held there much longer. Finally paint the aluminium leaf with *vegetable black* (Fig 18.23b). When it is dry, put it by the hole once again (Fig 18.23c). This time energy will be absorbed much more rapidly.

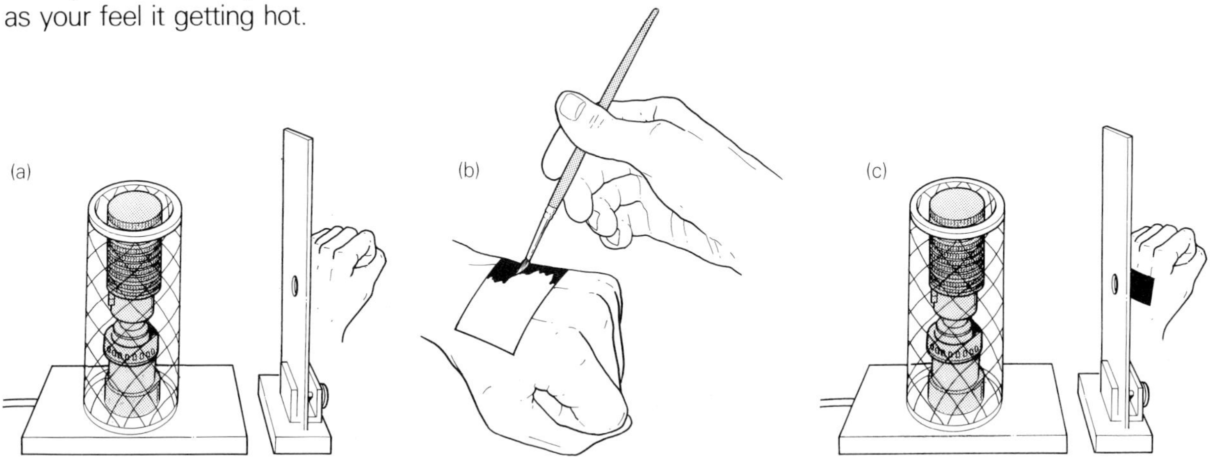

Fig 18.23

All surfaces radiate energy. How much energy is radiated every second depends on how hot the surface is and what the surface is like. Radiation rises very rapidly as the temperature goes up, and dull black surfaces radiate more energy than polished surfaces at the same temperature.

Surfaces also absorb some or all of the radiation which falls on them. The best absorber is a dull black surface and the poorest is a silvery polished surface. A solar heating panel is painted matt black so that it will absorb radiation at a faster rate than a white shiny surface would.

Questions for class discussion

1 A string vest is very good for keeping people warm, but it is full of holes. How do you explain this?

2 If you get out of bed with bare feet and stand on linoleum it feels cold, but standing on carpet makes you feel much warmer even though both the carpet and the linoleum are at the same temperature. What is the reason for this?

3a If an electric immersion heater is being put into a hot-water tank in the hot-water system of a house, should it be put at the top or at the bottom of the tank?

b Why is the freezing unit inside a refrigerator put near the top of the refrigerator?

4 In winter, a room usually feels warmer when the curtains are drawn. What is the reason for this?

5 The Earth is constantly receiving radiation energy from the Sun and it absorbs it. Will this mean that the Earth will go on getting hotter until it is at the same temperature as the Sun?

6 Why is a cloudy night usually much warmer than a clear cloudless one?

Background Reading

A vacuum flask or thermos flask

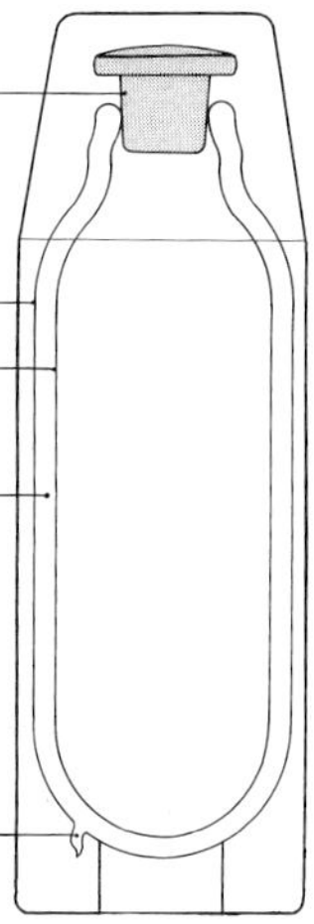

Fig 18.24

The flask is a double-walled glass vessel which has been evacuated and sealed. The inside surface of the double wall is silvered. Energy cannot be conducted away from the inner surface across to the outer because of the vacuum. Neither can there be any convection currents to carry internal energy away from the inner surface. Radiation does take place from the inner surface, but the silvering keeps this loss to a minimum. (A little energy is conducted away through the glass at the top of the flask.) In addition to loss of energy by conduction, convection and radiation, a liquid will also cool if evaporation occurs so that it is very important to cork the flask.

A vacuum flask (thermos flask) is often used to keep liquids hot on a picnic, but scientists usually use it to keep liquids very cold!

Chapter 19 Physics and the weather

The most common topic of conversation in the British Isles is the weather! Four very important processes in producing our weather patterns are conduction, convection, radiation and evaporation. The energy which 'drives' the weather systems arrives as radiation from the Sun. Some of this input is absorbed in the atmosphere (for example, an ozone layer high in the atmosphere absorbs most of the dangerous ultra-violet radiation, see page 319). Some is reflected back into space, particularly by clouds, but about $\frac{2}{3}$ is absorbed by the Earth, so warming it. If some parts are warmed more than others, then convection currents in the air can arise.

The effects of convection and evaporation

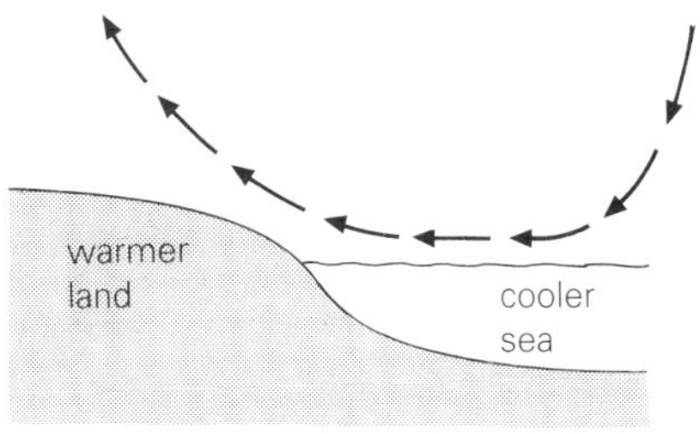

Fig 19.1

Think about a cloudless summer's day near the sea. The Sun's radiation raises the temperature of the land more quickly than the temperature of the sea because the specific heat capacity of water is much higher than that of the land. As the land gets hotter, the air near it receives energy by conduction and a convection current starts, rising over the land and circulating as shown in Fig 19.1, with cooler air moving in to take the place of the rising air. Thus, even on the calmest of summer days, there is usually a gentle in-shore breeze during the afternoons. Conditions are different at night, especially if there is no cloud. Now the land and the sea radiate energy into space, and the land cools faster than the sea. What might happen as a consequence?

Evaporation plays a major role in weather processes. The faster-moving molecules near the surface of water escape to become part of the air. Thus, because of this evaporation, air always has water vapour in it. However, there is a limit to how much water vapour a volume of air can hold before it becomes saturated. And the amount that can be held depends on the temperature of the air – the colder it is, the less is the mass of water vapour required to saturate the air. So, if unsaturated air is cooled, it eventually becomes saturated. Any further cooling results in the mass of water it can hold becoming less, and so some of the vapour has to become liquid again (that is, some vapour condenses). The water may appear as droplets on any surfaces present, or it may form a mist of droplets in the air.

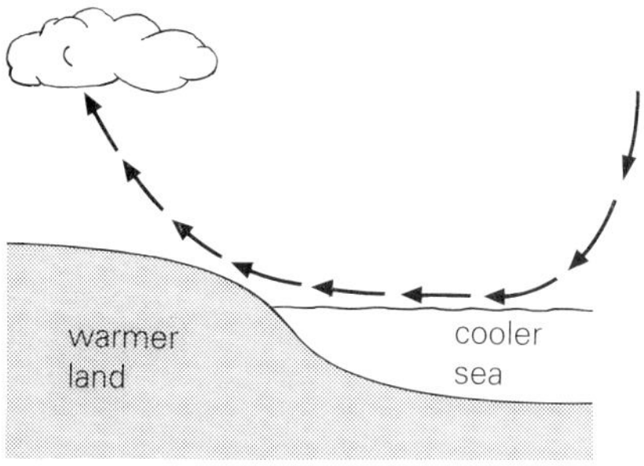

Fig 19.2

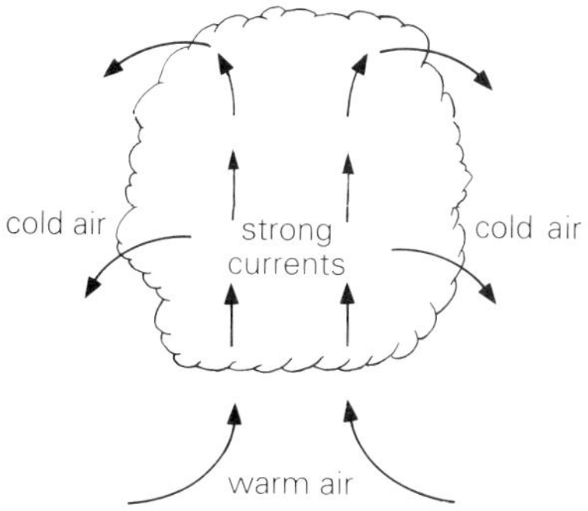

Fig 19.3

Have you seen the condensation on the bathroom window when you have had a hot bath?

In Fig 19.1, the air coming in from the sea will be moist because of evaporation, and, as it rises over the land and its pressure falls, it expands and cools. If it is cooled to the point of saturation, any further cooling will cause condensation as a mist; that is, a cloud of tiny water droplets starts to form. So, we should not be surprised if we find some cloud over the land (Fig 19.2).

The large 'cotton-wool' clouds (cumulus) are convection clouds. Warm moist air is swept up in a strong convection current rising over some warm place on the Earth's surface. As the air rises, the cloud starts to form at the level where the air becomes saturated, giving the cloud a well-defined base. But, the air goes on rising and cooling and more and more condensation into water droplets occurs (Fig 19.3).

These clouds can rise to great heights (several kilometres), when they are known as cumulo-nimbus clouds. The air currents inside are strong, which is why aircraft pilots avoid flying into them if possible. At the highest levels, the water may condense as ice crystals. There is a lot of energy being transferred in these clouds, and under certain conditions the cloud can act as an electrical generator. Then a thunderstorm results and the electrical energy is transferred partly to light and sound energy, but mostly to internal energy of the air.

Pressure and the weather

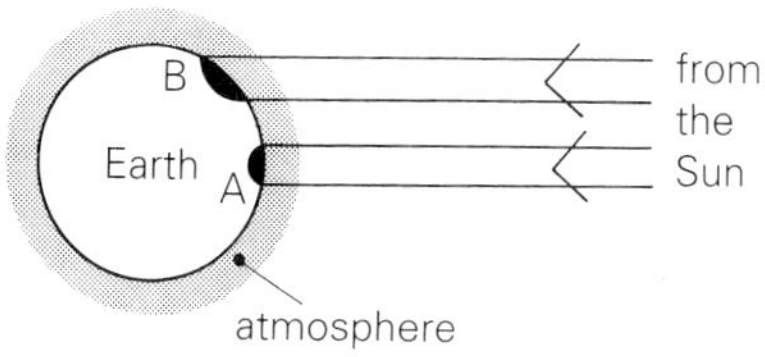

Fig 19.4

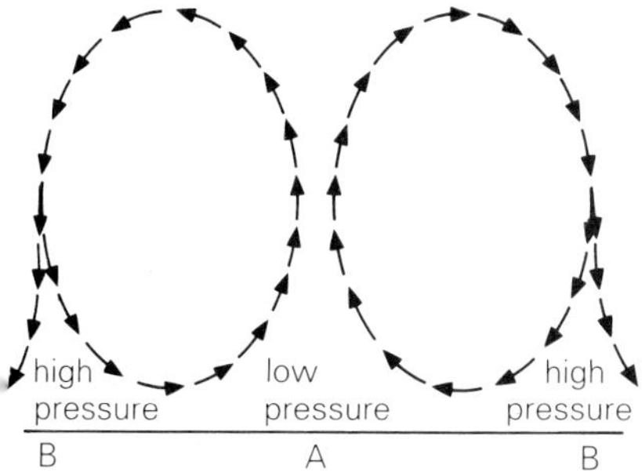

Fig 19.5

Convection is the main cause of wind patterns on the Earth. Fig 19.4 shows two regions, A and B, receiving supplies of energy from the Sun. At A where the Sun is overhead, the radiation falls on a smaller area than the same amount of radiation at B. What is more, the energy arriving at A has passed through a smaller depth of atmosphere so less is absorbed and reflected, and a bigger proportion reaches A than reaches B. Thus, the air over A will be hotter and therefore less dense than the air over B, and a convection current will occur. The convection arises because of a *pressure* difference. The expanded air over A is less dense than that over B and so the pressure of the air at A is less than the pressure of the air at B. The unequal heating of the Earth's surface results in regions of high pressure and regions of low pressure and the wind is the air moving from high to low pressure regions (Fig 19.5).

The full story is very complicated because the Earth is rotating and there are forces between the moving air and the ground and so on. But, to predict what our weather will be, we must know how the pressure varies over the Earth's surface. Weathermen (meteorologists) measure pressure in a unit called a *bar* which is a pressure equal to 100 000 N/m^2. Usually, pressures are quoted

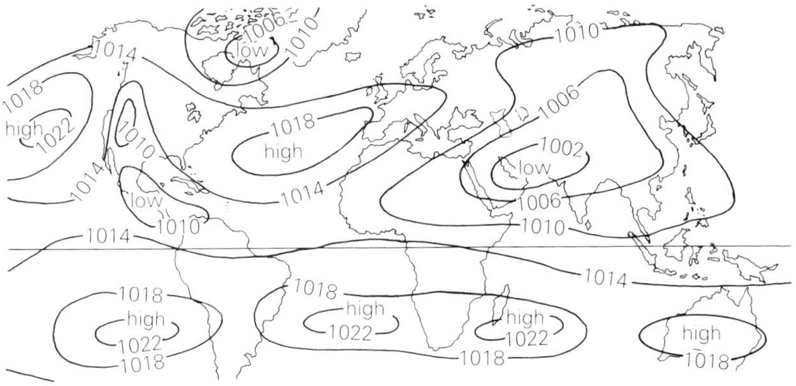

Fig 19.6

in millibars, a millibar being $\frac{1}{1000}$ of a bar, and air pressures are close to 1000 millibars. Meteorologists plot maps showing the pressure over regions of the Earth (Fig 19.6). To do this, they draw lines (called *isobars*) on a map joining places where the pressure is the same – much like a contour line joins places of equal height.

Notice that the isobars form closed loops round places of high pressure (an anticyclone) and round places of low pressure (a depression) just like contours round hilltops and round hollows. Indeed, weather forecasters use the same sort of language as geographers: they talk of a 'ridge of high pressure, a depression, a trough of low pressure'.

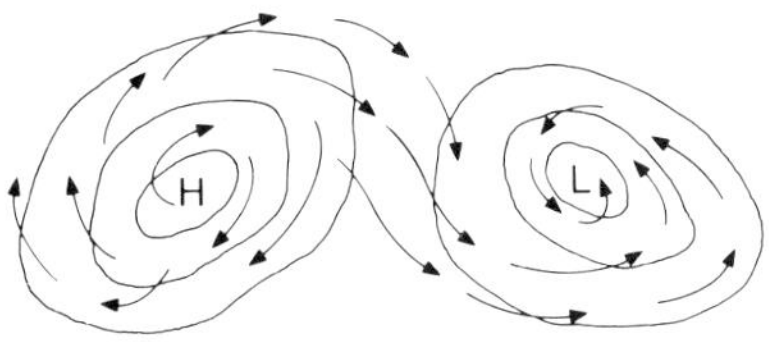

Fig 19.7

However, because of the Earth's rotation, the air does not move *directly* from places of high pressure to places of low pressure. In the nothern hemisphere, the air spirals out of an anticyclone clockwise and spirals into a depression anti-clockwise (Fig 19.7).

The wind directions are nearly parallel to the isobars (Fig 19.8). The more closely the isobars are grouped (that is, the greater the pressure gradient), the stronger are the winds. The bigger their velocity is, the more kinetic energy the air has and the greater may be the damage they can produce when they collide with a building or a tree.

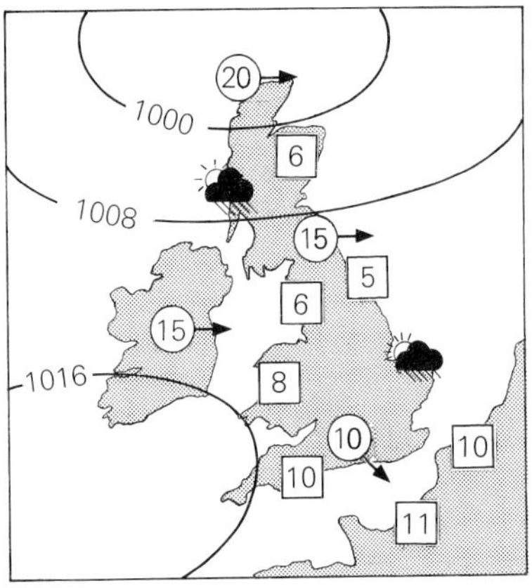

Fig 19.8 Wind speeds in mph, temperature in °C

Fronts

Weather systems arise where warm air and cold air meet, usually near the depression. If warm air moves into cold air, the warm air rides over the cold air because warm air is less dense than cold air and so tends to float on the cold air. The cold air forms a wedge under the warm air. At the boundary, the warm air is cooled and, if it is moist (as it will be if it originated over an ocean), clouds form along the boundary. Fig 19.9(a) shows what is called a **warm front** – warm air moving over cold. On weather maps, the junction between the two air masses along the ground is shown by a line as in Fig 19.9(b). The front moves towards the side on which the blobs are placed.

Fig 19.9

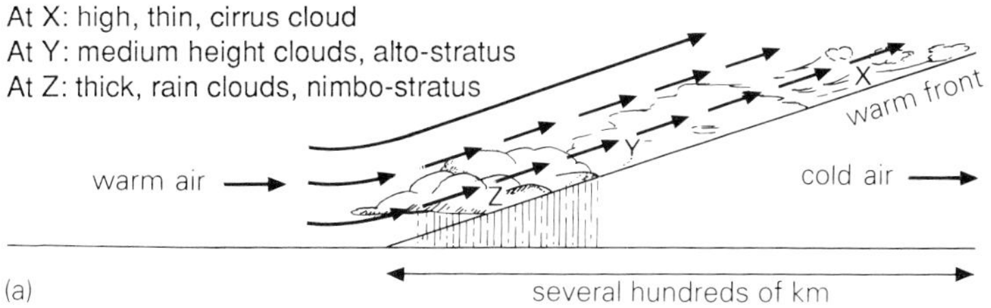

Often, colder, drier air pushes into warmer air. This is called a **cold front**. The cold air forces the warm air higher, and again clouds form along the front. The nose of the front becomes somewhat blunted, and it is here that large convection clouds may form, leading to heavy rain or hail and perhaps thunder and lightning (Fig 19.10(a)). A cold front is shown on a weather map by the symbol in Fig 19.10(b). The stormy weather of the front is usually followed by fine weather with small cumulus clouds.

Fig 19.10

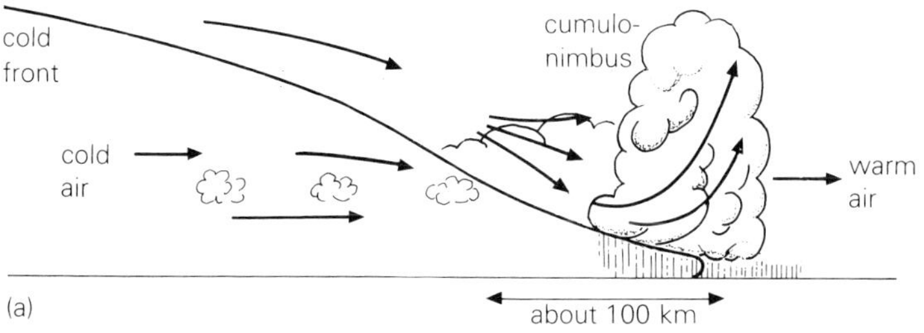

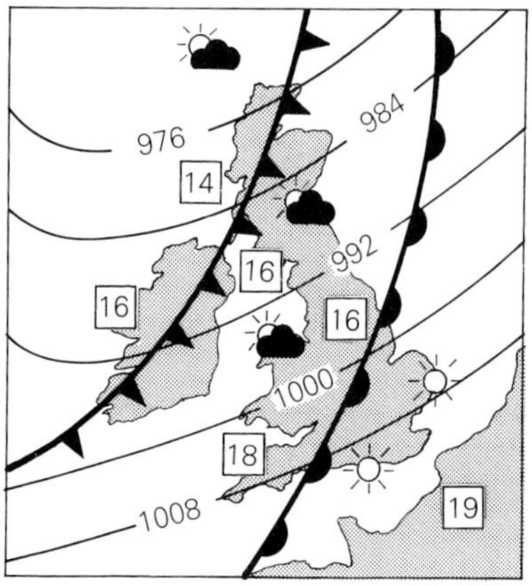

Fig 19.11

Fig 19.11 shows a weather map from a daily newspaper showing fronts over the British Isles during a spell of fine weather when the air masses were relatively dry. The weather associated with a front is easy to predict once it is known where the warm air and the cold air originated. That knowledge indicates how heavily laden with moisture the airs are. What makes forecasting so difficult is guessing how a front will move and where it will be in 24 hours time. Just a small error could mean that a front expected to cross these islands will miss them altogether.

Questions for homework or class discussion

1 At the start of this chapter sea breezes were mentioned. What effect would clouds have on the breezes?
2 What dangers would arise if the ozone layer were to disappear?
3 By what processes, other than evaporation from seas, lakes and rivers, does water vapour get into the atmosphere?
4 Prepare a short talk to your class on the various kinds of clouds. Try to give them their correct names.
5 Why does air 'cool as it expands'? (Think about energy.)

Chapter 20 Light

The photograph shows rays of light from the Sun passing through trees. This suggests that light travels in straight lines. You notice the same thing when a film projector sends light towards a screen through a dusty atmosphere.

In the first case light comes from the Sun; in the second it comes from the lamp inside the projector. Something which gives out light is called a **luminous source.**

Most objects are not luminous, as is obvious if you look for a book or a table in a darkened room; both are non-luminous. We can only see such things when light is reflected or scattered from them. In the drawing, light from the lamp shines on the book and the flowers and is then scattered into the person's eyes. The sunbeam streaming through the trees and the beam from a film projector are only visible because dust particles reflect or scatter some of the light to our eyes.

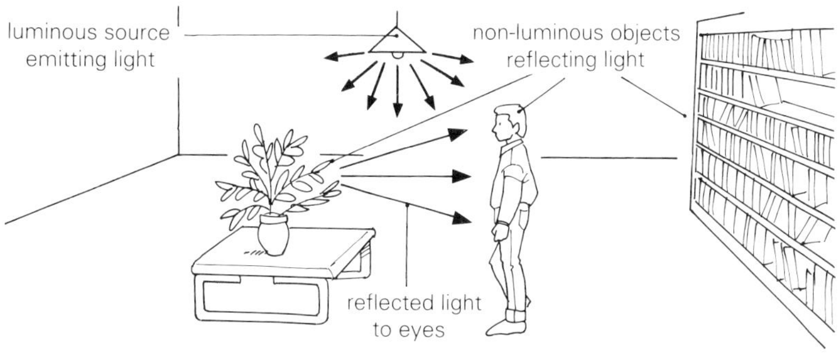

Experiment 20.1 Shadows

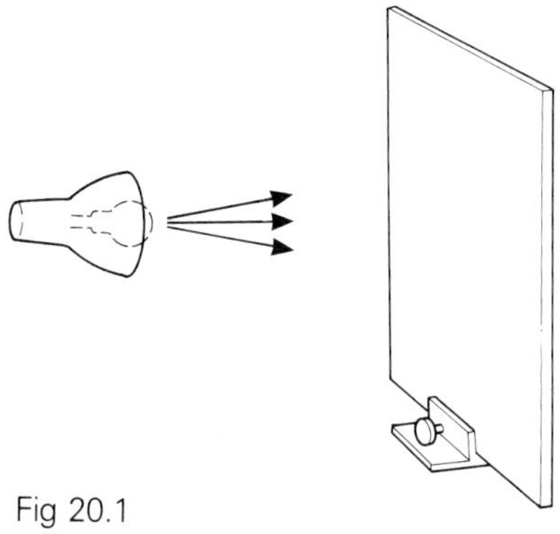

Fig 20.1

In this experiment, light from a lamp shines on to a white screen in a darkened room. A reading lamp makes a good source of light and the screen can be a sheet of paper pinned to a board.

a Fix a piece of card with a small hole in it (2 or 3 mm in diameter) in front of the lamp. Place various objects, such as a pencil or a finger, between the card and the screen. What do you see?

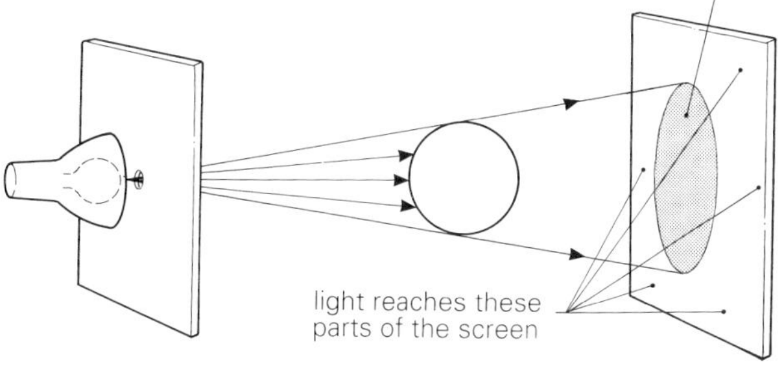

Fig 20.2

The object stops some of the light from reaching the screen and there is a shadow. The shadow is the same *shape* as the object because light travels in straight lines outwards from the hole. Fig 20.2 shows the shadow being formed. The straight lines which show the path of the light and its direction of travel are called **rays.**

b Find out what happens as you move the object nearer or further away from the source.

The size of the shadow changes. With the object near the screen, the shadow is only a little larger than the object, and the edges of the shadow are sharp and well defined. With the object nearer the source, the shadow is larger and a little less sharp.

c Now use a card with a larger hole in it, say about 1 or 2 cm in diameter. Look at the shadow again. What changes have occurred?

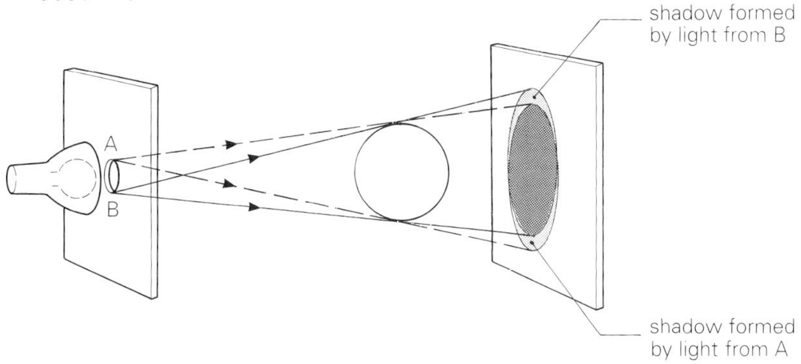

Fig 20.3

Around the shadow, the screen is brighter. The bigger hole lets more light through it. But the edge of the shadow is less sharp. It has become fuzzy! The bigger hole is like a collection of smaller ones, each casting a sharp shadow on the screen. But the shadows do not completely overlap and so the edges are not sharp (Fig 20.3).

d Now remove the card with the hole in it. The shadow will now be very fuzzy. Sharp shadows are produced when light comes from a *very* small hole, sometimes called a point source. An extended source throws shadows which have fuzzy edges.

Questions for homework

1 A filament lamp is a luminous source. Make a list of other types of luminous source and, in each case, say where the light energy comes from.

2 Light passes through a small hole and falls on a white screen 30 cm from it. A 2p piece is held up so that a circular shadow is formed on the screen. The coin is 10 cm from the hole and has a diameter of 2.5 cm. By drawing a scale diagram to show how the shadow is formed, find the diameter of the shadow. If the hole through which the light passes is made larger, what happens to the part of the shadow where there is no light at all?

Experiment 20.2 The pinhole camera

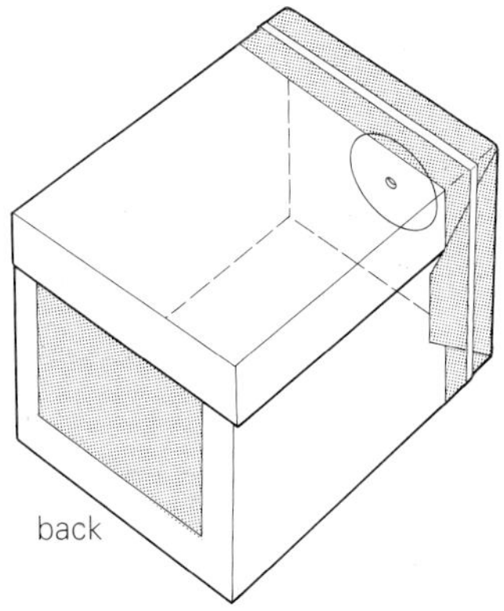

Fig 20.4

A pinhole camera can be made using a small rectangular cardboard box with a circular hole in one end (about 4 cm in diameter) and a large rectangular hole at the other end. The circular hole is covered with a sheet of black paper in which holes can be made with a pin. The rectangular hole at the back is covered with a screen of special paper like greaseproof paper which lets some light pass through it.

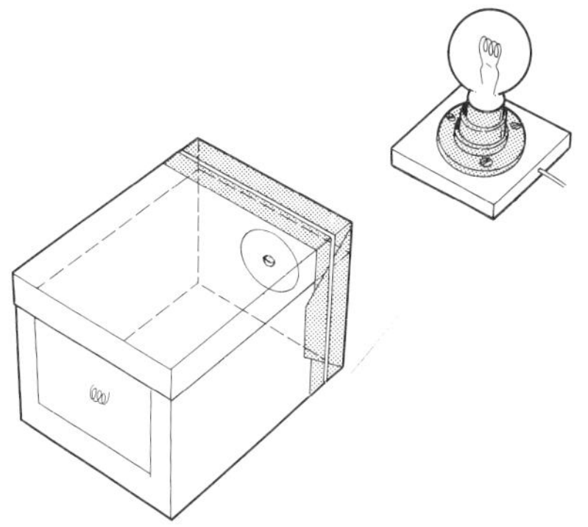

Fig 20.5

a Set up a clear filament lamp. Make a single hole with a pin in the centre of the black paper. Hold up the camera with the pinhole directed towards the lamp. Look at the screen from behind the camera. (The camera should be at least a metre away from the lamp.) What do you notice? What happens if you move the box closer to the lamp? What you see on the screen is usually called an image, and in this case, it is upside-down.

b Point the camera towards an open window and look at the image on the screen. Which way up is the image? Which way round is it? What happens to the image if you move closer to or further away from the window? Is the image coloured or black and white?

The image is formed on the screen by light travelling from the window through the pinhole in straight lines. The image is said to be **inverted**.

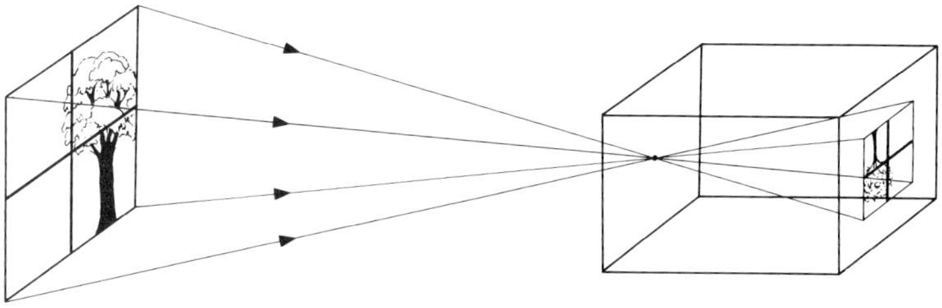

Fig 20.6

c Use the camera to view the filament lamp again. Make three or four pinholes in the black paper and see what happens.

There will be three or four inverted images in different places. What happens if you make a 'pepperpot' of pinholes?

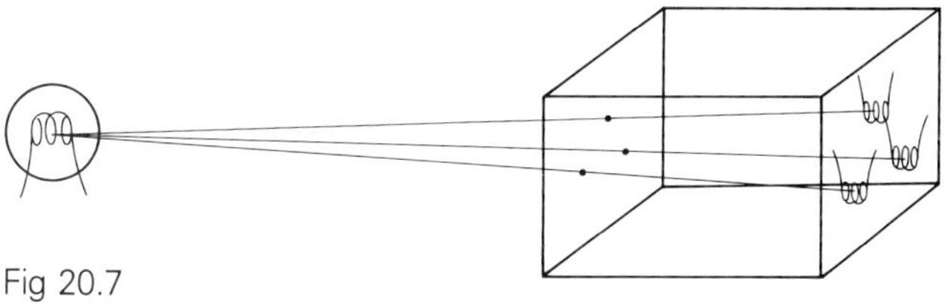

Fig 20.7

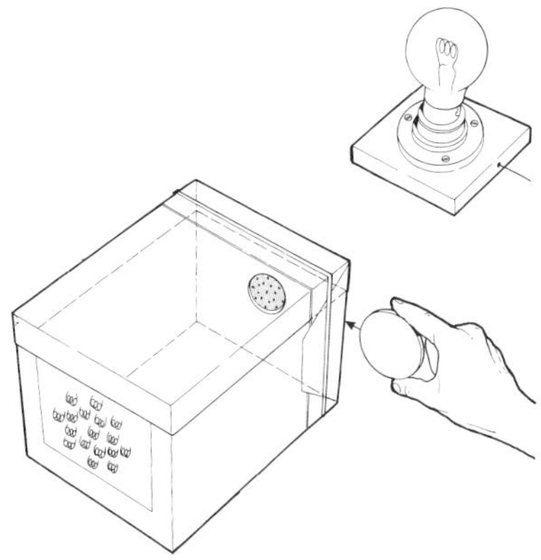

Fig 20.8

d Now hold a lens in front of the camera and find out what happens to the images.

The images come closer together and, for one special position of the lens, you can make the images overlap exactly. As you will see later, the lens does this by changing the direction of the light.

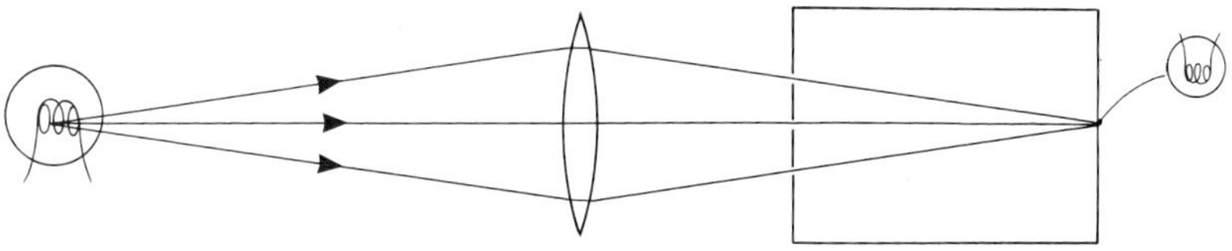

Fig 20.9

e Replace the black paper covering the circular hole in the camera box, and start again with one pinhole. Notice that the image is sharp but not very bright. To make it brighter, the hole could be made larger. Use the point of a pencil to do that.

Certainly more light gets through, but the image is not as clear as before; its edges have become fuzzy. If you have a large hole, it is like having a very large number of pinholes side by side, and instead of the image just getting brighter, it gets blurred too because the images do not overlap.

Again use the lens to make the multitude of images overlap exactly so that you have a clearly defined and bright image of the lamp. Now you can stick your finger through the black paper or remove it altogether; the lens will again produce a clear image for you if it is in the right position. Fig 20.10 shows the light beam passing through the lens to form what is called a **real** image on the screen.

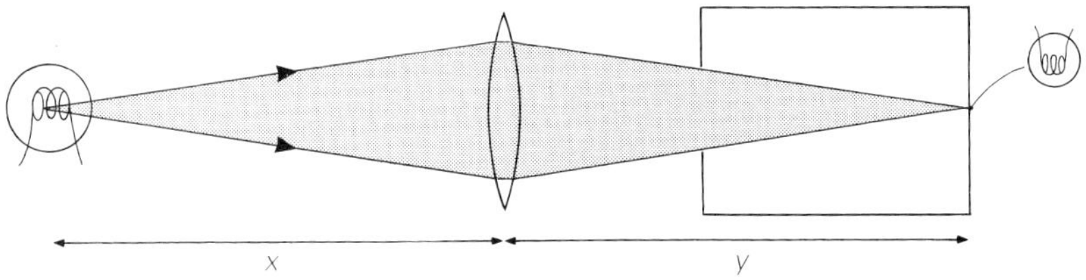

Fig 20.10

Experiment 20.3 The image in space

The last part of Experiment 20.2 can be done again but without the camera box.

a Using the same lamp, hold the lens in your left hand and a thin sheet of white paper in your right hand. Look at the paper as in Fig 20.11, and move the paper until you have found the one position for a sharp image.

You can now do a very surprising thing. Look at the image on the piece of paper and then take the piece of paper away. Provided you are directly behind the lens and not to one side, you can still see the image in space. The image is there whether there is a screen or not!

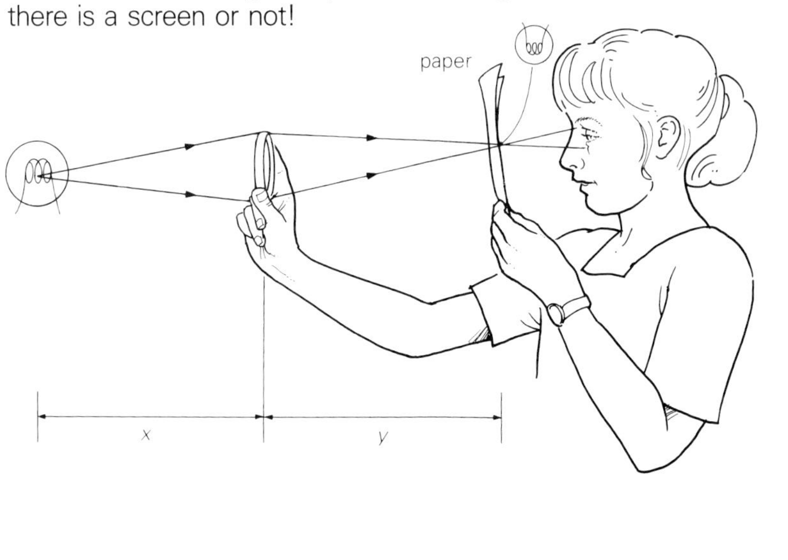

Fig 20.11

b Try the experiment again using the window as the object for the lens. If it is difficult, try holding up a piece of paper to catch half the image on the edge of the paper. Concentrate your attention on the image on the paper and then quickly take away the paper to see the image still there in space.

Fig 20.12

The next experiments show that a lens does change the direction of the light passing through it and that different lenses bend the light by different amounts.

Demonstration experiment 20.4
The smoke box

For this experiment a wooden-framed box with glass or Perspex sides is needed. A plate with holes in it should be placed at the end of the box. A strong lamp is set up about 30 cm from the end of the box.

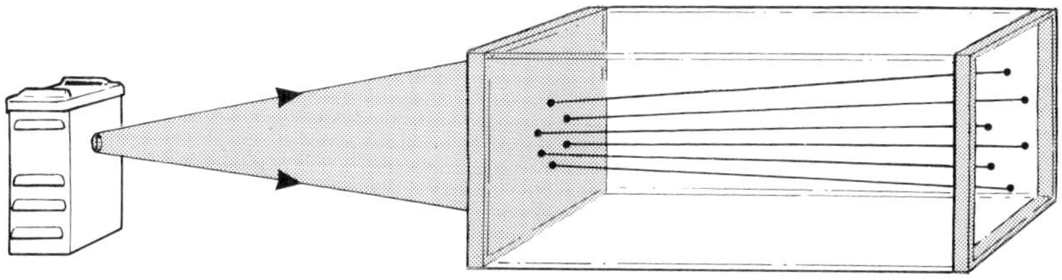

Fig 20.13

The path of the light inside the box will not be seen but if some smoke is put inside the box, the smoke particles will scatter some of the light and the light path will be visible.

If a large lens is put inside the box, the light paths will be bent by it and the light beams will cross at some point inside the box.

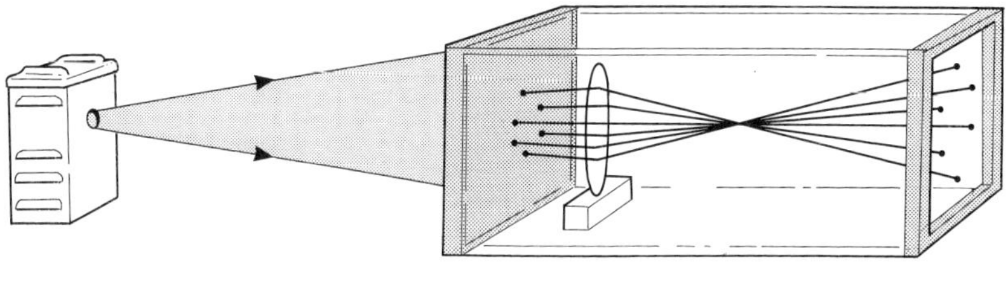

Fig 20.14

If the lamp is moved closer to the box, the point where the beams cross will move further away. It can be moved closer until the beams meet at the end of the box, just as in the experiment with the pinhole camera.

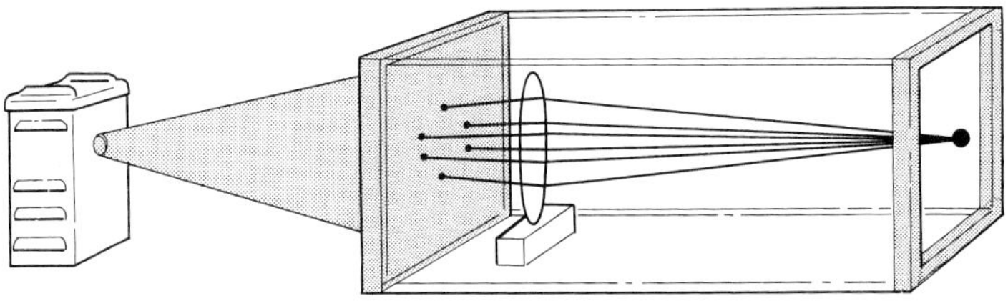

Fig 20.15

Finally take away the plate with the holes in it. There is now a broad beam of light, which is brought to a point at the end of the box by the lens.

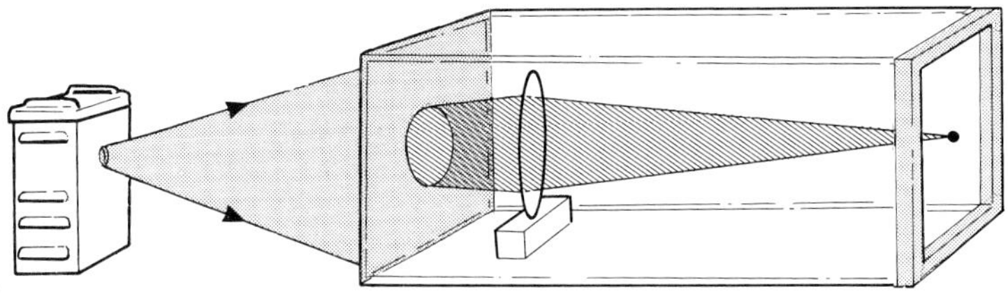

Fig 20.16

Experiment 20.5
Experiments with 'ray streaks'

You will need to work in a darkened room for these experiments. The light source is a lamp with a vertical 'line' filament with a shade placed so that a fan of light can spread out over a sheet of white paper. The paper should be placed on the bench about $\frac{1}{2}$ m from the lamp. A comb of slits, placed as shown in Fig 20.17, allows thin streaks of light to cross the paper.

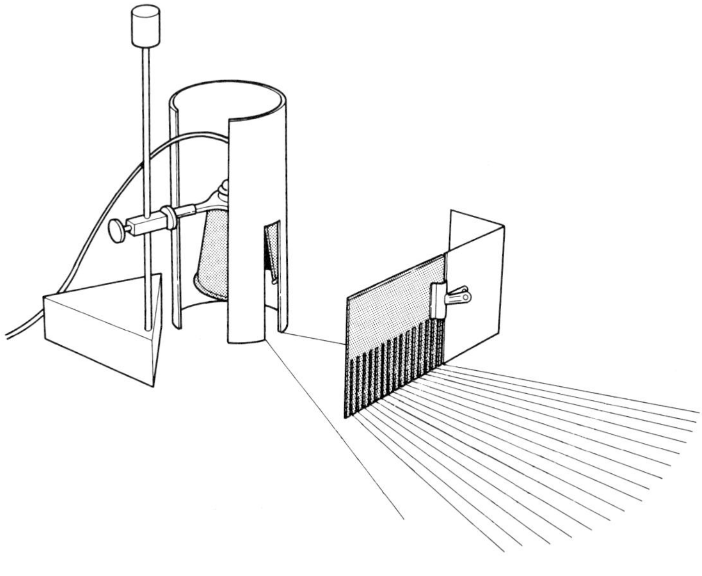

Fig 20.17

a Place a thin lens on the paper so that the ray streaks fall on the flat surface of the lens and see what happens. What is different for the outer ray streaks compared with the inner ones? What about the ray which goes through the centre of the lens? Try blocking off all the ray streaks except one (you can use small blocks of wood for this) and then see if your observations are correct if the ray streak hits the lens at different small angles. Does it make any difference if the lens is turned round so that the ray streaks fall on the curved surface first?

Fig 20.18

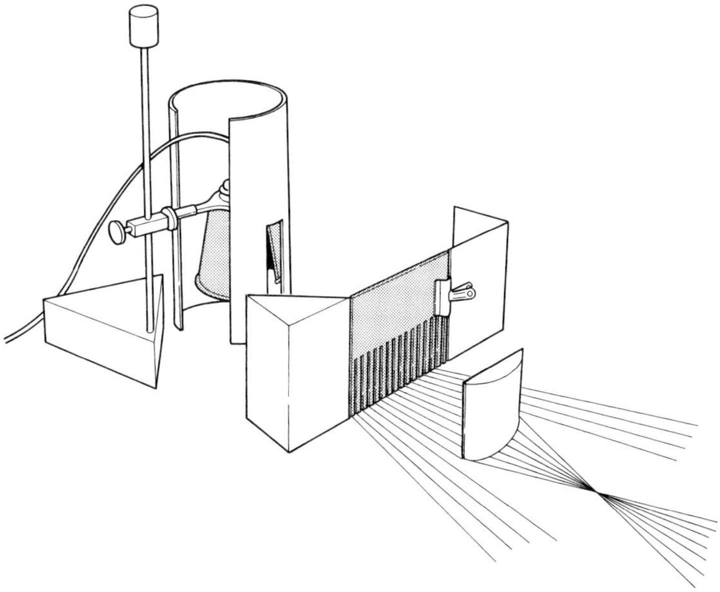

The lens bends the outer rays more than the inner ones so that all the rays go through one point (the image) on the other side of the lens (Fig 20.19a). The ray through the centre of the lens does not change its direction of travel (Figs 20.19b and c).

Fig 20.19

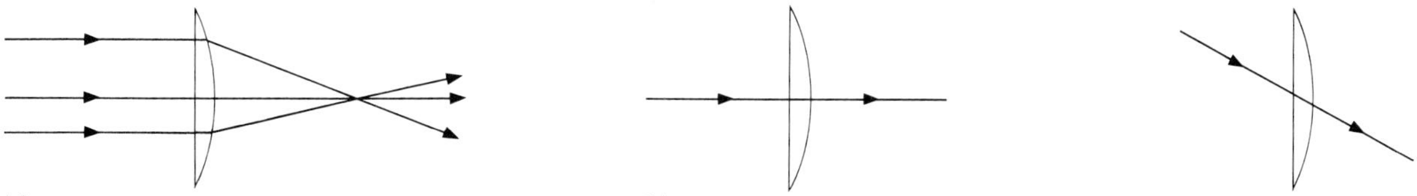

(a) (b) (c)

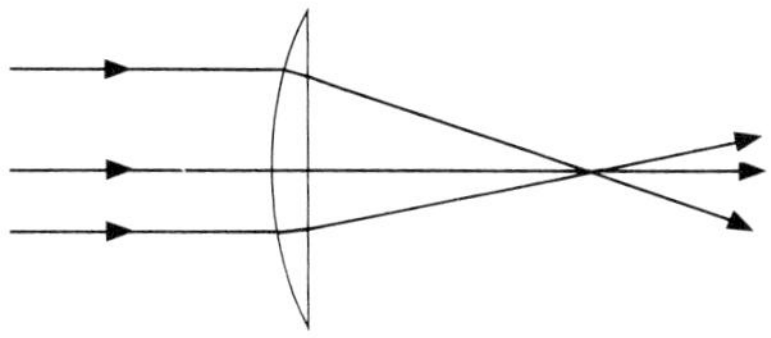

Fig 20:20

The same things happen if you turn the lens the other way round (Fig 20.20).

b Now use a lens whose cylindrical surface is more curved. (If the flat surface of the lens is the same size as the one used in part **a**, the lens will be fatter.) Find out how its behaviour is different from the lens used in part **a** when just three ray streaks fall on it.

The fatter lens bends the ray streaks more than the thinner one did when placed at the same position on the paper. The rays therefore cross at a place which is closer to the lens. The fatter lens is said to be stronger or more powerful than the thinner one.

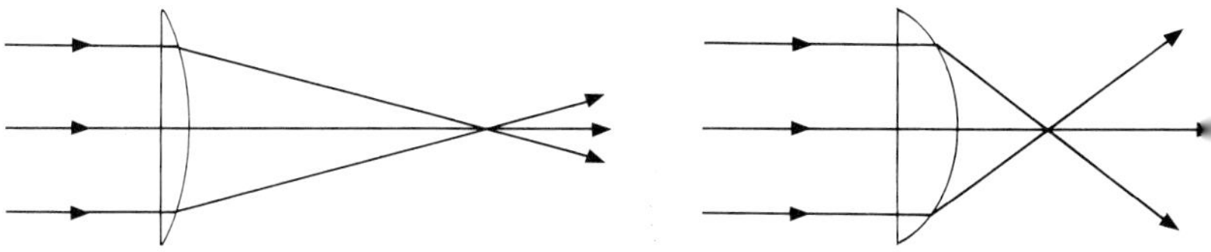
Fig 20.21

c Using the more powerful lens, see if you can find any difference in the way it behaves when there are more than three streaks falling on it.

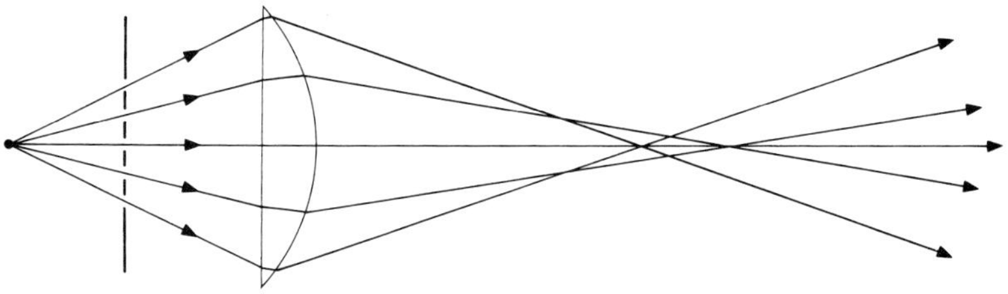
Fig 20.22

Now you should notice that it is not behaving 'perfectly'. The rays do not all pass through one image point. What happens if the lens is turned round? The defect is still there but even worse. See what happens if the comb of slits is removed. Does limiting the width of the lens have any effect?

All lenses suffer from this defect to some extent, but it is most evident with strong lenses. Limiting the width of the lens in use (the *aperture*) improves matters.

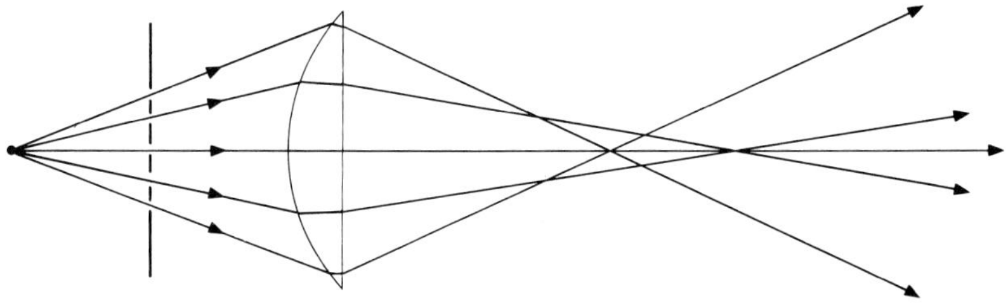
Fig 20.23

d Put the comb of slits back in place with the weaker lens close to it. Find out what happens when the light source is moved near to or further away from the lens (Fig 20.24).

As the source is moved *closer* to the lens, the image point moves *further away.* As the source is moved *away* from the lens, the image point gets closer to the lens. The nearest this real image point gets to the lens is when the lamp is at a great distance from the lens. That smallest distance, *f*, is called the **focal length** of the lens (see Fig 20.25). Strong lenses have small focal lengths and weak lenses have large focal lengths.

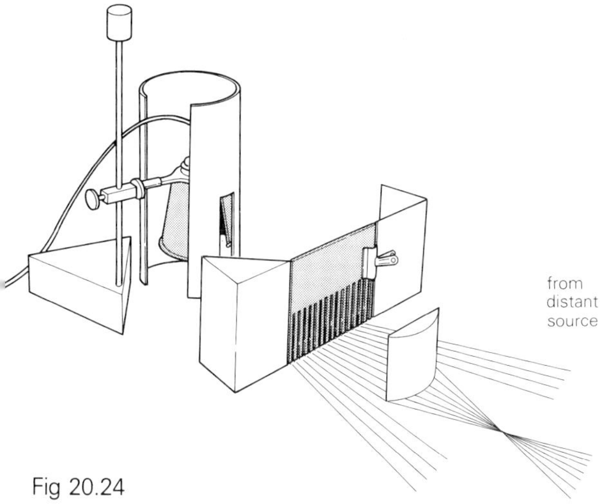

Fig 20.24

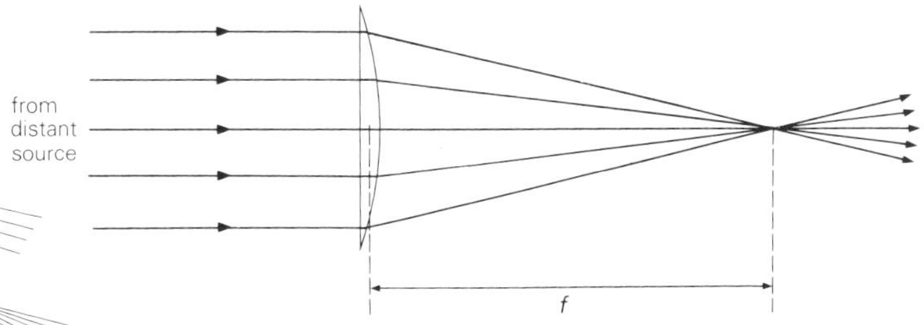

Fig 20.25

e The lenses you have tried up to now all cause the ray streaks to converge to an image point. They are all **converging** lenses; they are all fatter in the middle than at the edges. Opticians call these **positive** lenses. The other type of lens is thinner in the middle than at the edges. This is the **diverging** or **negative** lens. See what effect it has on the ray streaks.

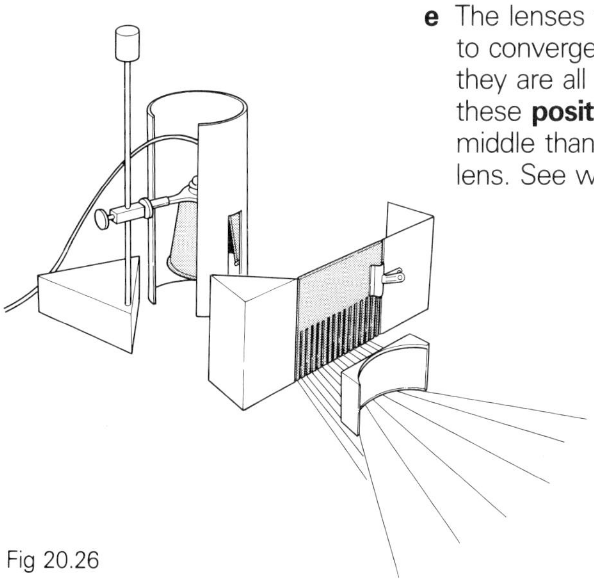

Fig 20.26

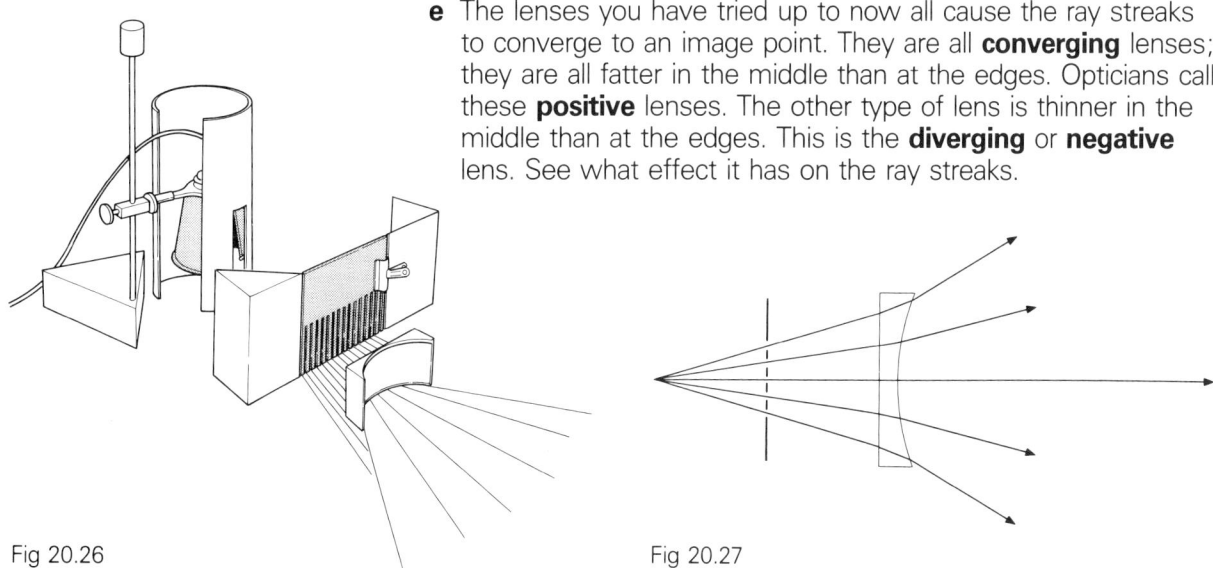

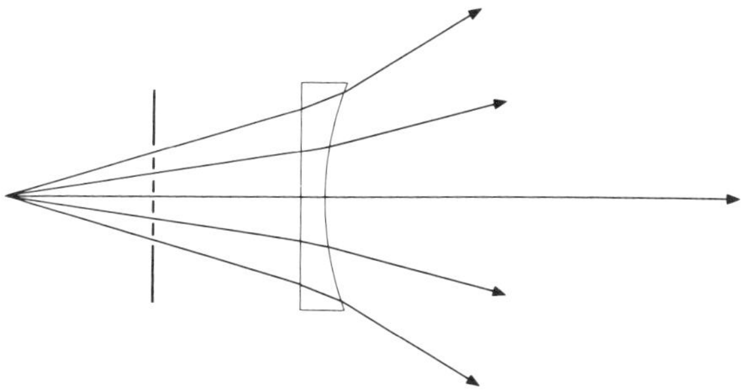

Fig 20.27

f Finally, find out what happens if you use two lenses at the same time.

Background Reading

The camera, the slide projector and the eye

The camera

In the camera, the screen is the photographic film. When light falls on the film it causes chemical changes within the film so that the image is captured. Only a small amount of energy is needed to do this so that it is necessary to have a shutter which will only admit light for a fraction of a second, usually $\frac{1}{30}$ th of a second or less. When the shutter is operated, light from a distant landscape passes through the lens which is positioned so that a sharp image falls on the film.

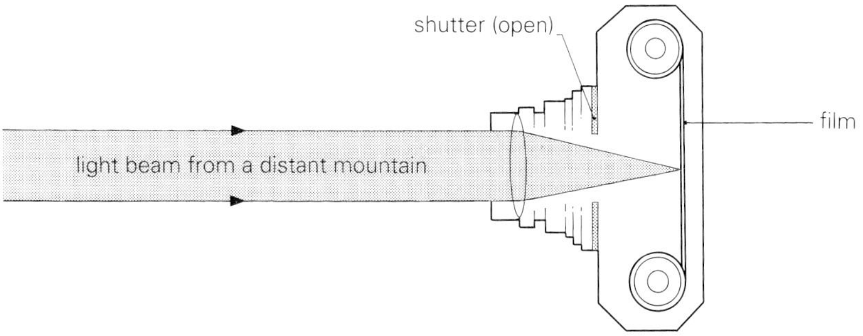

Fig 20.28

Strictly speaking, only objects at one distance from the camera will be in focus. Objects nearer or further away than that will produce images with fuzzy edges, that is to say, they will be out of focus. However if the diameter of the lens is not large, the 'fuzziness' is small, and we do not notice it when we look at the photograph, provided the object is not closer to the lens than a few metres. The diagrams in Fig 20.29 illustrate this.

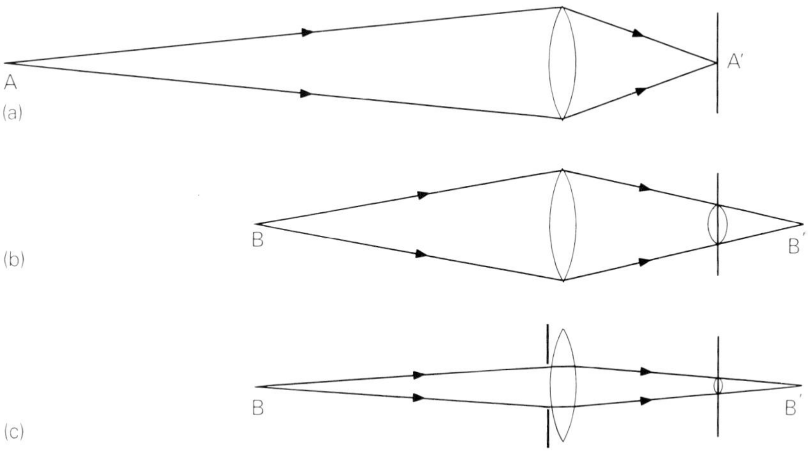

Fig 20.29

A and B are object points; A′ and B′ are their images. The image of A is in focus on the film but not the image of B. When the lens aperture is large (Fig 20.29b), the light from B forms a circular patch on the film. With a small aperture, the patch becomes very small and much more like a point (Fig 20.29c).

In the more expensive cameras, the lens can be moved so that objects over a wide range of distances can produce a sharp image on the film. Naturally, the lens is closest to the film when photographing objects at the greatest distance. It has to be moved further away from the film to focus objects near to the camera.

The slide projector

In the slide projector, an image of a transparent photograph (the slide) is thrown on to a screen by a lens. Naturally, a strong lamp is needed to illuminate the slide sufficiently for an image to be visible in a partly darkened room. The simplest arrangement is shown in Fig 20.30.

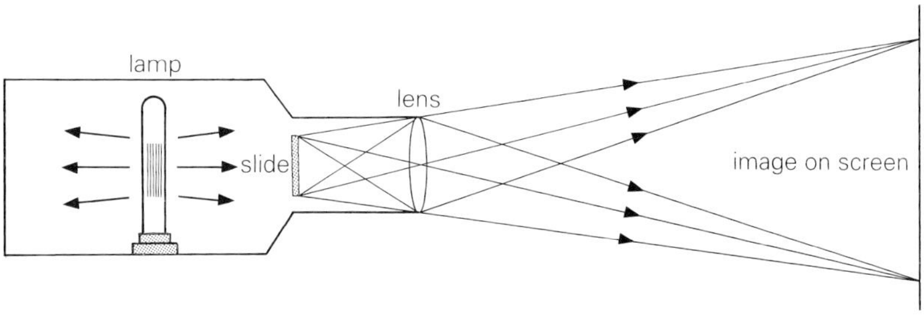

Fig 20.30

This arrangement is not very efficient. Only a small proportion of the light emitted by the lamp goes through the slide. To improve this, two lenses (called *condenser* lenses) are used to concentrate light evenly on the slide, and so give a brighter, evenly illuminated image. A curved reflector is often put behind the lamp to catch some of the light travelling away from the slide and send it back towards the slide.

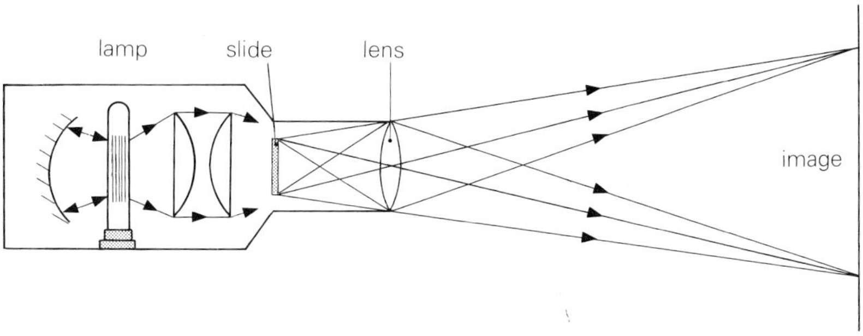

Fig 20.31

Filament lamps emit a lot of infra-red radiation (see page 319) as well as light and this can easily melt the slide. To stop this happening, a piece of special glass (the heat filter) which absorbs the infra-red radiation, is placed between the lamp and the slide, and the manufacturer usually incorporates a small fan too.

Notice that, with the object close to the lens, the image produced is larger than the object; it is magnified.

The eye

The eye is like a good camera in some ways. It has a lens to help form a real inverted image on a sensitive 'screen'; it automatically focuses the images of objects at different distances; it automatically adjusts the lens aperture to allow for different levels of brightness; and the inside surface is almost black so that light is not reflected about inside it. But there are differences too.

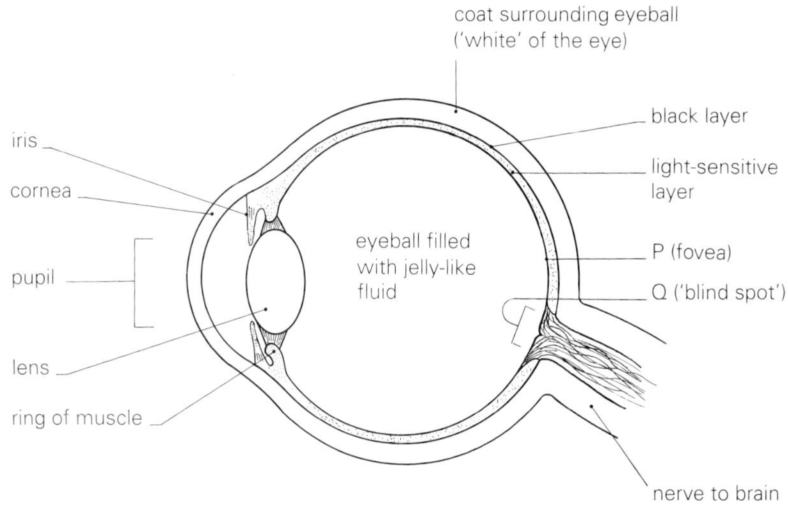

Fig 20.32

Fig 20.32 shows an eye with the names we give to its various parts. It has a tough, white coat which is spherical except for a transparent bulge (the *cornea*) at the front. Inside, a jelly-like lens is held in place by ligaments suspended from a ring of muscle. This divides the eye into two chambers; the first is filled with a watery fluid, the second a jelly-like fluid. In front of the eye lens, there is another ring of muscle (the *iris*) surrounding a hole (the *pupil*) through which the light passes. In dim light, the iris widens so the pupil gets bigger; in bright light, it shrinks so that the pupil gets smaller.

At the back of the eye there is a thin layer of light-sensitive cells (the *retina*). This layer receives the image and sends the messages to the brain via the main nerve cable (the *optic nerve*). The brain interprets these signals to give an upright image; it inverts the inverted image!

When light enters the eye, most of the bending occurs at the cornea, and then the eye lens is automatically adjusted to produce an inverted 'in focus' image on the retina. If the object moves closer to the eye, the eye lens needs to be a little more powerful in order to focus the image and so the muscles contract to allow the lens to become fatter. This process is called *accommodation*.

There is, of course, a limit to how fat the lens can become. So, there is a limit to how close to the eye an object can get and still form a sharp image. This distance (the *near distance*) gets larger as you get older, because as muscles and lens become less flexible, the lens does not fatten as much as it did in early childhood. Find out what your near distance is.

In middle age, a normal eye has a near distance of about 25 cm. Young children have near distances of about 10 cm, and, with older people, the distance can be many hundreds of centimetres.

For a normal eye, the lens of the eye is made as fat as possible to see a close object (Fig 20.33).

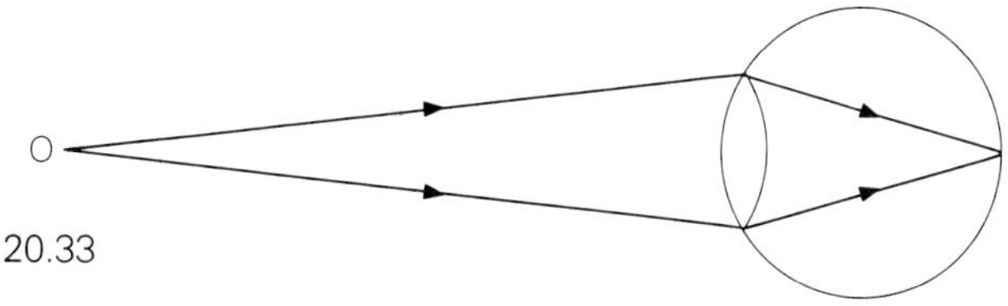

Fig 20.33

A short-sighted person can see closer objects (Fig 20.34).

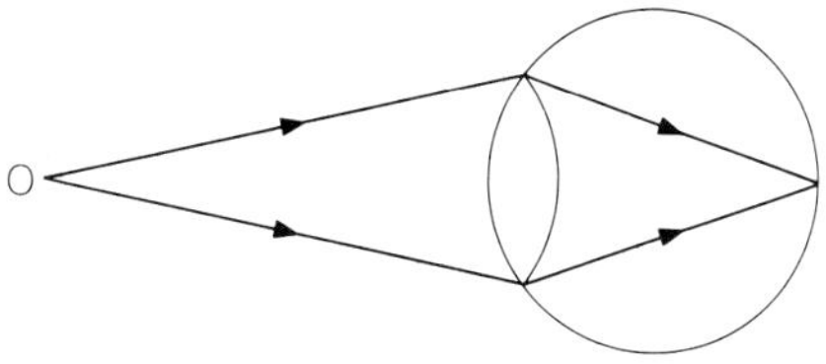

Fig 20.34

But such a short-sighted person cannot see distant objects clearly (Fig 20.35) because the eye lens is too strong.

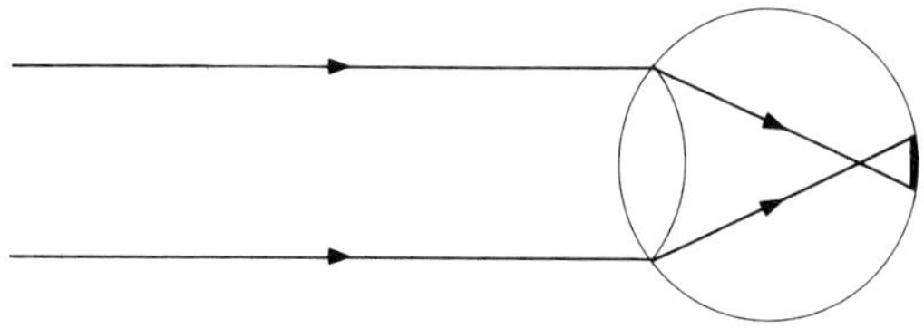

Fig 20.35

This can be corrected by putting a negative lens in front of the eye (Fig 20.36).

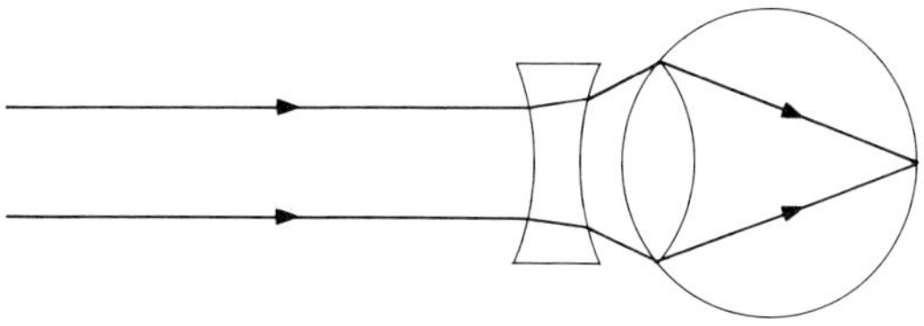

Fig 20.36

It is different for long-sighted people. They cannot see close objects clearly (Fig 20.37); the eye lens is not strong enough.

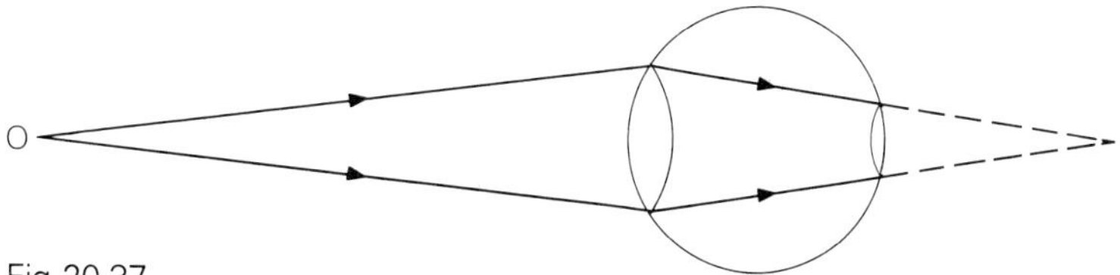

Fig 20.37

That situation can be corrected by putting a positive lens in front of the eye (Fig 20.38).

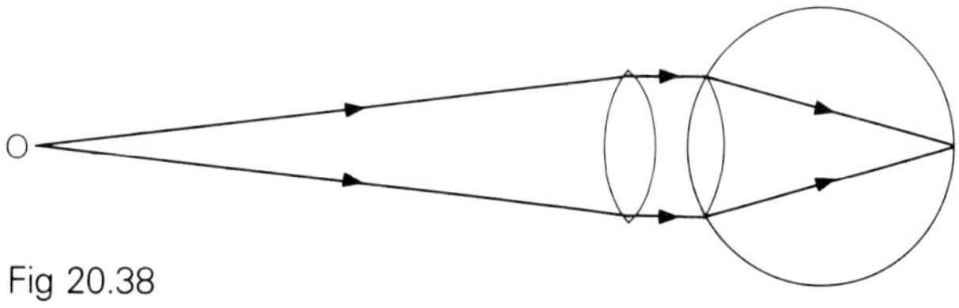

Fig 20.38

Questions for homework or class discussion

1 An illuminated sign has the word SALE on it. Draw what you would see on the screen of a pin-hole camera pointed towards it.

2 The best pin-hole cameras (and lens cameras too) are black on the inside. Why is this?

3 Architects sometimes use a pin-hole camera to photograph large buildings. What are the advantages and disadvantages of doing this?

4 In a pin-hole camera, what are the effects of
a moving the camera nearer to the object being viewed,
b increasing the distance between the pin-hole and the screen,
c *slightly* enlarging the pinhole,
d *greatly* enlarging the pinhole?

5 Fig 20.39 shows a pin-hole camera with four pin-holes, pointed at a lamp. Copy the diagram and draw light rays to show where the images appear on the screen. A lens is placed in front of the pinholes so causing the images to overlap exactly. Draw another diagram to show the paths of rays now.
What happens if one large hole replaces the four pin-holes with the lens in the same position?

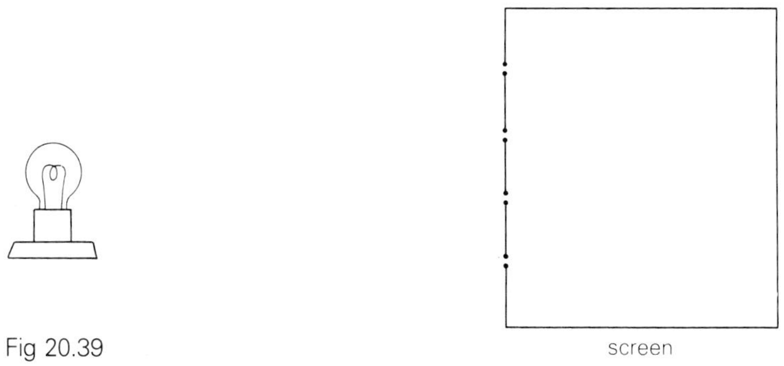

Fig 20.39

6 Fig 20.40 shows a set of positive lenses, all made of the same kind of glass. The focal length of each lens in this set is either 20 cm, 10 cm, 5 cm, or 2.5 cm. By looking at the drawings carefully, say what you think is the focal length of each lens. Explain how you got each answer.

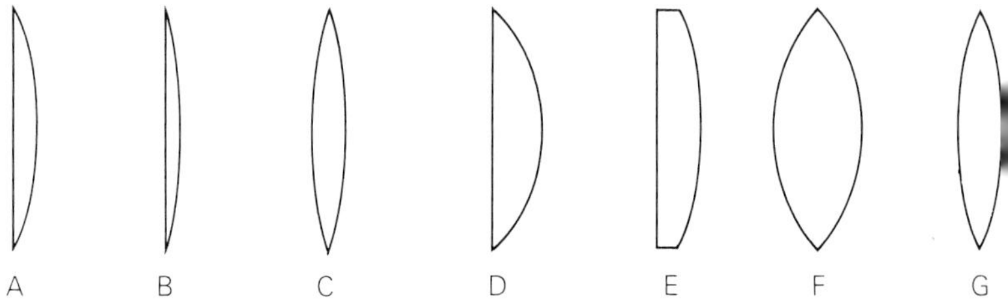

Fig. 20.40

7 A long-sighted person cannot see close objects clearly but finds it easier to read a book in a very bright light. Explain this. [*Hint*: think about the pupil of the eye.] What would you advise if the reader had to read small print and had no spectacles?

8 A good camera has an aperture (an opening) whose size can be changed, and a shutter which can be adjusted so that it is open for different times. Why does the manufacturer do this, and what advantages are to be gained?

How would you set the aperture and shutter time if you wanted to photograph a car moving at high speed?

Reflection of light

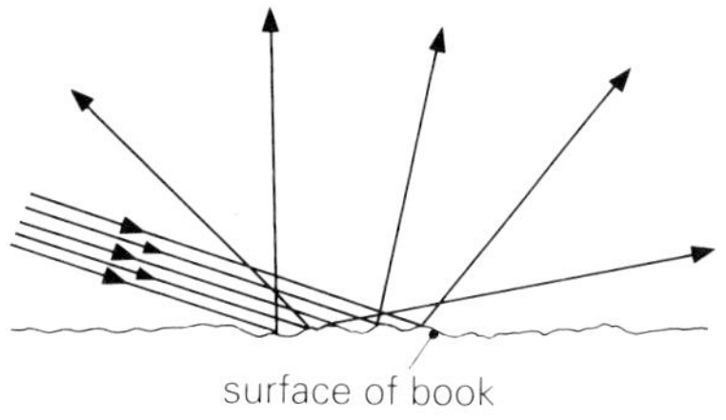

Fig 20.41

This chapter began by saying that non-luminous objects are visible because they reflect light, and if there is no light they cannot be seen. If light from a window or lamp falls on a book, the book can be seen in all parts of the room. This is because the surface of the book is actually quite rough and light bounces off it in all directions; in other words the book *scatters* light.

If the surface is very smooth like a sheet of glass or an unwrinkled piece of aluminium foil or a polished metal, it can reflect light in a regular way. A mirror provides such a surface as one of its faces has been silvered.

Experiment 20.6
Reflection by a mirror

Use the apparatus shown in Figure 20.42 to find out what happens when you place a mirror so that the ray streaks are reflected from it.

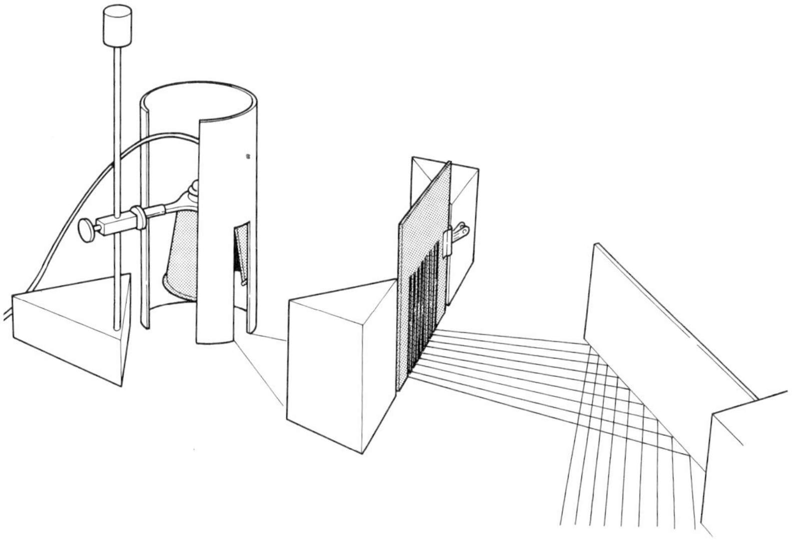

Fig 20.42

Try using only one ray streak by blocking off the other streaks with wooden blocks or by using a screen with just one slit in it. Turn the mirror at different angles to see what happens to the ray streak when it strikes the mirror.

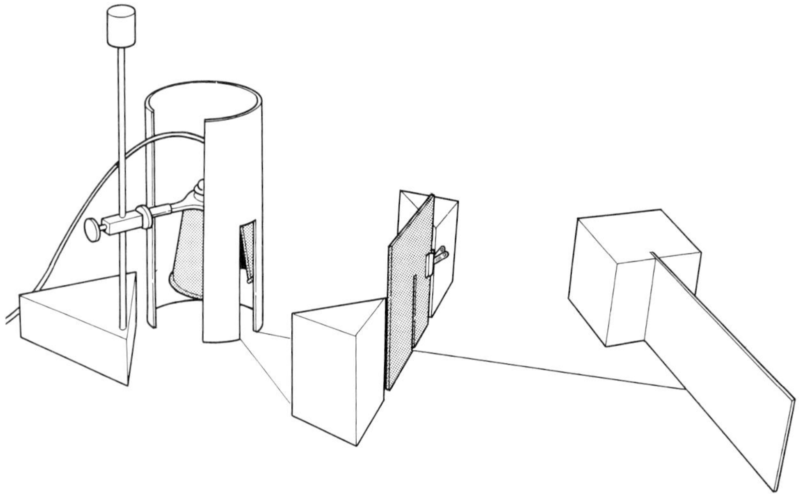

Fig 20.43

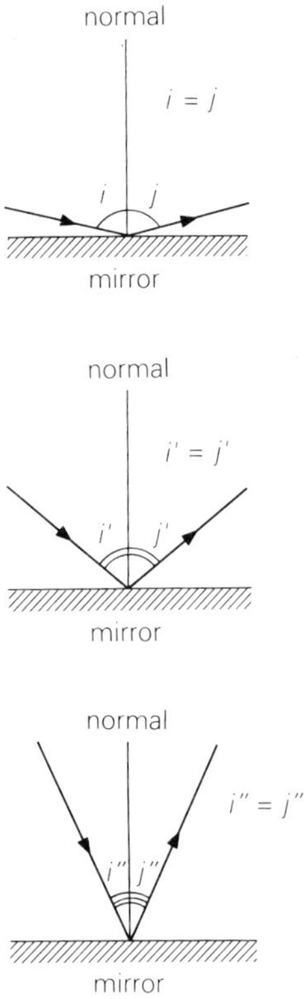

Fig 20.44

You will have noticed that the mirror reflects the ray at the same angle as that at which the light hits it.

Angles are usually measured between the rays and a line drawn at right-angles to the surface of the mirror (the *normal*).

Images formed in a mirror

Consider a candle in front of a mirror. Light from the candle will be reflected from the mirror as shown in Fig 20.45.

When you look into the mirror you see an image of the candle. It is an upright image and it is behind the mirror. It is not a real image: you cannot catch it on a screen or a piece of paper.

This kind of image is called a **virtual** image. It is quite different from a real image. Light actually passes through a real image so that we can get the image on a screen. With a virtual image, the light only appears to come from the image and we cannot catch it on a screen.

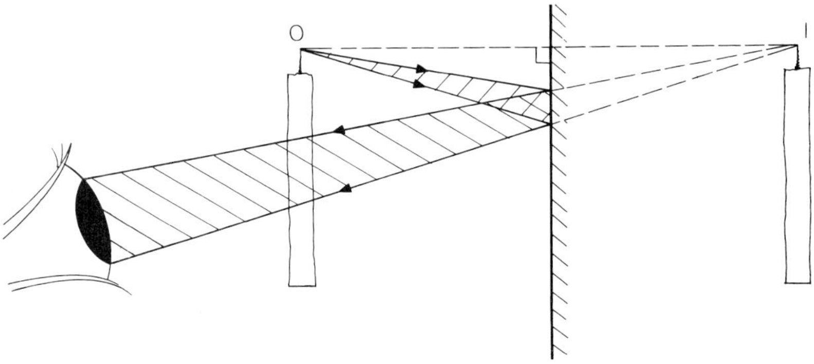

Fig 20.45

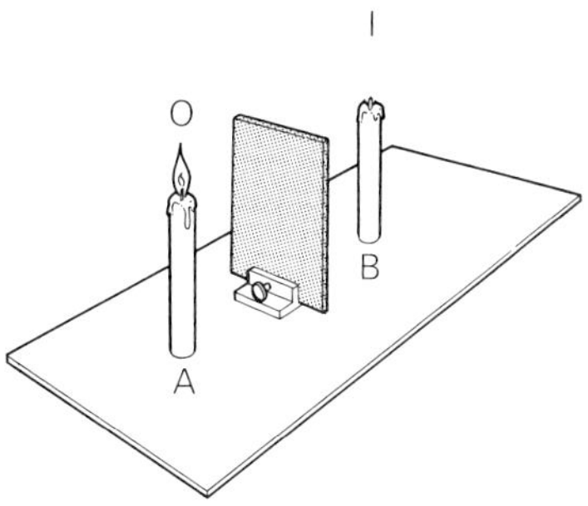

Fig 20.46

Experiment 20.7
Checking the position of the image

For this experiment, you need two candles and a large flat mirror. Set up the mirror vertically as shown in Fig 20.46 with one candle in front of the mirror and one behind.

Look at the image of candle A in the mirror and then partially looking over the top of the mirror adjust the position of candle B so that its position appears to coincide with the position of the image of candle A, from a range of viewpoints.

Measure the distances of the candles from the mirror and you should find they are the same distance from the silvered surface.

(It makes an entertaining effect if candle A is lit: the image of candle A is obviously lit. But when you raise your head candle B is seen to be out.)

Questions for homework or class discussion

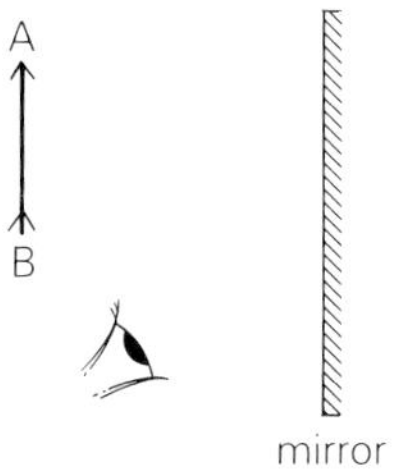

Fig 20.47

1 Fig 20.47 shows an arrow AB in front of a mirror, and an eye. Draw two rays which leave A and enter the eye. [*Hint*: you know where the image of A will be. Mark this position first.]

Now draw two rays from the tail B to the eye. What can you say about the length of the image?

2 Fig 20.48 shows a car in a very narrow lane where it joins a busy road. Two mirrors are to be placed at P so that approaching cars can be seen. Show by means of a diagram how the mirrors should be placed. Draw rays of light to show how the driver sees the approaching cars.

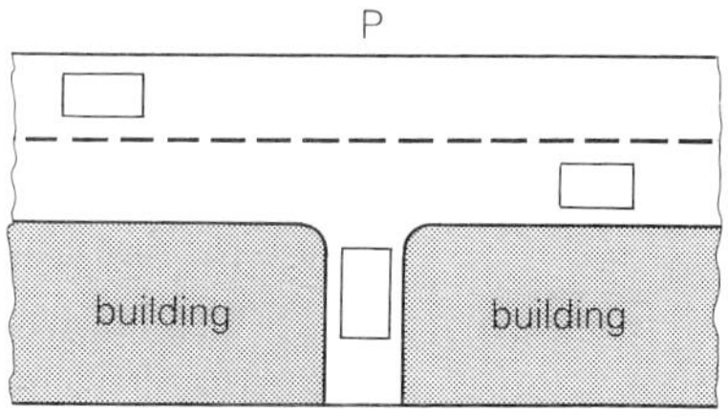

Fig 20.48

3 Find out what a periscope is and draw a diagram to show how you could make one using two small flat mirrors.

4 Explain why the light reflected from a mirror forms an image but the light reflected by a sheet of white paper does not.

The bending of light: refraction

We have already seen that lenses can bend rays of light. This 'bending' can occur in both glass and water as the following experiments show. The bending is usually referred to as *refraction*.

Experiment 20.8
The bending of light by water

Use the ray streak apparatus with a plastic box and water, as shown in Fig 20.49.

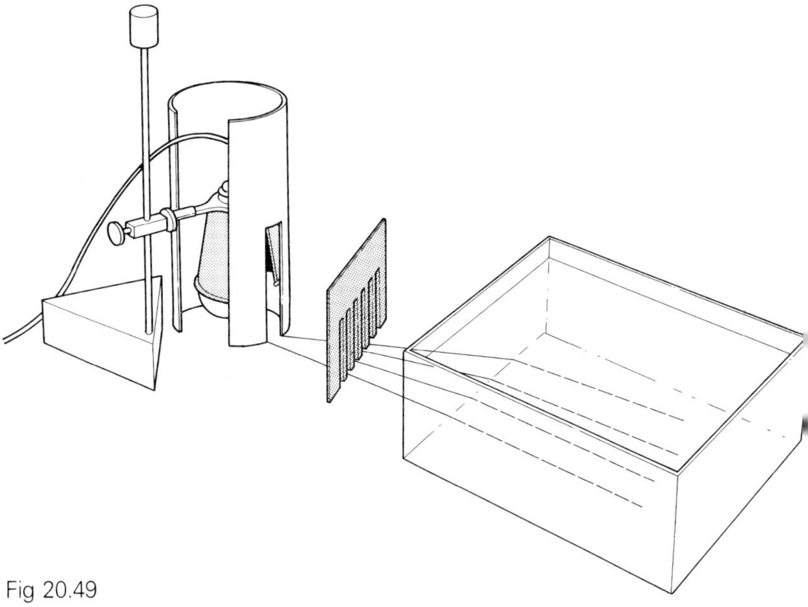

Fig 20.49

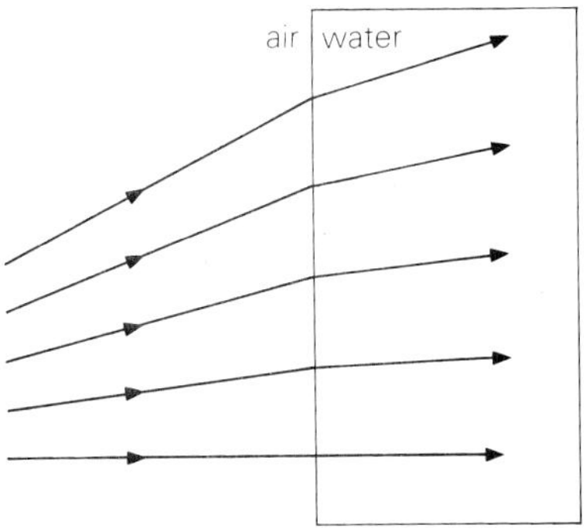

Fig 20.50

Bending occurs at the boundary between the air and the water (Figs 20.49 and 20.50).

In talking about refraction, we need to mention the angles between the light directions and the normal to the surface (the imaginary line at right-angles to the surface). When light is travelling from air into water, the angle *x* (usually called the *angle of incidence*) is bigger than the angle *y* (usually called the *angle of refraction*). We say that light has been bent towards the normal.

To see what happens when light comes from water into air, set up your apparatus as shown in Fig 20.53. Now angle *y* is larger than angle *x*; the light is bent away from the normal.

There is one other difference between this case and the case when light travels from air into water. The light does not emerge

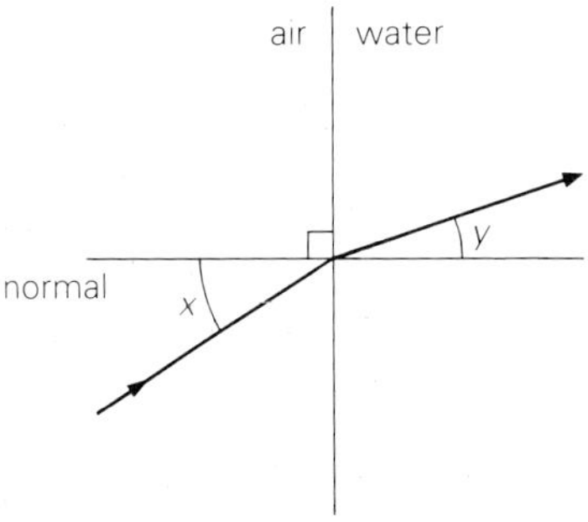

Fig 20.51

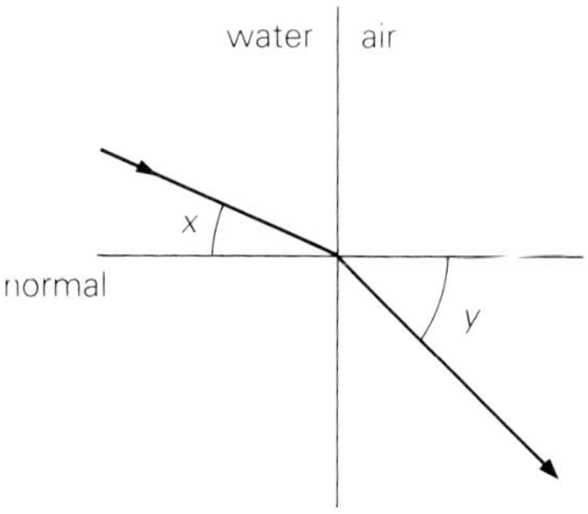

Fig 20.52

from the water if the angle x is larger than a certain value; all of it gets reflected instead (Fig 20.54).

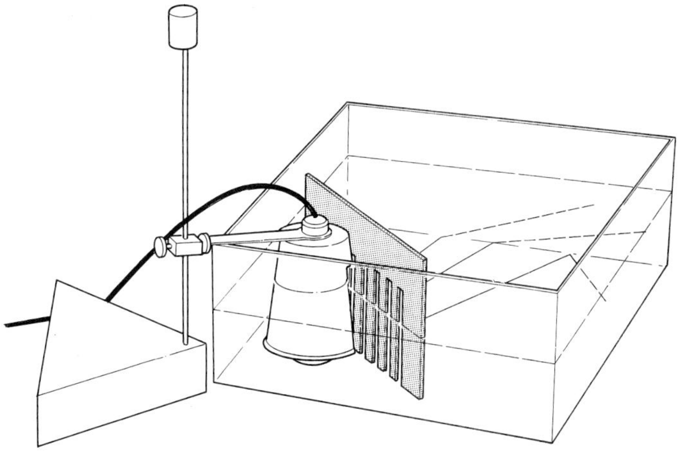

Fig 20.53

Did you notice that some of the light was reflected at other angles as well?

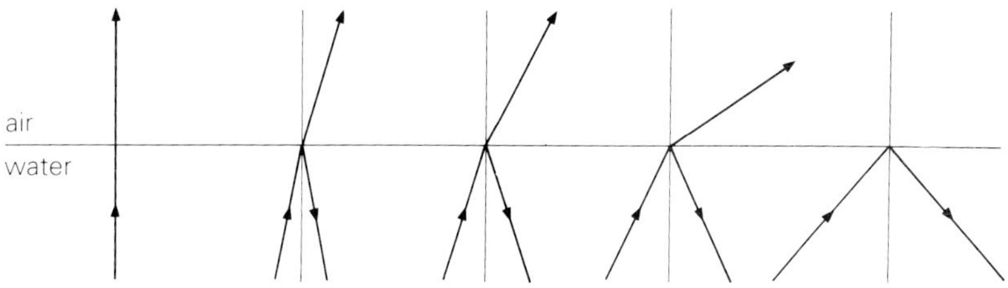

Fig 20.54

Experiment 20.9
Light passing through a block of glass

Use the ray streak apparatus and arrange it so that a single ray strikes the side of a parallel-sided block of glass placed flat on the sheet of paper.

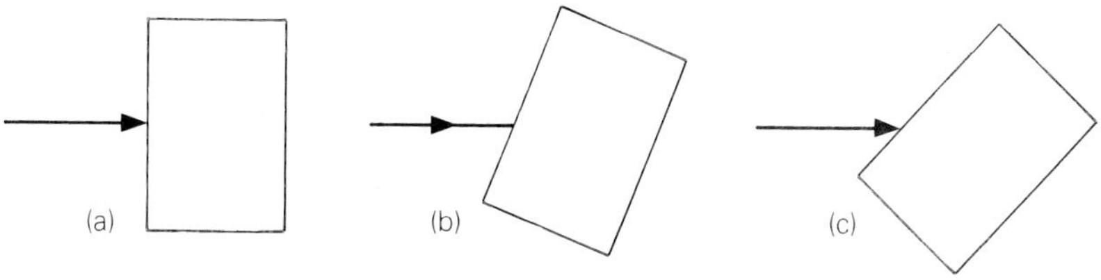

Fig 20.55

a First investigate what happens when the ray strikes the block at right angles. Does the ray go through the block? In what direction does it come out?

b Rotate the block slightly, as in (b). Does the ray go through the block? Does it go straight through? In what direction does it come out? Rotate the block more, as in (c). Does the ray come out parallel to the light entering, but moved sideways a bit? Notice which way the light bends at the first surface and which way it bends at the second. Did you also notice that some of the light at the second surface was also reflected?

When a ray of light passes through a parallel-sided block of glass, the direction of the light leaving the glass is parallel to the direction of the light entering, and the ray has been moved sideways a bit (Fig 20.56).

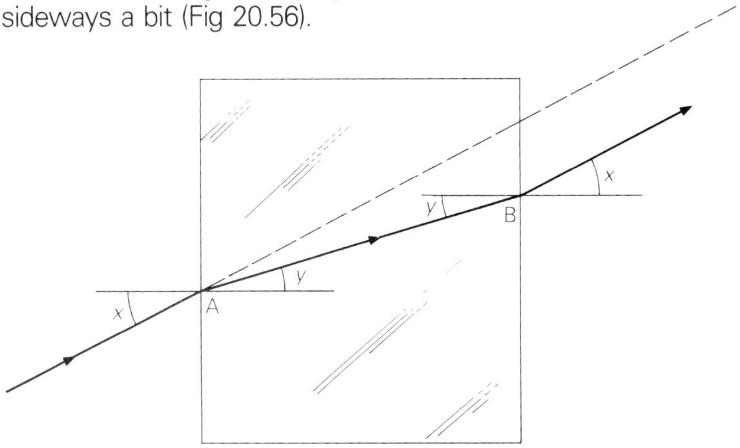

Fig 20.56

This happens when the sides of the block are parallel because the light is bent as much towards the normal at A as it is bent away from the normal at B.

This is why the centre part of a lens does not change the direction of the light: the centre part of the lens is just like a parallel-sided block. A good lens is thin and the slight 'moving sideways' of the ray is not noticeable.

The sides of the outer parts of a lens are not parallel and so there is some bending of the light. There is more bending near the edge because the angles are bigger.

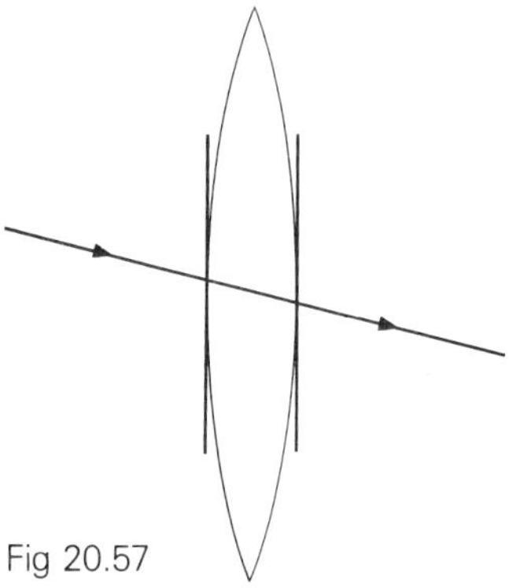

Fig 20.57

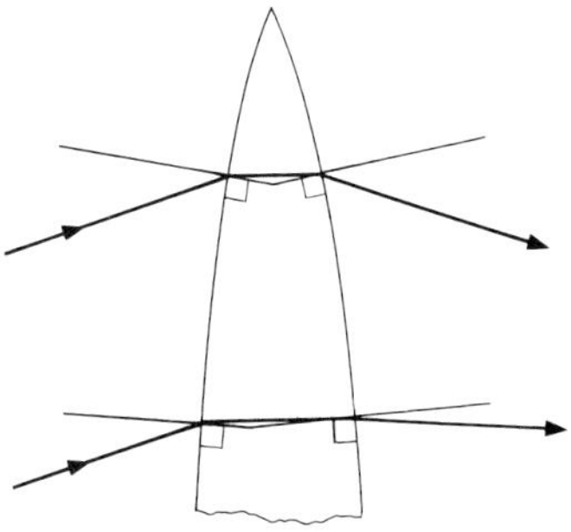

Fig 20.58

Later, you will learn that refraction occurs because light travels at a different speed in the transparent substance from its speed in air. The change of speed varies for different substances, so the amount of refraction produced is different. Water is not as 'good' at refracting as glass is, and glass is not as 'good' as diamond.

Does the light travel faster or slower in water? A difficult experiment has to be done to answer that and it was not until 1850 that Foucault, a Frenchman, was able to do it. He found that light travelled at a slower speed in water.

Experiment 20.10
Further experiments on refraction

a Put a coin in an empty bowl and stand at a place where you just cannot see the coin. Then get someone to fill the bowl with water. The coin suddenly appears!

Fig 20.59

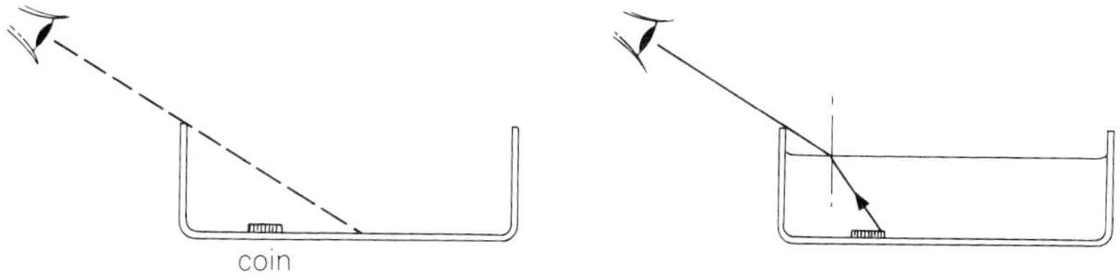

b Now put a straight rod into the water in the bowl. How does the rod appear? Is it bent towards or away from the surface?

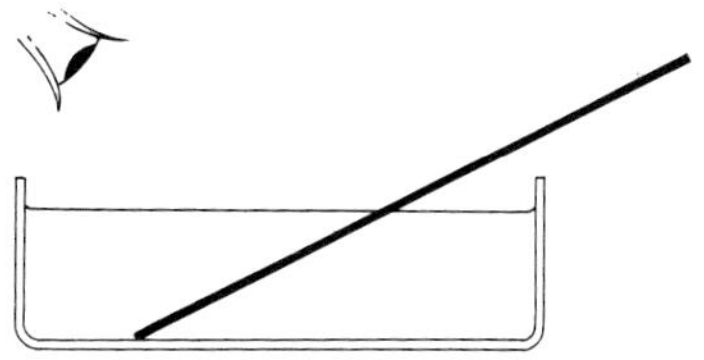

Fig 20.60

In both of the above cases you were looking at a virtual image which appeared to be nearer to the surface than the object causing the image. Indeed, the depth of the bowl itself appears to be less when there is water in it.

To understand how this happens, it is necessary to draw diagrams (Fig 20.61) to show how the light reflected from the object travels to the eye. To make the diagrams clear, the size of the eye's pupil has been exaggerated.

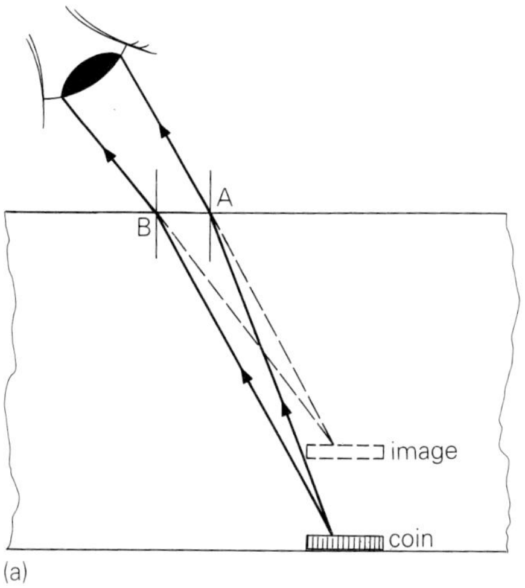

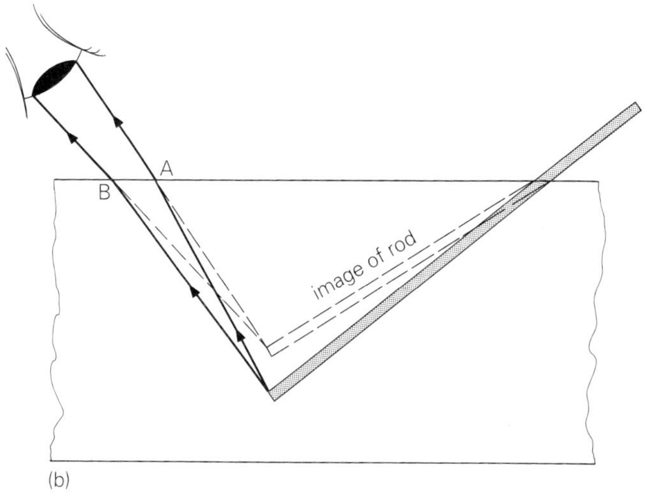

Fig 20.61

The light which travels to the eye via B is bent a little more than the light going via A. The light received by the eye appears to come from a place which is closer than the object, and lies in a slightly different direction.

Now you should be able to explain why a swimming bath appears to be less deep than it really is, and why hunters who spear fish have to aim a spear at a different place from where they see the fish.

Background Reading

Fibre optics

When light travels inside a transparent substance, like water or glass, towards a boundary with air, it cannot get out if the angle of incidence is larger than a certain value (which, in the case of glass, is about 42°). If the angle is bigger than this, then all the light is reflected back into the glass. Thus, light entering a glass rod can be reflected from side to side until it comes out at the other end with none escaping. Because all the light entering at one end comes out at the other end, the rod is often referred to as a 'light pipe'.

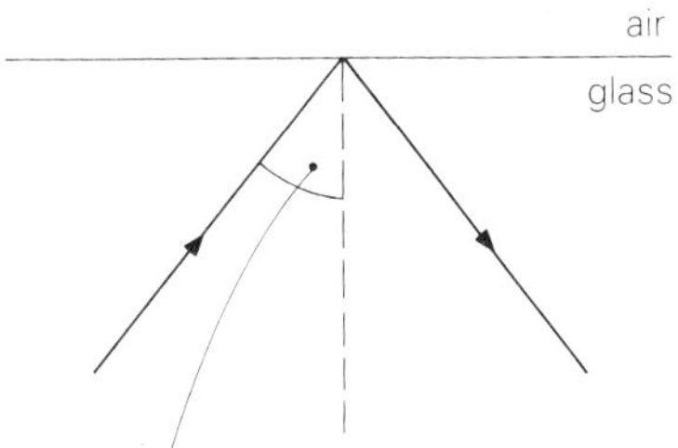

Fig 20.62

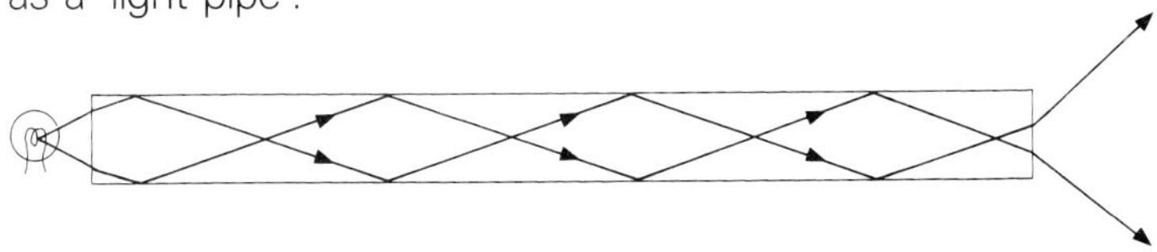

Fig 20.63

The glass (or plastic) rod need not be very thick for it to work in this way. Indeed, fibres of about 1 mm in diameter work just as well, and these are flexible enough for them to be bent, so allowing the light to be piped round corners. Perhaps you have seen lamps using these fibres.

Optical fibres are much used in medicine when it is necessary to look inside a patient's body, for example, at the stomach. Because the fibres are thin and flexible, a bundle of them can be passed into the patient via the gullet. Light is 'pumped' into the body along the fibres, and light reflected from inside then passes out again also by means of fibres. The doctor has an instrument for this and lenses are then used to form an image of the patient's stomach. Indeed, it has been found possible to 'operate' internally by using laser light 'piped' to the site of a growth. The photograph shows the equipment used by doctors for medical purposes.

Another very important use for optical fibres is in the telephone system. Messages can be sent down an optical fibre in the form of a series of light pulses. Of course, there has to be a special code for doing this. First, the electrical variations from the

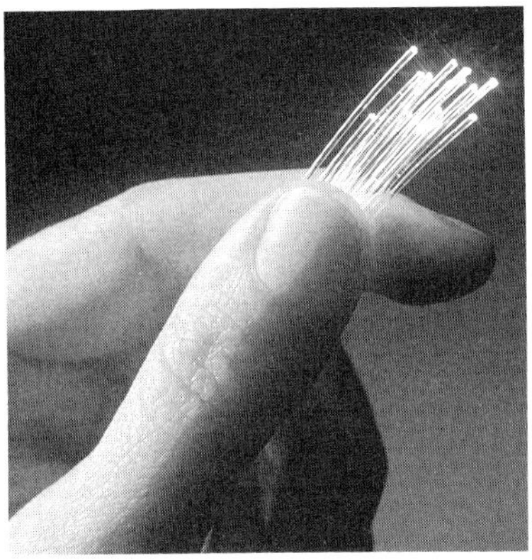

Optical fibres

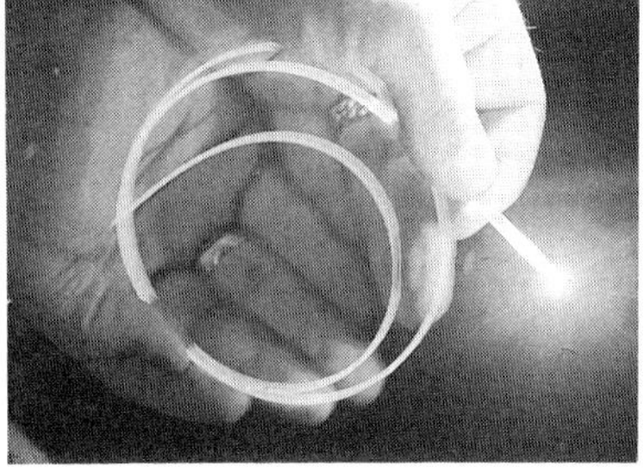

Fibre optic cable

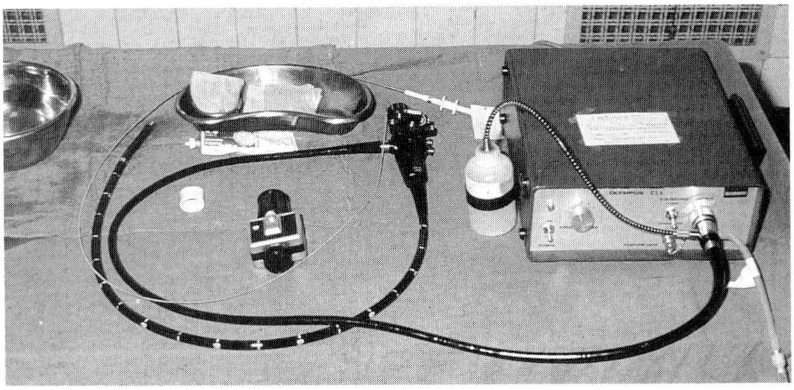

Equipment used by doctors to "see" inside the patient

microphone are converted, using the code, into a series of light flashes from an LED. This light is then sent along the optical fibre and converted back into electrical variations at the receiving end. See if you can find out what advantages this method has over the older method of using copper wires.

Refraction by a prism: colour

When light passes through the outer parts of a lens, its direction changes because it is refracted by the glass of the lens and the sides of the glass are not parallel. This bending can be seen clearly by passing a ray of light through a triangular glass prism.

Experiment 20.11
Refraction by a prism

Use the ray-streak apparatus with a prism.

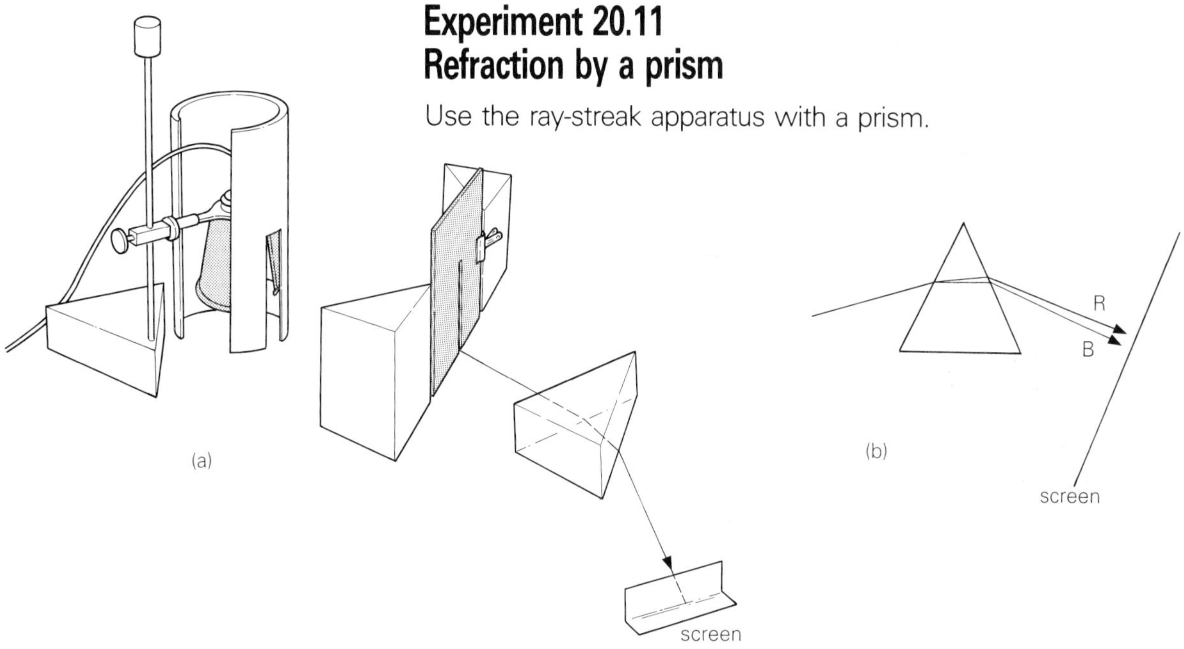

Fig 20.64

a The amount of bending depends on the angle at which the light meets the prism. Try turning the prism slowly to show this. You may find the light does not always go through the prism in the simple way illustrated. If it does not, can you see why?

b Place a screen made of white card so that light falls on it after passing through the prism as shown. You should see that there are now colours in the light. Turn the prism until the colours are clearest. Which colour light is bent most?

The colours, which are similar to the colours of the rainbow, range from deep blue to deep red, and change continuously from blue through green, yellow, and orange to red. This splitting up into colours is called *dispersion* and the spread of colours produced is called a *spectrum*. In the case of the rainbow, the dispersion is due to the refraction of the Sun's light in raindrops.

Newton's experiments with prisms and coloured light

Does the prism put the colours into the light? Or is white light just a mixture of all the colours of the spectrum which the prism has separated? Isaac Newton did some experiments to try to find the answer. He used a beam of sunlight passing into a darkened room through a small hole in the blind.

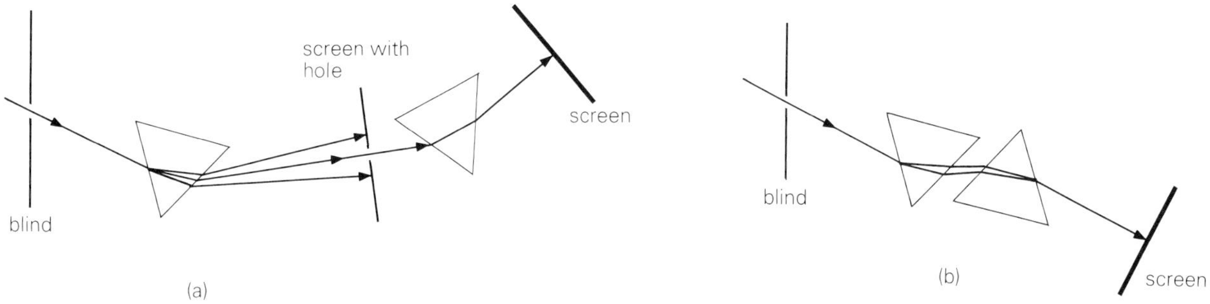

Fig 20.65

Fig 20.65 shows two of his experiments. In the first (a), he had a screen with a hole in it to allow through only one of the colours produced by the first prism. He then used a second prism to see if any more colours appeared. No new colours were produced. Green light, for example, could not be split up by the second prism into any other colours of the spectrum.

In the second experiment (b), he recombined all the colours produced by the first prism by using the second prism the other way round. He found that white light appeared on the screen.

As a result of these experiments he decided that the colours are there in the original light all the time and that the mixture of all the colours appears to us as white.

The prism sorts the various colours by bending each colour by a slightly different amount, the bluest colour being bent the most. We now know this happens because each colour has a slightly different speed in the glass.

The colour of objects

Why do objects have colours? If there is no light in a room, we cannot see any of the objects in it. We see an object because some of the light which falls on it is scattered into our eyes. Yet different objects can have different colours even though the light falling on them is the same.

The reason for this is that a red object has a surface which scatters the red parts of the spectrum but absorbs most of the green and blue parts. So we see it as red. With a blue object, the red and yellow parts of the spectrum are mostly absorbed, whilst the surface scatters the blue parts strongly.

This can be tested by holding lengths of coloured wool at various places in a white light spectrum. The spectrum needs to be larger and brighter than can be produced by using just a ray of light. A lens is needed, as in Fig 20.66, with the prism set to bend the light by the smallest amount.

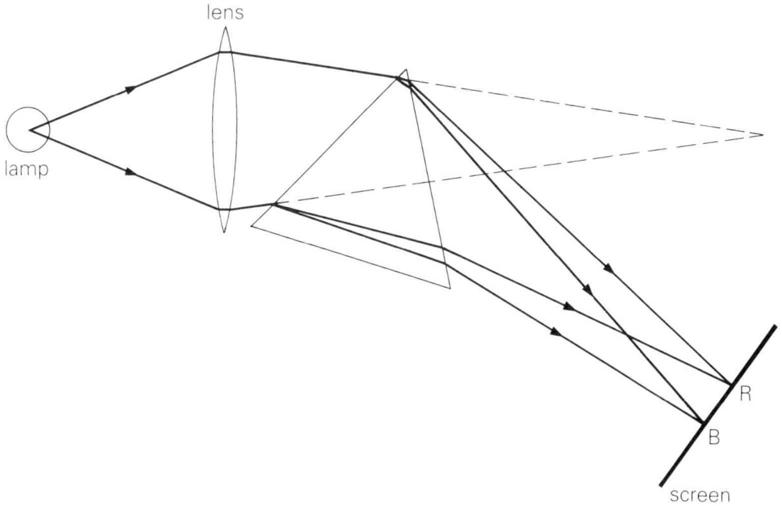

Fig 20.66

If the length of red wool is in the red part of the spectrum, it appears red because there is plenty of red light for it to scatter. In the blue part of the spectrum, it absorbs most of the light falling on it and scatters very little, so appearing to be black. If you set up this experiment try several coloured objects to see what happens.

Objects appear coloured in white light because they do not scatter the different parts of the spectrum equally.

Filters

Some materials which allow light to pass through (**transparent** materials as they are called) are coloured because some of the light is absorbed as it passes through. A filter is a thin piece of transparent plastic (or glass) which has a dye in it that allows some colours through but absorbs the others. A red filter appears red because red light passes through it whilst the blue/green/yellow parts of the spectrum are absorbed.

Notice that not all colours are present in the spectrum of white light. For example pink is not there, nor is magenta (a kind of purple). Pink is a mixture of white and red light, whilst magenta is a mixture of blue and red light. You can produce these colours by using two lamps with filters to illuminate a white screen as shown in Fig 20.67.

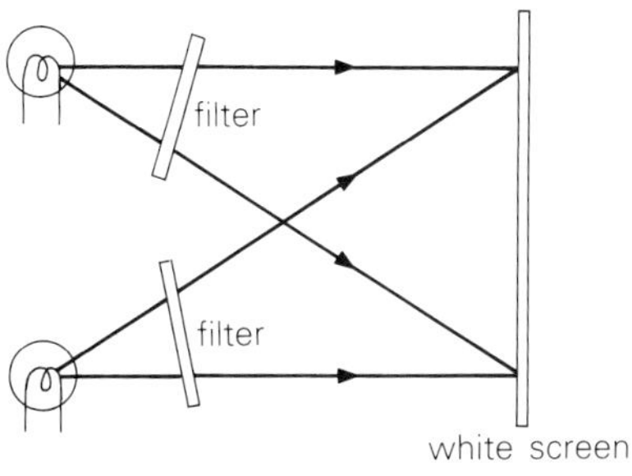

Fig 20.67

Questions for homework or class discussion

1 Fig 20.68 shows rays of light falling on glass blocks. Copy each diagram and then draw the path which each ray might follow.

Fig 20.68

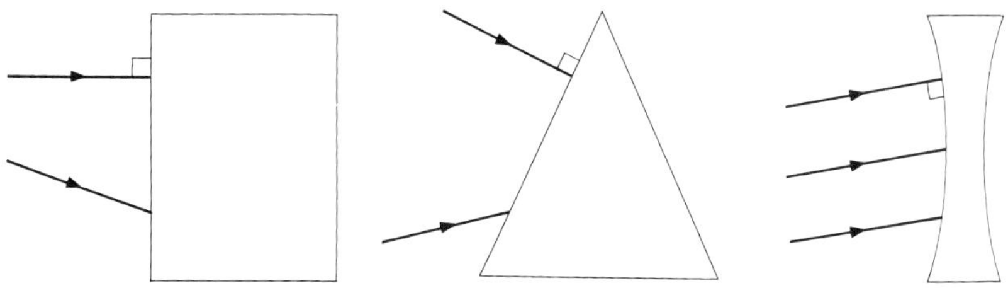

2 Draw a diagram to show why a swimming bath appears less deep than it really is.

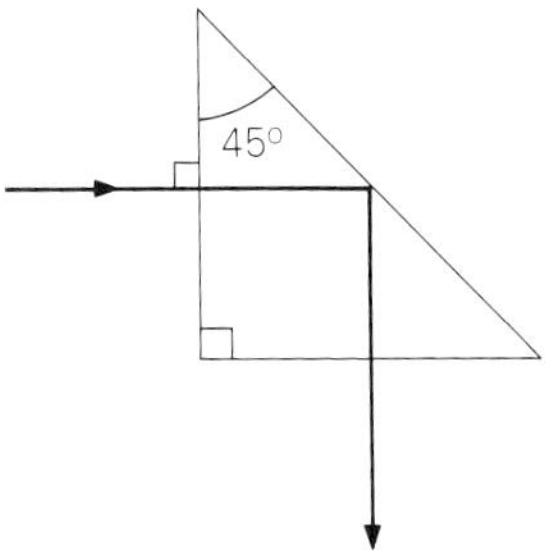

Fig 20.69

3 A prism can be used as a very good mirror if the light falls on it roughly as shown in Fig 20.69. Explain what is happening.

4 'Light is dispersed when it is refracted.' Why are colours not obvious when light passes through a glass block with parallel sides?

5 Why does a lens have a slightly different focal length according to the colour of the light falling on it? For which colour will the focal length be greatest?

6 Why do many things appear drab in the light of yellow street lamps?

Background Reading

Seeing things bigger

The magnifying glass

In order to see detail in an object, we bring it as close to the eye as possible as this makes it appear as big as possible. But there is a limit (the near distance) to how close the object can be brought and still be seen clearly. However, a lens can be used as a magnifier to make an object appear bigger still.

To use a lens to get a magnified image of the print on this page, you find that the lens needs to be fairly close to the page with your eye close to the lens. (The distance of the object to the lens has to be less than the focal length of the lens.)

When you look into the lens, you are looking at an image of the object. Fig 20.70 shows how the lens bends the light. Look at the light coming from the tip of the object, A. It is still spreading when it has passed through the lens.

If the eye is placed at, say, E, then the eye 'thinks' the beam from A has come from A′ and the light from B from B′. So it sees a large image A′B′. The lens has allowed you to have a large image to look at.

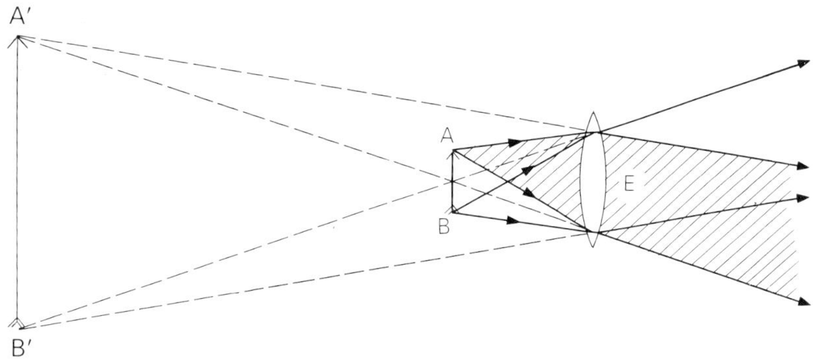

Fig 20.70

There is of course a difference between a real image and this kind of image, which is called a *virtual* image. Light actually passes through a real image so that we can get the image on a screen. With a virtual image, the light only appears to come from the image and we cannot get it on a screen.

The telescope

A magnifying glass on its own is of no use for looking at the Moon because we cannot get the Moon within the focal length of the lens. So it is necessary to use another lens which produces a real image of the Moon and then to use a magnifying glass to look at that image.

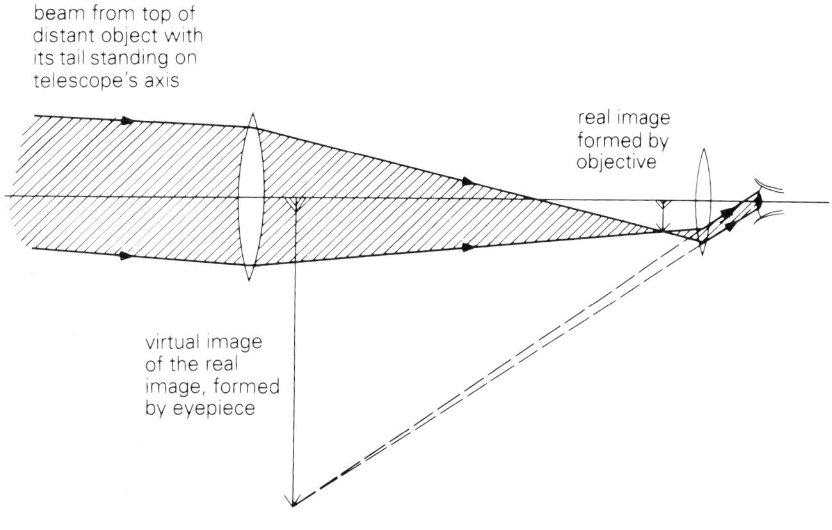

Fig 20.71

This is shown in Fig 20.71. The light from the distant object reaches the first lens (called the *objective* lens) and this forms a real image in the position shown. The second lens (called the *eyepiece*) acts as a magnifying glass and produces a large magnified virtual image. This system is called a telescope, and the photograph shows one used in an astronomical observatory.

It is worth pointing out that the final image is, of course, inverted. That does not matter if the telescope is being used to

Telescope used in an astronomical observatory

view the Moon, but it would matter if you were using it to look at ships at sea. Telescopes for that purpose have an extra lens between the objective lens and the eyepiece to invert the image formed by the objective.

The eye sees a much magnified version of the object because the image on the retina of the eye is so much bigger when the telescope is used.

The microscope

The microscope has a different job to do compared with the telescope. The thing to be looked at still needs to be magnified but it is now close at hand. However, as with a telescope, there are two stages: the first lens (the objective lens) again produces a real image and then the eyepiece produces a magnified image of that real image. This is illustrated in Fig. 20.72. As with the telescope the final image is inverted.

Fig 20.72

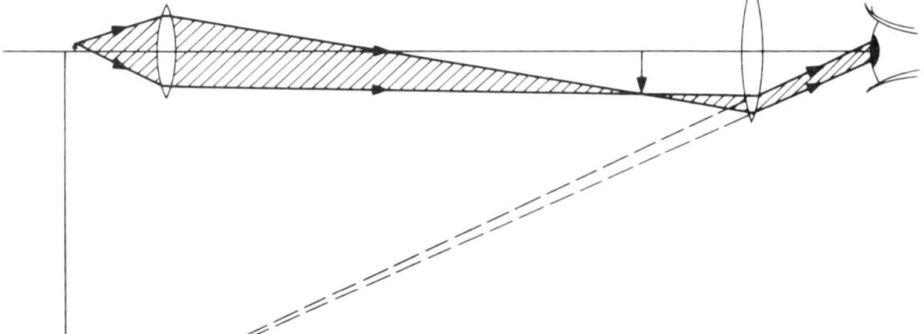

Chapter 21 The Earth in space

Man's view of the heavens

On a dark night, away from street lights, a myriad of points of light – the stars – will be seen in the sky. The stars form patterns which remain the same from year to year, but, each night, the patterns rotate about a point in the sky very close to the Pole Star.

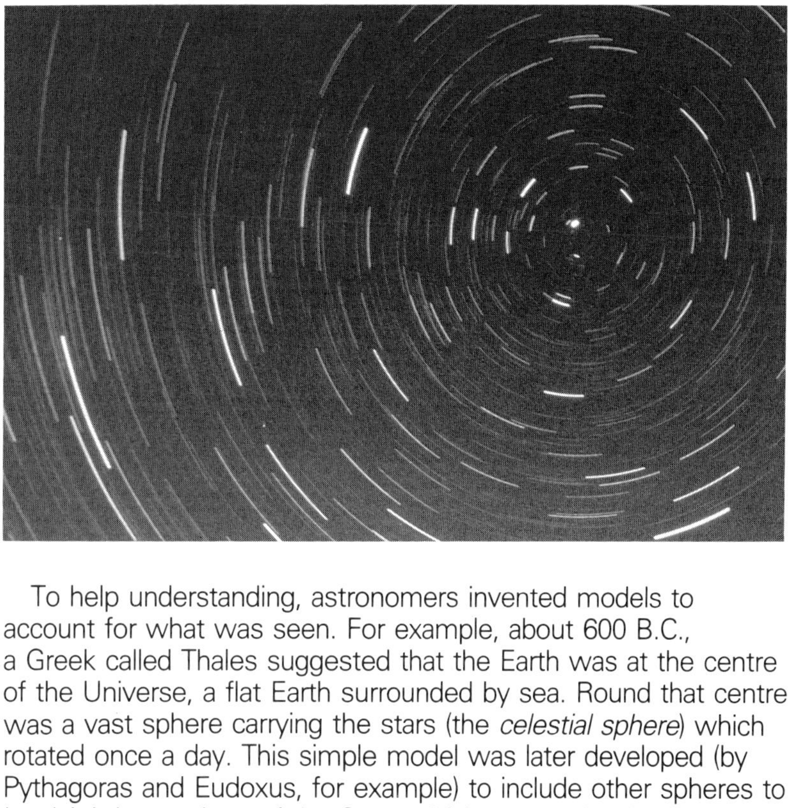

Rotation of stars around the pole

To help understanding, astronomers invented models to account for what was seen. For example, about 600 B.C., a Greek called Thales suggested that the Earth was at the centre of the Universe, a flat Earth surrounded by sea. Round that centre was a vast sphere carrying the stars (the *celestial sphere*) which rotated once a day. This simple model was later developed (by Pythagoras and Eudoxus, for example) to include other spheres to 'explain' the motions of the Sun and Moon – and to include the idea that the Earth was a sphere.

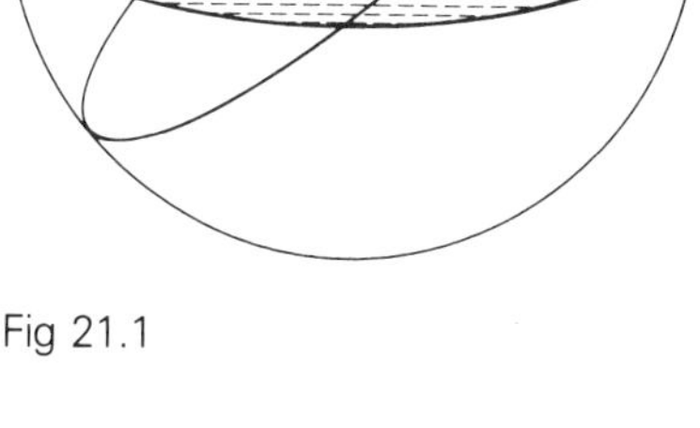

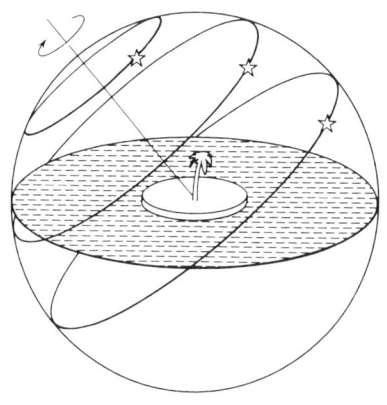

Fig 21.1

But, amongst all the points of light in the sky, there are a few which are not part of a fixed pattern and which wander through the stars. We call these *planets* (from the Greek word for *wanderer*). Their motion is not as regular as the motion of the Sun or the Moon and more spheres were added to the model to account for the motion of the planets, each having its own period of rotation.

Simpler models were suggested – there would be no need for the celestial sphere to rotate if the Earth itself were rotating, but astronomers were reluctant to accept that idea because it was thought that things would fly off a rotating Earth. Others suggested the Earth moved round a fixed Sun, but the idea found little favour until Copernicus (1473–1543) revived it. He believed that the Sun was the centre of our system and that the planets moved at different speeds in circular paths (*orbits*) round it. Nearly 70 years after the death of Copernicus, Galileo made a telescope and used it to view Jupiter, so discovering Jupiter's moons. He realised that the moons were in orbit around Jupiter and this was direct evidence that a system of orbiting planets was possible.

It was left to the genius of Newton to realise that the planets were held in orbit round the Sun by the same force, the force of gravitation, as that which causes an apple to fall to the ground. A force of gravitation acts between any two objects, but it is only noticeable if at least one of the objects is very massive like the Earth or the Sun. Just as the Earth pulls the apple towards the centre of the Earth, so the Sun pulls the Earth towards the centre of the Sun. It is this force which allows the Earth to go round the Sun in what is almost a circular orbit.

The Solar System

The stars are luminous sources (they give out light) and the Sun is a star. The planets (including the Earth) are non-luminous. They scatter light from the Sun and we see them as faint specks of light moving slowly against the background of the fixed stars.

The planets Mars and Venus, and the biggest planets, Jupiter and Saturn, can be seen by the naked eye, and a good pair of binoculars will show that they appear as discs. Those planets which are very far away are very faint and can be seen only by using a telescope. The stars are so far away that they appear as points of light even with the most powerful telescopes.

The orbits of the planets, drawn approximately to scale, are shown in Fig 21.2. The scale of drawing (b) is ten times larger than that of (a). The orbits are nearly circular. The time for Earth to complete one orbit of the Sun is one year (actually $365\frac{1}{4}$ days). The approximate time in years for each planet to complete one orbit is shown in brackets after the planet's name.

The Solar System as a whole is part of a galaxy which consists of an enormous number of stars. And our galaxy is merely one of an immense number of galaxies which are part of the Universe.

The light from the Sun lights up only one side of the Earth and all the places on that side are having day-time. No sunlight reaches the other side of the Earth; there it is night-time. Night turns into day and day into night because the Earth is rotating about an axis running between the North and South geographical Poles. On the night side of the Earth, the only light comes from

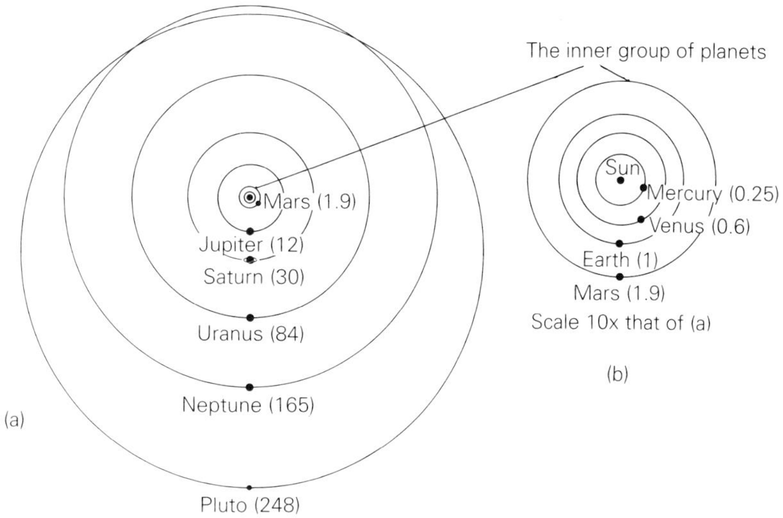

Fig 21.2

Fig 21.3

the luminous stars and perhaps from the Moon (see Fig 21.3).

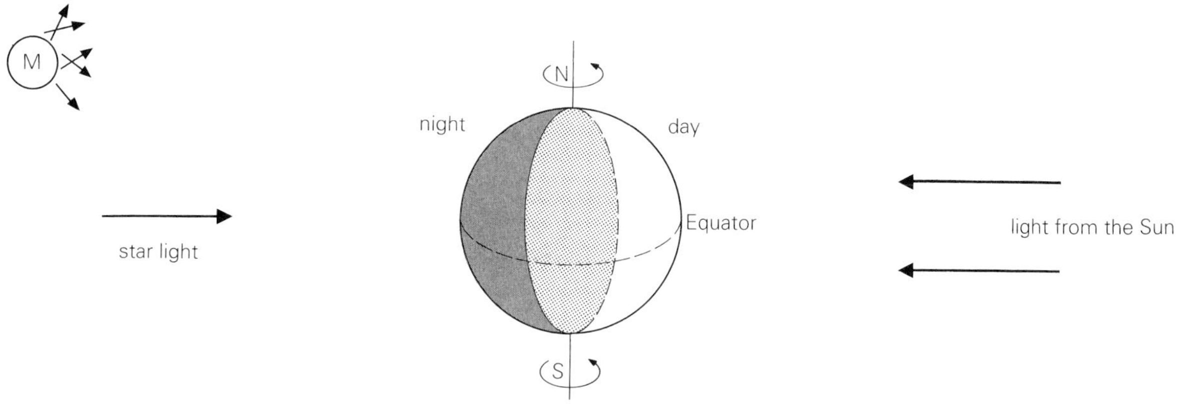

The Earth and the seasons

If the axis of rotation of the Earth were at right-angles to the direction of the Sun's light, everyone on the Earth would spend as much time on the sunlit side as on the dark side. The length of the day would always equal the length of the night. The height of the Sun above the horizon at noon would always be the same, and the Sun would rise due east and set due west. (See Fig 21.4.)

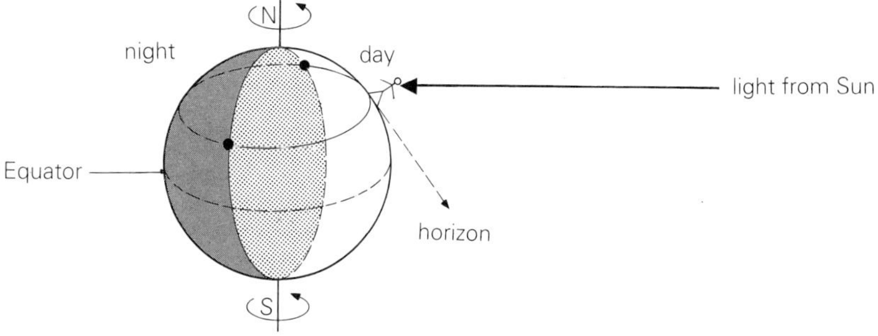

Fig 21.4

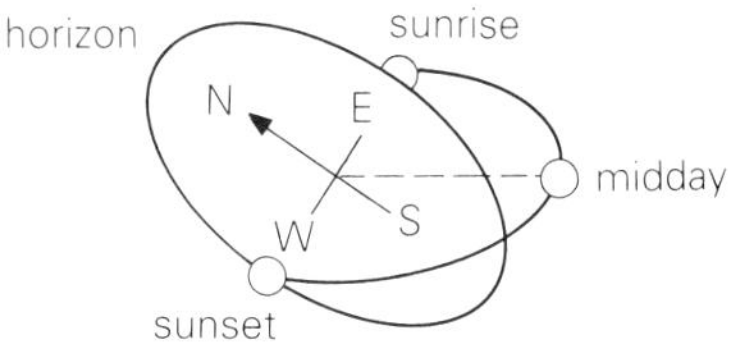

Fig 21.5

Fig 21.5 shows the path of the Sun as it would appear from a point on the Earth. In fact, this happens only on two days in each year. These are the *equinoxes* and occur about March 21st and September 21st. In the summer, the days are longer than the nights; the Sun is higher in the sky at midday; it rises north of due east and sets north of due west. And in the winter, the nights are longer than the days; the Sun is lower in the sky at midday; it rises and sets south of due east and south of due west.

This is because the axis of rotation is *not* at 90° to the direction of the Sun's light. As the Earth orbits the Sun, its axis of rotation is as shown in Fig 21.6.

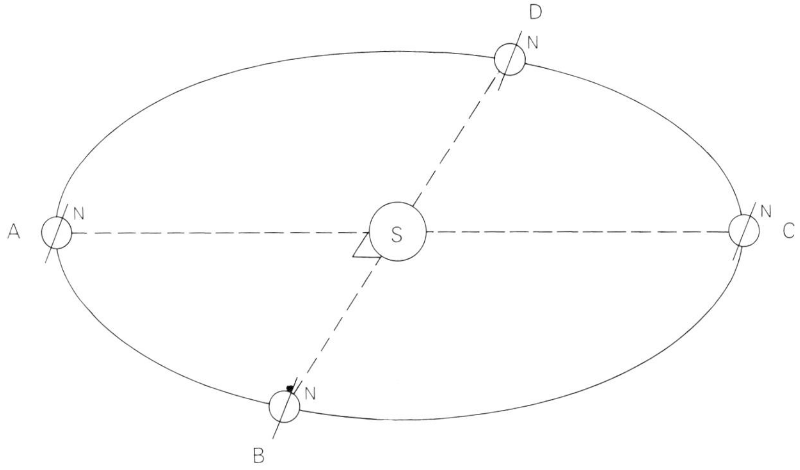

Fig 21.6

Fig 21.7 shows what happens when the Earth is at A in Fig 21.6.

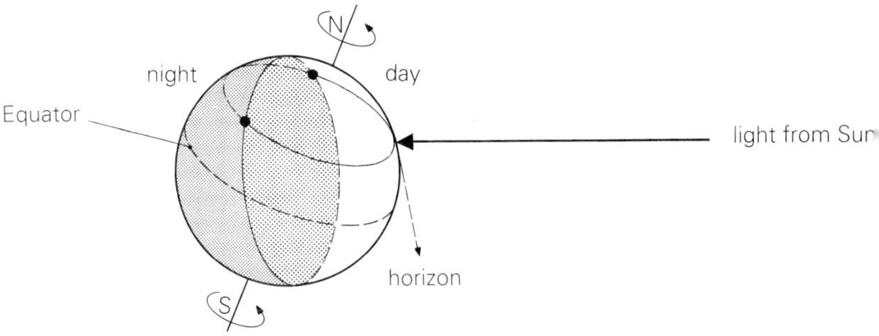

Fig 21.7

For someone in the northern hemisphere, the days are longer than the nights. It is summer, and at midday, the Sun is higher in the sky than in the winter.

At C in Fig 21.6, the situation is as shown in Fig 21.8. Now the northern hemisphere is having winter, the Sun is lower at midday and the nights are longer than the days. Can you explain why the temperatures are lower in winter than they are in summer? You will find a clue on page 249.

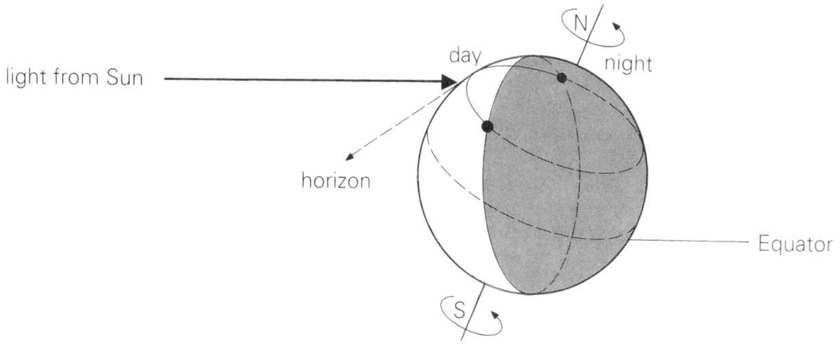

Fig 21.8

In positions B and D in Fig 21.6, the Sun's light is at right-angles to the axis of rotation. These are the positions of the Earth in its orbit at the equinoxes.

Questions for homework or class discussion

1 Suppose you were talking to Thales in 600 B.C. What arguments could you have produced to persuade him that the Earth is round?
2 Suppose Thales were alive today. What other evidence do we now have that the Earth is round?
3 Why are faint objects more easily seen by using a telescope? (Remember that a star appears as a point of light even in the largest telescopes.)
4 The time for the Earth to make one complete orbit is $365\frac{1}{4}$ days, yet the calendar year is a whole number of days. How do we manage that?
5 Explain why there are places on the Earth where, in summer, the Sun never sets and, in winter, never rises.

The Moon

Moon's surface

All the planets except Mercury and Venus have at least one satellite orbiting the planet. The Moon is the Earth's satellite and it has an orbital time of just under 28 days.

The Moon appears to be about the same size as the Sun, but that is because the Moon is much closer to us. In fact, the Sun is 400 times as far away as the Moon and its diameter is 400 times as big. The Earth has only about 4 times as large a diameter as the Moon.

As the Moon orbits the Earth, it rotates on its axis so that the same part of the Moon always faces the Earth. We always see the same surface markings ('the man in the Moon'), some parts being darker than others. With the invention of the telescope, it was soon discovered that the markings were due to its surface features. In general, the surface is pitted with craters similar to volcanic craters and the darker areas are relatively flat and free of craters. Astronomers have named them 'seas' for they were once thought to be water, though we now know that there is no water on the Moon.

It was natural that, as soon as the first artificial satellite had been launched, attention should be directed towards making a landing on the Moon. The table below shows some significant dates in the history of that.

Date	**Event**	
Oct. 4th 1957	USSR Sputnik I	First artificial satellite
Nov. 3rd 1957	USSR Sputnik II	First living creature in spac Laika, a mongrel bitch
Sept. 14th 1959	USSR Lunik II	First landing on the Moon (crash landing)
Oct. 7th 1959	USSR Lunik III	First pictures of far side of Moon obtained
Apr. 12th 1961	USSR Vostok I	First man in space – Yuri Gagarin (one orbit)
Feb. 20th 1962	USA Friendship VII	First American in space – John Glenn (three orbits)
June 16th 1963	USSR Vostok VI	First woman in space – Valentina Tereshkova-Nikolayev
Feb. 3rd 1966	USSR Luna IX	First soft landing on the Moon – unmanned craft
July 21st 1969	USA Apollo XI	First men on the Moon – Neil Armstrong and Edwi 'Buzz' Aldrin land from the lunar module, Eagle
Dec. 10th 1972	USA Apollo XVII	The last mission – three days spent on the Moon

Up to the end of the Apollo programme, the total estimated cost to the USA was £1×10^{10} (10 billion pounds). The Moon was found to be the inhospitable place we always thought it was – no air, no water, no life, no vegetation to provide a shield from the extreme temperatures of the day (more than 100°C) and of the night (less than −150°C). Even the other side of the Moon proved to be very similar to what we can see from Earth. However, the astronauts were able to bring back rock and sand samples – and to verify that the gravitational pull of the Moon is only about $\frac{1}{6}$ of that of the Earth. And it was all achieved using the ideas of gravitation first formulated by Newton 300 years earlier!

The Moon and its phases

The photographs show the Moon as it appears at different stages throughout a month. The reason for these phases is that the Moon is not a luminous body; it gives out no light of its own. We see it because it scatters light from the Sun. When we see it half-full, we are looking at that part of the Moon on which the Sun is shining, the part which is having day-time. The part which is having night cannot be seen (see Fig 21.9).

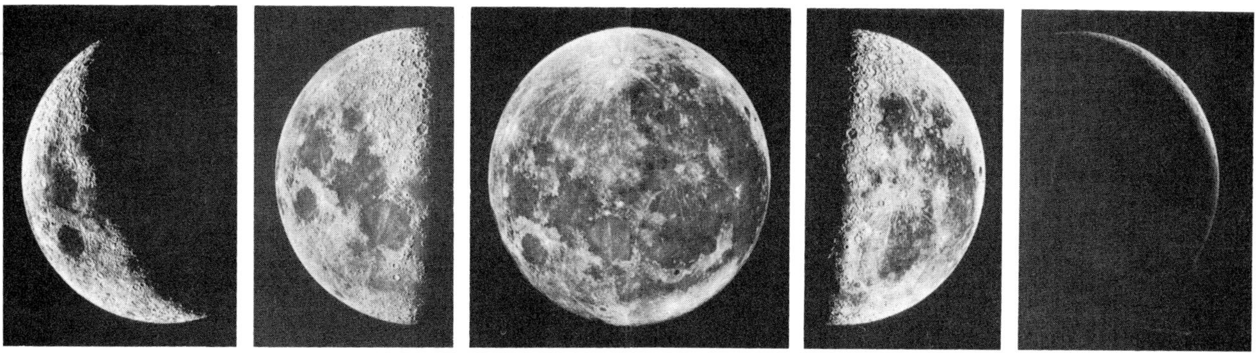

The Moon's phases

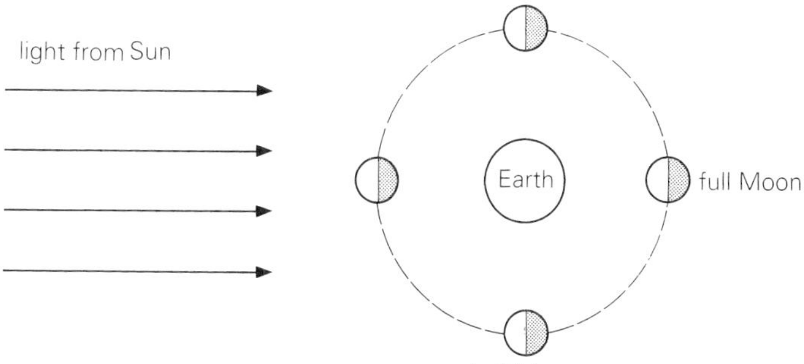

Fig 21.9

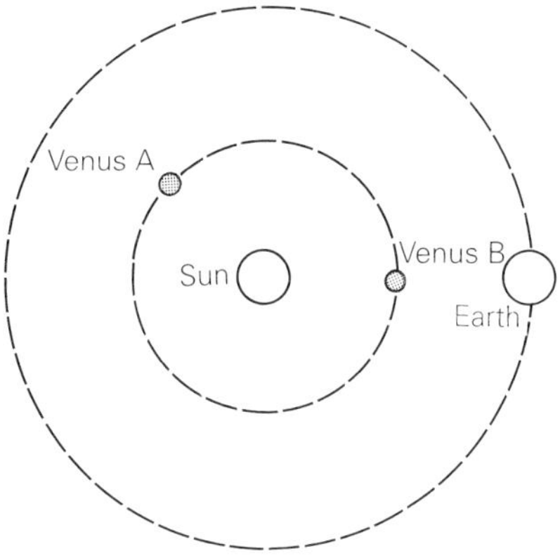

Fig 21.10

The planet Venus also shows phases. When it is in position A in Fig 21.10, it is almost full but appears small because it is far away. When it is in position B, it appears as a crescent because it is mainly the dark side which faces the Earth – and it appears much bigger because it is closer.

Background Reading

Eclipses

Eclipses are simply shadows on an astronomical scale. In an eclipse of the Moon, the Earth prevents light from the Sun from reaching the Moon; the Moon is in the shadow of the Earth (see the photograph).

The edge of the shadow is not very fuzzy because the Sun is so far away. Notice that the shadow shows the Earth is round. Fig 21.11 shows an eclipse of the Moon taking place, though the

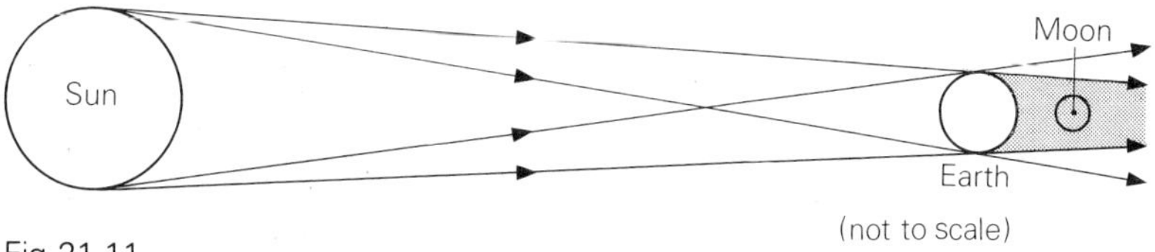

Fig 21.11

diagram is not to scale. If the sky is clear, anyone on the night side of the Earth ought to be able to see the eclipse. An eclipse of the Moon is not a very rare occurrence, but it does not happen every 28 days as you might expect.

An eclipse of the Sun *is* a rare occurrence. In such an eclipse, light from the Sun travelling to the Earth is cut off by the Moon. As it happens, the size of the Moon and its distance from the Earth are such that the part of the shadow where there is no light at all is very small (Fig 21.12).

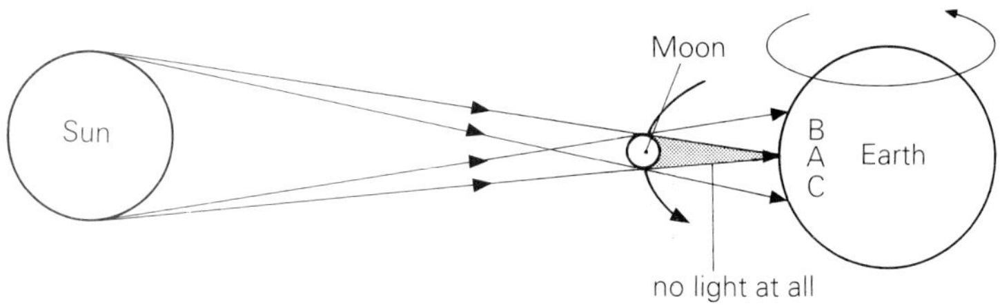

Fig 21.12 (not to scale)

To see a total eclipse, you have to be in the area marked A, and, since the Earth is rotating on its axis and the Moon is orbiting round the Earth, the total eclipse lasts only for a short time at any one place. If you are in the parts marked B and C, then you will see a partial eclipse (only part of the Sun is covered by the Moon). An eclipse of the Sun occurs during daytime, and lasts only for a few minutes at certain places on the Earth. And that is one reason why it is a rare occurrence.

During an eclipse of the Sun, the Moon covers the luminous disc of the Sun but not the hot gases which surround it, the *corona* as it is called. The photograph shows these luminous gases round the Sun and the plume of gas which has been expelled by an explosion on the Sun's surface.

Total eclipse of the Sun, 29 May 1919

Questions for homework or class discussion

1 How long do the Moon's day and night last?
2 'Astronauts wear space suits to protect themselves from sandstorms'. Comment on that statement.
3 In which way, relative to the Sun, do the horns of the crescent Moon point?
4 Why do we not have an eclipse of the Moon every 28 days?
5 How could you show that the Moon is just over 100 Moon diameters from the Earth?
6 Draw a diagram to show an eclipse of the Moon taking place on the Earth. What might an astronaut on the Moon see as this happened?
7 Explain what happens during an eclipse of the Sun. What would an astronaut on the dark side of the Moon see?
8 The cost of the Moon exploration has been immense. The Russians have spent even more than the Americans. Do you think it has been worthwhile?

The Galaxy and beyond

Distances in astronomy are immense. Astronomers know that the nearest star to the Sun is so far away that it takes light 4.3 years to travel to us. Light travels 300 000 km in 1 second and there are about 30 000 000 seconds in 1 year! Thus the distance travelled by light in 1 year

$= 30\,000\,000 \times 300\,000$
$= 9 \times 10^{12}$ km.

So that the distance to the nearest star

$= 9 \times 10^{12} \times 4.3$ km, which is about 4×10^{13} km.

It is an enormous number of kilometres, and that is the distance to the *nearest* star! If you spent every second of your life travelling at the speed of Concorde, it would take you 2 000 000 years to cover that distance, and you would be dead long before you got there. Even our fastest spacecraft would take 100 000 years!

Astronomers have found that the Universe contains groups of stars called galaxies. All the stars which we can see with the naked eye belong to one family, our own galaxy, which includes the Sun. A telescope will show that the Milky Way – the faint

Andromeda Galaxy

band of light which crosses the sky – is light from millions of stars (about 100 000 million) all of which belong to our galaxy. It is so enormous in size that it takes light 100 000 years to cross it; it is shaped like a flattish disc of diameter 100 000 light-years.

The strongest telescopes reveal the existence of many other galaxies in the Universe, like the Andromeda galaxy which is 2.2 million light-years away (see the photograph on page 289).

In all, astronomers estimate that there are more than 10 000 million galaxies in the Universe, each containing about 100 000 million stars. Surely, amongst all those stars, there must be millions with planets; and, amongst those, there might be at least a few million on which there is intelligent life.

Questions for homework or class discussion

1 How long will it take light **a** to reach the Earth from the Sun, **b** to travel across the Solar System?
The distance from the Earth to the Moon can be measured by sending a pulse of laser light to a reflector on the Moon and measuring the time for it to travel there and back. If the time is 2.56 s, how far away is the Moon?

2 The temperatures on Venus are much higher by day (and much lower by night) than they are on Earth. Why do you think this is?

3 How do you think our way of life would have developed on Earth if *we* had only one day and night in the time it takes the Earth to go round the Sun?

4 The largest telescopes use curved mirrors to collect the light and produce an image. Try to find out what advantages there are in using a mirror rather than a lens.

5 Astronomers are keen to have telescopes in orbit round the Earth. These telescopes are controlled from Earth, and send their results back, by radio. What advantages do you gain from having a telescope in orbit?

6 An athlete can jump 2 metres on Earth. On Mars, the weight of 1 kg is 4 N. On Jupiter, the weight of 1 kg is 26 N. How high would the athlete be able to jump on Mars if it had an atmosphere like ours? And what if the athlete visited Jupiter?

Chapter 22 Sound

'Keep quiet! Don't you dare move!' How many times have your parents said that to you? Did your parents know that objects have to move to produce sound?

Unless you are deaf, you know what a sound is. And you know that when a sound is made, it travels to your ears and you hear it. It must be your ears which are the sound receivers because, if you cover your ears, it 'cuts out' the sound.

What you hear depends mostly on whatever is producing the sound – the source of sound. The squeaky sound produced by a bird is quite different from the low booming sound of a cow. But what we hear also depends on the receiver, and in some circumstances, on what the sound travels through (the **medium**).

To begin, we shall look at what happens when something makes a sound.

Demonstration experiments 22.1 Making sounds

These experiments examine different ways of producing sound.

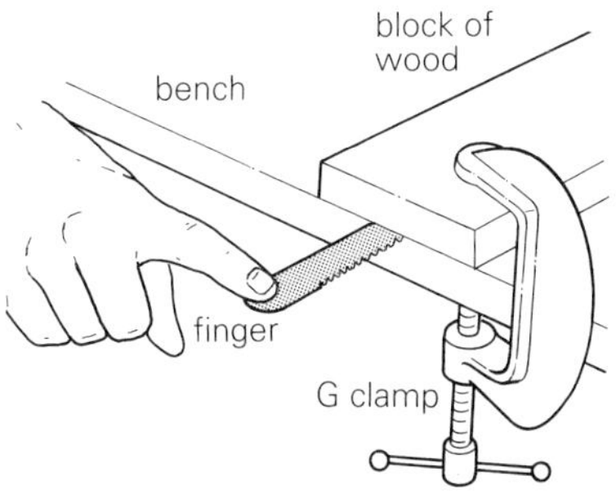

Fig 22.1

a Clamp a hacksaw blade or a plastic ruler at the edge of the bench with about 10 cm overhanging, as shown. Gently pluck the end of the blade or ruler downwards. It emits a sound briefly. How do you make it emit a louder sound? What happens if you shorten the length which overhangs the bench? If you lengthen it, what happens? What do you see if the overhanging length is large?

b Stretch a thin piece of rubber (a cut rubber band) and pluck it. What happens? How do you make the sound louder? What happens if you stretch a shorter length of the same band? What if the length is very short?

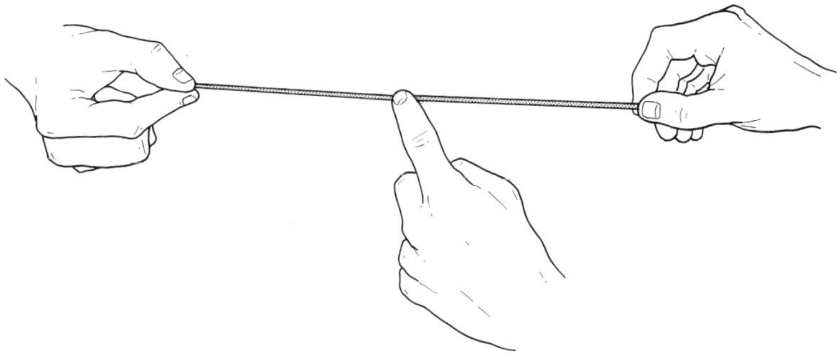

Fig 22.2

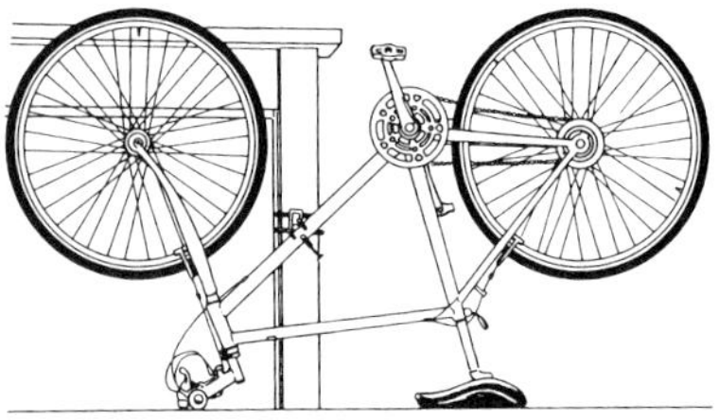

Fig 22.3

c You should do this part only if you can securely fix a bicycle so that its rear wheel is clear of the floor. This could be done by turning the bicycle upside-down and clamping the frame to a bench leg, or by hanging the bicycle from a beam with a rope firmly fixed round the saddle support.

When you are sure it cannot fall over, turn the pedals so that the rear wheel spins rapidly. Then hold a long piece of stiff card (about 5 cm wide) so that it just touches the rotating spokes near the rim. What happens if it just touches the spokes near the hub? What happens if the wheel speeds up or slows down? Can you think how you might make the sound louder?

All these experiments show that sound is emitted when an object is made to move to and fro. This to and fro motion is called **vibration.** Faster vibrations emit notes higher up the scale, and we say the sound is of higher **pitch**. Slower vibrations emit sounds of lower **pitch**. Faster vibrations occur with shorter lengths of the vibrator.

Experiment 22.2 The tuning fork

The tuning fork is used by holding the stem and hitting the prongs with a small rubber-faced hammer. Alternatively, the prongs can be squeezed together and suddenly released. Or the fork can be used to strike a large rubber bung or cork. It must *not* be struck against anything hard like the bench: it is bad for the fork.

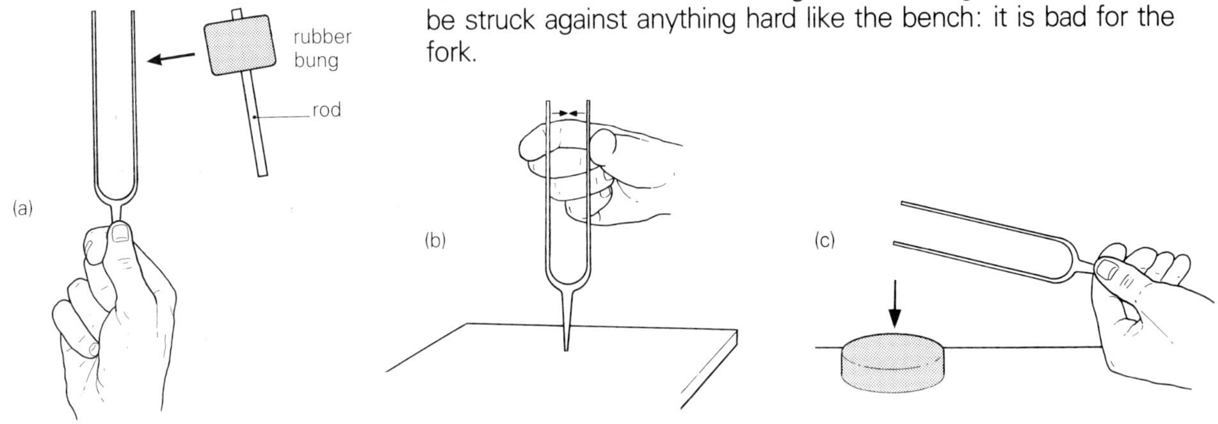

Fig 22.4

Strike the fork and hold it close to your ear. If you can, try a large fork and a small fork. Which emits the sound of higher pitch?

As in many other cases, the vibrations are too small or too rapid for them to be easily visible. But you can show the prongs are vibrating by allowing them to touch the surface of water in a shallow dish, or by allowing one of the vibrating prongs to touch the end of your nose.

A much louder sound can be obtained from a tuning fork if its stem is held against the top of the bench whilst the prongs are vibrating. This is because the fork causes the bench top to vibrate a little and the large bench can send a stronger sound to your ears.

Each of the forks has a number stamped on it. This number is the frequency of the fork. The smaller fork, which vibrates more rapidly and emits a sound of higher pitch, has a higher frequency.

Frequency

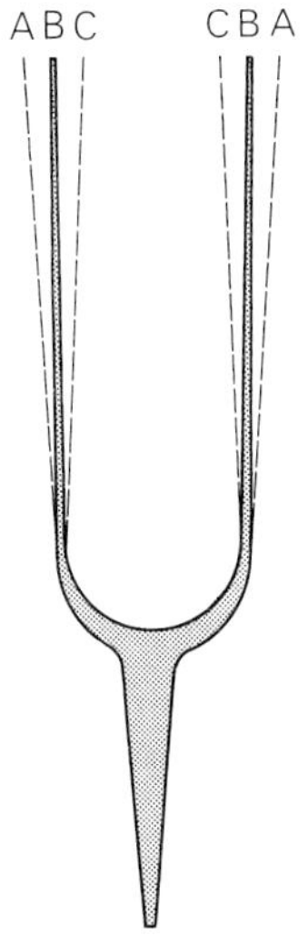

Fig 22.5

Fig 22.5 shows the prongs of a tuning fork.

Normally the prongs are in position B. When the fork sounds, the prongs move from the B position to the C position, back to B, then A, then back to B and so on. One vibration is B – C – B – A – B, or C – B – A – B – C, or any complete movement of the prongs. The number of complete vibrations in one second is called the **frequency** and it is measured in a unit called the **hertz** (Hz). A frequency of 100 Hz means that there are 100 complete vibrations every second. It is the frequency of the vibration which fixes the pitch of the musical sound emitted.

Experiment 22.3
Using Savart's Wheels to check the frequency of a tuning fork

Savart's Wheels are four toothed wheels of different diameters. There are a different number of teeth on each wheel. The wheels can be turned by a handle (or a motor) and a card held against the teeth to produce musical sounds. Without changing the speed of the wheels, hold a card against each wheel in turn, from the smallest to the largest.

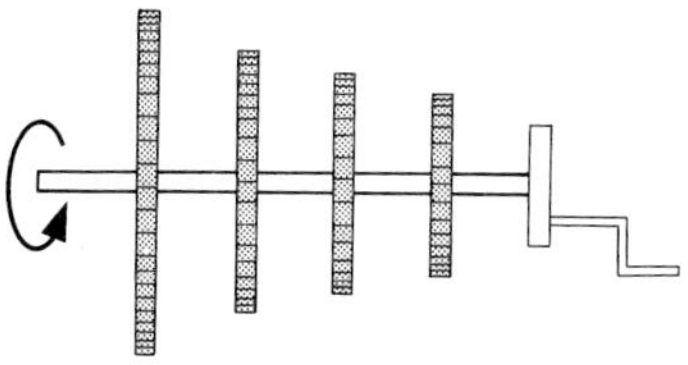

Fig 22.6

You should recognise the notes as doh–me–soh–doh. Count the number of teeth on the wheels. It is likely they will be 40, 50, 60 and 80. Can you see that the higher pitch 'doh' has a frequency just twice the lower pitch 'doh'? Notes an octave apart have frequencies in the ratio 2 : 1. And 'soh' has a frequency just $1\frac{1}{2}$ times the lower 'doh'.

The actual frequencies of the sounds depend on how quickly the wheels are turned. If you can measure the number of turns made in one second, then you could work out the frequency of the sound. Try to check the frequency of the larger tuning fork. Turn the wheels steadily, sound the fork and then find out which wheel gives a sound which most closely matches the pitch of the fork. Then adjust the speed of rotation until you have as good a match as you can get. Keeping the speed steady, measure how long it takes for 10 turns of the handle and use that to find out how many turns the wheel made in one second. What is your estimate of the frequency of the tuning fork?

Experiment 22.4
Experiments with a loudspeaker

Connect a loudspeaker to an audio oscillator as shown. An audio oscillator is a piece of equipment which produces an alternating voltage; when it is connected to a loudspeaker, the alternating current makes the loudspeaker vibrate.

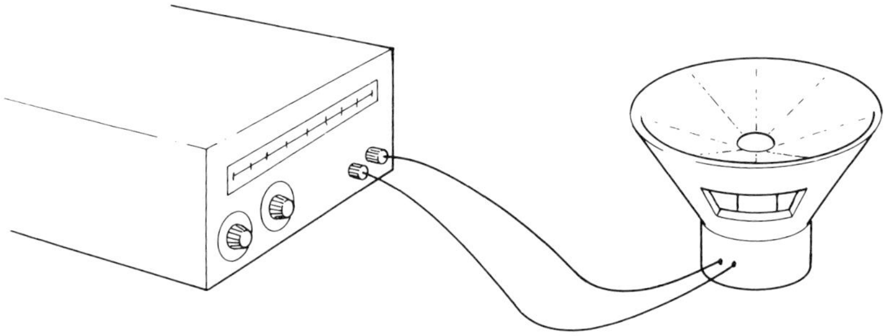

Fig 22.7

a Put a grain or two of rice on the loudspeaker to convince yourself that it is vibrating when a current flows.

b There is an *amplitude* control on the oscillator which makes the voltage bigger when it is turned up. Turn it up to see what happens using a fairly low frequency. Convince yourself that bigger vibrations cause louder sounds without changing the pitch.

c There is also a control on the oscillator which changes the frequency without changing the voltage. Turn it to see what happens. You can show that bigger frequencies cause sounds of higher pitch. As you go to higher and higher frequencies the note gets fainter until you cannot hear it any more. There may be several reasons for this. Firstly, there is a limit to how high a frequency your ears can detect. This upper limit is about 16 000 Hz (16 kHz) for average ears but it gets lower with increasing age. Secondly, the loudspeaker may not vibrate as much at higher frequencies, and how much sound is sent out depends on how much it vibrates. Thirdly, it may not be a very good oscillator – it may not be able to maintain the current through the loudspeaker as the frequency increases.

There is a lower limit of audibility too. How big is it for your ears? For most people it is somewhere near 30 Hz.

How the loudspeaker works

There are two main parts in a loudspeaker. First, a special cylindrical magnet (Fig 22.8a) provides a strong magnetic field. Secondly, a stiff paper cone is fixed to a card cylinder at its narrow end with a coil of copper wire wound on the cylinder (Fig 22.8b). The two parts fit together as in Fig 22.8c. The paper cone is held in a metal frame by a flexible mounting round its outer edge.

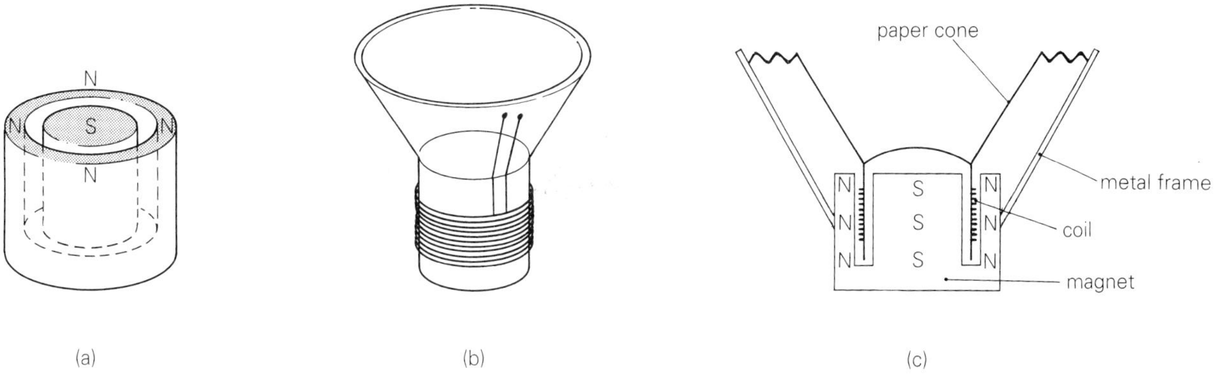

Fig 22.8

When a current passes through the coil, the coil is pushed out of or pulled into the gap according to the direction of the current. If a rapidly oscillating current goes through the coil, the cone vibrates at the same frequency as the oscillation in the current. The vibrating cone sends sound out into the surrounding air.

Microphones and the telephone

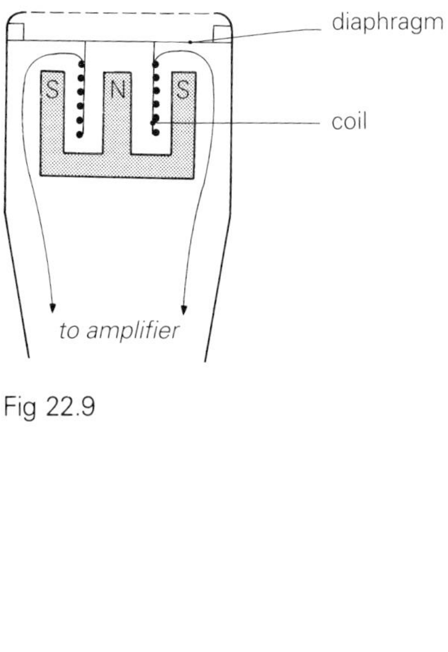

Fig 22.9

One kind of microphone (Fig 22.9) works by having a coil which moves in a magnetic field. Sound causes vibrations of the diaphragm which is made of a thin piece of springy metal. The coil is fixed to the back of the diaphragm and moves in the field of the specially-shaped magnet. Voltages are induced in the coil by this movement and these are then enlarged or amplified. High quality microphones are often of this type.

The mouthpiece in a telephone hand-set is a lower quality microphone (Fig 22.10). In this the diaphragm presses on the top of a small box packed with carbon granules. When someone is speaking with their mouth close to the diaphragm, it vibrates. This causes the pressure on the granules to change and that causes their electrical resistance to change. Higher pressure causes lower resistance, and *vice versa*. Thus, if the carbon box is connected to a battery, the current flowing will vary when sound falls on the diaphragm. (Notice that, in this case, sound is not transferred directly into electrical energy. This time, it is the cell which supplies the electrical energy and the sound energy merely causes variations in it.)

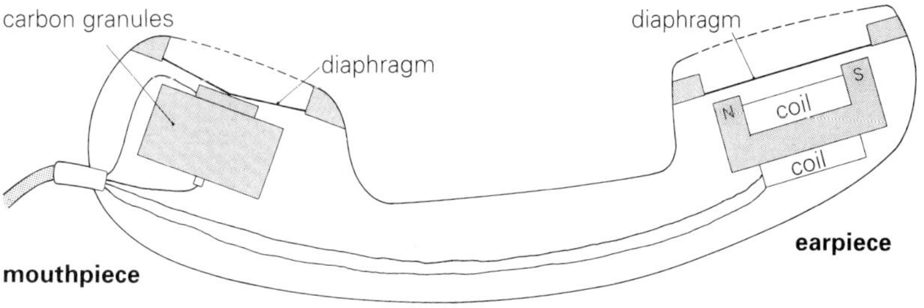

Fig 22.10

The earpiece of the telephone handset is a kind of loudspeaker (Fig 22.10). A coil of many turns is wound round a U-shaped magnet held near a diaphragm which is made of a springy magnetic metal. Variations in the current changes the strength of the magnet by small amounts. When the magnet is stronger, the diaphragm is more attracted, and, when it is weaker, it is less attracted. The diaphragm vibrates with the same frequencies as those present in the current variations, and so sends out sounds with those frequencies.

Experiment 22.5
Sounds produced by musical instruments

The sounds emitted by the loudspeaker in the last experiment were smooth and pure. Musical instruments vibrate in more complex ways and it is these complex ways that distinguish a piano sound from that of a violin, even though they are playing notes of the same pitch.

You can see these complex vibrations using a microphone and a cathode ray oscilloscope (CRO for short). The sound is picked up by the microphone and the sound is converted into changes of voltage which are displayed on the screen of the CRO (Fig 22.11). There is no need to understand how the CRO works at this stage.

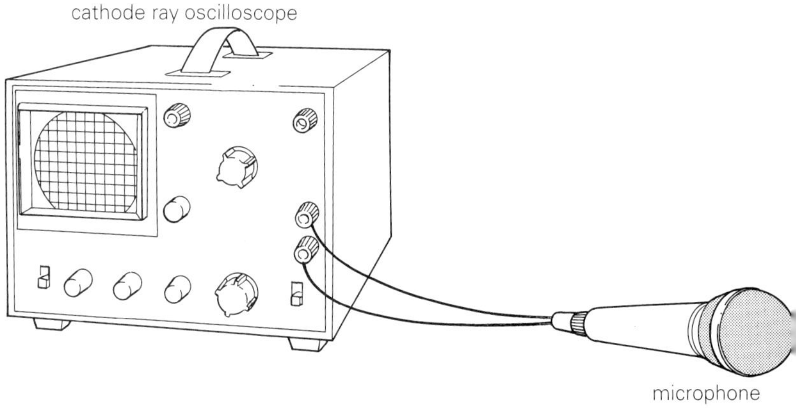

Fig 22.11

Try a tuning fork, a guitar, a violin and a piano – all sounding the same note in turn.

The tuning fork gives the simplest pattern on the screen (Fig 22.12). The patterns for other sources are different even though the basic frequency is the same and it is these different characteristic patterns which allow you to recognise different instruments.

Fig 22.12

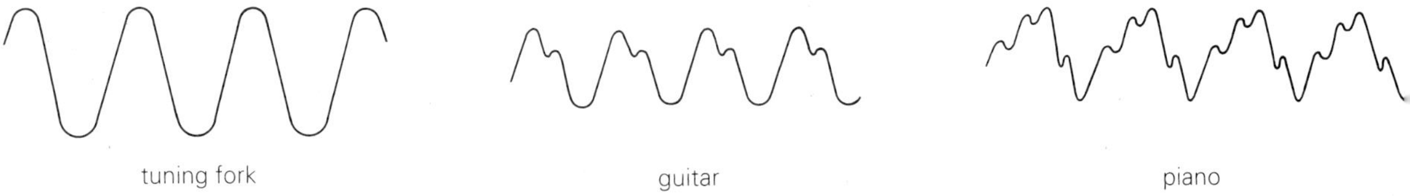

The medium

It is obvious that sound travels through air, but does it travel through a vacuum? What about other materials?

Demonstration experiment 22.6
The bell in a vacuum

A battery-driven bell is hung by two rubber bands from hooks in a rubber bung at the top of a bell jar. With the bell ringing, air is pumped out, preferably with a rotary vacuum pump. As the pressure falls, the sound of the bell fades away to nothing, even though the bell can be seen still to be working.

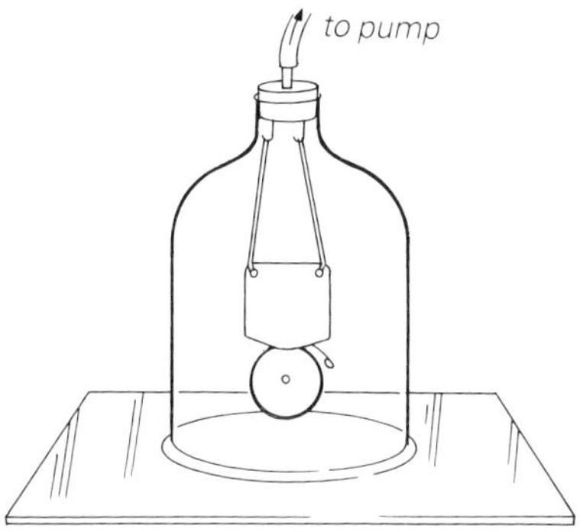

Fig 22.13

Sound will not travel through a vacuum; there must be air present in the jar to hear the bell.

Experiment 22.7
Can steel or wood be a medium for sound?

A test can be made to see if sound will travel through steel. A long length of railing is ideal for this.

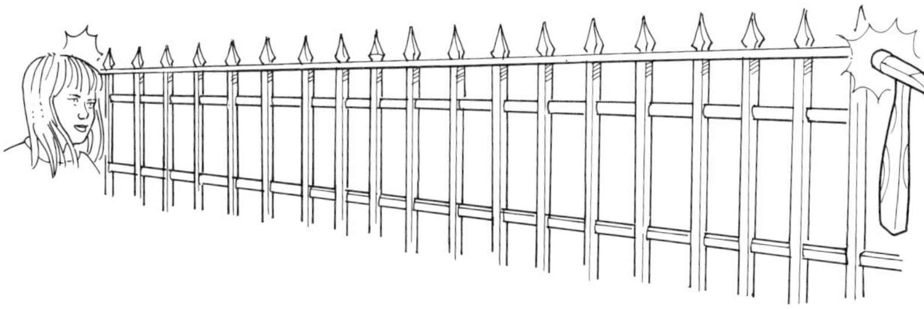

Fig 22.14

Place one ear against the steel and get someone to hammer the railings about 100 m away, say once every 5 seconds. Can you hear the sound which has travelled via the steel, as well as the sound reaching the other ear via the air? Can you *see* the hammer hit the railings?

It may be easier to see if sound will pass through wood. Place a ticking watch on the bench, or get someone to tap the bench with a pencil, and see if you can hear the sound coming through the wood.

Sound *does* travel through steel and wood. Its speed is greater than it is in air. In fact, sound travels through all materials so that there is no perfect sound insulator. And, incidentally, the experiment with the railings should also tell you that sound does not travel as quickly as light.

Does sound travel through water? Find out in the bath. With your head on one side and one ear in the air and one in the water, tap the bottom of the bath with your toe.

The speed of sound in air

Those who watch cricket or athletics should know that the speed of sound is less than the speed of light. When a batsman strikes a ball, you *see* the stroke first and then you *hear* the sound. At the start of a race, you *see* a puff of smoke from the starter's pistol, and then, a moment later, you *hear* the bang.

When the Cup Final is shown on television and the band plays the National Anthem, the crowd does not seem to be able to sing in step with the band. Of course the crowd sings in step with what it *hears* and the microphones high up in the stands pick up the singing. But it is out of step with the music picked up from the band because of the different distances and the fact that sound takes time to travel.

However, the National Anthem is still the same tune when it reaches the microphone. All the notes from the band arrive in the right order. None of the sounds overtakes others. All the different audible frequencies have the same speed.

If you listen to a band playing in the distance, the tune does not change but the sounds get weaker as you move further away. This is because the sound energy spreads over a larger and larger area as it moves away from the source, and the further the ear is from the source, the smaller is the energy received (Fig 22.15). Ultimately the sound is too small for our ears to

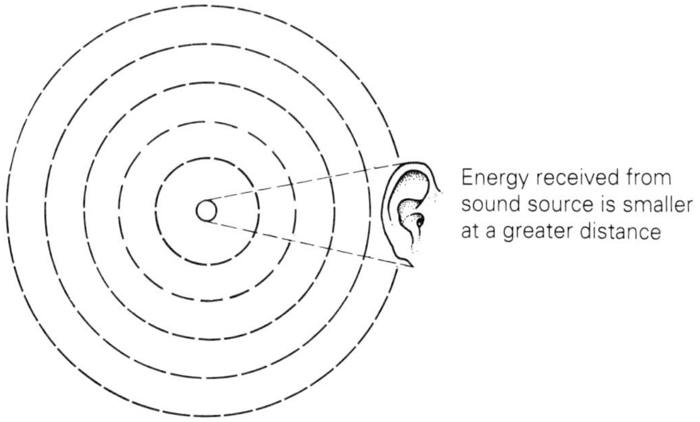

Energy received from sound source is smaller at a greater distance

Fig 22.15

detect it. This is why a jet aircraft flying at a height of 10 000 m cannot be heard.

But there is another effect you may have noticed about a band playing in the distance. The high-pitch notes become fainter more quickly than the low-pitch boom of a drum. This is because sound is absorbed as it passes through air, and high frequencies are absorbed more in a given distance than low frequencies. The low notes thus travel further than the high notes.

The speed of sound in air is about 350 m/s at ordinary temperatures. This is about 750 mph. How fast is Concorde flying if its speed is twice that of sound (Mach 2)? The answer is not 1500 mph! Concorde is very high up in the atmosphere where the pressure and the temperature are low. The low pressure does not affect the speed of sound, but the low temperature does. Can you think of any reason for that? At the height at which Concorde flies, the temperature is about −80 °C and the speed of sound only about 625 mph.

Questions for homework or class discussion

1 To emit sound, something has to move to and fro. Here is a list of musical instruments. For each one, say what is moving to and fro and how that movement is produced.
violin, drum, triangle, piano, harp, trumpet, oboe, flute

2 For each of the instruments in question **1**, say how you can change the pitch of the note which is emitted.

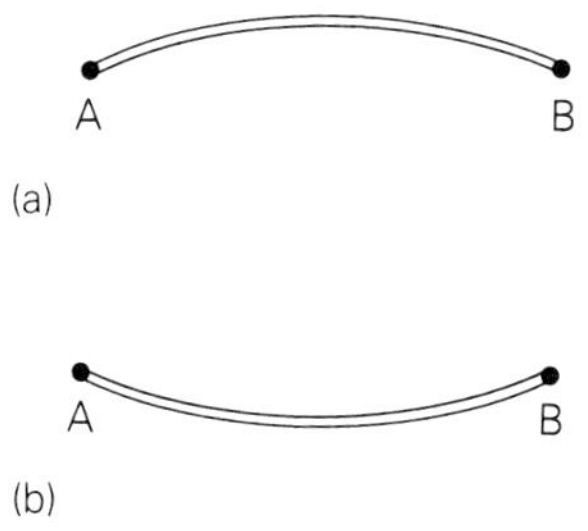

Fig 22.16

3 Fig 22.16 shows a rubber cord vibrating between two fixed points, A and B. At one moment its shape is as in (a). A hundredth of a second later, its shape is as in (b). In another hundredth of a second, it returns to the shape of (a) again. What is the frequency of the note it emits?

4 To start a rubber band vibrating you have to give it energy, but the energy soon disappears. Write a few sentences saying what has happened to the energy.

5 Sound travels at a speed of 350 metres per second in air. A flying bat avoids obstacles by emitting a high-pitched squeak and listening for echoes. What is the time between emitting the squeak and hearing the echo when it is 1 metre from a wall?

6 How many vibrations do the prongs of a 256 Hz tuning fork make in 5 seconds? How far does the sound from the fork travel in that time? [Speed of sound in air is 350 metres per second.]

7 Photographs of the Sun's corona at the time of an eclipse show that there are eruptions (explosions) occurring in the Sun's atmosphere. The Sun must be a very noisy place indeed. Why is it impossible for us to hear that noise?

Background Reading

Sound ranging

We can use the speed of sound to calculate how far away a thunderstorm is.

The lightning and the thunder occur at the same moment. The lightning is an electric spark and the energy which is transferred by it causes the air nearby to expand violently, just as gas expands when a gun is fired. The sudden crack of this explosion becomes the long rumble of a thunderclap as the sound is reflected many times from the clouds and the ground nearby (Fig 22.17).

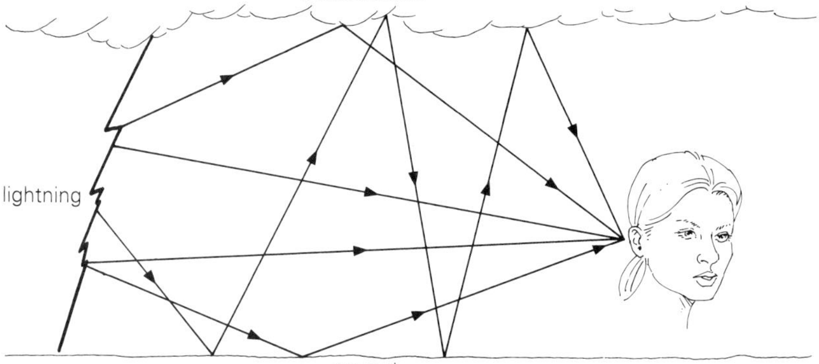

Fig 22.17

The *start* of the thunder is the sound which has travelled directly to the ears without being reflected. This sound travels at about 350 metres per second. So every 3 seconds of delay after the flash is seen corresponds to a distance of 1 kilometre approximately, or every 5 seconds a distance of about 1 mile.

Sound ranging is also used at sea to find the depth of the water below a ship.

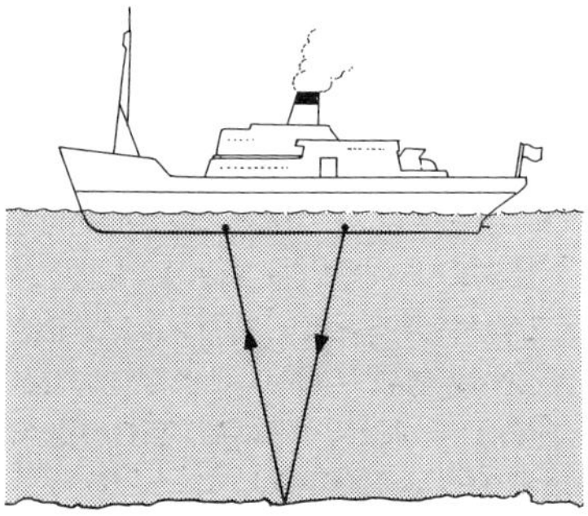

Fig 22.18

A source sends out a short burst of very high frequency sound (above the audible limit), some of which is reflected back to a receiver. The time which passes between emitting and receiving the sound is accurately measured and automatically plotted on a chart. (The reflected signal can also be shown on a CRO.) Sound travels at about 1500 m/s in water, and the times that have to be measured are very small. For example, even if the water is 75 m deep, the sound only takes about one-tenth of a second to travel the 150 m to the sea bed and back. The time measured can be used to calculate the depth of water below the ship.

This type of equipment is also used to detect submarines – and shoals of fish!

Communicating over large distances

Sound cannot be used for communicating over large distances. Even in a big hall it becomes difficult to hear what a speaker is saying and some *amplification* is desirable. A microphone may be

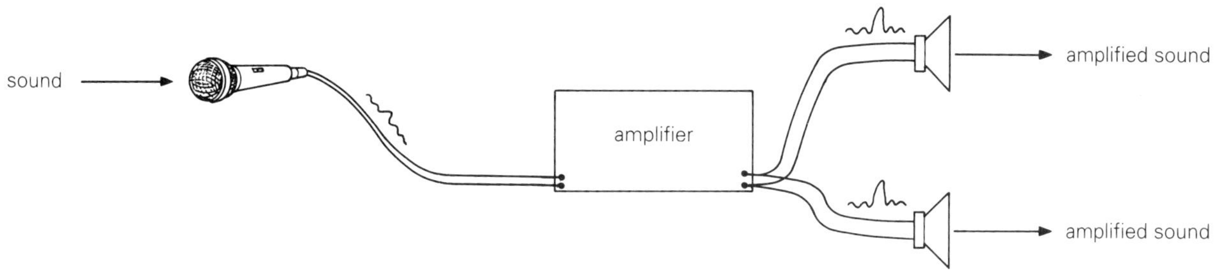

Fig 22.19

used to produce varying electric currents from the sound waves, and these can be enlarged by an amplifier in order to operate loudspeakers.

The leads from the microphone and to the loudspeakers can be quite long, so that the system is useful for large buildings, railway stations, local telephone networks, and so on. If the signals are to be sent over greater distances, leakages in the cables cause the signals to become gradually weaker and amplifiers are required every few kilometres. Nowadays, optical fibres are being used instead of wires. The electrical signals from the microphone are amplified and converted into 'light' pulses using a digital code (see page 312.) These pulses are sent down the fibre and are detected at the receiving end by a photosensitive device which converts them back to electrical pulses. An electronic circuit then decodes the pattern of pulses before passing the signals to the receiver. This type of system is cheaper than one using copper wires and requires fewer amplifiers.

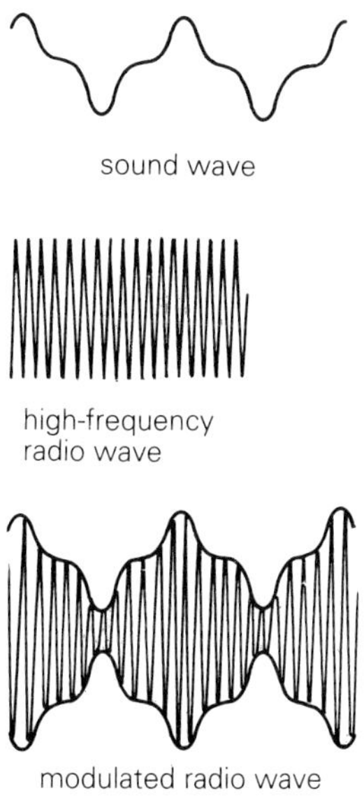

Fig 22.20

Another way of communicating over large distances is by means of radio waves. A radio wave will spread out from an aerial if electric charge in the aerial is made to oscillate at a very high frequency, usually at least 1 megahertz or 1 000 000 Hz. The problem is how to make the rapid oscillations of the radio wave carry the much slower variations due to the sound. One solution is to make the size of the wave (the amplitude) vary in the same pattern as the pattern of the sound signal. We call this *modulation* and say that the carrier wave has been *amplitude modulated* (AM).

When the radio wave reaches the aerial of the receiver, it causes similar variations in it. The receiving aerial will pick up radio waves from many different directions at different frequencies. So the receiver has to be tuned to pick out the required carrier wave frequency from all the others. The tiny voltage variations in the aerial circuit are then amplified and processed to extract the signal corresponding to the sound variations. Finally this is amplified again in order to drive the loudspeaker.

Fig 22.21

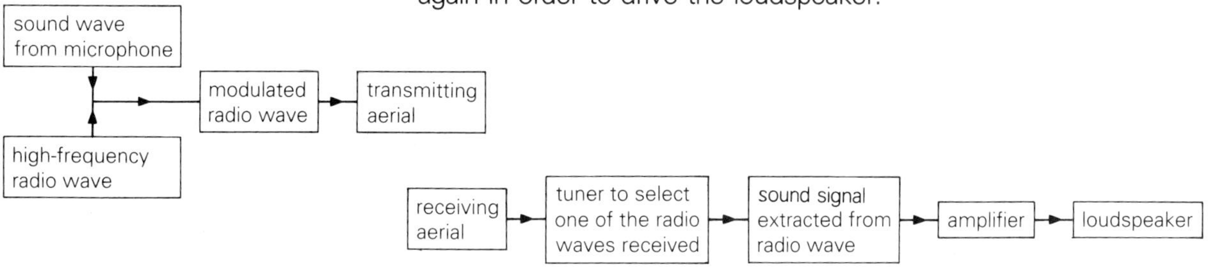

Chapter 23 Logic circuits and the bistable

You will have read at the end of Chapter 14 how switching circuits can now be made so small that more than 100 000 of them can be put on a chip of silicon a few millimetres square. These switching circuits, the basis of all computers, are not quite like those you have been studying – they use transistors as switches. But you can make a switching circuit which behaves in a similar way to one of the simplest microelectronic switches by using a reed relay with a change-over switch in it (see page 187). The circuit (Fig 23.1) has two sockets, A and B, called the *input* sockets, and an *output* socket, and it is connected to a suitable battery as shown.

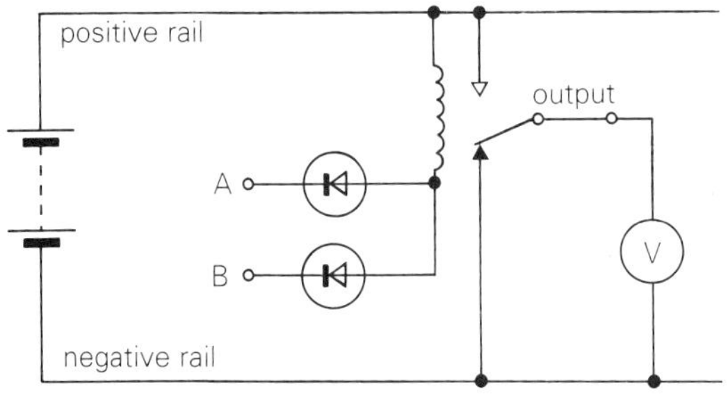

Fig 23.1

Normally, the moving reed rests against the contact connected to the negative rail so that a voltmeter connected between the negative rail and the output would read zero. If A or B (or both) is connected to the negative rail (the *low* 'pressure' side of the battery), a current flows through the coil, the reed moves to the other contact and the voltmeter gives a high reading (Fig 23.2(a)).

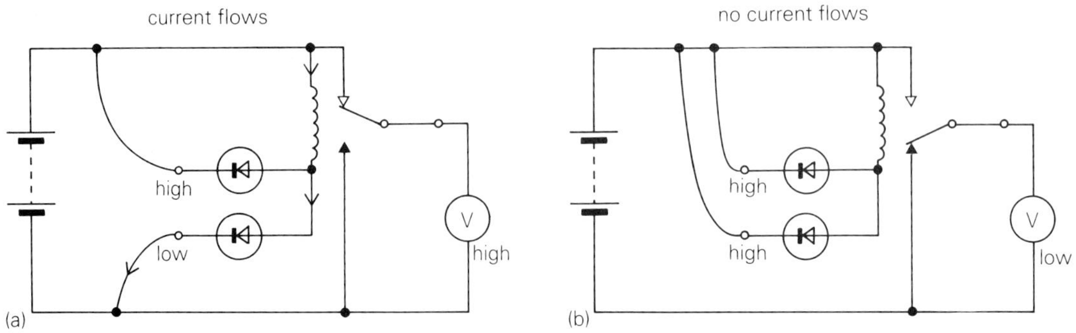

Fig 23.2

If *both* A and B are connected to the positive rail (the *high* 'pressure' side of the battery), no current flows through the coil and the output is connected to the low 'pressure' (Fig 23.2(b)).

B	A	Output
low	low	high
low	high	high
high	low	high
high	high	low

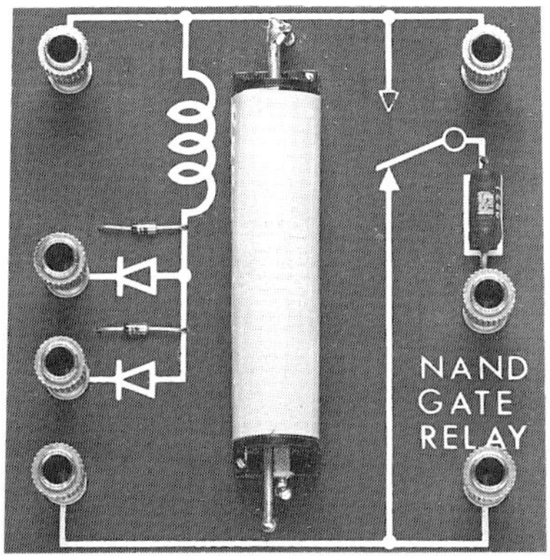

A module providing 4 NAND circuits

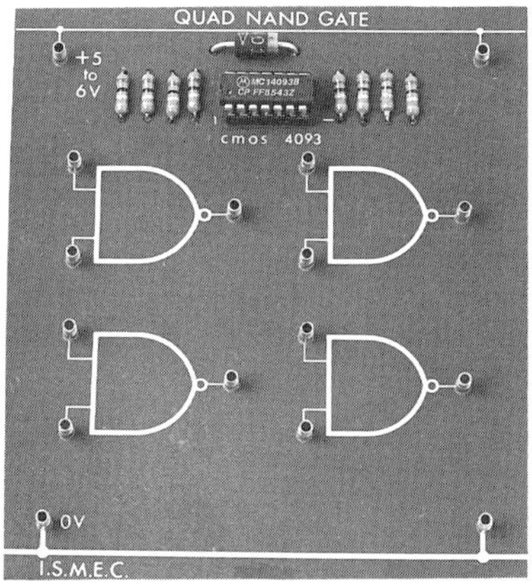

We can summarise this behaviour by using a table and the words 'low' and 'high' to show what happens. Tables which summarise switching behaviour are known as **truth tables**. Note that an LED module can be used to show the output voltage in place of the voltmeter. If the output is low the LED will not be lit; if it is high, the LED will glow.

We have not yet explained why the two diodes are needed. Think what would happen if A were made high and B low, or A low and B high. Draw the circuit without the diodes in and you will see that the battery is short-circuited. The diodes are needed to stop that happening.

One kind of small transistor switching circuit has the same truth table as the circuit above. The circuit is called a NAND circuit (for a reason which will be explained later) and the photographs show two NAND circuit modules, one using a relay and one using a chip.

In designing complicated circuits, such as television sets, tape recorders and computers, the engineers frequently use a number of smaller, simpler circuits which are connected together in a suitable way to do what is required. To do this, they must be familiar with the ways in which the simpler circuits behave, and they are then able to use the simpler circuits as 'building bricks' when constructing more complicated circuits. In the remainder of this chapter, you will explore some of the useful things the NAND circuit can do and you will use it as a building brick to do some useful jobs.

The NAND circuit

Experiment 23.1
Using the NAND circuit module as an INVERTER

In this and all the experiments which follow, we shall not show the circuit between the input sockets and the output socket inside the module. This is because you will not need to know what it is and it is not important in the work you will do.

a Set up the circuit below using a NAND module, an LED module and a battery.

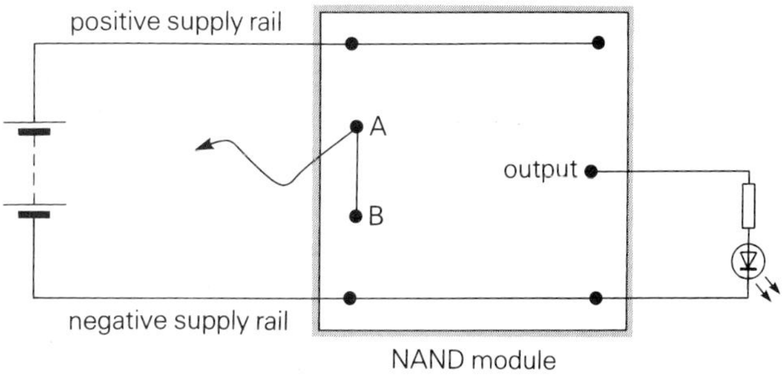

Fig 23.3

b Connect inputs A and B together and then plug another lead into socket A. Connect the other end of this lead to the positive supply rail (high). What does the LED tell you?

c Disconnect the lead to A from the positive rail and connect it to the negative rail. What does the LED tell you this time?

Input	Output
low	high
high	low

Because inputs A and B are connected together, there is in effect only one input, and the truth table for this circuit is as shown on the left. Notice that the circuit changes the voltage level over; if the input is LOW, the output is HIGH and *vice versa*. For this reason, the circuit is called an INVERTER ('invert' means 'turn upside down'). The circuit can be operated with a switch or an LDR at the input, and the output can be used to light an LED or operate a buzzer. The relay NAND circuit will also operate a motor directly.

Notice that, if the input lead is not connected to either power rail, the output is LOW. An unconnected input behaves as though it were HIGH.

Experiment 23.2
Operating an INVERTER with an LDR

a Use a NAND module, a buzzer module, an LDR module and a battery to set up this circuit.

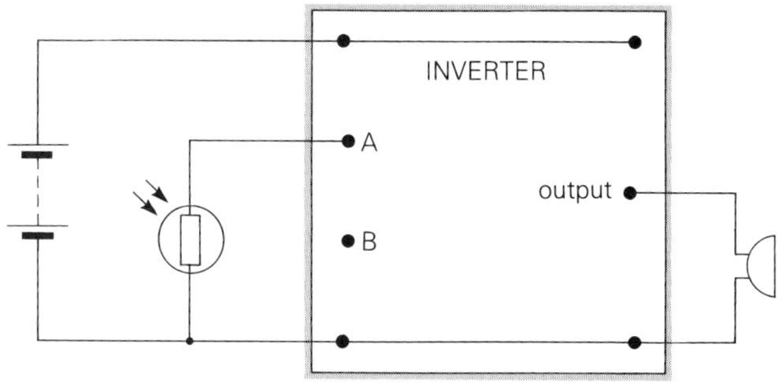

Fig 23.4

b What happens when you shine light from a torch on to the LDR?

When light is shone on the LDR its resistance becomes small and it behaves like a closed switch. Then the input is effectively connected to the negative supply rail. It is therefore LOW, the output goes HIGH and so the buzzer sounds.

In the dark, the resistance of the LDR is very high and it behaves like an open switch. The input is then unconnected and behaves as though it were HIGH, so the output is LOW and the buzzer is off.

Project: An automatic light

Use two NAND modules, an LDR module, an LED module and a battery to make a circuit which will light the LED automatically when it gets dark.

Experiment 23.3
Using both the inputs of the NAND module

a Use a NAND module, an LED module and a battery to set up the circuit below. Plug a lead into each of the inputs.

Fig 23.5

Input B	Input A	Output
low	low	
low	high	
high	low	
high	high	

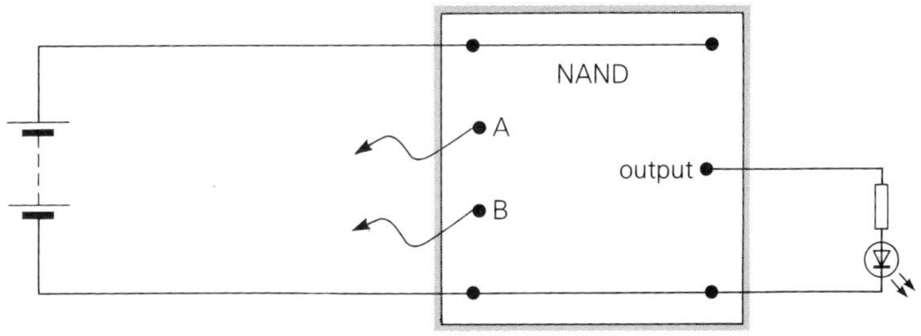

b Make a copy of the truth table. Complete the truth table by finding out what the output is for each of the input possibilities.

Truth table for the NAND module

Input B	Input A	Output
low	low	high
low	high	high
high	low	high
high	high	low

The truth table for the NAND module is as shown. Notice that the output is HIGH whenever one or both inputs is LOW. You could say that, with a NAND circuit, the output is

Not high only when input A **AND** input B are high.

That is why it is called a **NAND** circuit.

Experiment 23.4
Making an AND circuit

An AND circuit would have a different truth table. It would have a high output only if both inputs A **AND** B were high.

a Use two NAND circuits (one as an INVERTER), an LED module

Fig 23.6

Input B	Input A	Output
low	low	
low	high	
high	low	
high	high	

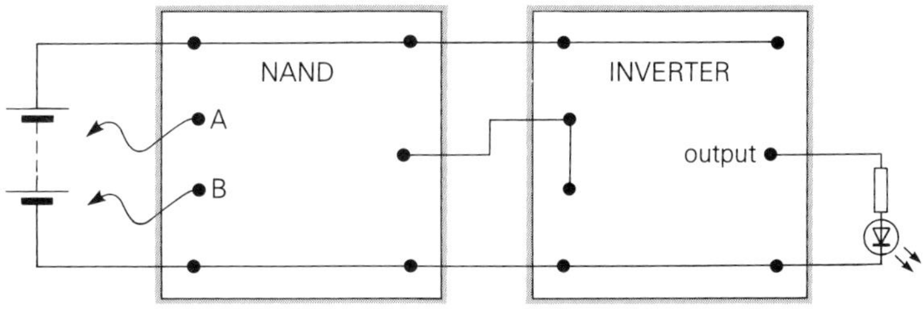

and a battery to set up the circuit above. Attach a lead to each of the inputs of the first NAND circuit.

b Copy the truth table. By connecting A and B to the appropriate supply rails, complete the truth table. Does your circuit behave like an AND circuit?

c Compare the truth table for this circuit with that of the NAND circuit. What has the INVERTER done?

Electronics uses circuits like AND, NAND and INVERTER circuits a great deal in designing larger circuits. These simpler circuits are called *logic circuits*. They all have inputs and an output. They are all switching circuits in which the voltage at the output at any time depends on the voltages at the inputs. They are 'decision-making' circuits.

Project: A 'length' detector

Use an AND circuit, two LDR modules, a buzzer module and a battery so that, if both LDRs are darkened, the buzzer will sound. This circuit could be used to detect objects of length greater than the distance between the LDRs.

Experiment 23.5
Using a NAND circuit to make a simple burglar alarm

a Use a NAND circuit, two push-button switches, a buzzer and a battery to connect up the circuit shown below.

Fig 23.7

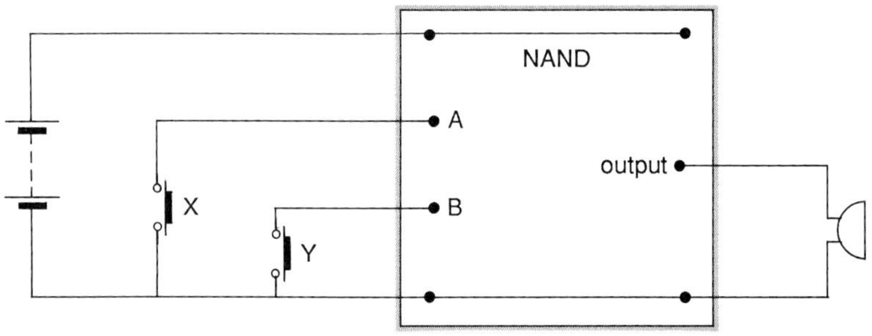

b What happens when switch X or switch Y is pressed? What happens when both are pressed together? Explain why this happens.

It makes a more realistic alarm if the switches are replaced by pressure pads. These could be placed under a door mat and so guard two doors. But, there is a serious weakness in this circuit as a burglar alarm. The alarm stops as soon as the intruder steps off the pressure pad. A better alarm would continue to sound once it had been set off, even if the switch were released. To do this requires the use of two NAND circuits in an arrangement known as a BISTABLE. Bistable circuits are used a great deal in modern electronics, particularly in computers.

Experiment 23.6
Making a bistable circuit

a Set up the circuit shown in Fig 23.8 using two NAND circuits, two LED modules and a battery. The LED modules are used to show whether the outputs are HIGH or LOW.

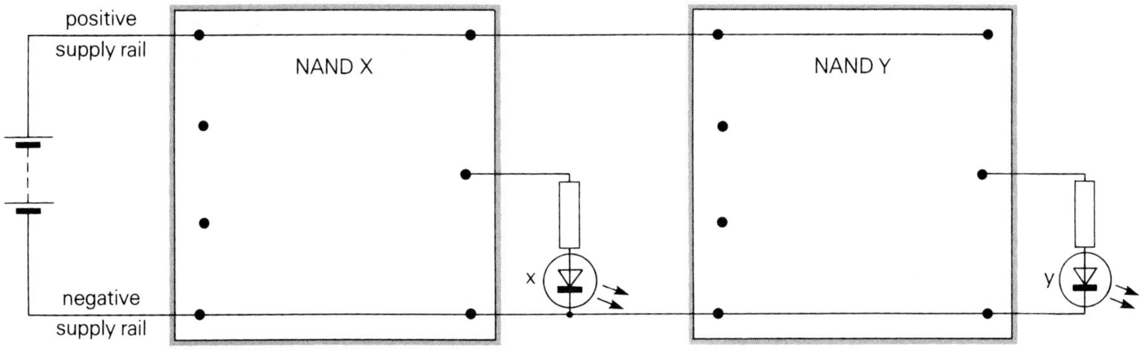

Fig 23.8

b Now connect the output of NAND X to one of the inputs of NAND Y. What do the LEDs show?

As there are no connections to the inputs of NAND X, the output of NAND X is LOW, and therefore the input of NAND Y must also be LOW. And that means that the output of NAND Y must be HIGH so that LED y is lit.

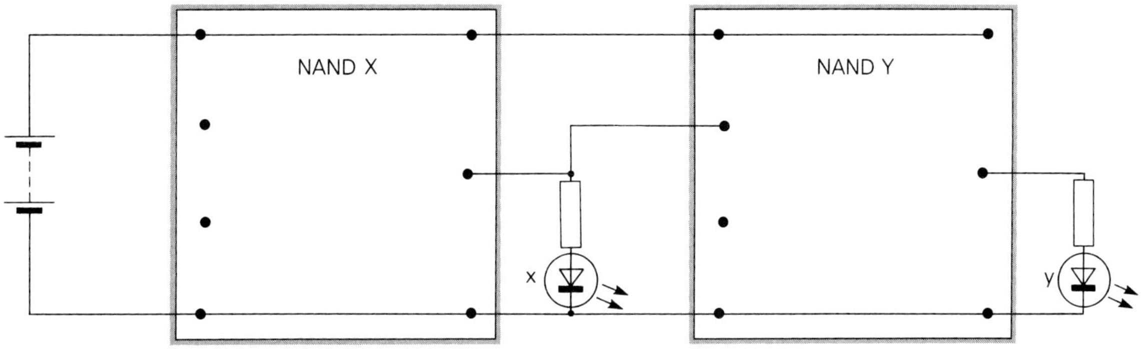

Fig 23.9

c Next, connect the output of NAND Y to one of the inputs of NAND X. (Note that in drawing this in the circuit diagram below, it is necessary for the lead from NAND Y to cross over the positive supply rail. This does not mean they make electrical contact. Electrical junctions are indicated with a dot.)

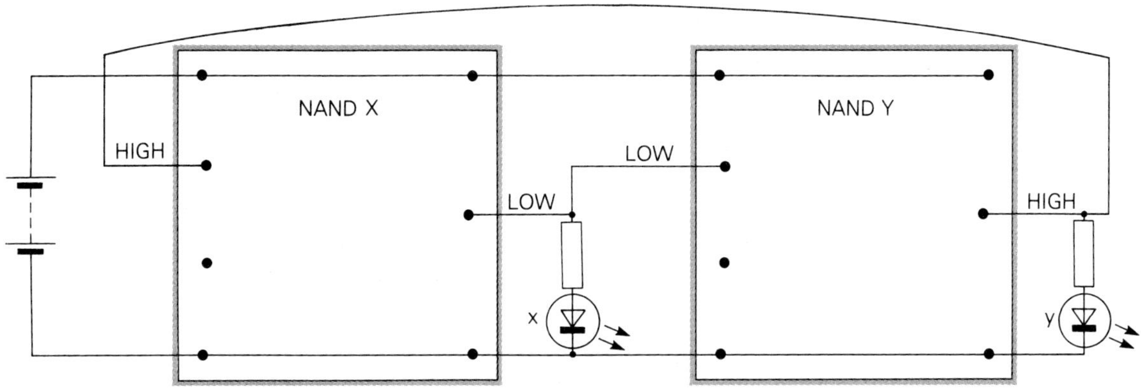

Fig 23.10

When NAND Y's output is connected to the input of NAND X, it makes that input HIGH. This does not change the output of NAND X which remains LOW. LED x remains unlit.

However long the connections are left, LED x remains unlit and LED y remains lit. It is a *stable* arrangement.

d But this is not the only stable arrangement. Remove the two leads connecting together the inputs and outputs of NAND X and NAND Y. Then connect first the output of NAND Y to the input of NAND X and afterwards connect the output of NAND X to the input of NAND Y. You will now have set up another stable arrangement. How does this second stable state of affairs differ from the first one?

In the first state, LED x is unlit, LED y is lit.
In the second state, LED x is lit, LED y is unlit.

As the circuit has two stable states, it is called a *bistable* circuit.

Fig 23.11

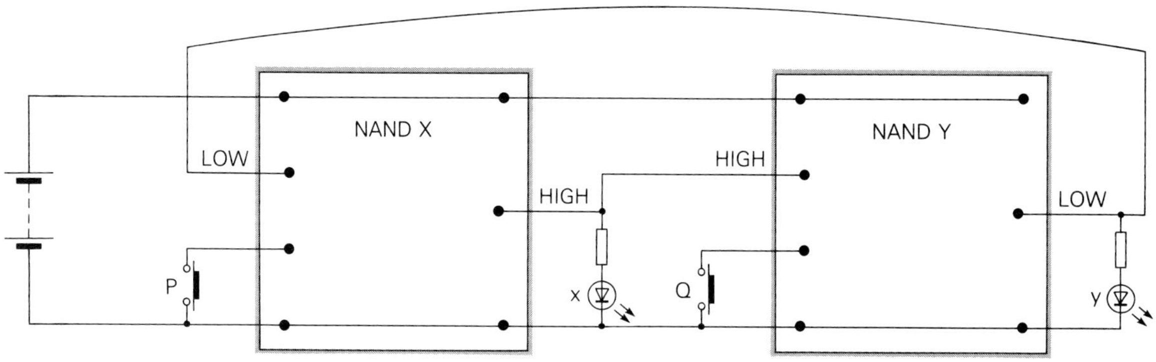

e Now add two push-button switches to the arrangement of modules, as shown above. What happens when you press switch Q? Try pressing it several times.

f What happens when you press switch P? Try pressing it several times.

g Now press Q again, and then P. Then press P and Q one after the other.

Pressing a switch causes the BISTABLE to switch from one of its stable states to the other, *provided* it is the switch connected to the NAND circuit whose output is LOW (that is, the one whose LED is unlit).

In the diagram above, it is only when you press Q that the circuit flips over to that shown in the diagram below. After it has flipped over, pressing Q again has no further effect.

Fig 23.12

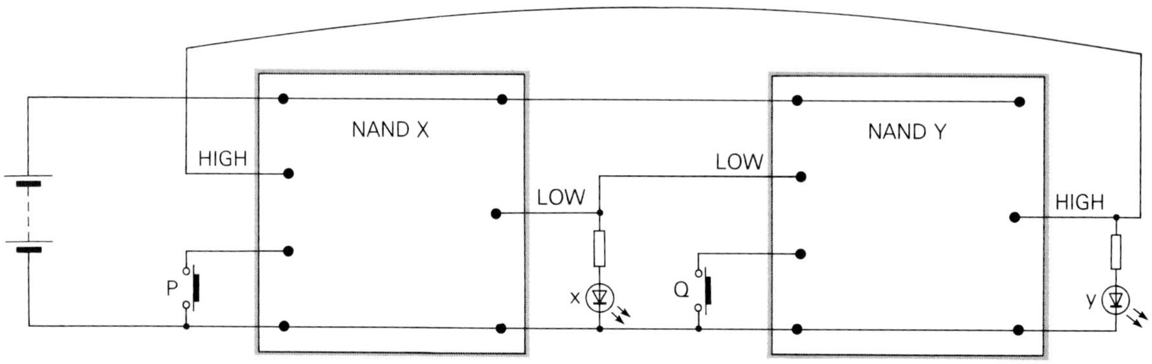

When Q is pressed for the first time, the input of NAND Y is connected to the negative supply rail and the output of NAND Y becomes HIGH immediately. That gives a HIGH input to NAND X, and with two HIGH inputs its output becomes LOW. Because that is connected to the input of NAND Y, NAND Y now has a LOW input (and therefore a HIGH output) whether Q is pressed again or not. The only way to get the BISTABLE to flop back again is to operate switch P. And now you can understand why bistable circuits are often called 'flip-flops'.

Experiment 23.7 A latched burglar alarm

Having built the BISTABLE, you are now in a position to make a really good alarm.

a First replace LED y with a buzzer. What happens when you switch from one stable state to the other?
b Now replace the push-button switch Q at the input of the NAND Y circuit by a pressure pad. What happens when the pressure pad is pressed? Why is the other push-button switch still necessary? How could you modify the circuit to have two pressure pads in use?

Project: Teacher warning system

A small variation of the arrangement in Experiment 23.7 will turn it into a 'teacher detector'. You could put a pressure pad under a mat in the corridor outside your classroom to let you know when your teacher is coming.

Of course, it would be better if the buzzer could be turned off before teacher arrives. Perhaps this could be done by putting a second pressure pad under another mat immediately outside the door to change the BISTABLE back when the teacher steps on the mat. The teacher will then automatically set off the alarm and silence it before entering the room. That is something for you to try.

Solving problems

Problem solving should be tackled systematically. First you have to decide what to use as input and output devices, LEDs, switches, buzzers for example. Then you must work out what the input voltage levels must be to produce the desired output effect and which circuit can be used for that. Next, you must draw a circuit diagram, check that it will work in the way you think it should, set it up and try it. If it does not do what you expect, try to work out why it is misbehaving. When you have done that, you should know how to put it right. The following are a few suggestions for further projects.

Project: Window alarm

Design a circuit which will sound an alarm if a window is opened. You may imagine that a magnet is hidden in the window sash (the moving part). A reed switch is buried in the frame so that the magnet is near it when the window is closed. Once the alarm sounds, it should continue even if the window is closed again; you will need to have a switch in the circuit to reset the alarm.

Project: Automatic signal for a railway

Use a BISTABLE to make an automatic signal for a model railway. The train should turn the signal to red as it passes and then back to green when it has travelled some distance further. Two reed switch modules can be placed between the rails a suitable distance apart, and a small magnet can be taped underneath the last coach of the train.

Project: Fire alarm

Design a circuit which will sound an alarm if a fire occurs. Use a piece of fine copper wire or solder which would melt in a fire. When you test your circuit, cut the fine wire with scissors to pretend that it has melted.

Project: Car doors warning light

Build a circuit for a two-door car which will light a warning indicator on the dash-board if both doors are not properly shut. Use two push-button switch modules to act as switches mounted in the doors, and an LED module as the warning light. The light should be on except when both switches are pressed (both doors are shut).

This circuit is an OR circuit – the output is HIGH when one door OR the other is open (input HIGH).

Project: Seat-belt warning light

A car is to have a pressure switch inside the front passenger seat and a switch in the seat-belt mechanism which opens when the seat-belt has been fastened. Design a circuit which will show a red light if the passenger has not fastened the seat-belt, but a green light otherwise. (Remember that there may not be a passenger in the car.)

Questions for homework

1 Write out the truth tables for **a** an INVERTER, **b** a NAND circuit, **c** an AND circuit, using 0 to stand for LOW and 1 to stand for HIGH.

2 Complete the following statements.

a An INVERTER is sometimes called a NOT circuit because the output is ...... when the input is NOT ......

b If both inputs of a NAND circuit are connected together and to the .... supply rail, the output is ....

c A NAND circuit is so called because the output is .... only when ....

d An AND circuit is so called because the output is .... only when ....

e An OR circuit is so called because the output is HIGH when ....

Fig 23.13

3 The diagram shows a bistable in one of its stable states.

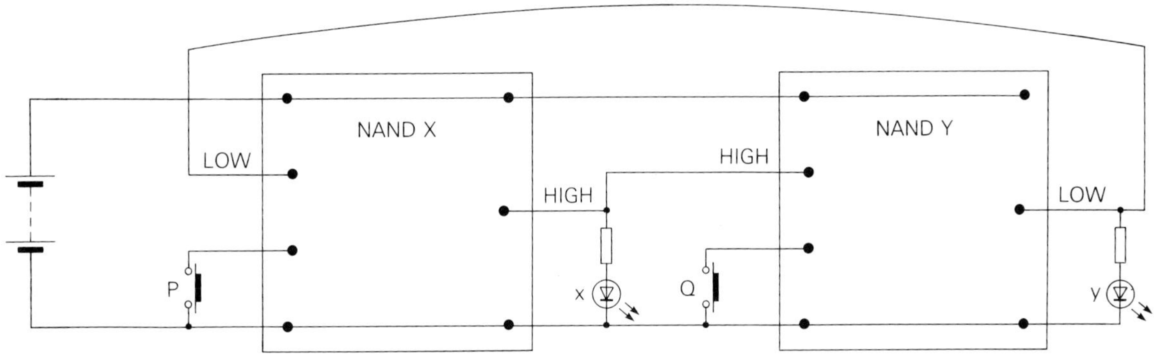

a What effect does pressing switch P have? Explain your answer.

b What effect does pressing switch Q have? Explain your answer.

Input B	Input A	C	D	Output
0	0			
0	1			
1	0			
1	1			

0 stands for LOW
1 stands for HIGH

4 Copy the truth table on the left and complete it for the circuit shown in Fig 23.14.

a In the column headed C, put down 0 or 1 for the output point marked C depending on what is written in the column headed A.

b In the column headed D, put 0 or 1 depending on what is in column B.

c Use columns C and D to decide what should be in the output column.

d What sort of circuit is this?

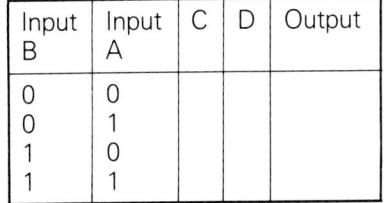

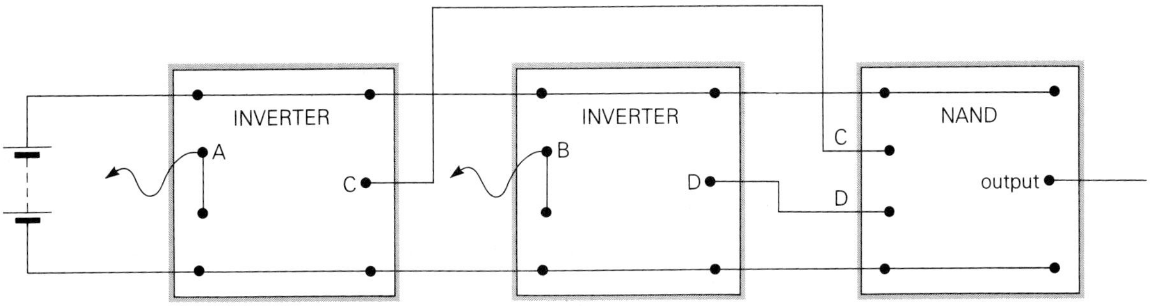

Fig 23.14

5 A thermistor is a special resistor whose resistance becomes smaller as it gets hotter. Draw circuit diagrams to show how it could be used to make a fire alarm using
a a battery and a buzzer alone,
b a battery, a buzzer and a reed relay,
c a battery, a buzzer and a NAND circuit.

Background Reading

Digital and analogue electronics

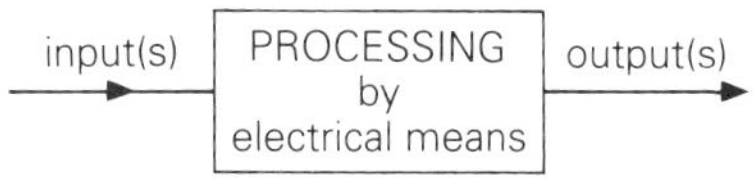

Fig 23.15

The electronic circuits you have met in this chapter have one thing in common. They need some sort of input which is changed by the circuit to produce an output. The changes, of course, are brought about by electrical methods and hence the name 'electronics'. The input and output information has been in the forms of 'low' or 'high', and there is no meaning to be given to anything in between. There are just the two voltage levels which the circuits can recognise and which are usually linked to the digits 0 and 1. So this kind of electronics is 'digital electronics'.

The digital method is not the only method by which information can be processed. Suppose you wish to make the sound from a radio louder. By adjusting the volume control the output gets steadily louder. There is a whole range of volumes, not just low and high. Most of the changes in the world around us are continuous changes: the change of light intensity from night to day, the rise of temperature of water in a pan when it is heated, the increase in speed of a falling body. And some electronic circuits process 'information' in its continuous form.

To do that, the information has to be converted into a corresponding value of an electrical quantity such as voltage, which is then called the analogue of the input ('analogue' means 'corresponding to'). Of course, the voltage has to be able to vary continuously at all stages of the processing, and this kind of electronics is called 'analogue electronics'.

A microphone amplifier is an example of an analogue circuit. When sound waves fall on the microphone, the microphone produces small voltages which change continuously according to the sounds. Those changes are then enlarged by the amplifier to drive the loudspeaker. A circuit which will produce an exact but magnified copy of the input variations is needed. The difficulty is to make it an exact copy — no amplifier or loudspeaker does it *exactly* at all frequencies so that there is always some distortion or lack of accuracy.

If electronic circuits are used in analogue systems, inaccuracies increase, but in digital systems that is not so. Compared with analogue circuits digital circuits offer greater precision, though they are usually more complicated.

(Adapted from *Electronics* by G E Foxcroft, J L Lewis and M K Summers, published by Longman.)

Explorations 4

A A tennis ball bounces on the floor several times. Each time it bounces it rises to a smaller height. Explore how much of its energy it loses at its first bounce, its second bounce and so on.

Does it matter what it is bouncing on?

Does it matter what height it falls from?

Does a golf ball behave differently?

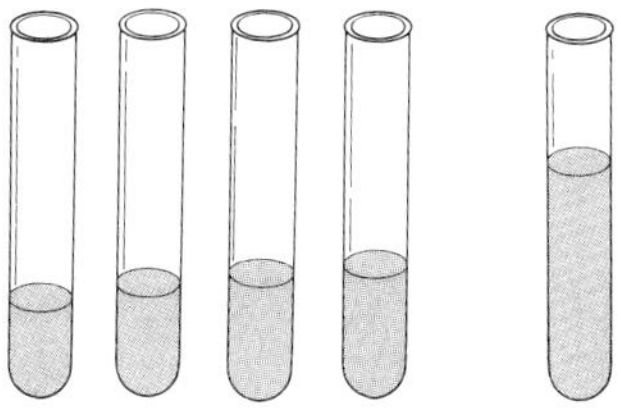

B If you partially fill a boiling tube with water and blow over the end of it, it will emit a note. (What do you think is vibrating to give this sound?)

Use eight boiling tubes to make an instrument which will play a musical scale.

Can you use the instrument to play a tune? 'Three blind mice' fits into your scale of eight notes.

Measure the lengths of the columns of air in the tubes. Do you notice anything about them?

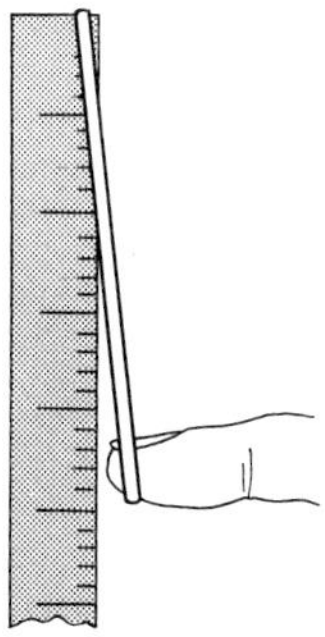

C Hook one end of a rubber band round a metre rule and stretch the band by pulling downwards on it with your finger in the free end.

If you release the band, it flies up into the air because of the original strain energy stored in it.

Explore how the height to which the band rises depends on how much it is stretched.

Does the gain in gravitational potential energy represent all the original stored energy?

D An open-air exploration can be done to measure the speed of sound. A drum or metal dustbin lid hit with a piece of wood can provide a source of sound.

Is it possible to watch the drum being hit from some distance away and to measure the time between seeing it hit and hearing the sound?

Why is this difficult to do?

Does the experiment really measure the speed of sound or is it measuring something else?

Explore whether there might be a way to measure the speed of sound using echoes reflected off a building.

E This exploration is to give you practice at observation.

When you boil water in a kettle you often hear the kettle 'singing', and after a time 'steam' appears from the spout. If you boil water in a beaker so that you can watch it, you will see other things happening too.

Place some *cold* water in a beaker and boil it using a Bunsen burner. Write down everything you see happening.

F Make a xylophone using wood.

You will have to experiment to find out the lengths of the bars for the various notes – and how to support them.

G This exploration involves finding out about a given circuit.

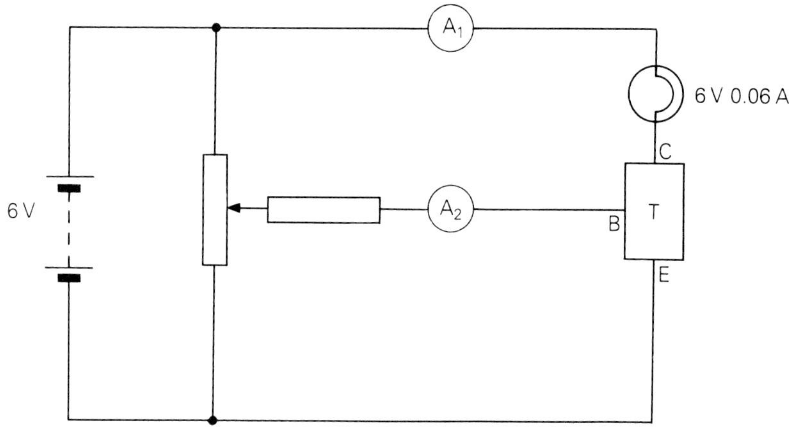

The circuit is a lamp dimmer for a small lamp and it includes a device T called a transistor.

Find out how the current through the lamp, as measured by the ammeter A_1, depends on the current shown by the ammeter A_2.

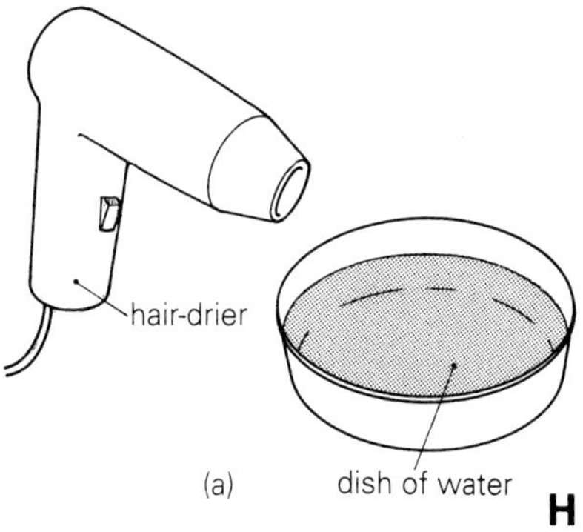

(a)

H This investigation is concerned with moisture in the air. It is in three parts.

First, put some water in a flat dish or saucer and leave it on a shelf to find out whether or not it evaporates and how long it takes. Does the time it takes to evaporate depend on whether the water was originally hot or cold? Does it make any difference whether the room is hot or cold?

Secondly, direct an electric hair-drier towards the dish of water. Does this affect the evaporation? Does it make any difference whether the hair-drier is blowing out hot or cold air?

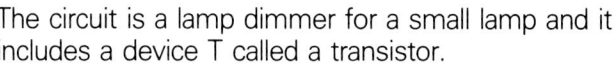

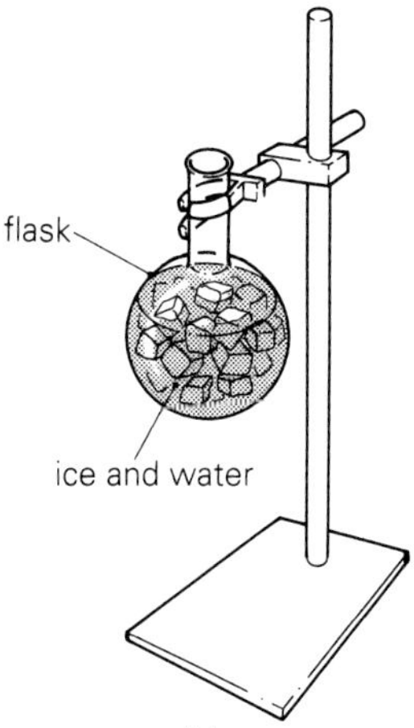

(b)

Thirdly, put some cold water and ice inside a large flask (as large as possible) and suspend the flask in a clamp. Arrange the apparatus so that the water vapour evaporated by the hair-drier is blown towards the flask. What do you notice about the outside of the flask? Explain what is happening.

I The Government encourages people to lag their hot water tanks in order to save energy.

Explore which is the best insulation to use, and what effect its thickness has. You could try cotton wool, expanded polystyrene, cork, sawdust, *etc.*

Try to make some measurements which would let you estimate how much could be saved in just keeping the water hot by lagging the tank.

J If a mirror reflecting a ray of light is turned, the direction of the reflected ray changes. Do some experiments to find out how the change of direction depends on the angle through which the mirror is turned.

K Use two lenses to set up a telescope to view a distant object such as a window. Try to measure the magnification of your telescope when looking at the window.

L The ancient Greeks and Romans were skilful at building engines for throwing projectiles. The trebuchet was a mediaeval weapon derived from the classical engines for hurling missiles into fortified towns and castles. The drawings below show a mediaeval trebuchet after and before firing.

Try building a device of your own for projecting missiles. Some mediaeval trebuchets used twisted cords. Would twisted elastic bands help? Investigate what are the best loads and the most suitable dimensions in the structure you build to get a projectile to travel the greatest distance.

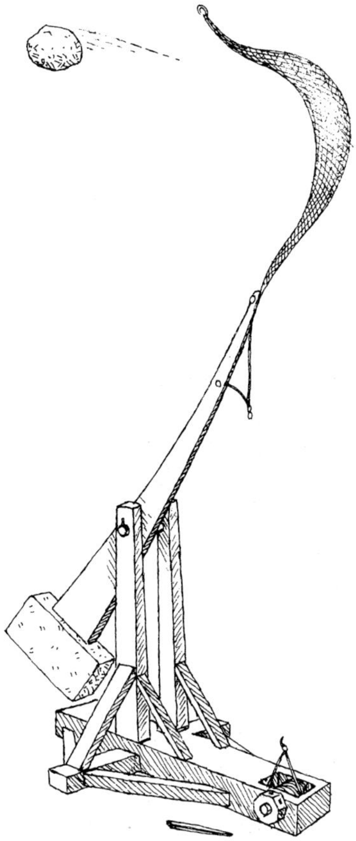

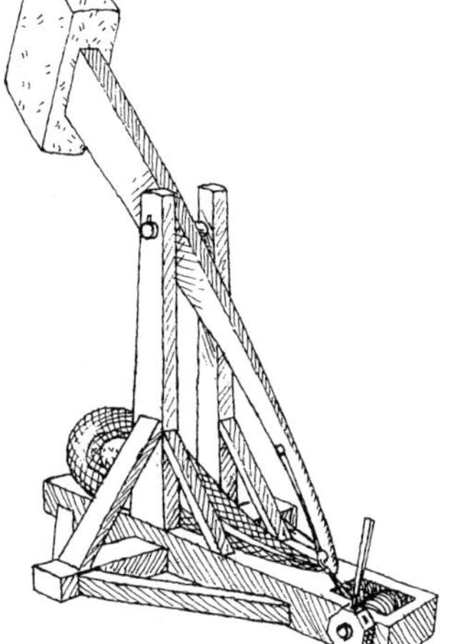

Chapter 24 The role of physics

In this book we have been considering those parts of science which are usually referred to as physics. In this final chapter we will consider some of its contributions and mention some of the problems that it will have to solve in the future. Of course other sciences, biology and chemistry for example, also have equally important contributions to make as we hope you have learned from your studies.

In all scientific work, it is important to observe, to think, to speculate, to experiment, to draw conclusions and then to test them. Physicists have always searched for patterns in their observations and in their experiments, and it is these patterns which enable them to formulate their laws of nature. In turn these laws have made it possible to apply the knowledge obtained to those engineering projects – electrical, mechanical, structural – which have added much to the quality of our lives.

Physicists have always been curious and model-making has played an important part when they were striving for understanding. The Greeks speculated on the nature of the Universe with their models of the Earth surrounded by celestial spheres carrying the Sun, the Moon and the planets. Physicists have always found it useful to make models to support their theories, and because of their interest in the nature of atoms, the development of a model of an atom has been very much one of their concerns. However, it is interesting to realise that a student giving what was considered a 'correct' description of an atom a hundred years ago would not have got any marks for it in an examination a few years later as the accepted model changed.

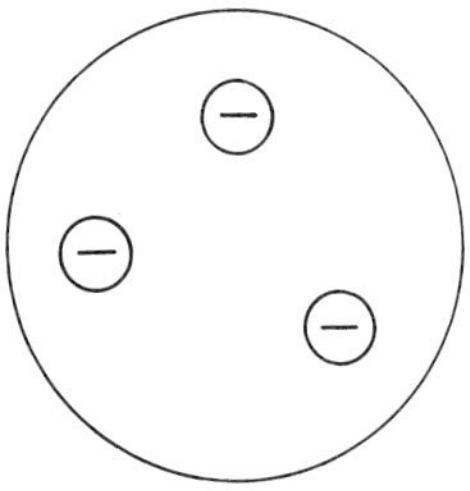

'plum-pudding model'

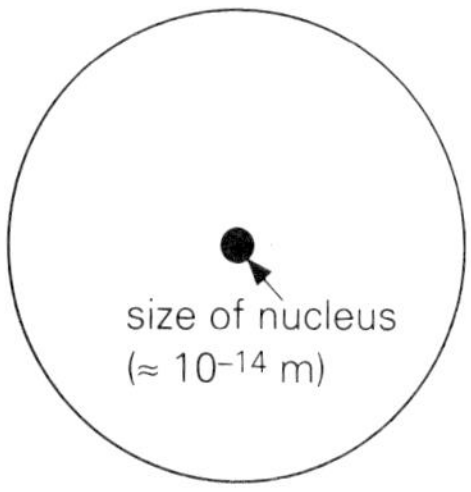

Rutherford's model

size of atom (≈ 10^{-10} m)

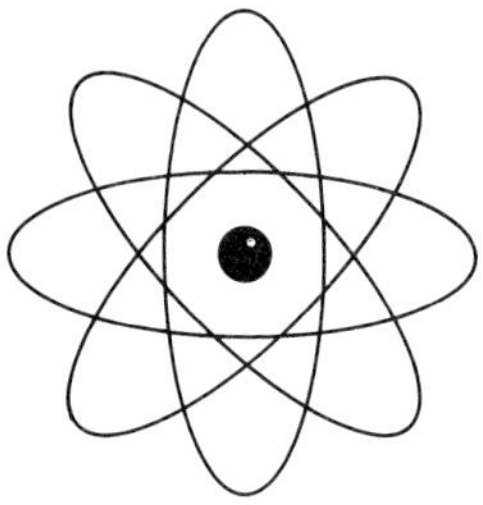

elliptic orbits model

After the discovery of the electron in 1895, the model proposed was the 'plum pudding model' in which the negatively charged electrons were embedded in positively charged material much like currants in a plum pudding. Even this model which was considered 'correct' for a number of years changed when Lord Rutherford in 1910 proposed the nuclear model of the atom, in which the atom consisted of a central nucleus of positive charge with electrons somehow around it. (This model was no speculation like some of the models of the Universe suggested by the ancient Greeks; it was the result of careful measurements conducted with the aid of instruments devised by physicists. Indeed, one of the most important contributions of physics has been the development of instruments which enable all the sciences to make measurements.)

In due course the Rutherford model was replaced by a new model with electrons moving in orbits around the nucleus. Even that model of the atom was replaced by a quite different model in 1926 when the electrons became clouds around the central nucleus.

The important thing to realise from all this is that physicists are always striving for greater understanding, and in the process produce better models as evidence shows that the previous model was not as good as they once thought it to be. They are also much more humble than they used to be; they now realise that their latest model may not be the final answer and may have to be abandoned for yet another model in the future.

Scientists have always enjoyed this process of 'finding out' and of producing better models. But is it perhaps a game? Is it just done for the fun of it (and scientific investigation is certainly fun) or does it have any useful purpose? We will return to this later. First we will consider some of the ways in which physics has contributed to our lives.

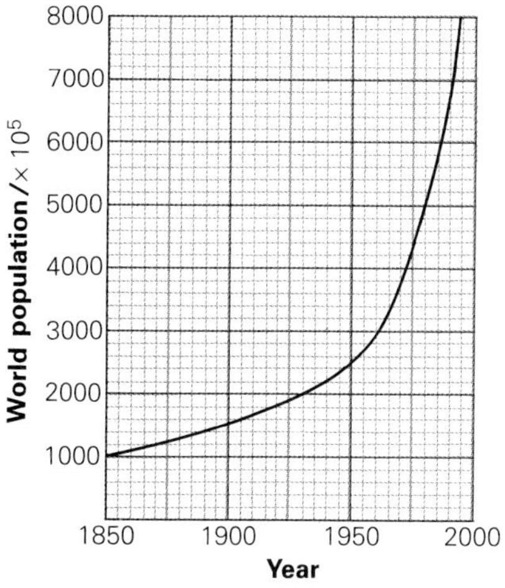

Fig 24.1

Energy usage in the world today

The photograph shows women in a developing country collecting firewood. They need this in order to cook their food and, in many developing countries, to provide warmth. At one time there were plenty of trees, but the demands for fuel have increased dramatically as the population has increased. At one time a woman in Bangalore in India would travel only a few miles to collect what she needed; now her successor may have to travel 30 miles each night to collect what she requires. Cutting down the trees means that the land is being devastated and for that reason it has been made illegal. But she still needs to cook and that means she has to do her collecting at night to avoid being arrested for breaking the law – and so the land deteriorates further.

Fig 24.1 shows how the world population has been increasing. Fig 24.2 shows how energy resources have been in greater and greater demand. Energy requirements are likely to increase in future, partly because expectations in developed countries seem always to increase and not least because the needs of developing countries will have to be met as their populations increase.

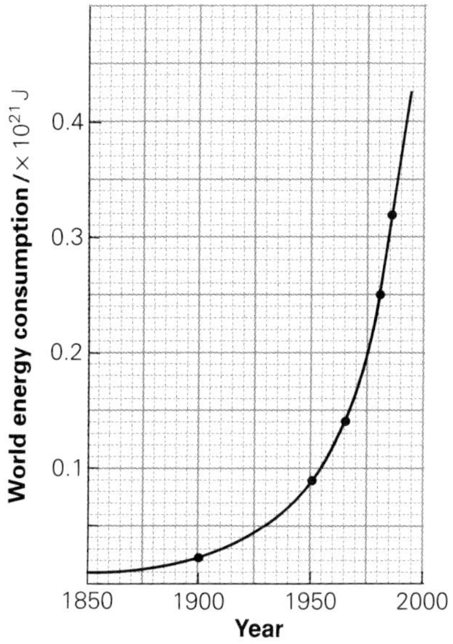

Fig 24.2

It should be remembered that the energy demands of the developing countries are nothing by comparison with the demands made by many developed countries. Fig 24.3 shows the average energy consumption per person per year in a number of countries and many African countries would not appear at all on such a plot. Is it any wonder that the developed countries go on getting richer whilst developing countries progress so slowly?

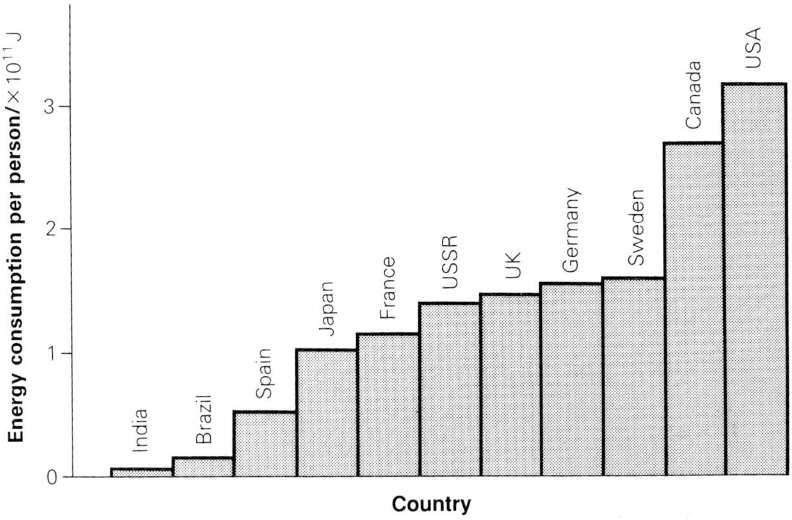

Fig 24.3

When one has grown up with electrical energy available everywhere at the throw of a switch, it is very easy to take it for granted. But, until the end of the 18th century, wood was the main source of our energy supply. Fortunately for us the population was much smaller and there were plenty of trees. Even more fortunately, scientists and engineers were beginning to exploit the use of coal and in 1811 gas was used for the first time for street lighting. Also very fortunate was the work done by Michael Faraday, working at the Royal Institution in London (see page 213) which led the way to the generation of electric power. It was out of the study of physics that many of those things which we take for granted today became possible, the warmth and light in our homes and the power to operate radios, television sets and a great variety of electrical devices. It is a supply of energy which enables us to drive our cars and makes possible those industrial activities which create the wealth to pay for our schools and hospitals and those social amenities which make our lives richer. It is a supply of energy which makes it possible for our crops to grow, as it is also a supply of energy which makes it possible to produce the fertilisers and pesticides which enhance their production. In this final chapter it is worth thinking about where all this energy comes from.

Sun as the source of energy

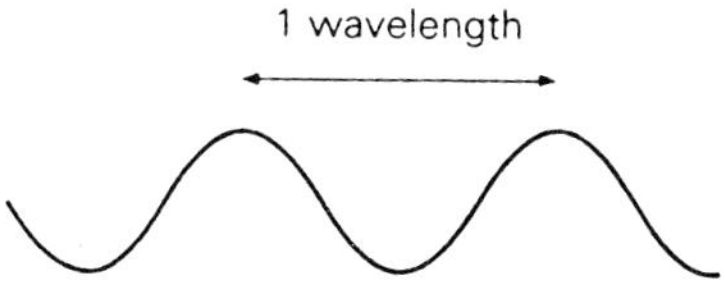

The Sun is the main source of all our energy. We receive radiation from it continuously in the form of waves and these extend over a wide spectrum from the long wavelength radio waves, through microwaves and the infra-red rays to the visible radiation in the form of light waves, then to ultra-violet waves and beyond to X-rays and gamma rays. Of this total radiation from the Sun, 98% is emitted in the infra-red region.

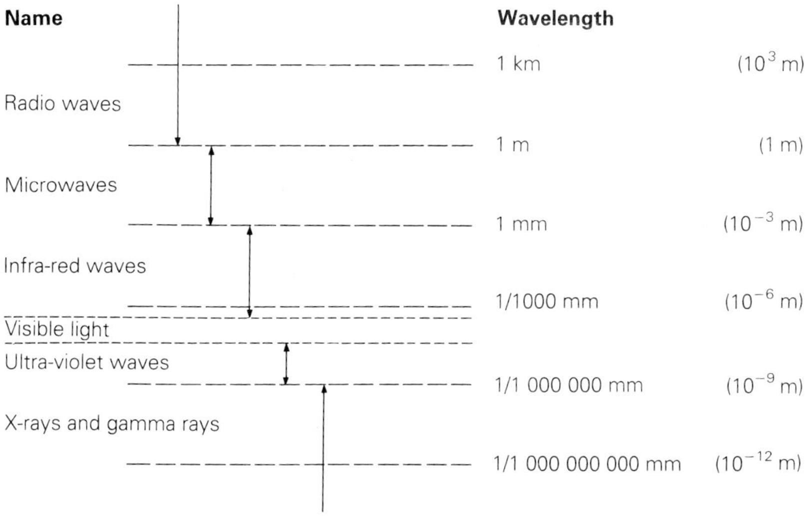

Fig 24.4

The Sun is the source of all food energy through the process of photosynthesis. It is the source of all coal and oil resources since they were produced through the burying of matter which once lived and owed its life to the Sun many years ago. It is the source of wind power through the convection currents it sets up in the air. It is the source of hydroelectric power through the evaporation of water which it causes and which eventually falls as rain, filling the lakes and rivers which make hydroelectric power stations possible (see page 141). And obviously it is the source of solar power when radiation from the Sun falls on solar cells which in themselves generate electricity.

Renewable and non-renewable sources

It is often convenient to classify energy sources as *renewable* or *non-renewable.* The renewable energy sources include solar energy, hydroelectric energy and wind energy, as they will be continually available as long as the Sun sends out radiation. Tidal energy is renewable energy because no sooner has the energy of an incoming tide been used to generate some electricity than further energy is available on another incoming tide. Likewise the energy of water waves is also renewable as long as waves are available on the seas.

The non-renewable sources include the fossil fuels – coal, oil and natural gas – and nuclear fuels such as uranium since they will not be replaced once they are used up. Strictly speaking oil, coal and gas are renewable in that the present deposits have been built up over a period of 600 million years. There is no reason to suppose deposits are not being built up now at the same rate as in the past, but this means that over the next 600 years only one millionth of the amount built up so far will be produced. What we have been using for the last two or three hundred years (and more particularly during this century) is being replaced at an extremely slow rate by comparison.

Primary fuel	UK	France
Oil	38.4%	43.6%
Natural gas	22.9%	12.0%
Coal	31.5%	8.8%
Nuclear energy	6.5%	27.7%
Hydroelectric energy	0.7%	7.9%

Electricity is often referred to as a *secondary* source of energy as it may be generated by burning the *primary* sources, oil, coal or gas or using uranium in a nuclear reactor. The amount of primary fuels used in a country will depend on its resources. The table compares the use of primary fuels in the United Kingdom and in France. What differences do you notice between them? What do you think are the reasons for those differences?

Some of the oil, gas and coal will be used for the generation of electricity in power stations (see page 140), but some of the oil will be used to produce petroleum, for example, or as a source for other chemical substances such as fertilisers. Coal is still the main fuel for generating electricity throughout the world, but during recent years there have been changes. The amount of electricity in the world produced from nuclear energy has increased 3 times, but in the same period the use of oil for the production of electricity has fallen from 22% to 9%. Why do you think that is?

Energy supplies in the future

We need to remember that many of our supplies of fuel will not last indefinitely. Estimates differ, but we all know that our North Sea supplies will not last much into the next century and that world supplies of oil and gas are likely to be used up during that century. Perhaps coal supplies might last another 400 years, but the fossil fuels will not last indefinitely. There is therefore an essential need for the scientists to develop alternative sources for the future.

One thing that this should remind us of is the need to 'save' energy, not to waste it. Fuel economy matters. It matters when energy is being transferred from one form to another, and nowhere is it more important than in our own homes.

Energy in the home

You have learned in this course that when energy is transferred from one form to another none is lost. Of course this is true in a home. Fuel is burned inside in order to warm the house, then the energy escapes through the roof, the windows and the walls. The energy is not 'lost' but it is spread out and ceases to be useful. A good example of '*spread out energy*' is the energy in the sea. There is a great deal of it in the ocean in the form of internal energy, but it is so spread out that it is 'useless'. When fuel is burned in a power station, the energy released does useful things like turning turbines which turn generators which produce electricity, but much energy is 'spread out' into a useless form. For example, in such a power station, some energy is transferred into the internal energy of the cooling water and of the air in the cooling towers. In fact, the efficiency of an energy transfer expressed as a percentage tells us how much of the input energy has been transferred to the form required as output. In the case of a power station, a more efficient process would produce more electrical energy from the same amount of fuel and waste less to the environment.

There is necessarily some energy 'lost' in a power station, but for many years countries such as Denmark have had combined-heat-and-power (CHP) schemes whereby the energy which might otherwise be wasted has been used to provide hot water for the local community. There is much in this direction that the scientists and engineers can develop in the future.

In a home, the object should be to prevent as much energy as possible getting spread out to the surroundings where it becomes useless. Fig 24.5 shows typical energy losses from a house. The sensible thing is to try to cut down on those losses by roof insulation, double glazing for the windows, the use of cavity walls and the insulation of such walls with foam. Draughts need to be reduced and too often people forget what is 'lost' through the floors. Of course adequate insulation may be expensive to install but the saving over the years can be considerable.

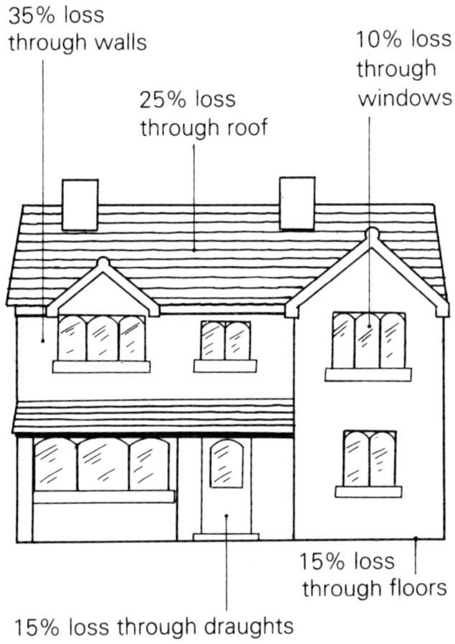

Fig 24.5

But are the effects of science all good?

We are all aware of the amount of pollution in developed countries, polluted rivers in many parts of Europe, polluted lakes in North America, the effects of acid rain on trees in Scandinavia and Germany, the harmful effects of exhaust fumes from cars, possible problems caused by the greenhouse effect. At least the scientific community is increasingly aware of these problems and physicists, chemists, biologists and engineers are all striving to find solutions to them. Problem solving is at the heart of scientific work, as it has always been.

But would we perhaps be better off if scientists had not, for example, shown so much interest in studying the atom? We discussed at the beginning of this chapter (page 316) how the physicist is fascinated by the process of 'finding out' and how this led to new models of the atom. It was Lord Rutherford's suggestion of a nuclear model of the atom which ultimately led to the nuclear physics work which in turn led to the discovery of fission and eventually to the atomic bomb. It was described on page 128 how the release of neutrons when fission takes place can lead to a chain reaction and the continuous release of energy (the process in a nuclear power station). But Fig 24.6 illustrates how there is an explosive build up in a nuclear bomb: in this diagram, two neutrons from each fission go on to produce further fissions and there is rapidly a great energy release.

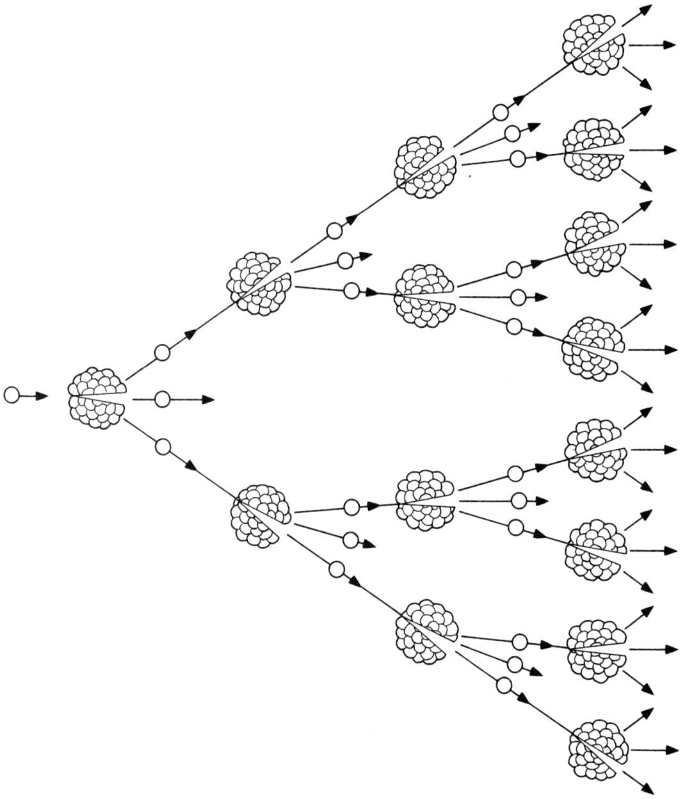

Fig 24.6

The type of explosion which occurs when a nuclear bomb explodes cannot possibly occur in a nuclear power station, but nevertheless it was the work of the physicists which led to the bomb.

Before deciding whether to accuse the physicists, it is important to remember that a time will come when the fossil fuels will be used up and other sources of energy will be needed. Renewable energy sources will need to be developed and scientists and engineers are at work on that, but it is not likely that such sources will be able to meet the ever growing need for more and more energy throughout the world. And now that it is realised that coal-fired power stations add considerably to the greenhouse effect, there may be a greater enthusiasm for more nuclear power stations which have no greenhouse effects.

Before trying to decide whether the work on nuclear physics has been beneficial or not, let us look a little more closely at the radiation which comes from the atom.

Radioactivity

Radioactivity was discovered at the end of the 19th century. We have already mentioned the energy which is stored in the nucleus of massive elements which is released in fission. With the atoms of some elements, such as radium and uranium, there is so much energy stored in the nucleus that the nucleus is liable to disintegrate of its own accord, and, when this happens, a bit of the nucleus is kicked out at high speed. Even a tiny sample of these elements contains so many atoms that there are always some disintegrating and sending out particles. This is the process we call *radioactivity*.

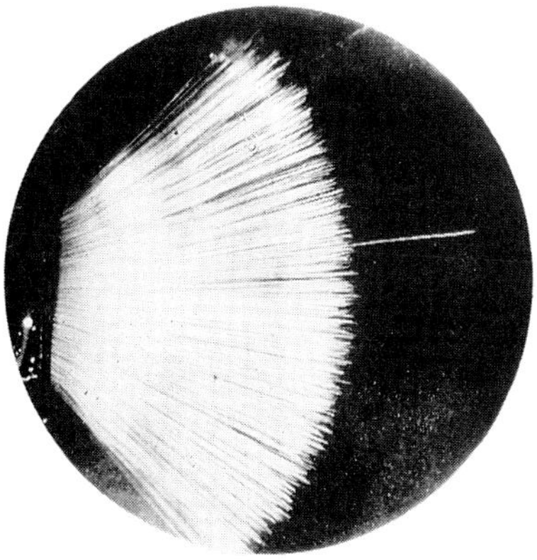

Alpha-particles

The cloud chamber gives evidence of the emission of particles. The cloud chamber is simply a space containing air which is made as moist as possible. When a particle travels through the air it leaves a vapour trail (a trail of drops of liquid, like the vapour trails left by high flying aircraft) which we can see. The photograph shows the tracks caused by *alpha particles* travelling from left to right. We now know that alpha particles are the nuclei of helium atoms travelling at speeds of about 15 000 km/s. Uranium and radium both emit alpha particles.

Some radioactive atoms emit *beta particles*. These have much less mass than alpha particles, but can have even greater speeds. You will learn later that these beta particles are electrons.

Yet another form of energy appears when particles are emitted. It is nearly always the case that when a particle shoots out of a disintegrating nucleus, there is a flash of radiation as well. We call this radiation from disintegrating atoms *gamma radiation*. We cannot see it, just as we cannot see radio waves or the warming radiation from the Sun. Gamma radiation is like X-rays in that it penetrates matter easily and too much exposure to it is dangerous.

Alpha particles cannot penetrate skin, but if a radioactive source gets inside you, these particles can do a great deal of damage and destroy living cells in your body. Beta particles and gamma radiation have much greater penetrating power. They can penetrate your skin and can kill living cells. It is because of the harmful effects of the radiations that radioactive substances have to be treated with great care. A consequence of using a nuclear bomb is that there is a considerable release of radioactive atoms into the environment. Radioactive atoms are also produced in a nuclear reactor in a power station, which is why special safety precautions have to be taken and why waste material from such power stations has to be handled very carefully.

This appears to suggest that radioactivity is entirely harmful. But there is another side to the story. Radioactivity is used very extensively in the cure of cancer. Just as it will kill living cells, it will also kill malignant (cancerous) cells. It has prolonged many lives and completely cured certain cancers. It has played an important part in diagnosing parts of the body which are not working effectively. It has helped in agricultural work to develop new strains of cereal crops which have proved of great economic benefit. And it is also important to realise that we are surrounded (and always have been) by radiations, which have nothing to do with nuclear power stations and nuclear bombs. There has always been radiation coming to us from outer space; there is radiation coming to us from the soil and from the rocks around us, of which many of our houses are built. This is referred to as *background radiation* and it will always be there. And it happens that a proportion of potassium nuclei are radioactive and potassium is present in the food we eat: there will always be radioactivity inside each one of us.

X-rays, which were discovered by a physicist at the end of the last century, can be very harmful, just like gamma radiation. Would we be better off if they had not been discovered? It is worth remembering the benefit which has been gained from them in diagnosing faults in the human body. Remember that when a bone was broken in someone's body a hundred years ago, all that could be done was to push the two parts together and hope for the best. However skilled a bone setter might have become, today the X-ray ensures that the limb is properly set. Perhaps that explains why there were so many deformed people before X-rays were discovered. Certainly some of the products of scientific work can be harmful, but they have also brought beneficial uses.

Of course the release of radioactive elements as a result of a nuclear bomb or of the accident at Chernobyl is a serious tragedy, but do you think it is right to blame the scientific work in pursuit of knowledge which has led to these things? There are beneficial results which have come from the study of the atom and the nucleus. Will we need one day to make much more use of nuclear energy?

The future

There are many global problems today and global problems require global action. Just as the threat of piracy on the high seas was finally overcome by international action, so today some problems can perhaps be solved only by international scientific action. 'At present Sweden blames Britain for some of its pollution, Canada blames the United States and everyone blames

Brazil.' There are serious problems to be solved, but casting blame solves little. It is very much the techniques of physics, discussing, exploring, testing theories, as well as using the instruments and tools developed by physicists, that will make possible an understanding of the world's problems and so help towards finding solutions. Just as physics has transformed the quality of life in Britain over the last 150 years, so it can do much to help the lot of those in the developing world in the years ahead.

a) Solar cells used for driving a submersible pump in Senegal

b) Bio-mass installation in Ouga, Senegal

c) Windmills for irrigation in Mozambique

d) A solar cooker at the Alternative Technology Station in Ouga, Senegal

Some final questions for class discussion or homework

1 It was said above 'It is a supply of energy ... which makes possible those industrial activities which create the wealth to pay for our schools and hospitals and those social amenities which make our lives richer'. What does this mean? What have industries to do with our schools and hospitals? Does wealth creation make our lives *richer* or *happier*?

2 It is stated above that the Sun is the source of food energy, of coal and oil resources, of wind power, of hydroelectric power. Describe the process in each of these cases.

3 a The two atomic bombs which were used at the end of World War II led to an enormous loss of life in Hiroshima and Nagasaki, as well as to a great number of people being maimed, and very great devastation. On the other hand it brought the war against Japan to an abrupt end. If it had not ended like that, there could have been a great loss of life if the ending of the war had required an invasion of Japan. Do you think the dropping of the bombs was justified?

b It has been said that the existence of the nuclear bombs has been the greatest deterrent from a war breaking out in Europe over the last 45 years. Do you think there is any truth in that statement?

4 'The pursuit of scientific knowledge is neither good nor bad. The important thing is the use that is made of it.' Do you agree?

5 Has physics anything to contribute to people in the developing world?

6 If you were appointed the Minister in charge of the Department of Energy, what policy or policies would you adopt for the nation in its use of energy?

Index